Table of Contents

First Edition

Pumping Apparatus DRIVER/OPERATOR Handbook

Michael A. Wieder, Writer
Carol Smith, Editor
Cynthia Brakhage, Editor

Validated by the International Fire Service Training Association
Published by Fire Protection Publications, Oklahoma State University

RECYCLABLE

The International Fire Service Training Association

The International Fire Service Training Association (IFSTA) was established in 1934 as a "nonprofit educational association of fire fighting personnel who are dedicated to upgrading fire fighting techniques and safety through training." To carry out the mission of IFSTA, Fire Protection Publications was established as an entity of Oklahoma State University. Fire Protection Publications' primary function is to publish and disseminate training texts as proposed and validated by IFSTA. As a secondary function, Fire Protection Publications researches, acquires, produces, and markets high-quality learning and teaching aids as consistent with IFSTA's mission.

The IFSTA Validation Conference is held the second full week in July. Committees of technical experts meet and work at the conference addressing the current standards of the National Fire Protection Association and other standard-making groups as applicable. The Validation Conference brings together individuals from several related and allied fields, such as:

- Key fire department executives and training officers
- Educators from colleges and universities
- Representatives from governmental agencies
- Delegates of firefighter associations and industrial organizations

Committee members are not paid nor are they reimbursed for their expenses by IFSTA or Fire Protection Publications. They participate because of commitment to the fire service and its future through training. Being on a committee is prestigious in the fire service community, and committee members are acknowledged leaders in their fields. This unique feature provides a close relationship between the International Fire Service Training Association and fire protection agencies which helps to correlate the efforts of all concerned.

IFSTA manuals are now the official teaching texts of most of the states and provinces of North America. Additionally, numerous U.S. and Canadian government agencies as well as other English-speaking countries have officially accepted the IFSTA manuals.

ISBN 0-87939-166-9   Library of Congress LC# 99-61722

First Edition, First Printing, May 1999   Printed in the United States of America
Second Printing, November 1999
Third Printing, May 2000
Fourth Printing, January 2001

If you need additional information concerning the International Fire Service Training Association (IFSTA) or Fire Protection Publications, contact:

Customer Service, Fire Protection Publications, Oklahoma State University
930 North Willis, Stillwater, OK 74078-8045
800-654-4055   Fax: 405-744-8204

For assistance with training materials, to recommend material for inclusion in an IFSTA manual, or to ask questions or comment on manual content, contact:

Editorial Department, Fire Protection Publications, Oklahoma State University
930 North Willis, Stillwater, OK 74078-8045
405-744-4111   Fax: 405-744-4112   E-mail: editors@ifstafpp.okstate.edu

This is the first edition of the **Pumping Apparatus Driver/Operator Handbook**. This text details the important responsibilities of firefighters who are assigned to drive and operate fire department vehicles that are equipped with a fire pump. It combines the information that was previously contained in three separate IFSTA manuals: **Fire Department Pumping Apparatus, Fire Streams Practices,** and **Water Supplies for Fire Protection.**

Acknowledgement and special thanks are extended to the members of the IFSTA validating committee who contributed their time, wisdom, and knowledge to this manual:

Chair
John Trenner
Carmel Fire Department
Carmel, California

Bob Guthrie
Ames Fire Department
Ames, Iowa

Tony Huemann
Mount Prospect Fire Department
Mount Prospect, Illinois

Bill Hulsey
Oklahoma State University
Stillwater, Oklahoma

Scott Kerwood
Orange County Emergency Services
Vidor, Texas

Matthew Manfredi
AAA Emergency Supply Company
White Plains, New York

Mark S. Pare
Providence Fire Department
Providence, Rhode Island

Michael Ridley
Elk Grove Fire Department
Elk Grove, California

Vice Chair
Arlen Gross
Louisiana State University
Baton Rouge, Louisiana

Jeff Rosenfeld
Virginia Beach Fire Department
Virginia Beach, Virginia

Mark D. Watts
Baton Rouge Fire Department
Baton Rouge, Louisiana

Michael Wilbur
Fire Department of New York
New York, New York

Matt Woodrow
Toronto Fire Services
Toronto, Ontario

Jeff Yaroch
Clinton Township Fire Department
Richmond, Michigan

Mary Beth Zampa
United States Air Force
Robins AFB, Georgia

Special thanks are extended to Mr. William Eckman of LaPlata, Maryland, whose years of experience in rural water supplies operations are partially reflected in Chapter 14 on water shuttles.

Special appreciation is also extended to the following fire departments that provided their personnel, resources, and time to assist our staff in shooting many of the new photographs that were necessary to complete this manual:

Stillwater (OK) Fire Department; Assistant Chief Rudee Cryer, B-Shift Personnel

Yukon (OK) Fire Department; Chief Robert Noll, Assistant Chief Jeff Lara, Members J.W. Nokes Jr., Larry Gossett, and Mike McGee

Tulsa (OK) Fire Department; Chief Tom Baker, Visual Communications Coordinator Frank Mason, Engine 16-A: Michael Ledbetter, Jeff Blackburn, and R.D. Harris

A manual of this scope would not be possible without the assistance of many other people and organizations who assisted us by providing pictures and information vital to the project's completion:

Waterous Company, Daniel L. Juntune, Sales and Marketing Support Manager

W.S. Darley & Co., Tom Darley, Sales Manager

Hale Fire Pump Company

Bob Esposito, Trucksville, Pennsylvania

Ron Jeffers, Union City, New Jersey

Warren Gleitsmann, Timonium, Maryland

Joel Woods, University of Maryland Fire & Rescue Institute, College Park, Maryland

Ron Bogardus, Albany, New York

Jeff Windham, Celanese Corporation, Clear Lake, Texas Plant

Rocky Hill, Connecticut Fire Department

Mount Shasta, California Fire Department

Pennsburg, Pennsylvania Fire Department

Oklahoma City, Oklahoma Fire Department

Class 1, Inc., Ocala, Florida

Akron Brass Company

Walter Kidde, Inc.

Conoco Oil Company, Ponca City, Oklahoma Refinery

KK Products, a division of Task Force Tips, Inc.

Houston, Texas Fire Department Station 22, Hazardous Materials Team

Rich Mahaney, Jackson, Michigan

Texas A&M University Fire Protection Training Division

Emergency One, Inc.

Williams Fire and Hazard Control, Inc.

3M Fire Protection Systems

3M Safety and Security

Volunteer Fireman's Insurance Service, York, Pennsylvania

National Interagency Fire Center, Boise, Idaho

Monterey County, California Fire Training Officer's Association

City of Phoenix, Arizona Development Services Department, Joe McElvaney and Steve Noblet

Eric Harlow, OSU School of Fire Protection and Safety

Harvey Eisner, Tenafly, New Jersey

Bill Tompkins, Bergenfield, New Jersey

Mount Prospect, Illinois Fire Department

Capt. Chris Mickal, New Orleans (LA) Fire Department Photo Unit

Bil Murphy, CFPS, Lancaster, California

Fire Marshal Andrew C. Mount, Plymouth Township, Pennsylvania

Plano, Texas Fire Department, William Peterson, Fire Chief

United States Air Force Academy Fire Department, Colorado Springs, Colorado, Asst. Chief James Rackl

Index provided by Kari Kells, Index West

Gratitude is also extended to the following members of the Fire Protection Publications staff, whose contributions made the final publication of the manual possible.

Barbara Adams, Associate Editor

Susan S. Walker, Instructional Development Coordinator

Tara Gladden, Editorial Administrative Assistant

Don Davis, Coordinator, Publications Production

Ann Moffat, Graphic Design Analyst

Desa Porter, Senior Graphic Designer

Connie Nicholson, Senior Graphic Designer

Don Burull, Graphics Assistant

Ben Brock, Graphic Assistant

Shelley Hollrah, Graphic Assistant

Introduction

A basic premise of the fire service is that when people need help, fire department personnel respond to provide assistance. The people who call may need medical, rescue, or fire fighting assistance. Unless the person who needs help happens to walk into the fire station or is directly in front of it, it is necessary for fire department personnel to board a piece of apparatus and ride it to the emergency scene. The apparatus must be appropriately positioned on the emergency scene so that equipment and/or hose may be efficiently deployed and the incident brought under control as quickly as possible.

The most basic type of fire apparatus operated in modern fire departments is the fire department pumper. Depending on local tradition, this piece of apparatus may also be called any of the following:

- Engine
- Triple
- Squad
- Wagon
- Pump

The fire pump, water tank, and fire hose are the three basic fire fighting components common to virtually every fire department pumper. These components allow hoselines to be deployed from the apparatus and water to be discharged. Again, depending on local preferences, the fire department pumper may carry a variety of other fire fighting, emergency medical, and rescue/extrication equipment.

In addition to pumpers, a variety of other types of fire apparatus may be equipped with fire pumps, water tanks, and fire hose. These include wildland fire apparatus, rescue vehicles, aerial apparatus, airport rescue and fire fighting (ARFF) vehicles, and water tenders (tankers). The addition of pumps and water tanks to these vehicles gives them added flexibility to handle a greater variety of incidents.

The *fire apparatus driver/operator*, herein simply referred to as the driver/operator, is responsible for safely driving the fire apparatus to and from the emergency scene and operating its pump and other components while on the emergency scene. The driver/operator is ultimately responsible for the safety of all personnel who are riding on the apparatus while it is in motion. He is also entrusted with the safety of personnel in positions of danger who are operating hoselines being supplied by the pumper. This is an enormous amount of responsibility and requires the driver/operator to complete an extensive training program in order to be able to satisfactorily handle these duties.

It should be the goal of every fire department to train its driver/operators to meet all the pertinent requirements contained in NFPA 1002, *Standard for Fire Apparatus Driver/Operator Professional Qualifications.* This standard is designed so that driver/operators may certify to only those portions that apply to the types of apparatus that their department operates. For example, many small volunteer and career fire departments do not have an aerial apparatus. Therefore, a driver/operator for that type of department would not have to meet the requirements of the chapters on operating aerial apparatus in order to be certified. As well, a driver/operator in a major urban fire department would not be required to be certified on a water tender/tanker.

One requirement of NFPA 1002 that has stirred some controversy over the years is the requirement that in order to be certified as a driver/operator, the person must also meet the requirements of and be certified to NFPA 1001, *Standard for Fire Fighter Professional Qualifications*, for Fire Fighter I. Obviously, this is not an issue for career fire departments. In their case, a person joins the fire department, goes through extensive recruit training, and usually serves as a firefighter for some period of time before being promoted to the position of driver/

operator. By the time the career firefighter becomes a driver/operator, he has extensive fire fighting experience. In many volunteer systems this progression is much the same.

However, in some systems, particularly volunteer fire departments in small communities, there are often potential fire apparatus driver/operator candidates who do *not* have any fire fighting background. The most common scenario is a professional truck driver or heavy equipment operator who would like to join the department as a driver/operator but has no inclination to go through basic firefighter training first. In today's climate of dwindling numbers of volunteers, this puts the fire chief in a difficult position. On one hand, there is a person who already has the training and experience necessary to drive the vehicle. Teaching this person to operate the fire pump or other equipment is certainly achievable, as the candidate is most likely very mechanically inclined. Having this person drive and operate the vehicle frees a trained member from being required to remain with the apparatus. Conversely, you have a standard saying that the driver/operator must first complete firefighter training before being certified as a driver/operator. For now the choice is clear: if certification is important to that department, the person must complete the firefighter training. IFSTA has long been an advocate and supporter of the NFPA standards and fire service certification.

Departments that do not choose to go the certification route should still use NFPA 1002 and this manual as the basis for their driver/operator training. This helps to ensure that the candidate is capable of driving and operating the vehicle in a safe and efficient manner.

Purpose and Scope

This first edition of the **Pumping Apparatus Driver/Operator Handbook** is designed to educate driver/operators who are responsible for operating apparatus equipped with fire pumps. This apparatus includes pumpers, initial fire attack apparatus, tenders (tankers), wildland fire apparatus, and aerial apparatus equipped with fire pumps. The information in this manual aids the driver/operator in meeting the objectives found in Chapters 1, 2, 3, 6, and 8 of NFPA 1002, *Standard for Fire Apparatus Driver/Operator Professional Qualifications* (1998 edition).

The information contained in this new manual is intended to replace that which was previously found in three separate IFSTA manuals. These manuals are the following:
- **Fire Department Pumping Apparatus**
- **Fire Stream Practices**
- **Water Supplies for Fire Protection**

By combining this information, we eliminate the need for driver/operator candidates to purchase three separate manuals. Also eliminated is the potential problem of conflicting information between the three manuals that sometimes occurs when different committees work on books related to the same topic.

The purpose of this manual is to present general principles of pump operation, with application of those principles wherever feasible. It is also meant to guide driver/operators in the proper operation and care of apparatus. This manual includes an overview of the qualities and skills needed by a driver/operator, safe driving techniques, types of pumping apparatus, positioning apparatus to maximize efficiency and water supply, fire pump theory and operation, hydraulic calculations, water supply considerations, relay pumping principles, water shuttle procedures, foam system operation, and apparatus maintenance and testing.

Driver/operators of apparatus other than standard fire department pumping apparatus may be required to obtain another manual other than this one. Operation of apparatus equipped with aerial devices (NFPA 1002 Chapters 4 and 5) is covered in IFSTA's **Fire Department Aerial Apparatus**. The operation of aircraft rescue and fire fighting (ARFF) apparatus (NFPA 1002 Chapter 7) is covered in IFSTA's **Aircraft Rescue and Fire Fighting** manual.

Notice on Gender Usage

In order to keep sentences uncluttered and easy to read, this text has been written using the masculine gender, rather than both the masculine and female gender pronouns. Years ago, it was traditional to use the masculine pronouns to refer to both sexes in a neutral manner. This usage is applied to this manual for the purpose of brevity and is not intended to address only one gender.

Chapter 1

The Driver/Operator

The fire apparatus driver/operator is responsible for safely transporting firefighters, apparatus, and equipment to and from the scene of an emergency or other call for service. Once on the scene, the driver/operator must be capable of operating the apparatus properly, swiftly, and safely. The driver/operator must also ensure that the apparatus and the equipment it carries are ready at all times.

In general, driver/operators must be mature, responsible, and safety conscious. Because of their wide array of responsibilities, often under stressful, emergency situations, driver/operators must be able to maintain a calm, "can do" attitude under pressure. Psychological profiles, drug and sobriety testing, and background investigations may be necessary to ensure that the driver/operator is ready to accept the high level of responsibility that comes with the job.

To perform their duties properly, all driver/operators must possess certain mental and physical skills. The required levels of these skills are usually determined by each jurisdiction. In addition, National Fire Protection Association (NFPA) 1002, *Standard for Fire Apparatus Driver/Operator Professional Qualifications* sets minimum qualifications for driver/operators. It requires any driver/operator who will be responsible for operating a fire pump to also meet the requirements of NFPA 1001, *Standard for Fire Fighter Professional Qualifications* for Fire Fighter I. The following sections discuss some of the basic mental and physical skills that may be required.

Skills and Physical Abilities Needed by the Driver/Operator

Anyone seeking the responsibility for operating emergency vehicles must possess a number of important cognitive and physical skills. Individuals who do not possess these traits may not be eligible to become a driver/operator unless they are able to obtain them. Some of the more important skills are highlighted in the following sections.

Reading Skills

Driver/operators must be able to read. They will be constantly required to understand the written word. Some examples of duties that call for good reading comprehension are as follows:

- Reading maps
- Reviewing manufacturer's operating instructions
- Studying prefire plans
- Reviewing printed computer dispatch instructions
- Reading and working on a mobile data terminal (MDT) (Figure 1.1)

Writing Skills

The driver/operator must also be able to convey information completely and accurately in writing. Some examples of job functions that require writing skills are completing maintenance reports, equipment repair requests, and fire reports. Each driver/operator should be evaluated for the ability to write clearly and concisely.

Mathematical Skills

Basic mathematical skills are important to the driver/operator. Every day the driver/operator uses math in hydraulic calculations and numerous other situations. The driver/operator should be able to add, subtract, multiply, divide fractions and whole numbers, and determine square roots. The driver/operator should also be able to solve simple equations such as those used in friction loss problems. It is not the purpose of this manual to review basic mathematical skills. If a prospective candidate is deficient in these skills, attempts should be made to provide the assistance necessary to correct this situation. Often, a local educational institution will offer such programs. These programs will not only be of benefit to prospective candidates but also to current driver/operators who need to brush up on their math skills.

Physical Fitness

The driver/operator often must perform rigorous physical activities while setting up the apparatus at a fire scene. These activities include tasks such as:

- Connecting to a hydrant with a suction hose
- Stretching a supply line to a hydrant by hand
- Deploying a portable water tank (Figure 1.2)

The driver/operator must be prepared to perform the lifts, bends, and strenuous actions needed to complete these tasks. The driver/operator must be subjected to a periodic medical evaluation in accordance with NFPA 1500, *Standard on Fire Department Occupational Safety and Health Program*.

Vision Requirements

The safe operation of an apparatus depends greatly upon the driver/operator's ability to see. NFPA 1582, *Standard on Medical Requirements for Fire Fighter*, requires that the firefighter have a corrected far visual acuity of 20/30 with contact lenses or spectacles. The standard contains further information on uncorrected vision and diseases of the eye. Consult the standard for specific details.

Hearing Requirements

Emergency vehicles en route to and on the emergency scene generate high levels of noise. Amid the noise of

Figure 1.1 Reading skills are important for functions such as understanding and operating a mobile data terminal (MDT).

Figure 1.2 The driver/operator may be required to perform any number of strenuous tasks, such as deploying a portable water tank from its storage location to a position of use on the fireground.

engines, sirens, air horns, and radio traffic, the driver/operator must be able to recognize different sounds and their importance. For example, he must be able to distinguish between the siren on his vehicle and that on another emergency vehicle.

The driver/operator must also be able to focus on particular sounds, such as radio instructions for placing apparatus. Failure to hear such orders may result in placing the apparatus in a less effective or even unsafe position. NFPA 1582 recommends rejecting the firefighter candidate who has a hearing loss of 25 decibels or more in 3 of 4 frequencies (500-1000-2000-3000 Hz) in the unaided worst ear. It also recommends rejecting a candidate who has a loss greater than 30 decibels in any one of three frequencies (500-1000-2000 Hz) and an average loss greater than 30 decibels for four frequencies (500-1000-2000-3000 Hz). Consult NFPA 1582 for more specific details.

Other Skills

Several other skills, although not required, will help the driver/operator perform well. Mechanical ability aids in understanding the operation and maintenance of the apparatus. Because the driver/operator is often in command of the apparatus while the officer is absent, basic supervisory skills will help in coordinating activities on the fireground.

Selection of Driver/Operators

In career fire departments, driver/operators are most often promoted from the rank of firefighter. Promotion is frequently based upon a required time of service with the department, written or performance tests, or a combination of service and tests. Whatever the method used, promotions should be based upon skill and ability rather than simply upon seniority or position.

Volunteer fire departments may base the "promotion" of a firefighter to driver/operator on the same criteria as career departments. More commonly, chief officers select members who they feel are ready for the responsibility of fire apparatus operation and train them toward that goal. Generally, the candidate will have to pass some form of examination before being approved as a driver/operator on emergency calls. Some volunteer fire departments may also allow lateral entry into a driver/operator position. This is common for members who have truck driving experience. To meet the intent of NFPA 1002, these candidates will also have to attend and complete a Firefighter I course.

Any fire department, whether it is career, volunteer, or industrial, must have an established and thorough training program for prospective fire apparatus driver/operators. Simply letting a firefighter drive the truck around the block a few times and showing him how to engage the fire pump is not adequate. An effective training program consists of appropriate amounts of classroom (theoretical) instruction, practical training in the field (application), and testing to ensure that the person is ready for the responsibility in a real-world setting. For more direction on establishing a driver/operator training program, consult NFPA 1451, *Standard for a Fire Service Vehicle Operations Training Program.*

Driving Regulations

Driver/operators of fire apparatus are regulated by state laws, city ordinances, and departmental standard operating procedures (SOPs). All driver/operators must be fully cognizant of all pertinent laws and SOPs. It is commonly known that ignorance of laws and ordinances is no defense if they are broken. More importantly, failure to know/follow department SOPs can have deadly consequences during emergency or even routine situations.

In general, a fire apparatus driver/operator is subject to all statutes, laws, and ordinances that govern any vehicle operator. Most laws and statutes concerning motor vehicle operation are maintained at the state or provincial level. Individual states or provinces may define what constitutes an emergency vehicle and exempt them from certain laws or statutes such as following posted speed limits and parking requirements. Driver/operators must understand these exemptions and their parameters. For example, a particular state or province may allow emergency vehicles to exceed the posted speed limit. However, it may also specify that this can only be done when responding to an emergency, under safe road conditions, and by no more than 10 mph (16 km/h). Driver/operators exceeding these parameters are subject to penalty and/or liability.

Most driving regulations pertain to dry, clear roads. Driver/operators must adjust their speed to compensate for wet roads, darkness, fog, or any other condition that makes normal emergency vehicle operation more hazardous. Under all circumstances, the fire apparatus driver/operator must exercise care for the safety of others and must maintain complete control of the vehicle.

In most jurisdictions, emergency vehicles are not exempt from laws that require vehicles to stop for school buses when their flashing red signal lights indicate that children are being loaded or unloaded. Fire apparatus should proceed only after the bus driver or a police officer gives the proper signal. Even then, the driver/operator should proceed with extreme caution as the children may not be aware of the approaching fire apparatus and the bus driver may panic and give a premature signal to proceed. Driver/operators must obey all traffic signals and rules when returning to quarters from an alarm.

Recent legal decisions have held that a driver/operator who does not obey state, local, or departmental driving regulations can be subject to criminal and civil prosecution if the apparatus is involved in a collision (Figure 1.3). If the driver/operator is negligent in the operation of an emergency vehicle and becomes involved in an accident, both the driver/operator and the fire department may be held responsible.

Licensing Requirements

In the United States, the federal Department of Transportation (DOT) establishes the basic requirements for licensing of drivers. In Canada, Transport Canada (TC) has the same authority. Other nations have similar organizations. Both the DOT and TC establish special requirements for licensing drivers of trucks and other large vehicles. While these are national guidelines, each state or province has the authority to alter them as it deems necessary for its jurisdiction. Some states and provinces require a fire apparatus driver/operator to obtain a com-

Figure 1.3 Driver/operators who operate the apparatus in a reckless manner that leads to a collision may be subject to criminal and civil penalties. *Courtesy of Ron Jeffers.*

mercial driver's license (CDL) for driving a large truck. Others have exempted fire service personnel from these licensing requirements. Each fire department must be aware of the requirements within its jurisdiction and make sure that its driver/operators are licensed accordingly.

Chapter 2

Types of Fire Apparatus Equipped with a Fire Pump

Fire apparatus are classified according to the functions for which they are designed. This chapter discusses different types of fire apparatus that may be equipped with a fire pump. Although many apparatus also have other systems, such as aerial devices or rescue tool systems, the focus of this manual will be on driving the vehicle and operating its pumping equipment. For information on operating aerial devices, see the IFSTA **Fire Department Aerial Apparatus** manual. For information on rescue apparatus and operating rescue tool systems, see the IFSTA **Principles of Extrication** manual.

Figure 2.1a A custom pumper.

Figure 2.1b A pumper built on a commercial truck chassis. *Courtesy of Warren Gleitsmann.*

Fire Department Pumpers

The main purpose of the fire department pumper (also called engine, wagon, triple, etc.) is to provide water at an adequate pressure for fire streams (Figures 2.1 a and b). The water supplied by the pumper may come from the apparatus water tank, a fire hydrant, or a static supply such as a lake, pond, or portable tank. NFPA 1901, *Standard for Automotive Fire Apparatus,* contains the requirements for pumper design. The standard specifies that the minimum pump capacity for these vehicles is 750 gpm (3 000 L/min). For pumps larger than 750 gpm (3 000 L/min), standard pump capacities are found in increments of 250 gpm (1 000 L/min). Municipal fire department pumpers rarely have pump capacities exceeding 2,000 gpm (8 000 L/min). However, industrial fire pumpers frequently have pump capacities in excess of 2,000 gpm (8 000 L/min).

In addition to the fire pump itself, a fire department pumper must also have intake and discharge pump connections, pump and engine controls, gauges, and other instruments to allow the driver/operator to use the pump. The pumper must also be equipped with a variety of hose sizes and types. These include intake hose, supply hose, and attack hose.

Fire department pumpers may carry a wide variety of portable equipment, in addition to the equipment associated with fire stream/water supply functions. NFPA 1901 specifies the minimum portable equipment that must be carried on all fire department pumpers. Local practices and procedures will dictate what equipment is carried on any particular pumper; however, the following is a list of types of equipment that may be found on pumpers.

- Ground ladders
- Self-contained breathing apparatus (SCBA)
- Rescue/extrication tools
- Forcible entry equipment
- Salvage equipment
- Portable water tanks
- First aid/medical equipment

In recent years, it has become increasingly popular to combine the functions of a rescue company with a fire department pumper. These apparatus, commonly called *rescue pumpers,* feature all the standard engine company equipment but also carry a larger than standard amount of rescue and extrication equipment (Figures 2.2 a and b). These apparatus typically are designed with more compartment space than a standard fire department pumper. Depending on the nature of the call, the personnel on the apparatus can function as either an engine company or a rescue company.

Pumpers with Foam Capability

Industrial, airport, and municipal fire departments commonly have pumpers capable of discharging fire fighting

Figure 2.2a A typical rescue pumper. *Courtesy of Joel Woods.*

Figure 2.2b Rescue pumpers have larger compartments than standard fire department pumpers.

foam on Class A (ordinary combustibles) and/or Class B (flammable and combustible liquids and gases) fires (Figures 2.3 a and b). Many industrial facilities that contain large quantities of flammable and combustible liquids are equipped with large-capacity foam pumpers. These foam pumpers are manned by trained members of the site fire brigade. Although they may be equipped to flow plain water on Class A fires, most industrial foam pumpers are primarily intended to produce large quantities of foam solution to attack Class B fires and spills. Industrial foam pumpers are built according to the standards provided in NFPA 11C, *Standard for Mobile Foam Apparatus,* and NFPA 1901.

Municipal and industrial foam pumpers may be equipped with either around-the-pump, direct injection, or balanced pressure foam proportioning systems. Most large-scale industrial foam pumpers use some form of balanced pressure proportioning system because of the reliability of the foam proportioning at large flows. The apparatus are equipped with fire pumps that range in capacity from 1,000 to 3,000 gpm (4 000 L/min to 12 000 L/min) or greater. Some have foam proportioning systems with capabilities that exceed the pumping capacity of that apparatus by itself. Most industrial foam pumpers have a large foam concentrate tank on board. These tanks range from 500 to 1,500 gallons (2 000 L to 6 000 L) of concentrate. The apparatus is typically equipped with a large fixed foam/water turret capable of flowing the entire capacity of the fire pump. The entire apparatus itself may be mounted on either a commercial or custom truck chassis.

Some municipal fire departments choose to equip fire department pumpers with fixed Class A and/or Class B foam systems. Class A foams are used on the standard types of fires to which municipal firefighters commonly respond (vehicles, brush, structures, etc.). Class B foam systems allow the firefighters to handle small-scale flammable/combustible liquids fires and spills. NFPA 1901 contains the requirements for these apparatus.

The Class A foam systems installed on municipal fire apparatus may be the high-energy or low-energy types described in Chapter 8. Regardless of the systems used, only a few discharge outlets are capable of flowing foam solution. Apparatus equipped with high-energy systems also require a sizable air compressor.

The foam proportioning systems that are used on municipal fire apparatus are typically scaled down versions of those described for industrial apparatus. Municipal fire apparatus equipped with foam systems also have foam concentrate tanks that supply the system. The most common size foam tanks for municipal fire apparatus range from 20 to 100 gallons (80 L to 400 L). These tanks are designed so that they can be refilled with 5 gallon (20 L) pails when necessary (Figure 2.4).

For more detailed information on foam systems and operations, see Chapter 15 of this manual.

Figure 2.3a Pumpers located at airports usually have foam capabilities. *Courtesy of Joel Woods.*

Figure 2.3b Pumpers that are maintained by industrial facilities usually have large foam concentrate tanks and proportioning systems. *Courtesy of Ron Jeffers.*

Figure 2.4 Most municipal pumpers that are equipped with foam tanks are refilled by using 5 gallon (20 L) pails.

Pumpers with Elevating Water Devices

Many fire departments choose to equip pumpers with elevating water devices (Figures 2.5 a and b). These devices provide a means for discharging fire streams from elevated nozzles. Through the use of articulating booms or telescoping pipes, hydraulically operated towers may be mounted on pumpers to form a combination unit. Elevating water devices can also be used to apply fire streams to the lower floors of a building. Elevating water devices typically range in height from 50 to 75 feet (15 m to 23 m). The operation of elevating water devices is covered in IFSTA's **Fire Department Aerial Apparatus**.

Figure 2.5a A pumper equipped with a telescoping elevating master stream device. *Courtesy of Dermot Scales.*

Initial Attack Fire Apparatus

Some fire departments use initial attack fire apparatus in their daily operations. These vehicles are basically scaled-down versions of the fire department pumpers described earlier in this chapter. Specific requirements for the design of initial attack apparatus are contained in Chapter 4 of NFPA 1901. Although they are not designations that are cited in the standard, manufacturers and fire departments typically use two categorical descriptions for initial attack fire apparatus: minipumpers and midipumpers.

Figure 2.5b A pumper equipped with an articulating elevating master stream device. *Courtesy of Joel Woods.*

Minipumpers

Smaller, quick-attack pumpers, known as *minipumpers*, are designed to handle small fires that do not require the capacity or personnel needed for a larger pumper (Figure 2.6). Using minipumpers enables a fire department to initiate a quicker attack than can be initiated using full-size pumpers. Many are equipped with four-wheel drive that allows them to be effectively driven over off-road terrain.

Minipumpers are most often mounted on one-ton chassis with custom-made bodies. Most of them have pumps with a capacity of no larger than 500 gpm (2 000 L/min), although some may have pumps rated up to 1,000 gpm (4 000 L/min). Most of the equipment carried on a larger pumper is also carried on a minipumper, although in smaller numbers. Some minipumpers also carry basic medical and extrication equipment, enabling them to serve as a rescue unit as well as a fire fighting unit. Some minipumpers are equipped with a turret gun that can be supplied directly from another pumper. The small size and maneuverability of the minipumper allows it to get into small spaces to set up a master stream.

Midipumpers

Midipumpers are used in the same types of situations as are minipumpers. A *midipumper* is well suited for such

Figure 2.6 Minipumpers are on light-duty truck chassis.

small fires as grass and dumpster fires and for service calls that do not require the capacity and personnel of a larger pumper (Figure 2.7). Midipumpers also have the ability to start an initial attack on larger fires.

Midipumpers are built on a chassis usually over 12,000 pounds (5 443 kg) Gross Vehicle Weight (GVW). The main differences between a midipumper and a minipumper are size, pump capacity, and the amount of equipment carried. Midipumpers routinely are equipped with pumps as large as 1,000 gpm (4 000 L/min). These units typically carry the same type of equipment as a full-size pumper,

Figure 2.7 A typical midipumper. *Courtesy of Ron Jeffers.*

Figure 2.8a Elliptical tenders are designed solely for hauling water to the fire scene. *Courtesy of Joel Woods.*

such as hose, ground ladders, and other fire fighting equipment. Some midipumpers also carry emergency basic extrication and medical equipment, so they can double as rescue and fire fighting vehicles.

Mobile Water Supply Apparatus

Mobile water supply apparatus, known as tenders or tankers, are widely used to transport water to areas beyond a water system or to areas where water supply is inadequate (Figures 2.8 a through c). Most attack pumpers carry water but not in large enough quantities to sustain an extended attack. Mobile water supply apparatus have water tanks that are larger than those generally found on standard pumpers. Specific requirements for the design of mobile water supply apparatus are contained in Chapter 5 of NFPA 1901.

The size of the water tank on mobile water supply apparatus depends upon a number of variables:

- Terrain — The mobile water supply apparatus may be required to climb steep hills or to operate on winding roads.

- Bridge weight limits — Bridges in the fire department's normal response district may be too old or may not be designed to bear the weight of heavy mobile water supply apparatus. This presents a danger to firefighters as they drive over these bridges if alternate routes are not available.

- Monetary constraints — The fire department may not have enough money to purchase a large mobile water supply apparatus.

- Size of other mobile water supply apparatus in the area — Water shuttles flow more easily when mobile water supply apparatus of the same or similar size are used.

According to NFPA 1901, the apparatus must carry at least 1,000 gallons (3 785 L) to be considered a mobile water supply apparatus. The weight distribution and

Figure 2.8b Some tenders have fire bodies that resemble a pumper. *Courtesy of Ron Bogardus.*

Figure 2.8c Jurisdictions with severe water problems often use large, semitrailer tenders. *Courtesy of Bob Esposito.*

load requirements generally limit tank capacity to 1,500 gallons (6 000 L) or less for single rear-axle vehicles. When tank capacities greater than 1,500 gallons (6 000 L) are desired, tandem rear axles, tri-axles, or a tractor-trailer design should be considered.

When designing a mobile water supply apparatus, the following construction requirements should be considered. These requirements allow the apparatus to move water safely and efficiently.

- Adequate but reasonable water tank capacity

- Adequate filling rate

- Adequate dump time

- Adequate suspension and steering

- Properly sized chassis
- Properly sized engine for tank size and terrain
- Sufficient braking ability
- Proper tank mounting
- Proper and safe tank baffling
- Ability to dump water from either side or the rear of the apparatus

Mobile water supply apparatus are used as support vehicles for pumpers that are attacking a fire. There are two basic methods in which a mobile water supply apparatus may be used. First, the apparatus can act as a reservoir or "nurse tanker/tender" for some fires. In this mode, the apparatus is parked at or near the fire scene and pumpers take water directly from the tank of the mobile water supply apparatus (Figure 2.9). The second method is to use the mobile water supply apparatus in a water shuttle operation. In these operations, mobile water supply apparatus dump their loads into a portable water tank or nurse tanker/tender and then go to a fill site to reload (Figure 2.10). For more information on the design and proper utilization of mobile water supply apparatus, see Chapter 14 of this manual.

Figure 2.9 In some jurisdictions, the tender and the attack pumper are directly connected together by a short section of hose. This is called a "nurse tender" operation.

Figure 2.10 Utilizing portable water tanks is the most efficient method for providing a continuous water supply when using water tenders.

Wildland Fire Apparatus

The control of ground cover fires often requires a light-weight, highly maneuverable vehicle that can go places inaccessible to larger apparatus. Fire apparatus specifically adapted for fighting ground cover fires are designed to fulfill these requirements. These units are usually built on a one-ton or larger vehicle chassis, and most have all-wheel drive. The majority of *wildland fire apparatus*, also known as *brush breakers* or *booster apparatus,* have pump capacities and water tank sizes of less than 500 gallons (2 000 L) (Figures 2.11 a and b). However, some jurisdictions have wildland fire apparatus with pump capacities of up to 1,000 gpm (4 000 L/min) and water tanks in excess of 1,000 gallons (4 000 L) (Figures 2.12 a and b).

The ability to "pump and roll" is a tremendous advantage when combating ground cover fires. Vehicles with the ability to pump and roll use a separate motor or a power take-off (PTO) to power the fire pump (Figure 2.13). This arrangement enables the apparatus to be driven and discharge water on the fire at the same time. There are two proper methods for making a moving fire attack. The first is to have firefighters using short sections of attack hose walk alongside the apparatus and extinguish fire as they go. The second is to use nozzles that are remotely controlled from inside the cab (Figure 2.14). Some jurisdictions design their apparatus so that firefighters may ride on the outside of the vehicle and

Figure 2.11a A typical small wildland fire apparatus.

Figure 2.11b Small wildland fire apparatus are usually constructed on a one-ton truck chassis.

Figure 2.12a Some jurisdictions use large wildland fire apparatus.

Figure 2.12b A typical large wildland fire apparatus.

Figure 2.13 Most wildland fire apparatus use a fire pump that is driven by an auxiliary engine.

discharge water as the vehicle is driven. This practice is strictly prohibited in NFPA 1500, *Standard for Fire Department Occupational Safety and Health Program.*

Most wildland fire vehicles carry booster hose, forestry hose, or small diameter attack lines (typically 1½-inch [38 mm]). In addition to the remote-controlled nozzles previously mentioned, apparatus may be equipped with ground sweep nozzles (Figure 2.15). Ground sweep nozzles are effective in protecting the front of the apparatus and extinguishing small fires in short vegetation as the apparatus is advanced.

Figure 2.14 Remote-controlled nozzles located on the front bumper eliminate the need for a firefighter to ride on the outside of the apparatus during pump-and-roll operations.

Figure 2.15 Ground sweep nozzles help protect the undercarriage of the apparatus.

In recent years, the practice of equipping wildland fire apparatus with Class A foam systems has become increasingly popular. Class A foam agents are extremely effective in attacking wildland fires and protecting exposures. Both high energy and low energy foam systems are used on these apparatus. More information on these foams systems can be found in Chapter 8 of this manual.

Booster tanks for wildland apparatus vary from approximately 50 gallons (200 L) on Jeeps to in excess of 1,000 gallons (4 000 L) on larger apparatus. Care should be taken to select a vehicle suitable for the terrain and ground conditions of the area that the vehicle will protect. Water tanks should be baffled into compartments to maintain vehicle stability when cornering or driving on rough terrain. Tanks and equipment should be mounted so that the vehicle is not top heavy.

For more information on the requirements for the design of wildland fire apparatus, consult NFPA 1906, *Standard for Wildland Fire Apparatus*. For more information on wildland fire fighting, see IFSTA's **Fundamentals of Wildland Fire Fighting** manual.

Aircraft Fire Apparatus

Aircraft rescue and fire fighting (ARFF) apparatus, formerly referred to as crash, fire, rescue (CFR) vehicles, are used to provide immediate suppression of flammable liquid fires and suppression of spill vapors on airport properties. In some instances, ARFF vehicles may respond off airport property to assist municipal firefighters at large-scale flammable liquid incidents.

Requirements for ARFF apparatus are contained in NFPA 414, *Standard for Aircraft Rescue and Fire Fighting Vehicles*. NFPA 414 divides ARFF apparatus into three general classifications:

- Major fire fighting vehicles (Figures 2.16 a and b)
- Rapid intervention vehicles (RIV) (Figures 2.17 a and b)
- Combined agent vehicles (Figures 2.18 a and b)

Although the information in this section is based on NFPA 414, there are other sources of information and requirements pertaining to ARFF apparatus. Additional requirements for airports in the United States are found in Federal Aviation regulations 14 CFR Part 139, *Certification and Operations; Land Airports Serving Certain Air Carriers*. Airports outside the United States follow the International Civil Aviation Organization (ICAO) Annex 14, *International Standards and Recommended Practices, Aerodromes*. Two other sources of information are helpful:

- Federal Aviation Administration (FAA), AC 150/5220-10, *Guide Specifications for Water/Foam Type Aircraft Fire and Rescue Trucks*

Figure 2.16a A major ARFF vehicle. *Courtesy of Joel Woods.*

Figure 2.16b A major ARFF vehicle. *Courtesy of Joel Woods.*

Figure 2.17a ARFF RIVs are smaller than major ARFF vehicles. *Courtesy of Joel Woods.*

Figure 2.17b A typical ARFF RIV.

Figure 2.18a Combined-agent vehicles carry dry chemical and foam extinguishing agents.

Figure 2.18b A combined-agent ARFF vehicle. *Courtesy of Joel Woods.*

- Federal Aviation Administration (FAA), AC 150/5220-14A, *Airport Fire and Rescue Vehicle Specification Guide*

For more information on ARFF apparatus and their operation, see the IFSTA **Aircraft Rescue and Fire Fighting** manual.

Fire Boat Apparatus

Fire boats are used in waterfront cities to protect docks, wharves, piers, and boats. A *fire boat* may be a small, high-speed, shallow-draft vessel, or it may be the size of a river, harbor, or ocean-going tug, depending upon its duties and the area to be covered (Figures 2.19 a and b). The number of personnel will vary with the size of the apparatus. The functions performed by a fire boat may include ice or water rescue, fire fighting, and relaying water to land-based apparatus. The two fire fighting operations for which fire boats are best suited are pumping through large master stream devices and providing additional water for onshore fire fighting operations. Fire boats have been built to deliver as much as 26,000 gpm (98 420 L/min). Individual master stream turrets that discharge 2,000 to 3,000 gpm (8 000 L/min to 11 500 L/min) are common.

Some smaller fire boats are propelled by water jets or are amphibious. Most heavy-duty fire boats are powered by marine-type diesel engines. Dual-purpose engines for propulsion and pumping have also been built.

Figure 2.19a Large fire boats can deploy powerful master streams. *Courtesy of Ron Jeffers.*

Figure 2.19b Small fire boats are able to maneuver in water that is too shallow for larger boats. *Courtesy of Ron Jeffers.*

Aerial Apparatus Equipped with Fire Pumps

Many fire departments choose to equip aerial apparatus with fire pumps (Figures 2.20 a through c). The following are reasons for equipping aerial apparatus with pumps:

- Depending on the nature of the call, the apparatus may be operated as a ladder company, engine company or both (if sufficient manpower is available). This provides the fire department with more flexibility. Keep in mind, however, that the apparatus should always be positioned with the aerial device in mind as the first priority. Simply stated, you have a limited amount of aerial device, but you can always pull an extra section of hose or two, if necessary.

- The apparatus will be capable of supplying its own elevated master stream.

- The apparatus may be used to extinguish small fires encountered when an engine company is not present.

- The apparatus may be used to protect itself in high radiant-heat situations.

Figure 2.20a Many full-sized aerial ladder apparatus are equipped with a fire pump.

Figure 2.20b Elevating platforms often have fire pumps to help the efficiency of elevated master stream operations.

Figure 2.20c Small quints have become increasingly popular in recent years.

Apparatus equipped with an aerial device, ground ladders, fire pump, water tank, and fire hose are commonly referred to as *quints.* Less common in today's fire service are *quads,* which are essentially engine companies that carry large quantities of ground ladders but do not have a main aerial device.

The size of the pump contained on the aerial apparatus will vary depending on the preference of the local jurisdiction. NFPA 1901 specifies that the minimum pump capacity for an aerial apparatus shall be 250 gpm (946 L/min) at 150 psi (1 035 kPa). If the apparatus is going to be considered a true quint, it must have at least a 750 gpm (3,000 L/min) pump. Departments that are only interested in having the aerial apparatus extinguish small fires or be able to protect itself will specify

these smaller pumps on their apparatus. Departments that want the aerial apparatus to have the same capabilities as a standard engine company may specify pumps as large as 2,000 gpm (8 000 L/min).

This manual will provide information on the operation of fire pumps that may be found on aerial apparatus. For information on operating aerial devices and their associated components, see IFSTA's **Fire Department Aerial Apparatus** manual.

Rescue Apparatus Equipped with Fire Pumps

Some fire departments choose to equip rescue vehicles with a small fire pump and water tank in order to handle small fires and provide protective hoselines at incident scenes (Figure 2.21). These vehicles differ from the rescue pumpers described earlier in this chapter in that they do not have the capabilities of a full-sized fire department pumper. Generally, these vehicles have a pump with a rated capacity of 500 gpm (2 000 L/min) or less. They also usually carry 500 gallons (2 000 L) of water or less. Some of these apparatus may also be equipped with foam proportioning systems and a foam concentrate tank.

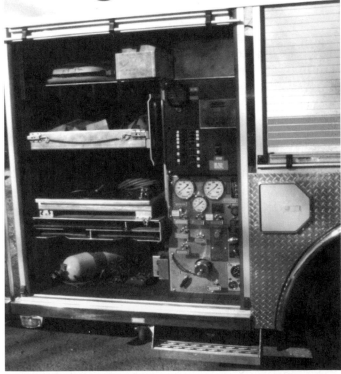

Figure 2.21 Some rescue apparatus are equipped with attack pumps and small water tanks that allow the apparatus to provide a lesser amount of fire protection than would a rescue pumper.

It is most common for the fire pump and water tank on these vehicles to be located inside one of the compartments. The compartment door will have to be opened in order to access the pump controls, intakes, and discharges. Preconnected attack hoselines may be found in the compartment, outside the vehicle in a cross-lay arrangement, or on the front bumper. Some vehicles may also have booster reels located somewhere on the apparatus.

Some jurisdictions choose to equip ambulances with small pumps and tanks. The design of these units is similar to that described in the previous paragraphs for rescue vehicles. Generally, ambulances with this extra equipment must be on a larger chassis than the standard one-ton vehicle used for most ambulances.

Apparatus-Mounted Special Systems

There are a variety of special systems commonly mounted on fire apparatus. In most jurisdictions, the driver/operator is the person responsible for operating these systems, in addition to the fire pump or aerial device. Although the exact procedures for operating these systems will vary depending on the manufacturer and design of the particular equipment found on the apparatus, the following sections provide general information on the types of special equipment that may be found on apparatus. Consult the owner's manual for exact operating instructions.

Electric Power Generation Equipment

Although once the bastion of rescue and specialty vehicles, it is becoming increasingly popular to equip pumping apparatus with electric power generation equipment. This equipment is used to power floodlights and other electrical tools or equipment that may be required on the emergency scene.

Inverters (alternators) are used on pumpers when the local jurisdiction determines that it is not necessary for the pumper to be able to generate large amounts of power (Figure 2.22). The *inverter* is a step-up transformer that converts the vehicle's 12- or 24-volt DC current into 110- or 220-volt AC current. Advantages of inverters are fuel efficiency and low or nonexistent noise during operation. Disadvantages include small capacities and limited mobility from the vehicle. These units are generally capable of providing approximately 1,500 watts (1.5 kW) of electric power. They are most commonly used to power vehicle-mounted floodlights.

Generators are the most common power source used for emergency services. They can be portable or fixed to the apparatus. Portable generators are powered by small gasoline or diesel engines and generally have 110- and/

or 220-volt capacities. They can be operated in the compartment of the apparatus, or they can be carried to a remote location (Figure 2.23). Most portable generators are designed to be carried by either one or two people. They are extremely useful when electrical power is needed in an area that is not accessible to the vehicle-mounted system. Portable generators are designed with a variety of power capabilities, with 5,000 watts (5 kW) of power being the largest.

Vehicle-mounted generators usually have a larger capacity than portable units. In addition to providing power for portable equipment, vehicle-mounted generators are responsible for providing power for the floodlighting

Figure 2.22 Inverters provide a limited amount of electrical power for emergency operations.

Figure 2.23 Many apparatus carry portable electric generators in a compartment.

Figure 2.24 Fixed generators can provide considerably more electrical power than their portable counterparts.

Figure 2.25 Handheld lights may be used anywhere on the emergency scene.

Figure 2.26 A tripod eliminates the need for someone to hold the floodlight.

Figure 2.27 These floodlights may be manually raised and turned.

system on the vehicle (Figure 2.24). Vehicle-mounted generators can be powered by gasoline, diesel, or propane engines or by hydraulic or power take-off systems. Fixed floodlights are usually wired directly to the unit through a switch, and outlets are also provided for other equipment. These power plants generally have 110- and 220-volt capabilities; capacities up to 12,000 watts (12 kW) are common on pumpers. Rescue vehicles commonly have larger generators than pumpers. Capacities of up to 50,000 watts (50 kW) are common. Mounted generators, particularly those powered by separate engines, are noisy and it is difficult to talk near them.

Scene Lighting and Electric Power Distribution Equipment

Most modern fire department pumpers have scene lighting and electric power distribution equipment on the apparatus. The driver/operator must be familiar with the location and proper operation of this equipment.

Lighting equipment can be divided into two categories: portable and fixed. Portable lights are used where fixed lights are not able to reach or when additional lighting is necessary. *Portable* lights generally range from 300 to 1,000 watts. They may be supplied by a cord from the power plant or may have a self-contained power unit. The lights usually have handles for safe carrying and large bases for stable setting and placement (Figure 2.25). Some portable lights are connected to telescoping stands that eliminate the need for personnel to either hold them or find something to set them on (Figure 2.26).

Fixed lights are mounted to the vehicle, and their main function is to provide overall lighting of the emergency scene. Fixed lights are usually mounted so that they can be raised, lowered, or turned to provide the best possible lighting. Often, these lights are mounted on telescoping poles that allow this movement (Figure 2.27). More elabo-

rate setups include electrically, pneumatically, or hydraulically operated booms with a bank of lights (Figure 2.28). The bank of lights generally has a capacity of 500 to 1,500 watts per light. The amount of lighting should be carefully matched with the amount of power available from the power-generating device carried on that vehicle. Overtaxing the power-generating device will give poor lighting, may damage the power-generating unit or the lights, and will restrict the operation of other electrical tools.

A variety of other electric distribution equipment may be used in conjunction with power generating and lighting equipment. Electrical cables or extension cords are necessary to conduct electric power to portable equipment. The most common size cable is a 12-gauge, 3-wire type. The cord may be stored in coils, on portable cord reels, or fixed automatic rewind reels (Figure 2.29). Twist-lock receptacles provide secure, safe connections. Electrical cable must be adequately insulated, waterproof, and have no exposed wires.

Junction boxes may be used when multiple connections are needed (Figure 2.30). The junction is supplied

Figure 2.28 Large banks of lights are usually raised by mechanical means.

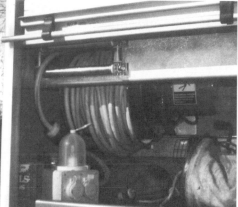

Figure 2.29 Some apparatus are equipped with electric-rewind power-cord reels.

Figure 2.30 A junction box allows several electrical devices to be supplied by one main power cord from the electric source.

by one inlet from the power plant and is fitted with several outlets. Junction boxes are commonly equipped with a small light on top to make them easier to find and plug into when it is dark.

In situations where mutual aid departments frequently work together and have different sizes or types of receptacles (for example, one has two prongs; the other has three), adapters should be carried so that equipment can be interchanged. Adapters should also be carried to allow rescuers to plug their equipment into standard electrical outlets.

Hydraulic Rescue Tool Systems

As mentioned earlier in this chapter, it is becoming increasingly common for fire department pumpers to carry a variety of extrication equipment. The most commonly used tools are powered hydraulic extrication tools. There are four basic types of powered hydraulic tools used by the rescue service: spreaders, shears, combination spreader/shears, and extension rams. The wide range of uses, the speed, and the superior power of these tools have made them the primary tools used in most extrication situations. These tools receive their power from hydraulic fluid supplied through special hoses from a pump. The pumps may receive their power from compressed air, electric motors, two- or four-cycle gas motors, or apparatus-mounted power take-off systems.

These hydraulic tool pumps may be portable and carried with the tool, or they may be mounted on the vehicle and supply the tool through long coiled hoses or a hose reel line (Figures 2.31 a and b). Driver/operators must know how the equipment on their apparatus operates and what limitations it has. Most pumps are not capable of supplying full power to the tool when the hose length between the pump and tool exceeds 100 feet (30 m). Apparatus-mounted power systems generally have

Figure 2.31a Portable hydraulic rescue tool motors generally use coiled hoses.

Figure 2.31b Hydraulic rescue tool motors that are affixed to the apparatus often supply hydraulic hoses that are stored on electric-rewind reels.

a standard method for engaging the device. This may be as simple as flipping an electric switch, or it could entail engaging the apparatus power take-off system.

The driver/operator should also be familiar with the number of tools that may be hooked to the system. Most manufacturers of powered hydraulic extrication tools have manifold blocks that can be connected to the end of a supply hose, from which multiple tools may be attached (Figure 2.32). The driver/operator should understand the limitations of this equipment and make sure that not too many tools are being attached to the system.

For more detailed information on powered hydraulic extrication equipment and its operation, see IFSTA's **Principles of Extrication** manual.

Figure 2.32 The hydraulic rescue tool manifold is similar in concept to the electric cord junction box.

ICS Typing of Pumping Apparatus

For jurisdictions that operate within the NIIMS Incident Command System (ICS), a method is employed to categorize pumping apparatus by capability. This method, called *apparatus typing*, is intended to make it easier for incident commanders to call for exactly the types of resources they need to handle an incident. The apparatus typing covered in Tables 2.1 and 2.2 is taken from the National Wildfire Coordinating Group (NWCG) *Fireline Handbook*. Individual states or jurisdictions may have their own method of typing apparatus. The driver/operator should know the method that is used in his jurisdiction.

The NWCG also has typing requirements for many other types of equipment, such as aircraft, bulldozers, and other heavy equipment, used for wildland fire fighting. Consult the *Fireline Handbook* for more information on apparatus typing.

Table 2.1 ICS Pumper Typing

Pumper Type	Minimum Pump Capacity in GPM (L/min)	Minimum Water Tank Capacity in Gallons (Liters)
Type 1	1,000 (4 000)	400 (1 600)
Type 2	500 (2 000)	400 (1 600)
Type 3	120 (480)	300 (1 200)
Type 4	50 (200)	200 (800)
Type 5	70 (280)	750 (3 000)
Type 6	50 (200)	500 (2 000)
Type 7	50 (200)	300 (1 200)
Type 8	20 (80)	125 (500)

Table 2.2 ICS Water Tender (Tanker) Typing

Tender/Tanker Type	Minimum Pump Capacity in GPM (L/min)	Minimum Water Tank Capacity in Gallons (Liters)
Type 1	300 (1 200)	5,000 (20 000)
Type 2	200 (800)	2,500 (10 000)
Type 3	200 (800)	1,000 (4 000)

Chapter 3

Introduction to Apparatus Inspection and Maintenance

Job Performance Requirements

This chapter provides information that will assist the reader in meeting the following job performance requirements from NFPA 1002, *Standard on Fire Apparatus Driver/Operator Professional Qualifications*, 1998 edition.

2-2.1* **Perform routine tests, inspections, and servicing functions on the systems and components specified in the following list, given a fire department vehicle and its manufacturer's specifications, so that the operational status of the vehicle is verified.**
- **Battery(ies)**
- **Braking system**
- **Coolant system**
- **Electrical system**
- **Fuel**
- **Hydraulic fluids**
- **Oil**
- **Tires**
- **Steering system**
- **Belts**
- **Tools, appliances, and equipment**

(a) *Requisite Knowledge:* Manufacturer specifications and requirements, policies, and procedures of the jurisdiction.

(b) *Requisite Skills:* The ability to use hand tools, **recognize system problems, and correct any deficiency noted according to policies and procedures.**

2-2.2 **Document the routine tests, inspections, and servicing functions, given maintenance and inspection forms, so that all items are checked for proper operation and deficiencies are reported.**

(a) *Requisite Knowledge:* Departmental requirements for documenting maintenance performed, understanding the importance of accurate record keeping.

(b) *Requisite Skills:* The ability to use tools and equipment and complete all related departmental forms.

3-1.1 **Perform the specified routine tests, inspections, and servicing functions specified in the following list in addition to those contained in the list in 2-2.1, given a fire department pumper and its manufacturer's specifications, so that the operational status of the pumper is verified.**
- **Water tank and other extinguishing agent levels (if applicable)**
- **Pumping systems**
- Foam systems

(a) *Requisite Knowledge:* Manufacturer specifications and requirements, policies, and procedures of the jurisdiction.

(b) *Requisite Skills:* The ability to use hand tools, **recognize system problems, and correct any deficiency noted according to policies and procedures.**

6-1.1 **Perform the specified routine tests, inspections, and servicing functions specified in the following list, in addition to those contained in 2-2.1, given a wildland fire apparatus and its manufacturer's specifications, so that the operational status is verified.**
- **Water tank and/or other extinguishing agent levels (if applicable)**
- **Pumping systems**
- Foam systems

(a) *Requisite Knowledge:* Manufacturer specifications and requirements, policies, and procedures of the jurisdiction.

(b) *Requisite Skills:* The ability to use hand tools, **recognize system problems, and correct any deficiency noted according to policies and procedures.**

Fire apparatus must always be ready to respond. Regardless of whether the truck responds to an emergency call once an hour or once a month, it must be capable of performing in the manner for which it was designed at a moment's notice. In order to ensure this, certain preventive maintenance functions must be performed on a regular basis. Most apparatus or equipment failures can be prevented by performing routine maintenance checks on a regular basis. Most fire departments require driver/operators to be able to perform these routine maintenance checks and functions. NFPA 1002 also requires the driver/operator to have certain preventive maintenance skills. This chapter covers the basic skills fire apparatus driver/operators should possess.

Before continuing, it is important to differentiate between the terms maintenance and repair. *Maintenance*, as used here, means keeping apparatus in a state of usefulness or readiness. *Repair* means to restore or replace that which has become inoperable. Apparatus or equipment that is said to be in a good state of repair has probably been well maintained. Preventive maintenance ensures apparatus reliability, reduces the frequency and cost of repairs, and lessens out-of-service time. The purpose of preventive maintenance is to try to eliminate unexpected and catastrophic failures that could be life and/or property threatening.

Preventive maintenance functions may be carried out by several different people. Fire departments may have a designated apparatus maintenance officer who routinely checks and services the apparatus. Fire departments should have separate mechanics who perform more detailed maintenance procedures. However, the driver/operator should be able to perform *basic* maintenance functions. In almost all cases, **repair** functions are carried out by qualified mechanics.

A Systematic Maintenance Program

The need for a planned fire apparatus maintenance program is obvious. Every fire department should have standard operating procedures (SOPs) for a systematic apparatus maintenance program. The SOPs should identify who performs certain maintenance functions, when they are to be performed, how problems that are detected are corrected or reported, and how the process is documented.

The SOP should clearly dictate those items that driver/operators are responsible for checking and which conditions they are allowed to correct on their own. Most departments allow the driver/operator to correct certain deficiencies such as low fluid levels and burned-out light bulbs. More detailed repairs need to be made by a certified mechanic. Large fire departments have their own repair shops and mechanics for this purpose (Figure 3.1). These mechanics may have their own vehicles and be able to come to a fire station or incident scene to perform a repair. Smaller fire departments may have a local automotive/truck repair business that assists them with these functions.

Figure 3.1 Larger-sized fire departments operate their own apparatus maintenance facilities.

The schedule for performing maintenance functions and checks varies from department to department. Typically, career fire departments require driver/operators to perform apparatus inspections and maintenance checks at the beginning of each tour of duty. They may also specify that more detailed work be completed on a weekly or monthly basis. Volunteer fire departments should establish a procedure by which all apparatus are inspected and maintained on at least a weekly or biweekly schedule.

Each fire department apparatus and equipment inspection and maintenance SOP should dictate how maintenance and inspection results should be documented and transmitted to the proper person in the fire department administrative system. Written forms or computer programs may be used to record the information. Appendix A contains several examples of apparatus inspection forms that may be used. Fire departments should maintain an effective filing system that allows the information on these reports to be reviewed, stored, and retrieved when required.

Apparatus maintenance and inspection records serve many functions. In a warranty claim, these records may be needed to document that the necessary maintenance was performed. In the event of an accident, maintenance records are likely to be scrutinized by the accident investigators. Proper documentation of recurrent repairs can also assist in deciding whether to purchase new apparatus in lieu of continued repairs on an older unit. All driver/operators must be trained to use their department's record-keeping system.

Cleanliness

One often overlooked but very important part of any apparatus inspection and maintenance program is the cleanliness of the vehicle and equipment. Many people look at apparatus cleanliness only from the standpoint of maintaining good public relations. The public sees a piece of fire apparatus as a unit of protection in which it has invested many thousands of dollars. Permitting apparatus to become marred with dirt, oil, and road grime damages public relations because individuals may feel their investment is not being properly protected.

Although the public relations aspect of cleaning apparatus is important, there are other more important reasons for keeping fire apparatus clean. A clean engine and clean functional parts permit proper inspection, thus helping to ensure efficient operation. Fire apparatus should be kept clean underneath as well as on top. Oil, moisture, dirt, and grime should not be permitted to collect. Some of the more vulnerable areas are the engine, wiring, carburetor or fuel injectors, and controls.

Keeping the apparatus body clean also helps promote a longer vehicle life. This is particularly true in jurisdictions where the use of road salt during inclement winter weather is prevalent (Figure 3.2). These chemicals have a corrosive effect on the steel components of the apparatus body and chassis. Frequent washing reduces the likelihood of damage caused by these chemicals.

On the other hand, overcleaning of the apparatus can also have negative effects. Fire departments whose members are fanatical about apparatus cleanliness often run into problems associated with the removal of lubrication from chassis, engine, pump, and aerial device components. This occurs when using any combination of degreasing agents, steam cleaners, and/or pressure washing equipment to clean the underside of the apparatus. Care should be taken not to strip all necessary lubrication from the apparatus when washing the apparatus body components. After a particularly heavy cleaning, it may be necessary to perform routine lubricating functions to ensure no unnecessary wear occurs on the apparatus.

Most apparatus manufacturers provide the fire department with specific instructions on how to clean its fire apparatus. If specific instructions are not available, use the following guidelines for apparatus cleaning.

Figure 3.2 Road grime, such as dirt and salt residue, is not only unsightly, it may also lead to corrosion on metal parts.

Washing

Washing the exterior of the apparatus is the most commonly performed function. Most apparatus manufacturers recommend slightly different procedures for cleaning the apparatus based on its age. Newer apparatus require gentler cleaning procedures than do older apparatus to avoid damage to new paint, detailing, and clear-coat protectants.

During the first six months after an apparatus is received, while the paint and protective coating are new and unseasoned, the vehicle should be washed frequently with cold water to harden the paint and keep it from spotting. To ensure the best appearance of the vehicle in the future and to reduce the chance of damaging new paint and protective coatings, the following washing instructions are recommended:

- Use a garden hose without a nozzle to apply water to the apparatus (Figure 3.3). The water pressure should be set so that the stream from the end of the hose is no more than 1 foot (0.3 m) in length. Higher pressures can drive grit and debris into the finish.

- Never remove dust or grit by dry rubbing.

- Wash the vehicle with a good automotive shampoo. Follow the shampoo instructions for proper use.

- Do not wash with extremely hot water or while the surface of the vehicle is hot.

- Rinse as much of the loose dirt from the vehicle as possible before applying the shampoo and water. This reduces the chance of scratching the surface when applying shampoo.

- Try to wash mud, dirt, insects, soot, tar, grease, and road salts off the vehicle before they have a chance to dry.

- Never use gasoline or other solvents to remove grease or tar from painted surfaces. Use only approved solvents to remove grease or tar from nonpainted surfaces.

- Dry the vehicle with a clean chamois rinsed frequently with clean water. Failure to dry the vehicle or the area around it completely will also encourage corrosion.

Once a new vehicle's finish is properly cured (according to the owner's manual), either a garden hose with a nozzle or a pressure washer may be used to speed cleaning of the apparatus. However, soapy water and hand washing on a regular basis are still required to assure proper cleanliness.

Glass Care

In general, warm soapy water or commercial glass cleaners should be used to clean automotive glass. These may be used in conjunction with paper towels or cloth rags (Figure 3.4). However, dry towels or rags should not be

Figure 3.3 Use a gentle stream when washing newer apparatus.

Figure 3.4 To clean automotive glass, use warm soapy water or commercial glass cleaners.

used by themselves because they may allow grit to scratch the surface of the glass. Do not use putty knives, razor blades, steel wool, or other metal objects to remove deposits from the glass.

Interior Cleaning

It is important to keep seat upholstcry, dashboard and engine compartment coverings, and floor finishes clean because an accumulation of dirt may cause deterioration of these finishes. Large, loose dirt particles should be swept or vacuumed first (Figure 3.5). Then, warm soapy water or commercial cleaning products may be used to clean the surfaces of these materials. Some manufacturers may specify particular cleaning agents or protective dressings that should be used on their materials.

> # WARNING
>
> **Many cleaners are toxic, are flammable, or cause damage to interior surfaces. Do not use volatile cleaning solvents, such as acetone, lacquer thinner, enamel reducer, nail polish remover, laundry soap, bleach, gasoline, naphtha, or carbon tetrachloride, to clean interior surfaces. Be sure that the vehicle is well ventilated when using any cleaning products inside the cab or crew-riding area.**

Figure 3.5 Sweep or vacuum the inside of the cab.

Waxing

Fire departments should follow the apparatus manufacturer's instructions regarding the application of wax or similar polishes to the exterior of the apparatus. On many newer apparatus, the application of these products is no longer necessary, and, in fact, may damage clear-coat protective-seal finishes that are applied over paints. If apparatus require waxes or polishes, they generally should not be applied until the paint is at least six months old. In general, before applying polish or wax the apparatus should be washed and dried. The wax or polish may then be applied with a soft cloth and buffed using a soft cloth or mechanical buffer.

Apparatus Inspection Procedures

The driver/operator should follow a systematic procedure for inspecting his apparatus. Having a systematic method helps ensure that all important items are checked every time the inspection is made. The information in the following sections is based on the requirements contained in NFPA 1002 and the government pretrip inspection requirements for obtaining a commercial driver's license (CDL). Even if your jurisdiction does not require that fire apparatus driver/operators obtain a CDL, these pretrip inspection principles provide a sound basis for the type of inspection that all fire apparatus driver/operators should be able to perform. These are the types of checks that career personnel should perform at the beginning of each tour of duty and volunteer personnel should do on a weekly or biweekly basis.

One particular method of performing an apparatus/pretrip inspection is referred to as the *circle* or *walk-around* method. This method involves the driver/operator starting at the driver's door on the cab and working around the apparatus in a clockwise pattern. As the apparatus is circled, important areas are checked along the way. The final step involves the driver/operator getting in the cab, starting the apparatus, and performing a functional check on apparatus systems. Figure 3.6 shows one method for performing the walk-around inspection. Exact procedures vary depending on departmental policies and vehicle design. For example, a vehicle with an

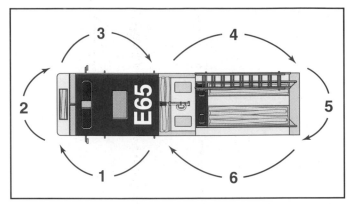

Figure 3.6 The driver/operator should use a systematic approach for checking the apparatus.

Figure 3.7 Check beneath the apparatus for signs of crucial fluid leaks.

engine ahead of the cab is checked in a different order than one with an engine behind or underneath the cab. The information contained in this section may be used during the course of the inspection, regardless of the exact order.

If the records from previous inspections are available, the driver/operator may wish to review them to see if any problems were noted at that time. This allows the current inspector to pay extra attention for reoccurrence of those concerns. If possible, the driver/operator may also talk to the last person to inspect or drive the apparatus to get any important information from him.

Approaching the Vehicle

The driver/operator should actually begin the inspection process by being observant when approaching the vehicle. Look for any general problems that are readily apparent by simply looking at the vehicle. Things such as vehicle damage or a severe leaning to one side are examples of things that could be readily apparent. Of course, be sure that a leaning vehicle is not doing so because it is resting on a sloped surface. Some apparatus bay floors are sloped to the extent that the vehicle appears to be leaning.

Look beneath the vehicle for spots that indicate leaking vehicle fluids such as water, coolant, oil, brake fluid, hydraulic fluid, or transmission fluid (Figure 3.7). These could be symptoms of serious problems. If the vehicle is indoors, make sure that proper ventilation equipment is in place or that doors are open to vent vehicle exhaust once the apparatus is started (Figure 3.8). If the weather permits, the apparatus should be parked outside for functional tests. Chock the wheels whenever the apparatus is parked.

CAUTION: Gasoline and diesel engines should not be run in unvented areas for any period of time. The buildup of carbon monoxide in that area could be harmful to

Figure 3.8 Exhaust venting equipment should be in place when the apparatus is parked inside the fire station.

personnel. In particular, diesel exhaust is known to emit benzene derivatives that have been shown to be carcinogenic in laboratory tests.

Left- and Right-Front Side Inspection

The first portion of the vehicle specifically checked by the driver/operator should be the left (driver's) side of the front, or cab, of the vehicle. After inspecting the left-front of the vehicle, the same procedure will be repeated on the right (passenger's or officer's) side. The driver/operator should begin this portion of the inspection by making a general observation of the cab on that side of the vehicle for any damage that was not noted in previous inspections.

The next thing the driver/operator should do is make sure various aspects of the cab doors are in proper order. Each door should close tightly, and the door latch should work as it was designed. There should be little or no play in the operation of the latch. Make sure that all door window glass is intact and clean. Make sure that all steps, platforms, handrails, and ladders are securely mounted and without deformation.

If the vehicle is equipped with saddle fuel tanks beneath the door opening, check to make sure that there are no apparent leaks or other problems with the tank. If the tank contains a fill cap, make sure it is tightly in place.

The driver/operator should then check the condition of the tire and wheel on that side of the vehicle (Figure 3.9). A quick visual check of the wheel itself should be made for any signs of missing, bent, or broken studs, lugs, or clamps. Check each lug nut by hand to feel if any are loose. The wheels should not be cracked or damaged because this prevents sealing of the tire to the rim. Look for unusual accumulations of brake dust on the wheel. This could signify a problem with the braking system. Inspect the wheel/tire assembly for other leaks. While seals that retain axle gear oil may show slight seepage and still be serviceable, trails of fluid on the wheel or tire are unacceptable. The driver/operator may also choose to make a quick visual inspection of the suspension components found behind the wheel and tire. Look for defects involving the springs, spring hangers, shackles, U-bolts, and shock absorbers. Springs should not have cracked or otherwise broken leaves. With the vehicle on a level surface, the springs on each side of the vehicle should have approximately the same amount of deflection.

There are a number of things that should be checked relative to the tires themselves. Some of the more important things that driver/operators should check include the following:

- *Proper tire inflation* — The tires should be inflated to the pressure noted by the manufacturer on the side of

Figure 3.9 Check the tires for damage and proper inflation.

the tire. Too much or too little pressure can damage the tire and cause poor road-handling characteristics.

- *Valve stem condition* — The valve stem should not be cut, cracked, or loose.

- *Tire condition* — Check for proper tire type, tread depth (varies according to state or provincial inspection requirements), tread separation, excessive wear to the sidewalls, cuts, or objects impaled in the tire.

The driver/operator may choose to check equipment in the rear portion of the cab at this time. This includes items such as portable fire extinguishers, self-contained breathing apparatus (SCBA), and emergency medical equipment. All of this equipment must be in proper working order and securely stowed. Consult IFSTA's **Essentials of Fire Fighting** manual for more information on how to properly inspect the fire fighting and rescue equipment carried on the apparatus. Once the left and right sides of the vehicle have been inspected, proceed to the front of the vehicle.

Front Inspection

As with the left- and right-side inspections, the first thing that should be noted when moving to the front of the vehicle is any significant body damage that was not present in previous inspections. The driver/operator may choose to take a quick look beneath the vehicle to note any obvious damage to the front axle, steering system, or pump piping (if present). Be observant for any loose, bent, worn, damaged, or missing parts. Some vehicle air systems have storage tanks that must be drained manually. Refer to the owner's manual for the location and procedures for manual drains.

Check the condition of the windshield and the wiper blades. The windshield should be free of defects and should be clean. The wipers should be held against the windshield with an appropriate amount of tension. The wiper blades should be intact and in good condition.

With the apparatus running or hooked into the electrical charging system, the condition of all lights, both running and emergency, should be checked for proper operation. It may be helpful to have a second firefighter in the cab operate the various light switches so that their operation can be properly checked. The lenses for all lights should be in place and not cracked or broken. Make sure all bulbs are working. Check the function of high-beam headlights and turn signals. It may be desirable to skip this check for now and wait until the apparatus operational checks are made. This is a matter of individual preference.

Any audible warning devices on the front of the vehicle should be checked for visible damage. This includes electronic siren speakers, mechanical sirens, and air horns.

CAUTION: Do not test the operation of audible warning devices with someone standing in front of the vehicle. This can cause hearing damage to that person. These devices may be tested by the driver/operator in the cab when no one else is in a position to be harmed.

Many modern fire department vehicles have functional emergency equipment on the front bumper area. The driver/operator should check this equipment to make sure it is in place and operable. The following is a list of some of the more common devices that are found on the front of fire apparatus and things about them that should be checked.

- *Pump intakes* — If the intake is normally capped, make sure the cap is tight enough to prevent air leaks when attempting to draft, but not so tight that it cannot be easily removed. If intake hose is normally preconnected, make sure the hose is firmly attached. If an intake valve is provided at that location, make sure it is fully closed (Figure 3.10). Intake hose should be in good physical condition and properly stowed.

- *Pump discharges* — Many apparatus have pump discharges and attack handlines located on the front of the apparatus (Figure 3.11). Check these in the same manner as described for intakes. Hose may be attached to the front bumper in a variety of ways. These include being in recessed wells, as cross lays, and on booster reels. Make sure the hose is loaded properly and is secure for road travel. Nozzles should be clean and in place.

- *Winches* — Many vehicles, especially rescue and wildland fire apparatus, are equipped with electric or hydraulic winches on the front bumper (Figure 3.12). The driver/operator should check them for proper operation. All associated components of the winch, including the remote control, cable, chains, hooks, and clasps, should be inspected for damage.

- *Hydraulic rescue tool systems* — It is becoming increasingly popular to mount hydraulic hose reels and small powered hydraulic rescue tools on the front bumper of the vehicle (Figure 3.13). These should be checked for proper operation, damage, and cleanliness.

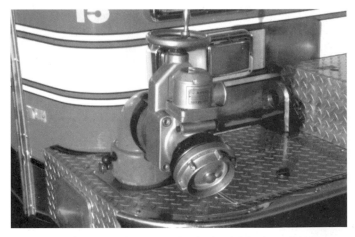

Figure 3.10 Check the front pump intake for debris that might be clogging the strainer.

Figure 3.11 Make sure that any attack lines that are on the front of the apparatus are properly stowed.

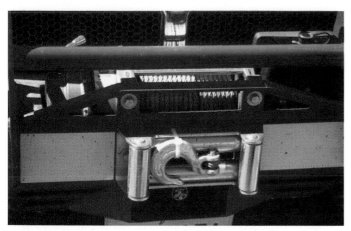

Figure 3.12 Check the front winch for proper operation and cable condition.

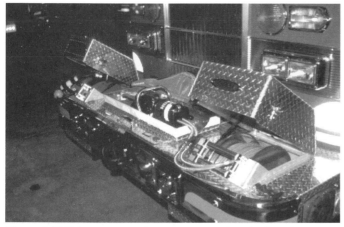

Figure 3.13 If the apparatus has a rescue tool on the front bumper, it should be checked for proper operation and stowing.

If the apparatus is equipped with a front-mounted fire pump, the driver/operator may choose to perform the pump check at this time. Information on performing pump checks is found later in this chapter. Once all of the front-end equipment has been checked, the driver/operator may proceed to the right-front side of the vehicle and follow the same procedure as listed for the left-front side. Once the right-front side inspection is complete, the driver/operator may continue down the right side of the vehicle using the same procedures as described for the left-front side.

Left- and Right-Rear Side Inspections

This part of the inspection should cover everything from the rear of the cab to the tailboard on each side of the apparatus (Figure 3.14). Again, note any obvious body damage that has occurred since the previous inspection. The same principles listed in the front-side inspection for checking the tires and suspension components may be followed here. Keep in mind that most fire apparatus have dual wheels on the rear axle. In addition to checking tire conditions as previously listed, the driver/operator

should also make sure that dual tires do not come in contact with each other or other parts of the vehicle. Rear wheels should be equipped with splash guards (mud flaps). Splash guards should be properly attached to the vehicle, be in good condition, and not be dragging on the ground.

Apparatus in cold weather climates may be equipped with automatic snow chains (Figure 3.15). These are mechanical devices that are activated by a switch in the cab. When activated, a rotating hub with numerous lengths of tire chain swings into place just ahead of the rear tires. The chains are swung around in a rotating motion so that they fall beneath the rear tires as they move forward. This provides better traction on snow- or ice-covered roads. The driver/operator should inspect the automatic snow chains to make sure that all of the chains are present and in good condition. During periods of inclement weather, it may be desirable to activate the chains and make sure they are operating properly.

All compartment doors should be opened and the equipment inside them inspected. Make sure all equipment that is supposed to be in each compartment is actually there and properly stowed. The driver/operator may choose to perform equipment inspections at this

Figure 3.14 This part of the inspection should cover everything from the rear of the cab to the tailboard on each side of the apparatus. *Courtesy of Joel Woods.*

Figure 3.15 Automatic tire chains swing into motion when a switch is activated in the cab of the apparatus.

time. As previously stated, information on inspecting equipment is contained in the IFSTA **Essentials of Fire Fighting** manual. The compartment and the equipment it contains should be neat and clean. When closing each door, make sure the door closes tightly and latches properly.

Any hose that is stored midship or on the side of the vehicle should be examined for proper stowing and security. This includes preconnected attack lines that traverse the midship area of the apparatus or are on top of the fender compartments (Figures 3.16 a and b). Top-mounted booster reels may also be checked at this time. The water level in the booster tank may also be visually checked through the top vent opening or sight glass (Figure 3.17).

Any equipment that is stored on the exterior of the vehicle should also be checked to make sure it is in good physical condition and properly stowed. This includes ladders, intake hose, hose in the hose bed, forcible entry tools, SCBA and/or spare cylinders, handlights, floodlights, cord reels, portable water tanks, portable fire extinguishers, and other portable equipment. Equipment that is stored above the pump panel area may also be checked at this time.

Note the condition of reflective striping on the side of the apparatus. Major losses of this striping affect the vehicle's visibility, particularly at night. Also check the operation of side-mounted warning lights.

Once everything on the right side of the vehicle has been checked, proceed to the rear of the vehicle.

Rear Inspection

When inspecting the rear of the vehicle, note the rear bumper or tailboard for new damage. Using the same procedure as that used on the front of the vehicle, make sure that all running and warning lights on the rear of the vehicle are in proper working order. Any equipment contained in the rear compartment should also be checked to make sure it is present, clean, operable, and properly stowed. Make sure the rear compartment door opens and closes properly. Any equipment stored on the outside of the rear of the apparatus should also be in proper working order and securely stowed. This includes portable fire extinguishers, spanner wrenches, hydrant wrenches, hydrant valves, portable master stream devices, etc. (Figure 3.18). Check any towing attachments for defects.

The driver/operator should also inspect the hose loads in the main hose bed. Make sure an adequate amount of hose is loaded on the apparatus and that it is loaded and finished correctly (Figure 3.19). If the apparatus is equipped with a hose bed cover, it should be in good

Figure 3.16a Preconnected handlines are commonly stowed in a cross lay arrangement above or below the pump panel.

Figure 3.16b Some departments have preconnected handlines in special compartments on the fender above the rear wheels.

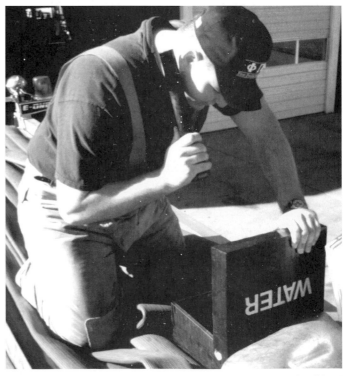
Figure 3.17 To be sure of the actual level in the water tank, the driver/operator should make a visual inspection through the tank vent.

Figure 3.18 Inspect any equipment that is stored on the apparatus tailboard. Make sure that it is securely fastened.

Figure 3.19 Check the hose load to make sure that it is finished according to standard operating procedures.

condition and in place. If solid hose bed doors are used to protect the hose, they should be checked to make sure they stay open when necessary (Figure 3.20). Any other equipment that is stored in the hose bed area should be checked for proper condition and stowing. This includes ground ladders, portable water tanks, intake hose, and pike poles (Figure 3.21).

Figure 3.20 Some departments equip their apparatus with hinged, metal doors to protect the hose in the bed. *Courtesy of Emergency One.*

Figure 3.21 Inspect any other equipment that may be stowed in the hose bed area.

Once the rear of the vehicle has been inspected, the driver/operator should proceed to the left rear side of the vehicle and proceed to check it following the information covered in the previous section.

In-Cab Inspection

Once the driver/operator has completed inspecting the outside of the apparatus, he should enter the cab and begin the mechanical check of the apparatus. Prior to starting the apparatus, if it is not already running, the driver/operator should make sure that the seat and mirrors are adjusted in a suitable manner. If the vehicle is not already running, it should be started at this time. Follow the manufacturer's instructions for starting the apparatus. All electrical switches should be in the off position when the apparatus is started to avoid an excessive load on the battery. Sudden temperature changes are not good for any engine. When an engine is started for nonemergency response, do not run it under full load until the engine has had time to warm up to its operating temperature. Likewise, anytime an engine has been running at full load, follow the engine manufacturer's recommendation for shutdown procedures. For more information on proper starting techniques, see Chapter 4 of this manual.

The seatbelts/restraints should be securely mounted and should operate freely without binding. Webbing must not be cut or frayed. Buckles must open and close freely. Mirrors should not be missing or broken. Check the tilt/telescopic steering wheel for proper position (Figure 3.22).

Once the vehicle is running, the driver/operator should assure that all of the gauges on the dashboard show the apparatus to be functioning in the normal operating range. Depending on the design of the apparatus, some or all of the following gauges may be found on the dashboard:

- Speedometer/Odometer
- Engine speed (measured in revolutions per minute [rpm])
- Oil pressure
- Fuel gauge
- Ammeter/Voltmeter
- Air pressure
- Coolant temperature
- Vacuum gauge
- Hydraulic pressure gauge

The speedometer should be at or very near zero with the truck parked (Figure 3.23). If it is showing anything above that, one of the two following possibilities exists:

1. The gauge is defective.

Figure 3.22 Make sure that the steering wheel is adjusted to fit the driver.

Figure 3.23 The speedometer should be at or very near zero with the truck parked.

2. The truck is in pump gear, rather than drive gear. Take the truck out of pump gear at this time. Information on how to do this is contained in Chapter 11 of this manual.

The fuel gauge should be checked to make sure an adequate amount of fuel is in the vehicle's fuel tank. Each department usually has its own policy on the minimum level of fuel that is permissible in the vehicle. As a rule of thumb, it is generally best to keep the fuel tank at least three-quarters full at all times. This assures that the apparatus can be operated for an extended period of time without the need to be refueled. The driver/operator should know the size of the fuel tank and how long it will last during prolonged pumping operations so that fuel can be requested in time when needed.

All other gauges should be checked to make sure they are operating within the intended limits specified by the apparatus manufacturer. These limits are typically graphically noted directly on the face of the gauge.

The driver/operator should also assure that all controls located in the cab are in proper operating condition. These would include the following:

- Electrical equipment switches.
- Turn signal switches.
- High beam headlight switches.
- Heating and air-conditioning controls.
- Radio controls (Figure 3.24).
- Audible warning device controls (sirens, auto warning horns, air horns, back-up alarms, etc.). **NOTE:** Appropriate hearing protection should be worn if any personnel will be exposed to noise levels in excess of 90 decibels.

Figure 3.24 Make sure that the radio and siren controls are set properly and in working order.

- Controls for any computer equipment in the cab (mobile data terminal [MDT], mobile computer terminal [MCT], etc.).
- Windshield wiper controls.
- Window defroster controls.
- Automatic snow chain control (if applicable).

The driver/operator should keep in mind that newer apparatus may be equipped with electrical load management systems. These devices are intended to prevent an overload of the vehicle's electrical generation system. Overload can be a problem because of the large amount of electrical equipment that tends to be added to modern apparatus. In general, these devices incorporate a load sequencer and load monitor into the same device. The *load sequencer* turns various lights on at specified intervals so that the start-up electrical load for all of the devices does not occur at the same time. The *load monitor* "watches" the system for added electrical loads that threaten to overload the system. When an overload condition occurs, the load monitor will shut down less important electrical equipment to prevent the overload. This is referred to as *load shedding.* For example, when activating the electric inverter to turn on two 500-watt floodlights, the load monitor may shut down the cab air-conditioning system or compartment lights. Driver/operators must understand the design of the electrical load management system on their apparatus so that they can determine if it is operating properly. They must also know the difference between load shedding and an electrical system malfunction.

If the apparatus is equipped with a manual shift transmission, the driver/operator should check the adjustment of the clutch pedal. The pedal should not have insufficient or excessive free play (also called *free travel*). *Free play* is the distance that the pedal must be pushed before the throw-out bearing actually contacts the clutch release fingers. Insufficient free play will cause the clutch to slip, overheat, and wear out sooner than necessary. The throw-out bearing will also have a shorter life. Excessive free play may result in the clutch not releasing completely. This can cause harsh shifting, gear clash, and damage to gear teeth. Driver/operators should be familiar with the normal amount of free play in vehicles they are assigned to drive. Any clutch that does not have the appropriate amount of free play should be checked as soon as possible by a certified mechanic.

The steering system should be checked for proper adjustment and reaction. The driver/operator can most effectively accomplish this by checking the steering wheel for excess play that does not result in the actual movement of the vehicle's front tires. In general, steering wheel play should be no more than about 10 degrees in either direction (Figure 3.25). On a steering wheel that has a 20-inch (500 mm) diameter, this will mean a play of about 2 inches (50 mm) in either direction. Play that exceeds these parameters could indicate a serious steering problem that might result in the driver/operator losing control of the apparatus under otherwise reasonable driving conditions.

Making sure that the vehicle's brakes are in proper operating order is an extremely important part of the vehicle inspection process. Many serious fire apparatus collisions have been caused by faulty brakes. There are a number of tests that can be used to check the function of the brakes. Federal, state, or provincial laws may dictate how and when brakes are tested. This section highlights the more common ones.

Most large, modern fire apparatus are equipped with air-operated braking systems. Smaller late-model apparatus and some older large apparatus are equipped with hydraulic braking systems. Most newer apparatus, regardless of whether they have air-operated or hydraulic brakes, are equipped with antilock braking systems (ABS) that reduce the possibility of the apparatus being thrown into a skid when the brakes are fully applied. Driver/operators must know the exact type of braking system on their apparatus so that the apparatus may be tested and driven in an appropriate manner. Procedures for using the different types of braking systems during road travel are covered in Chapter 4 of this manual.

An NFPA 1901 braking test requires that new apparatus be brought to a complete stop from a speed of 20 mph (32 km/h) in a distance not to exceed 35 feet (10.7 m). This test must be performed on a dry, paved surface. It is not recommended to conduct this test on a regular basis because it will cause excessive wear on the braking system components. The standard also requires the parking brake to hold the apparatus in place on a grade of 20 percent (Figure 3.26). In many jurisdictions, this type of grade may not be available, so testing by this method may not be practical.

There are several other NFPA requirements that driver/operators should be aware of and watch for on a regular basis. On apparatus equipped with air brakes, the standard requires the air pressure to build to a sufficient level to allow vehicle operations within 60 seconds of starting. If the driver/operator has to run the vehicle for more than 60 seconds to build sufficient air pressure, the apparatus should be checked by a certified mechanic. It may be necessary to add an electric air compressor or hook the apparatus into a fire station compressed air system when the apparatus is not in use in order to maintain adequate air pressures. This prevents delays in emergency responses.

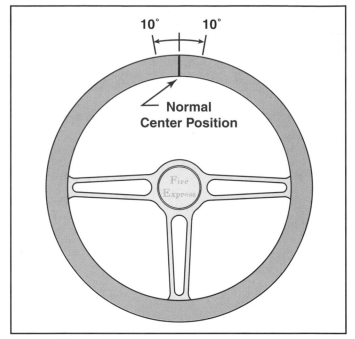

Figure 3.25 The steering wheel should not have excessive play in it.

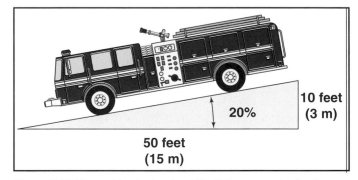

Figure 3.26 The parking brake should hold the apparatus in place on grades of up to 20 percent.

Apparatus with air brakes are to be equipped with an air pressure protection valve that prevents the air horns from being operated when the pressure in the air reservoir drops below 80 psi (552 kPa). If the driver/operator notices that this is a problem, it must be reported to the appropriate person and corrected.

To test the road brakes, allow the apparatus to move forward at about 5 mph (10 km/h). Then, push down on the brake pedal firmly. The apparatus should come to a complete stop within about 20 feet (6 m). Any of the following conditions may signify a brake problem and will require a mechanic to check the truck more thoroughly:

- The vehicle pulls to one side when the brakes are applied.

- The brake pedal has a mushy or otherwise unusual "feel" when the brakes are applied.

- The vehicle is not brought to a stop within about 20 feet (6 m).

The parking brake may also be tested in a similar manner. Allow the apparatus to move forward at about 5 mph (10 km/h). Then, activate the parking brake control (Figure 3.27). The apparatus should come to a complete stop within about 20 feet (6 m). Consult the apparatus manufacturer's operation manual or laws affecting the local jurisdiction for more detailed information on the braking system and brake tests.

Figure 3.27 On apparatus equipped with an air braking system, the parking brake control is most commonly a push-pull switch.

Figure 3.28 The tilt cab control mechanism must be activated before the cab may be lifted.

Engine Compartment Inspection

Once the entire exterior of the apparatus has been inspected and the in-cab checks have been completed, the driver/operator should shut down the vehicle and prepare to perform some routine checks and preventive maintenance procedures in the engine compartment. Depending on personal preference and departmental SOPs, the driver/operator may prefer to perform these checks before the apparatus has been started. Keep in mind that if this is the case, the readings of some fluid levels (crankcase oil, transmission fluid) have to be adjusted for the cold engine. Just remember that while it is acceptable to perform these checks either before or after the engine has been run, most (with the exception of automatic transmission fluid level) should not be done *while* the engine is running. For vehicles with a tilt cab, check to ensure that the level and/or control mechanism operates freely without binding (Figure 3.28). Check for the proper operation of cab lift motors and pumps (Figure 3.29). There is no established order in which to check the necessary items in the engine compartment. Each driver/operator will probably have his own preferred way of doing it. It is often advisable to follow the order of the required tasks listed on a department apparatus inspection form. As a minimum, the following items should be checked in the engine compartment:

Figure 3.29 Tilting the cab provides access to the apparatus engine and transmission.

Figure 3.30 Check the oil to make sure that it is maintained at an appropriate level.

- *Engine (crankcase) oil level*— Use the dipstick that is provided to assure that the oil level falls within the proper parameters (Figure 3.30). If the oil level is low, add a sufficient amount of the approved type of oil through the fill opening on the engine block. Consult the manufacturer's operation manual for direction on the proper type of oil use.

- *Engine air filter* — Inspect the air intake system for signs of damage (Figure 3.31). Some air intake systems include an air filter restriction gauge to show when it is time to change the filter. Refer to the owners manual for details.

- *Emergency shutdown* — Test according to manufacturer's instructions to ensure the emergency operating system operates properly and resets to the proper position. Some diesel engines incorporate an emergency shutdown switch and a manual reset.

- *Exhaust system* — Visually inspect the exhaust system for damage. Ensure that the rain cap on the exhaust system operates freely.

- *Radiator coolant (antifreeze) level* — Check by removing the cap on the antifreeze fill opening, commonly located on the coolant system overflow reservoir, or by viewing through the sight glass if one is provided (Figure 3.32). There will generally be at least one mark on the inside of the reservoir indicating the proper level for the antifreeze. Some vehicles' coolant reservoirs have two marks: one for when the engine is hot and the other for when it is cool. Add approved antifreeze when the level is low. The driver/operator should also check the condition of radiator hoses to make sure no leaks are evident. Make sure no debris, such as leaves or trash, is resting against the radiator intake because it reduces the cooling efficiency of the unit.

Figure 3.31 Inspect the air filter and other intake system components for damage.

Figure 3.32 It is best to inspect the coolant level when the engine is cool.

WARNING

Use caution when removing the radiator fill cap on a vehicle that is currently running or has been recently running. Boiling antifreeze and/or steam may be emitted, causing severe injury to the person removing the cap. It is most desirable to check this item when the engine and radiator system are cool.

- *Cooling fan* — Inspect the cooling fan for cracks or missing blades (Figure 3.33).

WARNING

Some engine cooling fans are activated automatically without warning. Use caution when working around the fan.

- *Windshield washer fluid level* — Check level of fluid, which is usually contained in a semitransparent tank or pouch that is appropriately labeled (Figure 3.34).

Figure 3.33 Inspect the cooling fan for cracks or missing blades.

Figure 3.34 It is recommended that the windshield washer fluid reservoir be refilled any time it is less than one-half full.

Figure 3.35a Some batteries are located in the engine compartment.

Figure 3.35b Custom apparatus frequently have their batteries in a location that is remote from the engine.

There is no specified level of the fluid in this reservoir. The driver/operator will simply add fluid as it becomes depleted. It is recommended that the reservoir be refilled any time it is less than one-half full. These fluids are commercially available, and compatibility from one brand to another is usually not a concern. Plain water will work; however, it has a tendency to freeze in cold weather conditions.

• *Battery condition* — Check various battery components. Depending on the design of the vehicle, the batteries may be located in the engine compartment or in a separate compartment elsewhere on the vehicle (Figures 3.35 a and b). Most modern vehicle batteries are the sealed type that require no internal inspection by the driver/operator. Unsealed batteries have caps on them that must be carefully removed so that the electrolyte (water) level in the batteries may be checked. Add water if the internal level is low. Also check all the battery connections for tightness and excessive corrosion (Figure 3.36). Corrosion around the battery terminal connections may be cleaned by mixing baking soda (which is chemically a *base*) with water and pouring it on the connections. This will cause a neutralization reaction with the corrosive (*acidic*) buildup. The terminal connection may then be cleaned with a wire brush and rinsed with clear water. Lastly, check the battery tie-downs to make

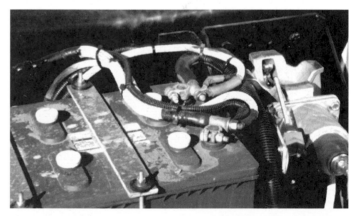

Figure 3.36 Make sure that the battery terminal connections are free from corrosion.

sure the battery is held firmly in place. Information on charging or jump-starting dead batteries is contained later in this chapter. If the apparatus is equipped with a built-in battery charger, make sure this unit is operating properly.

CAUTION: Wear appropriate personal protective equipment, including eye protection, when working with batteries. Contact with battery acid may damage the skin or eyes. Also work in a well-vented area so that any fumes that are developed will dissipate.

- *Automatic transmission fluid level* — Check fluid level in a similar manner to that of crankcase oil in many vehicles. A special dipstick is provided for this purpose. The correct level of the transmission fluid is noted on the dipstick. Some newer vehicles may be equipped with transmissions that provide an electronic readout of the transmission fluid level. Be familiar with the proper method for checking the fluid level in your vehicle. Add fluid if the reading on the dipstick indicates it is necessary. Depending on the manufacturer's recommendations, it may be necessary to check the transmission fluid level while the vehicle is *running*. If this is the case, the driver/operator may wish to perform this check first, before shutting off the vehicle.

- *Power steering fluid level* — Check the level of this fluid using the indicator marks provided by the manufacturer. Add approved power steering fluid if the level is low. Some power steering systems require the fluid level to be checked while the engine is running at normal operating temperature. Be careful not to overfill the reservoir because damage can occur to the system.

- *Brake fluid (hydraulic brake systems)* — Check the level of the brake fluid in the master brake cylinder using the procedure specified by the manufacturer, and add fluid if needed.

- *Air system* — Check for leaks in the system. With the air system at normal operating pressure and the engine shut off, walk around the apparatus and listen for leaks. Some apparatus may have an air dryer filter that must be inspected.

- *Belts* — Check all belts (water pump, air compressor, fan, alternator, etc.) in the engine compartment for tightness and excessive wear. The driver/operator must be familiar with the proper feel for tightness of each belt when it is properly adjusted. Most engines have multiple drive pulleys.

> ## WARNING
> Never attempt to check the belts while the vehicle's engine is running. This could result is severe injuries or entrapment in the belt and pulley(s).

- *Leaks* — Look for leaks of any of the fluids used in the vehicle's engine, including antifreeze, water, windshield wiper fluid, oil, transmission fluid, hydraulic fluid, power steering fluid, and/or battery fluid. Inspect the condition of all hoses and hydraulic lines for the presence of fluids.

- *Electrical Wiring* — Check the general condition of all electrical wiring in the engine compartment. Look for wires that are frayed, cracked, loose, or otherwise worn. Have a mechanic correct any wiring problems that are detected.

Some departments require driver/operators to check chassis lubrication. Proper lubrication saves maintenance and repair dollars and reduces out-of-service time. Effective lubrication depends upon the use of a proper grade of recommended lubricant, the frequency of lubrication, the amount used, and the method of lubrication. To select the proper lubricant, consideration must be given to the requirements of the unit to be lubricated, the characteristics of lubricants, and the manufacturer's recommendations. The manufacturer's manual will "recommend" the Society of Automotive Engineers (SAE) numbers for the engine oil. The SAE number indicates only the viscosity. Some other essential characteristics of oil are corrosion protection, foaming, sludging, and carbon accumulation, which may be controlled by the refiner. Do not mix different types of engine oil. Always consult the owner's manual for the type of oil and the location of fill ports and grease fittings.

If the driver/operator is expected to lubricate the chassis components, he must be familiar with all of the lubrication fill connections. These have an appearance that is similar to the valve stem on a tire (Figure 3.37). When located, the driver/operator should press the end of the lubrication gun fill hose onto the inlet. Operate the pump handle on the gun until no more lubricant enters the inlet (lubricant squeezes out between the hose outlet and the inlet). Continue this process around the apparatus until all of the inlets have been filled. Newer apparatus may be equipped with automatic vehicle lubrication systems. If your apparatus is equipped with such a system, follow the manufacturer's directions for operating the system.

Figure 3.37 Lubricate the chassis at every available grease fitting.

In addition to those listed, each fire department may specify other parts of the engine or apparatus that the driver/operator is responsible for checking. The driver/operator should ensure that all necessary items are checked and documented according to departmental policies.

Posttrip Inspections

All of the functions described to this point in the chapter are intended to be performed prior to the operation of the vehicle. It is also a good idea to perform this type of an inspection after the vehicle has been operated for a prolonged period. This could include after operating at a fire for a long period or after an unusually long road trip. Long road trips may occur in remote areas that have large response districts, on mutual aid responses, following parades, or following trips to distant service centers. Each department should have a policy for performing posttrip inspections. IFSTA recommends that the same procedures described for pretrip inspections be used for posttrip inspections. However, some jurisdictions may choose to use an abbreviated procedure. Each driver/operator must know what is expected of him in these circumstances.

Charging Batteries

Historically, the charging of vehicle batteries is a function that is commonly performed by fire apparatus driver/operators. Apparatus batteries often require charging or jump-starting because of long periods of inactivity or improper drains on the electrical system. However, apparatus that regularly require charging or jump-starting should be reported to the department mechanic so that corrective measures may be taken.

In the station, these charging functions are generally performed with a battery charger. Because batteries produce explosive hydrogen gas when being charged, chargers must be used correctly to prevent needless accidents.

Driver/operators should be proficient in the use of battery chargers because they often need to be used in a rapid manner to start an apparatus prior to an emergency response. Follow the procedure outlined below when attaching battery charger cables to any vehicle battery. The driver/operator should be wearing safety eye protection whenever performing this evolution.

Step 1: Make sure that the battery and ignition switch(es) are in their off positions.

Step 2: Identify the polarity of the battery to be charged (positive or negative ground).

Step 3: Attach the red (positive or "+") charger cable to the red (positive or "+") battery post (Figure 3.38).

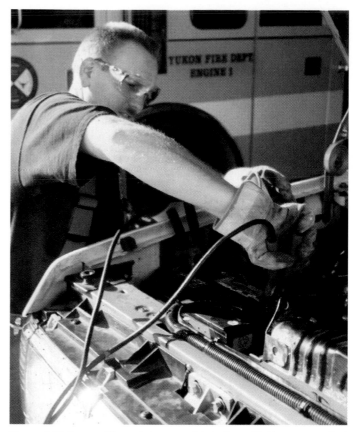

Figure 3.38 Connect the red charger cable to the red (+) battery terminal post.

Step 4: Attach the black (negative or "-") charger cable to the black (negative or "-") battery post.

Step 5: Connect the battery charger to a reliable power source (away from gasoline and other flammable vapors).

Step 6: Set the desired battery charging voltage and charging rate if so equipped. (**NOTE:** Switches on battery charger should be in the **OFF** position when not in use.)

Step 7: Reverse the procedure to disconnect the battery charger.

General Fire Suppression Equipment Maintenance

Although fire pumps are tested at regularly scheduled times, these tests are designed to compare performance to specific standards. In addition, certain preventive maintenance procedures should be regularly performed to detect pump failure or other apparatus fire suppression equipment deficiencies. Some departments require the driver/operator to perform these functions. Others may have mechanics or pump technicians perform these functions. Not all of the maintenance functions need to be conducted on a daily basis.

Daily Inspections

Although a complete apparatus inspection is not needed on a daily (or in the case of some departments, weekly) basis, there are some things that the driver/operator should check/perform daily. These items include the following:

- Operate the pump drive control and make sure that the pump can be engaged (Figure 3.39). Depending on the design of the apparatus the pump may be powered by a power take-off (PTO), split shaft transmission, or separate engine. Whatever the case, make sure the pump is in working order. See Chapter 9 of this manual for information on how to operate fire pumps.

- Make sure the auxiliary fuel tank is full in the case of separate engine-driven pumps with fuel supplies independent of the main apparatus fuel tank.

- Make sure all gauges and valves on the pump panel are in working order (Figure 3.40). Any gauges that are duplicated on the pump panel and dashboard should be checked to make sure they are in agreement. You may wish to open and close each valve several times to make sure it operates smoothly. Make sure that all pump drains are closed.

- Make sure that the fire pump and booster lines are completely drained of water to prevent damage from freezing water in cold climate conditions.

- Operate the controls to check or inspect a fire pump. It is not necessary to pump full capacity or even to deliver any more water than a booster line will convey.

- Inspect the water and foam tanks (if applicable) for proper fluid level.

- Check the underside of the apparatus and inside compartments for evidence of water or foam leaks.

- Check for damage, leaks, or obstructions in any auxiliary winterization system used to prevent fire suppression water from freezing. It may be necessary to run the booster heater on a periodic basis, even during summer months, to prevent corrosion. Refer to the operator's manual for detailed instructions.

- Test roof and bumper turrets (if applicable) for proper operation and full range of motion (Figures 3.41 a and b). Test to ensure that the length and pattern of discharge conform to the specifications in the operator's manual. Follow your department's SOPs regarding frequency for checking the agent dispensing system.

- Check all components of the auxiliary fire suppression systems on board (halon, dry chemical, etc., if applicable) for damage, leaks, or corrosion. Ensure that all connections are secure. Check that all the valves are in

Figure 3.39 Operate the pump drive control and make sure that the pump can be engaged.

Figure 3.40 Make sure that all the pump panel controls and indicators are in proper working order. *Courtesy of Joel Woods.*

the normal position for operation. Check the agent level in the tank by means of an agent level or sight gauge. Check dry chemical systems for signs of lumping. If the system is equipped with a hose reel, check it for proper operation.

Weekly Inspections

The following items should be checked on at least a weekly basis:

- Flush the pump with clear water if it is a department policy to carry the pump full of water. This may be

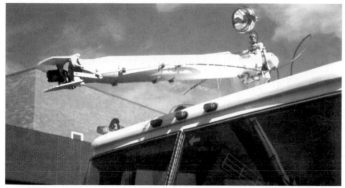

Figure 3.41a Some pumpers are equipped with roof turrets, such as this foam pumper.

Figure 3.41b Check the bumper turret for proper operation.

Figure 3.42 Flush the pump with clean water from a municipal water supply system.

Figure 3.43 Remove the intake cap to check the strainer for debris.

done while the pump is out of gear. Simply open all of the valves and drains and push water through the system until it runs clear and no debris is being discharged (Figure 3.42). Some departments prefer to thoroughly flush the pump and piping by pumping water into the system through both an intake and a discharge connection (at different times). Water for flushing the pump and piping is usually supplied from a fire hydrant. It may be necessary to briefly flush the hydrant to remove debris before connecting to the pumper.

- Check and clean the intake strainers (Figure 3.43). After flushing, it may be necessary to remove strainers to remove debris that has lodged against them.

- Check the pump gear box for proper oil level and traces of water. Consult the manufacturer's operations manual for more information on the proper oil level and corrective measures.

- Operate the pump primer with all pump valves closed.

- Operate the changeover valve while pumping from booster tank in the case of a multistage (two- or three-stage) pump (Figure 3.44).

Figure 3.44 If the apparatus is a two-stage pump, operate the changeover valve to make sure that it works properly.

- Check the packing glands for excessive leaks. Each pump manufacturer provides information on the acceptable amount of water dripping from the packing.

- Recalibrate the flowmeter according to manufacturer's instructions.

- Operate the pump pressure control device(s).

- Test the accuracy of the foam proportioning system. Consult IFSTA's **Principles of Foam Fire Fighting** manual for more information on foam system testing.

- Refer to the pump manufacturer's recommendations for additional instructions, if any.

Some of these functions may also need to be performed on an "as-needed" basis. For example, anytime a pump has been operated at draft from a static water supply, such as a lake or pond, or in areas that have old water mains or poorly designed water systems, the pump and piping should be thoroughly flushed before the apparatus is placed back in service. If untreated water has been used to fill the apparatus water tank, it should be drained, flushed, and refilled with clean water as soon as possible. For more detailed information on fire pump testing, consult Chapter 12 of this manual.

Chapter 4

Operating Emergency Vehicles

Job Performance Requirements

This chapter provides information that will assist the reader in meeting the following job performance requirements from NFPA 1002, *Standard on Fire Apparatus Driver/Operator Professional Qualifications*, 1998 edition. Particular portions of the job performance requirements (JPRs) that are met in this chapter are noted in bold text.

2-3.1* **Operate a fire department vehicle, given a vehicle and a predetermined route on a public way that incorporates the maneuvers and features specified in the following list that the driver/operator is expected to encounter during normal operations, so that the vehicle is safely operated in compliance with all applicable state and local laws, departmental rules and regulations, and the requirements of NFPA 1500,** *Standard on Fire Department Occupational Safety and Health Program,* **Section 4-2.**

- **Four left and four right turns**
- **A straight section of urban business street or a two-lane rural road at least 1 mile (1.6 km) in length**
- **One through-intersection and two intersections where a stop has to be made**
- **One railroad crossing**
- **One curve, either left or right**
- **A section of limited-access highway that includes a conventional ramp entrance and exit and a section of road long enough to allow two lane changes**
- **A downgrade steep enough and long enough to require down-shifting and braking**
- **An upgrade steep enough and long enough to require gear changing to maintain speed**
- **One underpass or a low clearance or bridge**

(a) *Requisite Knowledge:* **The effects on vehicle control of liquid surge, braking reaction time, load factors, general steering reactions, speed, and centrifugal force; applicable laws and regulations; principles of skid avoidance, night driving, shifting, and gear patterns; negotiating intersections, railroad crossings, and bridges; weight and height limitations for both roads and bridges; identification and operation of automotive gauges; and proper operation limits.**

(b) *Requisite Skills:* **The ability to operate passenger restraint devices, maintain safe following distances, maintain control of the vehicle while accelerating, decelerating, and turning, maintain reasonable speed for road, weather, and traffic conditions, operate safely during nonemergency conditions, operate under adverse environmental or driving surface conditions, and use automotive gauges and controls**.

2-3.2* **Back a vehicle from a roadway into restricted spaces on both the right and left sides of the vehicle, given a fire department vehicle, a spotter, and restricted spaces 12 ft (3.66 m) in width, requiring 90-degree right-hand and left-hand turns from the roadway, so that the vehicle is parked within the restricted areas without having to stop and pull forward and without striking obstructions**.

(a) *Requisite Knowledge:* Vehicle dimensions, turning characteristics, spotter signaling, and principles of safe vehicle operation.

(b) *Requisite Skills:* The ability to use mirrors, judge vehicle clearance, and operate the vehicle safely.

2-3.3* **Maneuver a vehicle around obstructions on a roadway while moving forward and in reverse, given a fire department vehicle, a spotter for backing, and a roadway with obstructions, so that the vehicle is maneuvered through the obstructions without stopping to change the direction of travel and without striking the obstructions**.

(a) *Requisite Knowledge:* Vehicle dimensions, turning characteristics, the effects of liquid surge, spotter signaling, and principles of safe vehicle operation.

(b) *Requisite Skills:* The ability to use mirrors, judge vehicle clearance, and operate the vehicle safely.

2-3.4* **Turn a fire department vehicle 180 degrees within a confined space, given a fire department vehicle, a spotter for backing, and an area in which the vehicle cannot perform a U-turn without stopping and backing up, so that the vehicle is turned 180 degrees without striking obstructions within the given space**.

(a) *Requisite Knowledge:* Vehicle dimensions, turning characteristics, the effects of liquid surge, spotter signaling, and principles of safe vehicle operation.

(b) *Requisite Skills:* The ability to use mirrors, judge vehicle clearance, and operate the vehicle safely.

2-3.5* Maneuver a fire department vehicle in areas with restricted horizontal and vertical clearances, given a fire department vehicle and a course that requires the operator to move through areas of restricted horizontal and vertical clearances, so that the operator accurately judges the ability of the vehicle to pass through the openings and so that no obstructions are struck.

(a) *Requisite Knowledge:* Vehicle dimensions, turning characteristics, the effects of liquid surge, spotter signaling, and principles of safe vehicle operation.

(b) *Requisite Skills:* The ability to use mirrors, judge vehicle clearance, and operate the vehicle safely.

2-3.6* Operate a vehicle using defensive driving techniques under emergency conditions, given a fire department vehicle and emergency conditions, so that control of the vehicle is maintained.

(a) *Requisite Knowledge:* **The effects on vehicle control of liquid surge, braking reaction time, load factors, general steering reactions, speed, and centrifugal force; applicable laws and regulations; principles of skid avoidance, night driving, shifting, and gear patterns; negotiating intersections, railroad crossings, and bridges; weight and height limitations for both roads and bridges; identification and operation of automotive gauges; and proper operation limits.**

(b) *Requisite Skills:* **The ability to operate passenger restraint devices, maintain safe following distances, maintain control of the vehicle while accelerating, decelerating, and turning, maintain reasonable speed for road, weather, and traffic conditions, operate safely during nonemergency conditions, operate under adverse environmental or driving surface conditions, and use automotive gauges and controls.**

3-1.2* Perform the practical driving exercises specified in 2-3.2 through 2-3.5, given a fire department pumper and a spotter for backing, so that each exercise is performed safely without striking the vehicle or obstructions.

(a) *Requisite Knowledge:* Vehicle dimensions, turning characteristics, the effects of liquid surge, spotter signals, and principles of safe vehicle operation.

(b) *Requisite Skills:* The ability to use mirrors, judge vehicle clearance, and operate the vehicle safely.

3-1.3* Operate a fire department pumper over a predetermined route on a public way that incorporates the maneuvers and features specified in the list in 2-3.1, so that the vehicle is safely operated in compliance with all applicable state and local laws, departmental rules and regulations, and the requirements of NFPA 1500, *Standard on Fire Department Occupational Safety and Health Program,* Section 4-2.

(a) *Requisite Knowledge:* **The effects on vehicle control of liquid surge, braking reaction time, load factors, general steering reactions, speed, and centrifugal force; applicable laws and regulations; principles of skid avoidance, night driving, shifting, and gear patterns; negotiating intersections, railroad crossings, and bridges; weight and height limitations for both roads and bridges; identification and operation of automotive gauges; and proper operation limits.**

(b) *Requisite Skills:* **The ability to operate passenger restraint devices, maintain safe following distances, maintain control of the vehicle while accelerating, decelerating, and turning, maintain reasonable speed for road, weather, and traffic conditions, operate safely during nonemergency conditions, operate under adverse environmental or driving surface conditions, and use automotive gauges and controls.**

6-1.2* Perform the practical driving exercises specified in 2-3.2 through 2-3.5, given a wildland fire apparatus, so that each exercise is performed safely without striking the vehicle or obstructions.

(a) *Requisite Knowledge:* Vehicle dimensions, turning characteristics, the effects of liquid surge, spotter signaling, and principles of safe vehicle operation.

(b) *Requisite Skills:* The ability to use mirrors, judge vehicle clearance, and operate the vehicle safely.

6-1.3* Operate a wildland fire apparatus over a predetermined route on a public way that incorporates the maneuvers and features specified in the list in 2-3.1, so that the vehicle is safely operated in compliance with all applicable state and local laws, departmental rules and regulations, and the requirements of NFPA 1500, *Standard on Fire Department Occupational Safety and Health Program,* Section 4-2.

(a) *Requisite Knowledge:* **The effects on vehicle control of liquid surge, braking reaction time, load factors, general steering reactions, speed, and centrifugal force; applicable laws and regulations; principles of skid avoidance, night driving, shifting, and gear patterns; negotiating intersections, railroad crossings, and bridges; weight and height limitations for both roads and bridges; identification and operation of automotive gauges; and proper operation limits.**

(b) *Requisite Skills:* **The ability to operate passenger restraint devices, maintain safe following distances, maintain control of the vehicle while accelerating, decelerating, and turning, maintain reasonable speed for road, weather, and traffic conditions, operate safely during nonemergency conditions, operate under adverse environmental or driving surface conditions, and use automotive gauges and controls.**

6-1.4* Operate a wildland fire apparatus, given a predetermined route off of a public way that incorporates the maneuvers and features specified in the following list that the driver/ operator is expected to encounter during normal operations, so that the vehicle is safely operated in compliance with all applicable departmental rules and regulations, the requirements of NFPA 1500, *Standard on Fire Department Occupational Safety and Health Program*, Section 4-2, and the design limitations of the vehicle.

- Loose or wet soil
- Steep grades (30 percent fore and aft)
- Limited sight distance
- Blind curve
- Vehicle clearance obstacles (height, width, undercarriage, angle of approach, angle of departure)
- Limited space for turnaround
- Side slopes (20 percent side to side)

(a) *Requisite Knowledge:* **The effects on vehicle control of braking reaction time, load factors, general steering reactions, speed, and centrifugal force; applicable laws and regulations; principles of skid avoidance, night driving, shifting, and gear patterns; negotiating intersections, railroad crossings, and bridges; weight and height limitations for both roads and bridges; identification and operation of automotive gauges; and proper operation limits.**

(b) *Requisite Skills:* **The ability to operate passenger restraint devices, maintain safe following distances, maintain control of the vehicle while accelerating, decelerating, and turning, maintain reasonable speed for road, weather, and traffic conditions, operate safely during nonemergency conditions, operate under adverse environmental or driving surface conditions, and use automotive gauges and controls.**

8-1.2*   Perform the practical driving exercises specified in 2-5.2 through 2-3.5, given a fire department mobile water supply apparatus and a spotter for backing, so that each exercise is performed safely without striking the vehicle or obstructions.

(a) *Requisite Knowledge:* Vehicle dimensions, turning characteristics, the effects of liquid surge, spotter signals, and principles of safe vehicle operation.

(b) *Requisite Skills:* The ability to use mirrors, judge vehicle clearance, and operate the vehicle safely.

8-1.3*   Operate a fire department mobile water supply apparatus over a predetermined route on a public way, using the maneuvers specified in the list in 2-3.1, so that the vehicle is safely operated in compliance with all applicable state and local laws, department rules and regulations, and the requirements of NFPA 1500, *Standard on Fire Department Occupational Safety and Health Program*, Section 4-2.

(a) *Requisite Knowledge:* **The effects on vehicle control of liquid surge, braking reaction time, load factors, general steering reactions, speed, and centrifugal force; applicable laws and regulations; principles of skid avoidance, night driving, shifting, and gear patterns; negotiating intersections, railroad crossings, and bridges; weight and height limitations for both roads and bridges; identification and operation of automotive gauges; and proper operation limits.**

(b) *Requisite Skills:* **The ability to operate passenger restraint devices, maintain safe following distances, maintain control of the vehicle while accelerating, decelerating, and turning, maintain reasonable speed for road, weather, and traffic conditions, operate safely during nonemergency conditions, operate under adverse environmental or driving surface conditions, and use automotive gauges and controls.**

The ability to safely control and maneuver fire apparatus is one of the most critical aspects of a driver/operator's responsibilities. Quite simply stated, the first goal of the driver/operator is to get the apparatus and its crew to the scene in an expedient yet safe and efficient manner. You cannot perform the necessary emergency operations until you arrive on the scene, and you cannot perform them at all if you never arrive.

Consider the impact on a response system when an emergency vehicle is involved in a collision while responding to an emergency. Suppose that a single-engine company is dispatched to a car fire, drives through a red light, and strikes a passenger car in an intersection while en route. At least one more engine company is going to be dispatched to respond to the collision (and likely even

more resources than that), and someone still has to be dispatched to extinguish the car fire. What started out as a single-company response now involves at least three engine companies, chief officers, ambulances, police, and other vital resources. If the driver/operator had only come to a complete stop at the intersection, all of these other resources would still be available for other potential emergencies. The impact on people's lives, including those who were hit by the fire truck and those who awaited someone to come put out their car fire, goes without saying.

Statistics compiled annually by the National Fire Protection Association (NFPA) historically show that 15 percent to 20 percent of all firefighter injuries and deaths are caused by vehicle collisions while responding to or returning from emergency calls (Figure 4.1). This equates

Figure 4.1 A significant percentage of firefighter injuries and deaths result from traffic collisions.

to about 25 firefighter deaths per year. United States Department of Transportation studies have indicated that about an equal number of civilians are killed annually in apparatus-related collisions. As we will learn in this chapter, most of these collisions are avoidable if safe driving principles are used. Sound driving principles also reduce wear and tear and extend the life cycle of the apparatus.

This chapter discusses the many elements of safe fire apparatus operation. First, it is important to understand the common causes of collisions and how we can avoid them. The driver/operator must also understand the proper techniques for starting and driving the vehicle, driving in adverse conditions, and using the warning and traffic control devices. The latter portion of the chapter provides information on performing practical driving exercises required of the driver/operator to meet NFPA 1002.

It is also important to note that while this chapter primarily addresses the operation of fire apparatus, all of the safe driving practices contained here also apply to the operation of privately owned vehicles (POVs) by volunteer firefighters. A significant number of collisions involving volunteer firefighters and POVs occur each year. Adherence to the information in this chapter can help reduce the incidence and severity of POV collisions.

Collision Statistics and Causes

It is a commonly held belief that those who fail to recognize history are doomed to repeat it. With that in mind, it is important that we review emergency vehicle collision statistics and causes. Reviewing this information helps us realize where our problems lie and what we can do to correct them.

While the NFPA keeps detailed information on firefighter injuries and deaths, its studies do not provide extensive background on collision causes or conditions.

The studies do typically note that things such as failure to wear seatbelts, poor road conditions, poor vehicle conditions, and failure to obey traffic rules are often factors in apparatus-related collisions. Outside of the work done by individual fire departments relative to their own situations, extensive research information on fire apparatus collision causes is not available. However, two statewide studies on collisions involving emergency medical service (EMS) vehicles do provide keen insight into the problem. Because EMS and fire vehicles operate under similar conditions, the information from these studies is relevant to fire apparatus.

The first study was conducted by Indiana University of Pennsylvania (IUP). It was centered on the 1,079 known providers of EMS in the state of Pennsylvania. During a one-year period, 212 collisions involving EMS vehicles were documented. As indicated by Tables 4.1 and 4.2, most collisions occurred in broad daylight on dry roads.

The second study was conducted by the New York State Department of Health EMS Program. This study looked at 1,412 EMS vehicle collisions that occurred over a four-year period. The results were very similar to the IUP study (Tables 4.3 and 4.4).

What these two studies indicate is that although we must be trained to drive in adverse weather conditions, collisions are most likely to occur during ideal vision

Table 4.1 Times that Collisions Occur (IUP Study)	
Time of Day	**Number of Collisions**
Daylight	108 (51%)
Dawn/Dusk	23 (11%)
Night	58 (27%)
Unknown	23 (11%)

Table 4.2 Road Conditions when Collisions Occur (IUP Study)	
Road Conditions	**Number of Collisions**
Dry Road	130 (61%)
Wet Road	22 (10.5%)
Snow/Ice	28 (13%)
Muddy Road	1 (0.5%)
Unknown	32 (15%)

Table 4.3	
Times that Collisions Occur (NY Study)	
Time of Day	**Number of Collisions**
Daylight	825 (70%)
Dawn/Dusk	52 (5%)
Night	283 (24%)
Unknown	12 (1%)

Table 4.4	
Road Conditions when Collisions Occur (NY Study)	
Road Conditions	**Number of Collisions**
Dry Road	891 (63%)
Wet Road	352 (25%)
Snow/Ice	90 (6%)
Muddy Road	4 (1%)
Unknown	77 (5%)

and road conditions. Thus, we must look to other reasons as the major causes of fire apparatus collisions. In general, fire apparatus collisions can be grouped into the following five basic causes:

1. Improper backing of the apparatus

2. Reckless driving by the public

3. Excessive speed by the fire apparatus driver/operator

4. Lack of driving skill and experience by the fire apparatus driver/operator

5. Poor apparatus design or maintenance

A large percentage of collisions occur while backing the vehicle. While they are seldom serious in terms of injury or death, they do account for a significant portion of overall damage costs (Figures 4.2 a and b). Backing collisions occur in a variety of locations. On the emergency scene, it is often necessary to back the apparatus into position for use or to back it out when the assignment is complete. Backing collisions also occur in parking lots while the apparatus is either on routine or emergency duties. Lastly, backing collisions commonly occur when backing the apparatus into the fire station.

Reckless driving by the public occurs in many forms. Some of the more common problems include:

• Failure to obey posted traffic regulations or directions

Figure 4.2a Backing accidents are the most common cause of apparatus damage. *Courtesy of Chris Mickal.*

Figure 4.2b The result of an apparatus backing accident. *Courtesy of Ron Jeffers.*

• Failure to yield to emergency vehicles

• Excessive speed

• Unpredictable behavior created by a panic reaction to an approaching emergency vehicle

• Inattentiveness

Fire apparatus driver/operators must always be cognizant of the fact that they have little control over the way members of the public react toward them. With this in mind, driver/operators must never put themselves, or the public, in a situation where there is no alternative (other than crashing into each other).

The urgency of the emergency often leads to the driver/operator driving the fire apparatus at speeds faster than should reasonably be used. Excessive speed may lead to one of the two following types of collisions occurring:

• Control of the apparatus is lost on a curve or adverse road surface, which may cause the vehicle to leave the road surface, roll over, or strike another vehicle or object (Figure 4.3).

• The driver/operator is unable to stop the apparatus in time to avoid a collision with another vehicle or object.

Figure 4.3 Many apparatus that leave the roadway in an uncontrolled manner eventually roll over. *Courtesy of the National Interagency Fire Center.*

Driver/operators must remember that the fire apparatus they are driving does not handle the same, or stop as fast, as the privately owned vehicle they drove to the fire station. It takes a much greater distance for a fire apparatus to stop than does a smaller passenger vehicle because fire apparatus weigh substantially more than standard passenger vehicles. As well, the air brake systems commonly used on fire apparatus take a little longer to activate and stop a vehicle than do the hydraulic/mechanical brake systems on smaller passenger vehicles. More detailed information on stopping distances and braking times is contained later in this chapter. Fire apparatus also tend to be top-heavy and more severely impacted by quick turns and maneuvers than are smaller vehicles.

Lack of driving skill by the driver/operator may be attributed to a number of factors, including insufficient training and unfamiliarity with the vehicle. Fire departments must ensure that all driver/operator candidates complete a thorough training program *before* they are allowed to drive fire apparatus under emergency conditions. Never assign someone who has not been trained on a particular vehicle to drive it. This unfamiliarity with the driving characteristics of the vehicle can lead to a serious collision.

There are a number of other factors that may contribute to collisions that involve driver/operator error as the cause. These factors include the following:

- *Overconfidence in one's driving ability* — The driver/operator may have an inflated opinion of his capabilities during the adrenaline rush of an emergency response. This may cause him to operate the vehicle in a manner in which neither he nor the vehicle is capable.

- *Inability to recognize a dangerous situation* — Poorly trained driver/operators do not have the ability to sense when they are approaching a hazardous situation. In a study of commercial truck drivers by the Society of Automotive Engineers (SAE), it was determined that in 42 percent of all collisions, the driver/operator was not aware of a problem until it was too late to correct it.

- *False sense of security because of a good driving record* — This manifests itself in the attitude "I've never had an collision before, why should I worry about one now."

- *Misunderstanding of apparatus capabilities* — Many driver/operators are ignorant of the potential for disaster the emergency vehicle presents. They do not know the capabilities of the apparatus that can be used to avoid collisions. They are also unaware of simple vehicle characteristics such as the fact that a fire apparatus is lighter and will travel faster when the water tank is empty, which makes the vehicle more likely to skid under certain conditions.

- *Lack of knowledge about how to operate the controls of the apparatus in an emergency* — Again, this is a problem associated with insufficient training.

Poor vehicle design and maintenance have been attributed to many serious fire apparatus collisions. Poor vehicle design problems are typically not as serious with vehicles that have been built by a fire apparatus manufacturer as they are with "homebuilt" vehicles that have been constructed by members of a fire department or by local mechanics. These vehicles are often built on government-surplus or otherwise-used vehicle chassis. Commonly, these vehicles are overweight, have high centers of gravity, and are on chassis that were worn out before they were ever made into a fire apparatus (Figure 4.4).

In particular, homebuilt water tenders (tankers) have had a high incidence of serious collisions. They are commonly constructed on old military 6×6 chassis, as well as converted fuel oil or gasoline tankers (Figure 4.5). These chassis frequently are not designed for the weight of the water that will be carried on them. Keep in mind that a gallon (liter) of water (8.33 pounds [1 kg]) weighs more than a gallon (liter) of gasoline or fuel oil (5.6 pounds [0.67 kg] and 7.12 pounds [0.85 kg] respectively). Multiplied by a thousand gallons (4 000 L) or more, this difference becomes significant.

Another problem is that in many cases the water tanks are improperly baffled. Thus, when the vehicle is being driven with a partially filled tank, liquid surges within the tank can result in the vehicle becoming out of control. More information on tank baffles and liquid surges is contained in Chapter 14 of this manual.

Figure 4.4 An apparatus with a high center of gravity is especially prone to rollover accidents.

Figure 4.5 Water tenders that are constructed from old milk or petroleum trucks must be driven with special care.

Poor maintenance of apparatus can also result in vehicle system failures that lead to collisions. This is particularly true of braking systems. Several fatal fire apparatus collisions have been traced back to improperly maintained apparatus braking systems. By following an effective apparatus maintenance program, the likelihood of mechanical failure leading to collision can be reduced.

Driving Regulations

Driver/operators of fire apparatus are regulated by federal laws, state or provincial motor vehicle codes, city ordinances, NFPA standards, and departmental policies. Copies of these laws and rules should be studied by all members of the fire department. Because the regulations vary from state to state (province to province), those discussed here are general in nature.

Unless specifically exempt, fire apparatus driver/operators are subject to any statute, rule, regulation, or ordinance that governs any vehicle operator. Statutes usually describe those vehicles that are in the emergency category; this classification usually covers all fire department vehicles when they are responding to an emergency. In some jurisdictions, the statutes may exempt emergency vehicles from driving regulations that apply to the general public concerning the speed, direction of travel, direction of turns, and parking if they are responding to a reported emergency. Under these circumstances, the driver/operator must exercise care for the safety of others and must maintain complete control of the vehicle. All traffic signals and rules must be obeyed when returning to quarters from an alarm or during any other nonemergency driving.

Keep in mind that most driving regulations pertain to dry, clear roads during daylight conditions. Driver/operators should adjust their speeds to compensate for conditions such as wet roads, darkness, fog, or any other condition that makes normal emergency vehicle operation more hazardous (Figure 4.6).

Emergency vehicles are generally not exempt from laws that require vehicles to stop for school buses that are flashing signal lights to indicate that children are boarding or disembarking. Fire apparatus should proceed only after a proper signal is given by the bus driver or police officer. The driver/operator should proceed slowly, watching for any children that may not be aware of the apparatus' approach.

A driver/operator who does not obey state, local, or departmental driving regulations can be subject to criminal and civil prosecution if the apparatus is involved in a collision (Figure 4.7). If the driver/operator is negligent in the operation of an emergency vehicle and becomes involved in a collision, both the driver/operator and the fire department may be held responsible.

Figure 4.6 Driver/operators must use extreme caution when driving in adverse weather conditions.

Figure 4.7 The driver/operator must obey all traffic laws. Most manuevers, such as passing a stopped vehicle on its right, are just as illegal in an emergency vehicle as they are in a passenger vehicle.

Starting and Driving the Vehicle

Obviously, before the driver/operator can perform virtually any other part of his duties, he must be able to start the vehicle and drive it in a safe and efficient manner. The following sections detail the standard procedures that should be used for starting and driving fire apparatus. Keep in mind that each apparatus manufacturer has specific directions that apply to its vehicles. Consult the manufacturer's operator's manual, supplied with each vehicle, for detailed instructions specific to the vehicle.

The driver/operator should start the vehicle as soon as possible so that it is warmed up when the rest of the crew is assembled and ready to respond. Let it idle as long as possible before putting it into road gear — for nonemergency response this could be 3 to 5 minutes; for an emergency response it may be only a few seconds.

When starting the apparatus under any conditions, but especially emergency response conditions, the first thing the driver/operator needs to know is where the apparatus is going. By taking the time to review the incident location, the driver/operator can consider important factors that may affect the response such as road closings and traffic conditions. The vehicle should not be moved until all occupants are within the cab, in a seated position, and wearing seat belts (Figure 4.8).

Starting the Vehicle

The following procedure may be used to start the vehicle and begin the response:

Step 1: Disconnect all ground shore lines. Generally, the first thing the driver/operator will do when preparing to board and start the apparatus is disconnect any external electrical cords, air hoses, or exhaust system hoses from the apparatus (Figure 4.9). Electrical cords are used to keep the vehicle's batteries charged at all times. Air hoses may be connected to the apparatus to keep an adequate amount of air in the vehicle's air brake system at all times. Exhaust hoses are used to vent diesel fumes to the exterior of the station. The benzene derivatives contained in diesel emissions have been found to be carcinogenic in laboratory studies. Some apparatus are equipped with variations of ground shore lines that are designed to pop off automatically when the apparatus is started or moved (Figure 4.10). Make sure that they release before the apparatus is driven from the station.

Step 2: Turn on the vehicle battery(ies). Most fire apparatus are equipped with a battery switch that is intended to turn off all vehicle electrical systems

Figure 4.8 All firefighters should be seated and belted before the apparatus is put into motion.

Figure 4.9 Disconnect the electric cord before entering the cab.

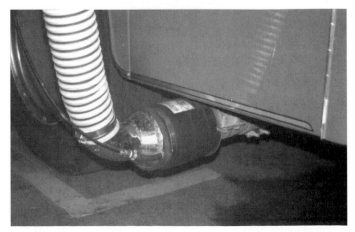

Figure 4.10 Some exhaust systems automatically eject the hose when the apparatus begins to pull away from its parking spot.

when the apparatus is parked and shutdown. The purpose of this switch is to prevent unwanted electrical drains on the battery that might result in a dead battery when the driver/operator attempts to start the vehicle.

Most fire apparatus are equipped with two vehicle batteries. This is a back-up measure in the event that one of the batteries is dead. It is also necessary because of the larger-than-normal amount of electrical equipment that is attached to the fire apparatus. On apparatus equipped with two batteries, the battery switch may have four settings: *Off, Battery 1, Battery 2,* and *Both* (Figure 4.11). Obviously, this gives the driver/operator the ability to use either or both of the batteries when starting and operating the vehicle. Newer apparatus may have a simple on/off switch (Figure 4.12). Regardless of which type of switch the apparatus is equipped with, the battery switch should never be operated while the engine is running. Follow the apparatus manufacturer's directions and departmental SOPs as to whether one or both batteries should be used. Depending on the location of this switch, the driver/operator may choose to operate this switch prior to entering the cab or immediately upon sitting in the driver's seat.

Step 3a: Start the engine (Manual Transmission). On an apparatus equipped with a manual shift transmission, the driver/operator should start the engine with the drive transmission in Neutral (N), and the vehicle's parking brake should be set. The driver/operator begins the process by turning on the ignition switches (Figure 4.13). These are usually located on the dashboard. Once these switches are on, press down on the clutch pedal to disengage the clutch. Once the pedal is depressed, you may operate the starter control. There are several designs for fire apparatus starter controls. Some utilize a single key, similar to a standard passenger vehicle key (Figure 4.14). Others utilize single or dual toggle switches or push buttons (Figures 4.15 a and b). The dual starter controls are used on apparatus equipped with two batteries. Each control is connected to a particular battery. If your SOP is to turn on only one battery, make sure that you use the appropriate starter control to start the apparatus. If you use both batteries, operate both starter controls at the same time. The starter

Figure 4.11 The multiposition, rotary battery switch is most common on modern fire apparatus.

Figure 4.14 Some apparatus, particularly those on commercial truck chassis, have key-operated starters.

Figure 4.12 Some apparatus may be equipped with a simple on/off battery switch.

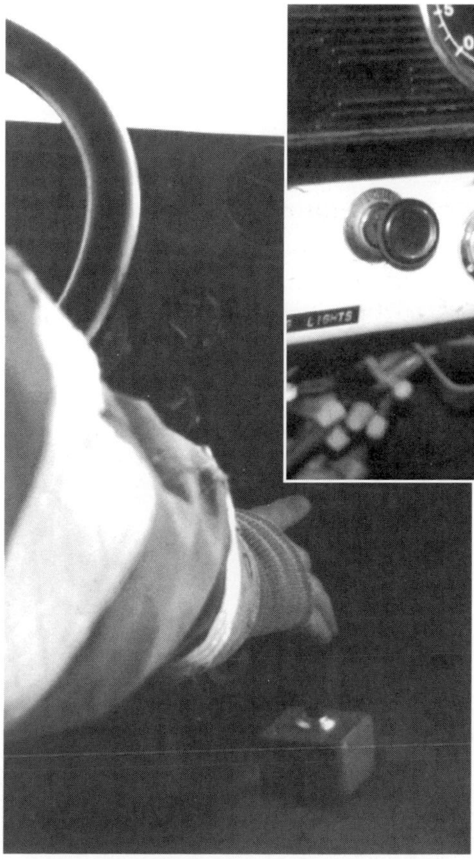

Figure 4.13 The driver/operator begins the process by turning on the ignition switches.

control should be operated in intervals of no more than 30 seconds, with a rest of 60 seconds between each try if the vehicle does not start sooner. The starter may overheat if the controls are operated for longer periods. (**NOTE:** On gasoline-powered apparatus, it may be necessary to operate a manual choke control before operating the starter control [Figure 4.16]. Use the manual choke sparingly in warm weather or after the apparatus is already warm.)

Step 3b: Start the engine (Automatic Transmission). On an apparatus equipped with an automatic transmission, the driver/operator should start the engine with the drive transmission in Neutral (N) or Park (P). The vehicle's parking brake should be set. The driver/operator begins the process by turning on the ignition switches (see Figure 4.13). These are usually located on the dashboard. You may then start the apparatus

using the starter control(s). There are several designs for fire apparatus starter controls. Some utilize a single key, similar to a standard passenger vehicle key (see Figure 4.14). Others utilize single or dual toggle switches or push buttons (see Figures 4.15 a and b). The dual starter controls are used on apparatus equipped with two batteries. Each control is connected to a particular battery. If your SOP is to turn on only one battery, make sure that you use the appropriate starter control to start the apparatus. If you use both batteries, operate both starter controls at the same time. The starter controls should be operated in intervals of no more than 30 seconds, with a rest of 60 seconds between each try if the vehicle does not start sooner. The starter may overheat if the controls are operated for longer periods. (**NOTE:** On gasoline-powered apparatus, it may be necessary to operate a manual choke control before operating the starter control [see Figure 4.16]. Use the manual choke sparingly in warm weather or after the apparatus is already warm.)

Step 4: Observe the apparatus gauges. Make sure that all gauges on the dashboard move into their normal operative ranges. In particular, pay attention to the oil pressure and air pressure gauges (Figure 4.17). If the oil pressure gauge does not indicate any reasonable amount of oil pressure within 5 to 10 seconds of starting the apparatus, stop the engine immediately and have the lubricating system checked by a trained mechanic. The air pressure gauge should be checked to make sure that adequate pressure is built up to

Figure 4.15a Some starter switches are of the toggle switch variety.

Figure 4.15b Some custom apparatus have push-button starter switches.

Figure 4.16 Apparatus that are powered by a gasoline engine may have a manual choke control.

Figure 4.17 Most modern apparatus is equipped with oil and air pressure gauges on the dashboard.

release the parking brake. An interlock should prohibit the parking brake from being disengaged before there is enough air pressure in the system to operate the service brakes. If there is not enough air pressure in the system to release the parking brake, allow the engine to idle until air pressure is built up to an appropriate level. It may speed the buildup of air pressure if the throttle is increased to a fast idle. The driver/operator should also check the ammeter to make sure the electrical system is operating/charging properly.

Step 5: Adjust the seat, mirrors, and steering wheel. If the driver/operator was not the last person to drive this vehicle, he should take a moment while the engine is idling/warming to properly adjust the seat and mirrors. It is generally best to adjust the seat first. The seat may be adjusted for height as well as distance from the steering wheel, foot pedals, and other controls. Once the seat is in the desired location, adjust the mirrors so that you can clearly see to the rear of the apparatus. Newer apparatus may have adjustable or telescoping steering wheels that may be adjusted to fit the driver/operator.

The apparatus is now ready to be placed into a road gear and driven from its present location.

Driving the Vehicle

Once the apparatus is running and the air system (if so equipped) has sufficient air pressure, the driver/operator is ready to release the parking brake and place the apparatus into a road gear (Figure 4.18). The following sections discuss the procedures for driving, stopping, engine idling, and shutting down the apparatus.

Driving Manual Transmission Apparatus

NOTE: The information in this section pertains to the typical 4- or 5-speed manual shift transmission with a single-speed rear axle. For information on driving apparatus with 2-speed rear axles or more than a 5-speed transmission (10, 12, or more gears are common in some jurisdictions), consult the apparatus/transmission manufacturer's operational manual for specific information.

Once you are ready to move the vehicle, depress the clutch pedal with your left foot, depress the service brake pedal with your right foot, and release the parking brake (Figure 4.19). Place the gear shifter into a low gear that will allow the vehicle to move without an inordinate amount of wear on the engine. Never attempt to start the apparatus moving while it is in a high or drive gear. This action causes the clutch (on both manual and automatic transmissions) to slip, which may damage the clutch facing. The clutch should be released slowly when starting from a standstill because a sudden, rapid release of

Figure 4.18 The parking brake must be released before the vehicle is ready to be driven.

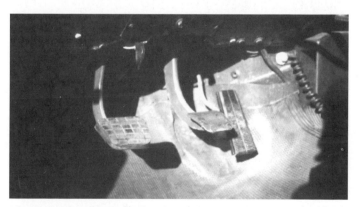

Figure 4.19 The clutch pedal is operated with the driver's left foot.

the clutch throws a heavy load upon the engine, clutch, transmission, and drive components. Take care to avoid vehicle rollback before engaging the clutch.

The apparatus should be kept in low gear until the proper speed or revolutions per minute (rpm) is reached for shifting to a higher gear. An engine develops its maximum power up to a certain speed, and excessive engine speed may result in decreased power and excessive use of fuel. It is good practice to keep the transmission in low gear until the apparatus is clear of the station and the driver/operator has an unobstructed view of the street and traffic conditions. When the driver/operator shifts gears, the clutch should be entirely disengaged (pedal pushed in all the way). The gear shift lever should be carefully moved into proper position, not jammed.

When climbing a hill, shift the transmission to a lower gear. This practice provides adequate driving power and enables the driver/operator to keep the apparatus under control. On sharp curves or when turning corners, shift standard transmissions into a lower gear *before* entering the curve or intersection. This action maintains peak engine power and apparatus control. When fire apparatus must be driven over rough or rugged terrain, use the lower gears. If the apparatus becomes stuck, such as in mud or snow, do not race the engine or jump the clutch. These actions only cause the apparatus to become further stuck and can lead to mechanical failure. Always maintain front wheels in line with the chassis of the vehicle.

When driving downhill, select a lower gear. Remain in gear at all times. The engine provides braking power when the vehicle is in gear. To prevent engine damage, limit downhill speed to lower than maximum governed rpm. The engine governor cannot control engine speed downhill; the wheels turn the driveshaft and engine. Engine rotation faster than rated rpm can result in valves hitting pistons, increased oil consumption, damage requiring overhaul, and injector plug seizures.

Table 4.5 contains information for the appropriate gear selections for both manual shift and automatic transmissions. These gear selections are based on the listed conditions.

Driving Automatic Transmission Apparatus

Apparatus equipped with automatic transmissions eliminates a lot of the decision making by the driver/operator regarding when to shift gears. Once the apparatus is ready to move, the driver/operator should depress the interlock on the shifter and move it to the appropriate gear selection (Figure 4.20). This selection varies depending on the manufacturer of the apparatus or trans-

mission. Some shifters have a normal driving (D) designation on the shifter similar to that found on smaller passenger vehicles. Others have a series of numbers on the shifter. They may be single numbers (1, 2, 3, etc.) or range numbers (1-3, 1-4, 1-5, etc.). Newer apparatus may have push-button transmission selectors (Figure 4.21). In this case, the appropriate gear is

Table 4.5
Transmission Gear Selection Criteria

Automatic	Manual	Operating Conditions
Shift Selector Position	Gear Shift Position	
Drive (D), 5, or 1-5	5 or Overdrive	Normal load, grade, and traffic conditions with open road ahead
4 or 1-4	4	Moderate grades and over-the-road operation with moderate speeds
3 or 1-3	3	Operating in heavy traffic
2 or 1-2	2	Need for speed control requiring a hold condition such as descending a steep grade or operation on rough terrain
1	1	Starting gear for initial forward movement

Figure 4.20 Some automatic transmissions are operated with a stick shift.

Figure 4.21 This automatic transmission is operated by a push-button pad.

chosen simply by pushing the appropriate button. Consult the manufacturer's operator's manual for recommendations on which gear to select for normal operation.

Driver/operators of apparatus equipped with automatic transmissions should be aware that the pressure placed upon the accelerator influences automatic shifting. When the pedal is fully depressed, the transmission automatically upshifts near the governed speed of the engine. This may result in reduced power and excessive fuel consumption. If the accelerator is partially depressed, the upshift occurs at a lower engine speed.

Driver/operators have the option of manually selecting a particular gear for operation on apparatus equipped with an automatic transmission. This may be desirable when operating the apparatus at a slow speed for a long period of time or when driving up a steep hill. Simply move the shifter to the lower gear when this change is desired. Do not attempt to jump more than one gear at a time. Consult Table 4.5 for information on the appropriate gears to select for various conditions.

Cruising

Once the apparatus is moving, accelerate the vehicle gradually. Do not try to reach rated speed in the low gears. Going to rated speed in the low gears can be very noisy and increases engine wear. Stay in the highest gear that allows the apparatus to keep up with traffic and still have some power in reserve for acceleration.

Operators of fire service vehicles can significantly reduce drivetrain damage and extend apparatus service life by adopting proper operating habits. When driving a fire service vehicle, attempt to maintain engine rpm control through correct throttling. Keeping the engine operating within its power curve ensures adequate power and optimum fuel economy for a given set of conditions.

Whenever possible, avoid overthrottling, which results in lugging. *Lugging* occurs when the throttle application is greater than necessary for a given set of conditions. When this happens, the engine cannot respond to the amount of work being asked for by the operator. Do not allow the engine rpm to drop below peak torque speed if lugging does occur. Automatic transmissions downshift automatically to prevent lugging. Standard transmissions must be downshifted to avoid stalls and prevent lugging. If the apparatus is equipped with a tachometer, it is easier to control the transmission shift points.

When ascending a steep grade, momentary unavoidable lugging takes place. Although this brief lugging is unavoidable, time spent in the lugging state should be minimized to avoid possible damage to the power plant. Engine rpm drops even with heavy throttle settings. Progressively lower gears should be selected until the combination of available power and the torque multiplication of the transmission gears allow the hill to be climbed easily.

When overthrottling occurs with a diesel engine, more fuel is being injected than can be burned. This results in an excessive amount of carbon particles issuing from the exhaust (black smoke), oil dilution, and additional fuel consumption.

Another point that requires consideration is maximum engine rpm. Fire service apparatus are maintained for a much longer period of time than are commercial vehicles. Over this period of time, the valve springs can become weakened. For this reason, allowing an engine to overspeed as the result of improper downshifting or hill descent should be avoided in an effort to prolong engine life. Choose a gear that cruises the engine at 200 or 300 rpm lower than recommended rpm. This reduces engine wear; power losses caused by the fan, driveline, and accessories; noise; and fuel consumption.

Stopping the Apparatus

The process of braking fire apparatus to a standstill should be performed smoothly so that the apparatus will come to an even stop. Before braking, the driver/operator should consider the weight of the apparatus and the condition of the brakes, tires, and road surface (Figure 4.22). An abrupt halt can cause a skid, injury to firefighters, and mechanical failure. Some apparatus employ engine brakes, or other types of retarding devices, that assist in braking. The engine brake and retarder are activated when pressure is released from the accelerator. Because they provide most of the necessary slowing action, these devices allow the driver/operator to limit the use of service brakes to emergency stops and final stops. Both devices save wear on the service brakes and make the apparatus easier to manage on hills and slippery roads.

Figure 4.22 The driver/operator must use extra care when braking heavy fire apparatus. *Courtesy of Joel Woods.*

Driver/operators of units having retarders should become thoroughly familiar with the manufacturer's recommendations regarding their operation prior to use.

Some air brake systems have limiting systems for varying road conditions. The clutch should not be disengaged while braking until the last few feet (meters) of travel. This practice is particularly important on slippery surfaces because an engaged engine allows more control of the apparatus. More information on proper braking techniques is discussed later in this chapter.

Engine Idling

Shut the engine down rather than leave it idling for long periods of time. This applies to apparatus at an emergency scene that are not being used (Figure 4.23). Long idling periods can result in the use of ½-gallon (2 L) of fuel per hour; the buildup of carbon in injectors, valves, pistons, and valve seats; misfiring because of injector carboning; and damage to the turbocharger shaft seals. When the engine **must** be left to idle for an extended period of time because of extremely cold weather or during floodlight operations, set it to idle at 900 to 1,100 rpm rather than at lower speeds. Most departments have standard operating procedures for times when the apparatus may be forced to idle for an extended period of time. The driver/operator should be familiar with the SOP and follow it accordingly.

Engine Shutdown

Never attempt to shut down the engine while the apparatus is in motion because this action cuts off fuel flow from the injectors. Fuel flow through the injectors is required for lubrication anytime the injector plunger is moving. Fuel pressure can build up behind the shutoff valve and prevent the valve from opening.

Never shut down immediately after full-load operation. Shutting down the engine without a cooling-off period results in immediate increase of engine temperature from lack of coolant circulation, oil film "burning" on hot surfaces, possible damage to heads and exhaust manifolds, and possible damage to the turbocharger that may result in turbo seizure. Allow the engine temperature to stabilize before shutdown. A hot engine should be idled until it has cooled. Generally, an idle period of 3 to 5 minutes is recommended.

The procedures for shutting down the apparatus are as follows:

Step 1: Place the transmission in the Park (P) or Neutral (N) position once the apparatus is parked in or near the desired shutdown location.

Step 2: Set the parking brake.

Figure 4.23 Apparatus that are assigned to a staging area for an extended period of time should be shut down, rather than allowed to idle.

Step 3: Allow the engine to idle and cool down for 3 to 5 minutes. If the truck is being prepared to park in a fire station that is not equipped with an exhaust vent system, this cooldown idle should be performed on the front apron. The apparatus may then be backed into the station before proceeding.

Step 4: Shut off the engine by moving the ignition key or switch(es) to the *off* position.

Step 5: Turn the battery switch to the *off* position.

Step 6: Reconnect all ground shore lines (electric, air, exhaust) if parking in the station.

Safe Driving Techniques

The driver/operator's job is to keep the fire apparatus under control at all times. In the following sections, we will review some of the more important issues involving the safe operation of fire apparatus during emergency and nonemergency operations.

Attitude

The first element in learning to drive safely is to develop a safety-conscious attitude. It is critically important that the driver/operator remain calm and drive in a safe manner. Reckless driving, even in response to an emergency, is never acceptable. The driver/operator who drives aggressively, failing to observe safety precautions, is a menace to other vehicles, pedestrians, and other firefighters on the apparatus.

Driver/operators must realize that they cannot *demand* the right-of-way, although they may legally have it. Actions such as ignoring approaching apparatus, refusing to yield, and driving erratically due to panic may be expected from the public. At all times, the driver/operator must be prepared to yield the right-of-way in the interest of safety. One manner of looking at it is to drive as you would during nonemergency situations and take advantage of the room that clears for you on the road.

In addition to the safety aspects of having a proper attitude, the driver/operator should also consider fire department public image aspects. All fire department members should strive to present a positive fire department image at all times. Reckless operation of the vehicle, degrading gestures, and verbal assaults toward members of the public will not assist in maintaining the positive image you seek. Other actions, such as blaring sirens and air horns at 3 a.m. on deserted roads that wake homeowners from their sleep (unless required by law or department SOP), show the fire department in a negative light. Remember that these are the same people who vote on bond issues and tax referendums and, in the case of volunteer departments, who donate money to your department. They are less apt to do this if they have negative impressions of your organization.

Apparatus Rider Safety

The driver/operator must always assure the safety of all personnel riding on the apparatus. It is most desirable for riders of emergency vehicles to don their protective gear before getting in the apparatus (Figure 4.24). The one possible exception to this is the driver/operator himself. Some driver/operators are not comfortable driving the apparatus wearing fire boots or bulky protective coats. In this case, the driver/operator should don his protective clothing at the scene.

All riders on the apparatus should be seated within the cab or body (rescue apparatus) and wearing their seat belts before the apparatus is put into motion (Figure 4.25). Currently, NFPA 1901, *Standard for Automotive*

Fire Apparatus, requires that a seat and seat belt be provided within the cab or body of the apparatus for every firefighter who is expected to ride the truck. NFPA 1500, *Standard on Fire Department Occupational Safety and Health Program,* also specifically states that all riders must be seated and belted. The standard does, however, provide three exceptions to the seated and belted requirement:

- When providing patient care in the back of an ambulance that makes it impractical to be seated and belted
- When loading hose back onto a fire apparatus
- When performing training for personnel learning to drive the tiller portion of a tractor-drawn aerial apparatus

Providing emergency care and operating aerial apparatus are not within the scope of this manual. However, driver/operators who drive apparatus equipped with fire pumps are likely to find themselves in the situation of having to load fire hose back onto the vehicle. Loading fire hose while driving the apparatus is particularly common when loading large diameter (4-inch [100 mm] or larger) supply hose. NFPA 1500 provides the following specific directions on how these operations should be performed to maximize safety:

- The procedure must be contained in the department's written standard operating procedures (SOPs), and all members must be trained specifically on how to perform the moving hose-load operation.
- At least one member, other than one actually loading the hose, must be assigned as a safety observer to the operation. He must have complete visual contact with the hose-loading operation, as well as visual and voice communications (usually via a portable radio) with the driver/operator (Figure 4.26).

Figure 4.24 Firefighters should don their protective clothing before boarding the apparatus.

Figure 4.25 Before the apparatus is put into motion, all riders on the apparatus should be seated within the cab or body (rescue apparatus) and wearing their seat belts.

Figure 4.26 One firefighter, equipped with a portable radio, should monitor the hose loading operation at all times.

- The area in which the hose loading is being performed must be closed to other vehicular traffic.

- The apparatus must be driven in a forward direction (straddling or to one side of the hose) at a speed no greater than 5 mph (8 km/h) (Figure 4.27).

- No members are allowed to stand on any portion of the apparatus.

- Members in the hose bed must sit or kneel while the apparatus is moving (Figure 4.28).

Although a common practice in the past, firefighters should *never* be allowed to ride the tailboard or running boards of any moving apparatus. This is specifically prohibited by NFPA 1500. As well, the practice of attacking wildland fires while riding on the outside of a moving wildland fire apparatus is specifically prohibited by the standard (Figure 4.29). Many departments still choose to construct riding platforms on the front, sides, or rear of wildland fire apparatus. These are used by firefighters during pump-and-roll fire fighting operations. These riding positions offer little protection if the vehicle is involved in a collision or rollover.

Figure 4.27 The apparatus is driven in a forward direction either next to or straddling the hose.

> ## WARNING
>
> **Firefighters should never ride on the outside of a moving fire apparatus for any reason, other than those exceptions noted in NFPA 1500. Serious injury or death could occur if the apparatus is involved in a collision or rollover or if the rider falls from the moving apparatus.**

Although all newer apparatus are designed with fully enclosed cabs, many older apparatus with jump seat riding areas not totally enclosed are still in service. Some of these are equipped with safety bars or gates that are intended to prevent a firefighter from falling out of a jump seat (Figures 4.30 a and b). These devices are not substitutes for safety procedures that require firefighters to ride in safe, enclosed positions wearing their seat belts. Safety devices that are held in an upright or open position by straps or ropes provide no extra security for the firefighter riding in the jump seat.

Figure 4.28 Firefighters on the apparatus should either be seated or kneeling in the hose bed area.

Backing the Vehicle

As stated earlier in this chapter, a significant portion of fire apparatus collisions occur while the apparatus is being driven in reverse. All fire departments should have firmly established procedures for backing the vehicle, and these procedures must always be followed by fire apparatus driver/operators.

Whenever possible, the driver/operator should avoid backing the fire apparatus. It is normally safer and some-

Figure 4.29 Firefighters should not be riding on the outside of a moving wildland fire apparatus.

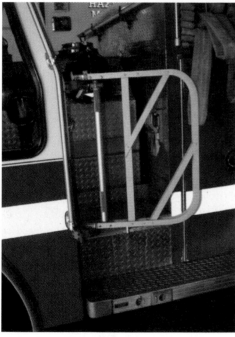

Figure 4.30b Safety gates provide somewhat more secure protection than do safety bars.

a large percentage of the collisions that occur during backing operations. Very simply, if you are the driver/operator and you do not have or cannot see the spotters behind you, **do not back the apparatus!** All fire apparatus should be equipped with an alarm system that warns others when the apparatus is backing up.

Defensive Driving Techniques

Sound defensive driving skills are one of the most important aspects of safe driving. Every driver/operator should be familiar with the basic concepts of defensive driving. They include anticipating other drivers' actions, estimating visual lead time, knowing braking and reaction times, combating skids, knowing evasive tactics, and having knowledge of weight transfer.

The driver/operator should know the rules that govern the general public when emergency vehicles are on the road. Most laws or ordinances provide that other vehicles must pull toward the right and remain at a standstill until the emergency vehicle has passed. These laws, however, do not mean that the apparatus driver/

Figure 4.31 Drive-through apparatus bays eliminate the need for backing apparatus into their parking position.

times quicker to drive around the block and start again. It is most desirable that new fire stations be designed with drive-through apparatus bays that negate the necessity to back the apparatus into them (Figure 4.31). However, there are situations when it is necessary to back fire apparatus. This operation should be performed very carefully. When backing, there should be at least one firefighter —and preferably two—with a portable radio assigned to clear the way and to warn the driver/operator of any obstacles obscured by blind spots (Figure 4.32). If two spotters are used, only one should communicate with the driver/operator. The second spotter should assist the first one. This is a very simple procedure that can prevent

Figure 4.32 When possible, two firefighters should assist the driver/operator in backing the apparatus.

operator can ignore stopped vehicles. People may panic at the sound of an approaching siren and accidentally move into the lane of traffic or stop suddenly. Others may not hear warning signals because of car radios, closed windows, or air-conditioning. It must also be remembered that some individuals simply ignore emergency warning signals.

Intersections are the most likely place for a collision involving an emergency vehicle (Figure 4.33). When approaching an intersection, the driver/operator should slow the apparatus to a speed that allows a stop at the intersection if necessary. Even if faced with a green signal light, or no signal at all, the apparatus should be brought to a complete stop if there are any obstructions, such as buildings or trucks, that block the driver/operator's view of the intersection.

Depending on the motor vehicle statutes and departmental SOPs within a particular jurisdiction, fire apparatus on an emergency response may proceed through a red traffic signal or stop sign after coming to a **complete** stop and assuring that all lanes of traffic are accounted for and yielding to the apparatus. The determination of what is an emergency response is covered later in this chapter (Warning Devices and Clearing Traffic). Do not proceed into the intersection until you are certain that every other driver sees you and is allowing you to proceed. Simply slowing when approaching the intersection and then coasting through is not an acceptable substitute for coming to a complete stop. When proceeding through the intersection, attempt to make eye contact with each of the other drivers to ensure they know you are there and about to proceed.

Traffic waiting to make a left-hand turn may pull to the right or left, depending upon the driver. In situations where all lanes of traffic in the same direction as the responding apparatus are blocked, the apparatus driver/operator should move the apparatus into the opposing lane of traffic and proceed through the intersection at an extremely reduced speed (Figure 4.34). Oncoming traffic must be able to see the approaching apparatus. Full use of warning devices is essential. Driving in the oncoming lane is **not recommended** in situations where oncoming traffic is unable to see the apparatus such as on a freeway underpass. Be alert for traffic that may enter from access roads and driveways. Approaching traffic on the crest of a hill, slow-moving traffic, and other emergency apparatus must be closely monitored.

Even though use of warning sirens, lights, and signals is essential, fire apparatus driver/operators should realize that these signals may be blanketed by other warning devices and by street noises. Serious collisions and fatalities have been caused by overreliance on warning signals.

Anticipating Other Drivers' Actions

Never assume what another driver's actions will be; expect the unexpected.   Anticipation is the key to safe driving. Always remember the following control factors:

- *Aim high in steering*: Find a safe path well ahead.
- *Get the big picture*: Stay back and see it all.
- *Keep your eyes moving*: Scan — do not stare.
- *Leave yourself an "out"*: Do not expect other drivers to leave you an out. Be prepared by expecting the unexpected.
- *Make sure others can see and hear you*: Use lights, horn, and signals in combination.

Visual Lead Time

The concept of visual lead time refers to the driver/operator scanning far enough ahead of the apparatus, for the speed it is being driven, to assure that appropriate action can be taken if it becomes necessary. For example, if the operator is concentrating on vehicles that are 100 feet (30 m) in front of the apparatus and based on the speed the apparatus is being driven it would take 200 feet (60 m) for it to stop or perform an evasive maneuver, a collision is likely to occur. The driver needs to learn to match the speed he is traveling with the distance ahead of the vehicle he is surveying. Visual lead time interacts directly with reaction time and stopping distances. By "aiming high in steering" and "getting the big picture," it is possible to become more keenly aware of conditions that may require slowing or stopping.

Braking and Reaction Time

A driver/operator should know the total stopping distance for a particular fire apparatus. The *total stopping distance* is the sum of the driver/operator reaction distance and the vehicle braking distance (Figure 4.35). The

Figure 4.33 Intersections are the most likely place for a collision involving an emergency vehicle. *Courtesy of Ron Jeffers.*

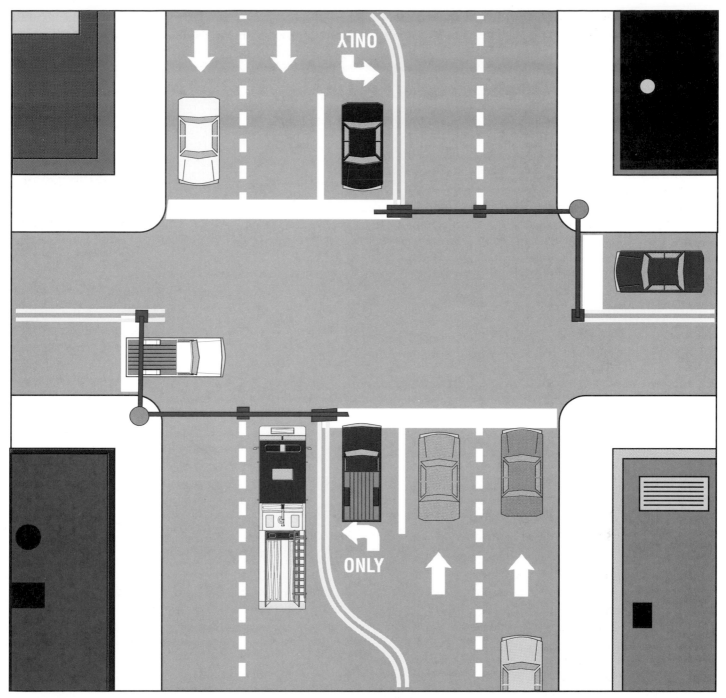

Figure 4.34 Use caution when entering the opposing lane of traffic at an intersection.

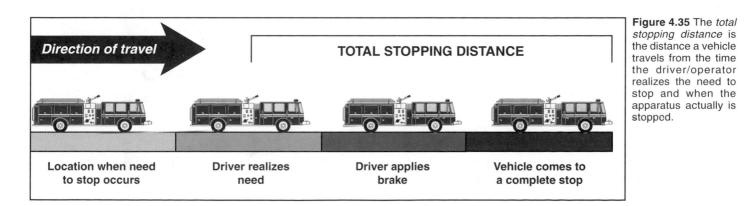

Direction of travel

TOTAL STOPPING DISTANCE

| Location when need to stop occurs | Driver realizes need | Driver applies brake | Vehicle comes to a complete stop |

Figure 4.35 The *total stopping distance* is the distance a vehicle travels from the time the driver/operator realizes the need to stop and when the apparatus actually is stopped.

driver/operator *reaction distance* is the distance a vehicle travels while a driver is transferring the foot from the accelerator to the brake pedal after perceiving the need for stopping. The *braking distance* is the distance the vehicle travels from the time the brakes are applied until the apparatus comes to a complete stop. Tables 4.6 a and b show driver reaction distances, vehicle braking distances, and total stopping distances for different sizes of vehicles. These tables indicate approximates for vehicles, and the statistics may vary for different fire apparatus. Each department should conduct braking distance tests with its own apparatus. Apparatus manufacturers may also be able to provide this information for specific apparatus.

There are a number of factors that influence the driver/operator's ability to stop the apparatus:

- Condition of the driving surface
- Speed being traveled
- Weight of the vehicle
- Type and condition of the vehicle's braking system

A dry, paved road provides the optimal stopping ability from a driving surface standpoint. The ability of the apparatus to stop is negatively affected by wet, snowy,

icy, or unpaved roads. Driver/operators must compensate for these conditions by reducing their speeds by an appropriate amount to match the conditions.

The correlation between vehicle weight and speed and stopping distance should be obvious to anyone. At an equal speed, it will take a greater distance to stop a three-axle water tender (tanker) than it will a brush pumper (Figures 4.36 a and b). It will also take a greater distance to stop a vehicle that is going 50 mph (80 km/h) than the same vehicle when it is traveling 30 mph (48 km/h).

The type and condition of the braking system have a tremendous impact on the ability to stop the fire apparatus. Several serious fire apparatus accidents have been traced to poor maintenance of the braking system. Obviously, a vehicle that has a properly maintained braking system will stop faster than one that has a system in disrepair.

Weight Transfer
The effects of weight transfer must be considered in the safe operation of fire apparatus. Weight transfer occurs as the result of physical laws that state that objects in motion tend to stay in motion; objects at rest tend to remain at rest. Whenever a vehicle undergoes a change in

		Braking Distance (feet)				Total Stopping Distance (feet)			
Speed (mph)	**Average Driver Reaction Distance (feet)**	**Vehicle A**	**Vehicle B**	**Vehicle C**	**Vehicle D**	**Vehicle A**	**Vehicle B**	**Vehicle C**	**Vehicle D**
10	11		7	10	13		18	21	24
15	17		17	22	29		34	39	46
20	22	22	30	40	50	44	52	62	72
25	28	31	46	64	80	59	74	92	108
30	33	45	67	92	115	78	100	125	148
35	39	58	92	125	160	97	131	164	199
40	44	80	125	165	205	124	169	209	249
45	50	103	165	210	260	153	215	260	310
50	55	131	225	255	320	186	280	310	375
55	61	165	275	310	390	226	336	371	451
60	66	202	350	370	465	268	426	436	531

Table 4.6a (US)
Braking and Stopping Distances (dry, level pavement)

Typical Brake Performance
A–Average automobile
B–Light two-axle trucks
C–Heavy two-axle trucks
D–Three-axle trucks and trailers

Table 4.6b (metric)
Braking and Stopping Distances (dry, level pavement)

Speed (km/h)	Average Driver Reaction Distance (meters)	Braking Distance (meters)				Total Stopping Distance (meters)			
		Vehicle A	Vehicle B	Vehicle C	Vehicle D	Vehicle A	Vehicle B	Vehicle C	Vehicle D
16	3.4		2.1	3	4		5.5	6.4	7.3
24	5.2		5.2	6.7	8.8		10.4	11.9	14
32	6.7	6.7	9.1	12.2	15.2	13.4	15.8	18.9	21.9
40	8.5	9.4	14	19.5	24.4	18	22.6	28	32.9
48	10.1	13.7	20.4	28	35.1	23.8	30.5	38.1	45.1
56	11.9	17.7	28	38.1	48.8	29.6	39.9	50	60.7
64	13.4	24.4	38.1	50.3	62.5	37.8	51.5	63.7	75.9
72	15.2	31.4	50.3	64	79.2	46.6	65.5	79.2	94.5
80	16.8	39.9	68.6	77.7	97.5	56.7	85.3	94.5	114.3
88	18.6	50.3	83.8	94.5	118.9	68.9	102.4	113.1	137.5
96	20.1	61.6	106.7	112.8	141.7	81.7	129.8	133	161.8

Typical Brake Performance
A–Average automobile
B–Light two-axle trucks
C–Heavy two-axle trucks
D–Three-axle trucks and trailers

Figures 4.36a and b Obviously, it will take a greater distance to stop this large tender than it will the brush pumper, given the same speed and road conditions.

velocity or direction, weight transfer takes place relative to the severity of change. Apparatus driver/operators must be aware that the excessive weight carried on most fire apparatus can contribute to skidding or possible rollover due to excessive weight transfer. These hazardous conditions can result from too much speed in turns, harsh or abrupt steering action, or driving on slopes too steep for a particular apparatus. This is of particular concern with apparatus that have large water tanks that are improperly baffled and partially filled with liquid (water or foam concentrate).

Use only as much steering as needed to keep weight transfer to a minimum. Steering should be smooth and continuous. Also, maintain a speed that is slow enough to prevent severe weight transfer from occurring. This is particularly important on curves.

Combating Skids
Avoiding conditions that lead to skidding is as important as knowing how to correct skids once they occur. The most common causes of skids involve driver error:

- Driving too fast for road conditions

- Failing to properly appreciate weight shifts of heavy apparatus
- Failing to anticipate obstacles (these range from other vehicles to animals)
- Improper use of auxiliary braking devices
- Improper maintenance of tire air pressure and adequate tread depth

Tires that are overinflated or lacking in reasonable tread depth make the apparatus more susceptible to skids.

Most newer, large fire apparatus are equipped with an all-wheel, antilock braking system (ABS). The power for the braking ability comes from air pressure. These systems are effective in that they minimize the chance of the vehicle being put into a skid when the brakes are applied forcefully. ABS works using digital technology in an onboard computer that monitors each wheel and controls air pressure to the brakes, maintaining optimal braking ability. A sensing device located in the axle monitors the speed of each wheel. The wheel speed is converted into a digital signal that is sent to the onboard computer. When the driver/operator begins to brake and the wheel begins to lock up, the sensing device sends a signal to the computer that the wheel is not turning. The computer analyzes this signal against the signals from the other wheels to determine if this particular wheel should still be turning. If it is determined that it should be turning, a signal is sent to the air modulation valve at that wheel, reducing the air brake pressure and allowing the wheel to turn. Once the wheel turns, it is braked again. The computer makes these decisions many times a second, until the vehicle is brought to a halt. Thus, when driving a vehicle equipped with an ABS, maintain a steady pressure on the brake pedal (rather than pumping the pedal) until the apparatus is brought to a complete halt.

Keep in mind that in the case of air brakes, there is a slight delay in the time from which the driver/operator pushes down on the brake pedal until sufficient air pressure is sent to the brake to operate. This must be considered when determining total stopping distance.

When an apparatus that is **not** equipped with an antilock braking system goes into a skid, release the brakes, allowing the wheels to rotate freely. Turn the apparatus steering wheel so that the front wheels face in the direction of the skid. If using a standard transmission, do not release the clutch (push in the clutch pedal) until the vehicle is under control and just before stopping the vehicle. Once the skid is controllable, gradually apply power to the wheels to further control the vehicle by giving traction.

Proficiency in skid control may be gained through practice at facilities having skid pads. These are smooth surface driving areas that have water directed onto them to make skids likely (Figure 4.37). All training should be done at slow speeds to avoid damaging the vehicle or injuring participants. Some jurisdictions choose to use reserve apparatus or other older vehicles for this part of the training process (Figure 4.38). If a skid pad is not available, it may be possible to use an open parking lot.

Auxiliary Braking Systems

In addition to the engine retarders mentioned earlier in this chapter, the apparatus may be equipped with one or more other types of auxiliary braking systems. The driver/operator needs to understand these systems so that they can be used properly; in some cases, their use should be avoided completely.

The first type of auxiliary braking system is the front brake-limiting valve type. These were commonly installed

Figure 4.37 The skid pad has a smooth surface, and water is floated across the top of it.

Figure 4.38 Some jurisdictions choose to use surplus vehicles to learn the principles of vehicle control on the skid pad.

on apparatus built before the mid-1970s, but are also found on some newer apparatus (Figure 4.39). These were more commonly known as the "dry road/slippery road" switches. These devices were intended to help the driver/operator maintain control of the apparatus on slippery surfaces. This was accomplished by reducing the air pressure on the front steering axle by 50 percent when the switch was in the slippery-road position. This would prevent the front wheels from locking up, allowing the driver/operator to steer the vehicle even when the rear wheels were locked into a skid.

In reality, these systems were not overly effective or safe. With the switch in the slippery-road position, the braking capabilities were actually reduced by 25 percent. If the braking system was not in optimum condition to begin with, say it was only working at 80 percent of its designed capability, then using this switch would drop it 25 percent more or to 55 percent of the designed braking ability. After the adoption of the Federal Motor Vehicle Safety Standard 121 of 1975 by the United States government, few trucks were built with this switch installed. *IFSTA recommends that apparatus equipped with this switch have it placed in the dry-road position and disconnected.*

Another type of auxiliary braking system is the interaxle differential lock, also known as a power divider or third differential (Figure 4.40). This is another type of switch

Figure 4.39 Dry/slippery road switches were more common on older apparatus.

Figure 4.40 Interaxle differential lock switches may be found on apparatus with tandem rear axles.

that may be activated from the cab of an apparatus that has tandem rear axles. It allows for a difference in speed between the two rear axles, while providing pulling power from each axle. This is intended to provide greater traction for each axle.

Under normal operating conditions, the interaxle differential switch should be in the unlocked position. Move the switch to the locked position when approaching or anticipating slippery-road conditions to provide improved traction. Always unlock the switch again when road conditions improve. You must lift your foot from the accelerator when activating the interaxle differential lock. Do not activate this switch while one or more of the wheels are actually slipping or spinning because damage to the axle could result. Also, do not spin the wheels with the interaxle differential locked because damage to the axle could result.

Some vehicles that are equipped with an ABS are also equipped with automatic traction control (ATC). ATC turns itself on and off; there is no switch for the operator to select. A green indicator light on the dash illuminates when the ATC is engaged. The engine speed also decreases, as needed, until traction is acquired to move the chassis.

ATC helps improve traction on slippery roads by reducing drive wheel overspin. ATC works automatically in two ways: First, when a drive wheel starts to spin, the ATC applies air pressure to brake the wheel. This transfers engine torque to the wheels with better traction. Second, when all drive wheels begin to spin, the ATC reduces the engine torque to provide improved traction.

Some vehicles equipped with ATC have a snow-and-mud switch. This function increases available traction on extra soft surfaces. When this switch is activated, the ATC indicator light flashes continuously. Deactivate this feature when normal traction is regained. To deactivate this switch, press the switch a second time and turn off the vehicle ignition. If you desire to "rock" an apparatus out of a particular spot and the ATC has cut the throttle, activate the mud-and-snow switch. Use caution when activating this switch because if the apparatus regains traction suddenly, axle damage may occur.

Consult the apparatus manufacturer's operations manual for detailed information on operating the auxiliary braking system(s) on your apparatus.

Passing Other Vehicles

In general, it is best to avoid passing vehicles that are not pulling over to yield the right-of-way to the fire apparatus. However, in some instances, the need to pass will occur, and the driver/operator must be prepared to do it in the safest manner possible. The following guidelines should be used to ensure safe passing:

- Always travel on the innermost lane on multilane roads. Wait for vehicles in front of you to move to the right before proceeding (Figure 4.41).

- Avoid passing vehicles on their right sides. Most civilian drivers' natural tendency is to move to the right when an emergency vehicle is approaching (Figure 4.42). Thus, they could turn into your path if you are passing on the right. Some departments have strict SOPs prohibiting this practice.

- Make sure you can see that the opposing lanes of traffic are clear of oncoming traffic if you must move in that direction.

- Avoid passing other emergency vehicles if at all possible. However, in some cases, it may be desirable for a smaller, faster vehicle (such as a chief's vehicle) to pass a larger, slower vehicle (such as an aerial apparatus). In these cases, the lead vehicle should slow down and move to the right to allow the other vehicle to pass. This maneuver should be coordinated by radio if possible.

- Flash your high beam lights to get the driver's attention when passing.

Adverse Weather

Weather is another factor to consider in terms of safe driving. Rain, snow, ice, and mud make roads slippery. A driver/operator must recognize these dangers and adjust apparatus speed according to the crown of the road, the sharpness of curves, and the condition of road surfaces. The driver/operator should decrease speed gradually, slow down while approaching curves, keep off low or soft shoulders, and avoid sudden turns. The driver/operator should recognize areas that first become slippery such as bridge surfaces, northern slopes of hills, shaded spots, and areas where snow is blowing across the roadway.

Because the stopping distance is greatly increased on slippery-road surfaces, it is sometimes a good policy to try the brakes while in an area free of traffic to find out how slippery the road is. Speed must be adjusted to road and weather conditions so that the apparatus can be stopped or maneuvered safely. Good windshield wipers and defrosters should keep the windshield clean and clear.

Snow tires or tire chains will reduce the stopping distance and considerably increase starting and hill-climbing traction on snow or ice. Apparatus may be equipped with the traditional, manually applied tire chains or the

Figure 4.41 When driving on a multilane road, stay in the inside (left) lane.

Figure 4.42 Avoid passing other vehicles on their right side.

newer automatic variety. Automatic tire chains consist of short lengths of chain that are on a rotating hub in front of each rear wheel (Figure 4.43). The hubs swing down into place when a switch on the dashboard is activated (Figure 4.44). The rotation of the hub throws the chains underneath the rolling tires. These chains tend to lose their effectiveness in snow that is deeper than 8 inches (200 mm).

During slippery-road conditions, the safe following distance between vehicles increases dramatically. Remember that it takes 3 to 15 times more distance for a vehicle to come to a complete stop on snow and ice than it does on dry concrete.

Warning Devices and Clearing Traffic

All fire apparatus are equipped with some combination of audible and visible warning devices. Audible warning devices may include air horns, bells, mechanical sirens, or electronic sirens (Figures 4.45 a through d). Studies have shown that civilian drivers respond better to sounds that change pitch often. Short bursts with the air horns and the constant up-and-down oscillation of a mechani-

Figure 4.43 Many newer apparatus are equipped with automatic tire chain devices.

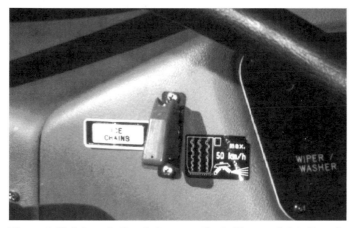

Figure 4.44 Automatic tire chains are activated by a switch in the cab.

cal or electronic siren are the surest ways to catch a driver's attention. At speeds above 50 mph (80 km/h), an emergency vehicle may "outrun" the effective range of its audible warning device. A study conducted by the staff of Driver's Reaction Course concluded that a siren operating on an emergency vehicle moving at 40 mph (64 km/h) can project 300 feet (90 m) in front of the vehicle. At a speed of 60 mph (97 km/h), however, the siren is only audible 12 feet (3.7 m) or less in front of the vehicle. Driver/operators must operate within the effective range of their audible warning devices.

Warning devices are of no value to the driver/operator or the public if they are not used. Warning devices should be operated from the time the apparatus begins its response until it arrives on the scene. The driver/operator needs to use some discretion in the use of sirens. When responding to sensitive situations, such as psychiatric emergencies, it is often better to turn off the siren as the apparatus nears the destination. The sudden use of an audible warning device immediately behind another vehicle may unnerve the driver. If this happens, the driver may suddenly stop or swerve, causing a collision. The use of warning devices is essential, but it does not give the driver/operator the right or permission to disregard other drivers.

Figure 4.45a An air horn.

Figure 4.45b Out of tradition, some apparatus are still equipped with bells.

Figure 4.45c Mechanical sirens are traditionally associated with fire apparatus.

Figure 4.45d An electronic siren speaker.

The use of warning devices should be limited to true emergency response situations. Each department should have standard operating procedures on what types of calls are considered emergencies and which are not. In general, certain types of calls for service, such as water shutoffs or covering an empty fire station, are not really emergencies. Warning devices and emergency driving tactics should not be used on these types of responses. Some fire departments have further extended their list of nonemergency responses to include such calls as automatic fire alarms, carbon monoxide detector alarms, smoke odor investigations, and trash fires. Be aware that emergency organizations have lost lawsuits when they were deemed responsible for causing collisions while driving at an emergency rate to nonemergency calls.

Some fire departments have standard operating procedures that require driver/operators to turn off all warning devices and proceed with the normal flow of traffic while driving on limited-access highways and turnpikes. In reality, the apparatus will probably not be capable of driving as fast as many of the vehicles on that roadway and the use of warning devices is unnecessary. Warning lights may be turned back on when the apparatus reaches the scene and parks.

There have been numerous collisions involving fire apparatus and other emergency vehicles. It is not always possible to hear the warning devices of other emergency vehicles when the audible warning device on your apparatus is being sounded. When more than one emergency vehicle is responding along the same route, units should travel at least 300 to 500 feet (90 m to 150 m) apart (Figure 4.46). Some fire departments rely upon designated response routes. This practice can be hazardous if a company is delayed or detoured for some reason. Standard operating procedures should call for radio reports of location and status, particularly when you are certain that you are approaching the same intersection as another emergency vehicle. Regardless of the system or pattern used, always take precautions to ensure a safe, collision-free response. This includes coming to a complete stop at any intersection that has a stop sign or a red signal light.

White lights can be readily distinguished during daylight hours. For this reason, headlights should be turned on while responding (Figure 4.47). Some departments use white warning lights in conjunction with red or other colored lights, but some state and provincial laws prohibit the use of flashing white lights. A spotlight moving across the back window of a vehicle rapidly gains the driver's attention. The spotlight should not be left shining on the vehicle, however, because this blinds the driver. Headlights should be dimmed and spotlights turned off in situations where they may blind oncoming drivers, including the drivers of other apparatus that are approaching the scene. Headlight flashers are an inexpensive and effective warning device; however, some states (provinces) do not allow or consider them warning lights. Do not drive with high beam headlights on constantly because they tend to drown out the other warning lights.

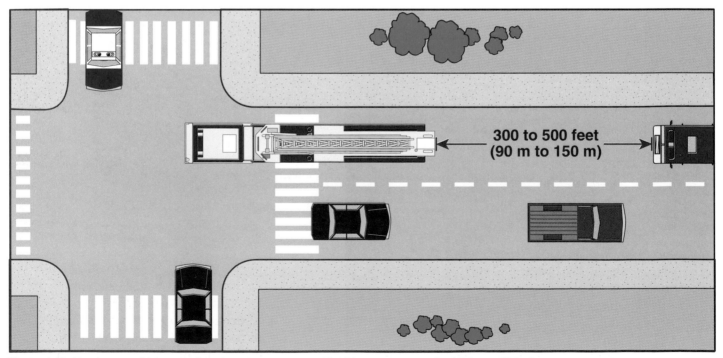

Figure 4.46 Emergency vehicles traveling along the same route should stay 300 to 500 feet (90 m to 150 m) apart.

Figure 4.47 Headlights should be on when making an emergency response, even during daylight hours.

The trend in recent years has been to equip apparatus with a large amount of visual warning devices. While this quantity has proven to be effective during the response, studies have shown that they may actually pose hazards for the firefighters once the apparatus is parked on the scene during nighttime operations. The large quantity of warning lights, combined with on scene floodlights overpower the effectiveness of the reflective trim on the firefighters' protective clothing or vests. This makes the drivers of approaching vehicles unable to see firefighters who are standing in the street. In these situations, it is desirable to turn off some of the warning lights on the apparatus once it is parked. Some fire departments have equipped their apparatus with one or two small, yellow warning lights that are turned on when the apparatus is parked on the scene. This allows the approaching vehicles' headlights to more effectively illuminate the reflective trim worn by firefighters.

Traffic Control Devices

Some jurisdictions use traffic control devices to assist emergency vehicles during their response. The driver/operator must be aware of the traffic control devices used in his jurisdiction and how they operate. One of the simplest involves placing a traffic signal in front of the fire station to stop the flow of traffic in front of the station so that the apparatus can exit safely (Figure 4.48). This signal may be controlled by a button in the station or by the dispatcher. It may also be activated when the station is toned out. Some jurisdictions have systems that control one or more traffic lights in the normal route of travel for fire apparatus. Again, these may be controlled from the fire station, remote controls on the fire apparatus, or from the dispatch center.

Another common system for controlling traffic signals for fire apparatus is the Opticom™ system, which involves the use of special strobe lights, called emitters, on the fire apparatus and sensors mounted on the traffic lights. The emitter generates an optical signal that is received by the sensor on the traffic light as the apparatus approaches (Figure 4.49). The sensor converts this signal

Figure 4.48 Some stations have a traffic light out front to stop cross traffic when apparatus leave the station.

Figure 4.49 The Opticom™ emitter is located on the apparatus. It may be contained within a light bar, or it may be a separate unit.

to an electronic impulse that is routed to the phase selector in the traffic light control cabinet (Figure 4.50). The phase selector then provides a green light for the direction that the apparatus is traveling and red signals in all other directions (Figure 4.51). In some jurisdictions, the traffic light standard may be equipped with a white light that indicates to the driver that the signal has been received and a green light is forthcoming. On some apparatus, the emitter is wired into the parking brake system. When the parking brake is set, the emitter will be turned off. On apparatus that do not have this feature, the driver/operator should remember to turn off the emitter when the apparatus is parked on the scene of an emergency. Otherwise, the emitter could affect any traffic signals that are within reach and disrupt the normal flow of traffic.

A newer type of traffic control system is the SONEM 2000 system, which is activated by the emergency vehicle's siren as it approaches an intersection. A microphone on the traffic signal "hears" the siren and sends a signal to the traffic signal controller, ordering a preemption of the current traffic signal. The microphone may be adjusted to order the preemption from distances of anywhere from a few hundred feet (meters) to about one-half mile (1 km). Intersections equipped with this system will have 3-inch (76 mm) white and blue lights in each direction of travel, somewhere to the side of the regular traffic signals. As soon as the microphone sends the preemption signal to the signal controller, the direc-

Figure 4.50 The Opticom™ sensor is located somewhere on the traffic light standard.

tion of travel for the emergency vehicle gets a white light indicating that the signal was received and that a green traffic light is forthcoming. All other directions of travel get a blue light that indicates an emergency vehicle coming from one of the other directions has gained control of the signal first. This is extremely important when emergency vehicles are approaching the intersection from more than one direction. Vehicles getting the blue light know that they will have to come to a stop because a green signal is not immediately forthcoming in their direction of travel.

Regardless of which types of traffic control devices are used in any jurisdiction, they are not substitutes for using proper defensive driving techniques. When traversing an intersection with a green signal, the driver/operator must maintain a speed that will allow for evasive actions in the event another vehicle enters the intersection. If for any reason the fire apparatus does not get a green signal, the driver/operator should bring the vehicle to a complete stop at a red signal. Keep in mind that if two apparatus equipped with Opticom™ devices approach the same traffic signal from different directions, only the apparatus whose sensor affects the signal first will get a green light. The later-approaching apparatus gets a red signal. Do not assume that just because you did not get a green light that the system is not working. Approach the intersection with caution and come to a complete stop.

Driving Exercises and Evaluation Methods

After driver/operators have been selected and trained, their performance should be evaluated by some standard method. These evaluations should occur before the driver/operator is allowed to operate the apparatus under emergency conditions. NFPA 1002 provides some specific directions on how driver/operator candidates should be tested. These directions need to be followed by

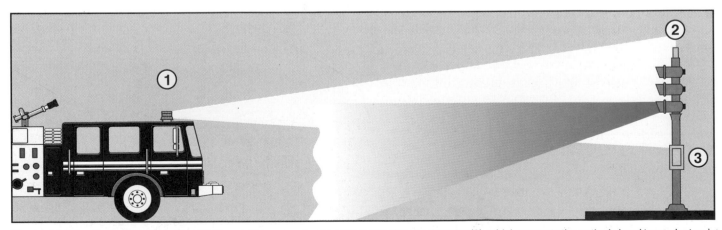

Figure 4.51 How the Opticom™ works: (1) The emitter generates an optical signal to the detector (2), which converts the optical signal to an electronic impulse. The phase selector (3) in the control cabinet processes and then manipulates the controller to provide a green signal for the emergency vehicle and red signals in all other directions. *Courtesy of 3M Safety and Security.*

agencies that certify their personnel to the standard. Other agencies should at least follow the standard to avoid possible civil law liabilities should the driver/operator be involved in a collision. Most agencies use a combination of written and practical testing for driver/operator candidates. On volunteer fire departments, successful completion of this testing may simply result in making the person qualified to drive the fire apparatus. On career fire departments, this may be a promotional exam.

All fire apparatus training and testing should follow the requirements contained in NFPA 1451, *Standard for a Fire Service Vehicle Operations Training Program.*

Written Test

There are some facets of the driver/operator's job that are most easily tested through the use of a written exam. The written exam for driver/operators may include questions pertaining to the following areas:

- State and local driving regulations for emergency and nonemergency situations
- Departmental regulations
- Hydraulic calculations
- Specific operational questions regarding pumping
- Department standard operating procedures

Depending on local preference, the test may be open or closed book. The style of questions also vary according to local preference.

Practical Driving Exercises

NFPA 1002 specifies a number of practical driving exercises that the driver/operator candidate should be able to successfully complete before being certified to drive the apparatus. The standard requires that driver/operators be able to perform these exercises with each type of apparatus they are expected to drive. Some jurisdictions prefer to have driver/operators complete these evolutions before allowing them to complete the road test. This assures that driver/operators are competent in controlling the vehicle before they are allowed to drive it in public. The exercises that follow are those that are specifically required in the standard. Individual jurisdictions may choose to add other exercises that simulate local conditions. However, as a minimum, all of these exercises should be completed.

NOTE: The descriptions for the exercises listed contain minimum dimensions for setting up these exercises. NFPA 1002 notes that these dimensions may not be reasonable for extremely large fire apparatus. This will re-

quire the local agency to modify the dimensions. The authority having jurisdiction should be able to justify any modifications that are made as being reasonable for local conditions.

Alley Dock

The alley dock exercise tests the driver/operator's ability to move the vehicle backward within a restricted area and into an alley, dock, or fire station without striking the walls and to bring the vehicle to a smooth stop close to the rear wall. The boundary lines for the restricted area should be 40 feet (12.2 m) wide, similar to curb-to-curb distance (Figure 4.52). Along one side and perpendicular is another simulated area 12 feet (3.66 m) wide and 20 feet (6.1 m) deep. The test procedure has the driver/operator moving past the alley (which is on the left), backing the apparatus, making a left turn in reverse into the defined area, and stopping. This exercise should then be repeated from the opposite direction. The driver/operator is considered to have successfully completed this exercise when able to back into the restricted area without having to stop and pull forward and without striking any obstructions or markers.

As an alternative to the traditional alley dock exercise, the local jurisdiction may choose to substitute the apparatus station parking maneuver (Figure 4.53). In this exercise, an apparatus bay is simulated by allowing a 20-foot (6.1 m) minimum setback from a street that is 30 feet (9 m) wide. A set of barricades are positioned 12 feet (3.66 m) apart at the end of the setback to simulate a garage door opening. A simulated apparatus bay should be con-

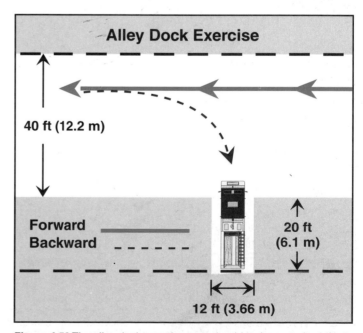

Figure 4.52 The alley dock exercise tests the driver/operator's ability to back into a tight spot.

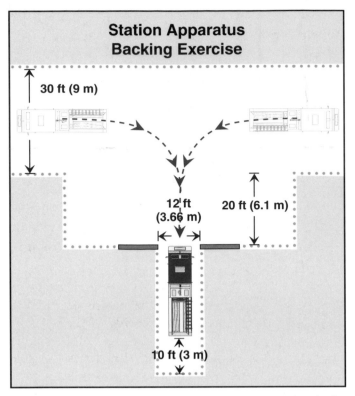

Station Apparatus Backing Exercise

30 ft (9 m)

12 ft (3.66 m)

20 ft (6.1 m)

10 ft (3 m)

Figure 4.53 This exercise simulates backing the apparatus into the fire station.

and then forward. The course must be traveled in each direction in one continuous motion without touching any of the course markers.

First, the driver/operator is required to drive the apparatus along the left side of the markers in a straight line and stop just beyond the last marker. Then the driver/operator must back the apparatus between the markers by passing to the left of No. 1, to the right of No. 2, and to the left of No. 3. At this point, the driver/operator must stop the vehicle and then drive it forward between the markers by passing to the right of No. 3, to the left of No. 2, and to the right of No. 1.

Confined Space Turnaround

The confined space turnaround exercise tests the driver/operator's ability to turn the vehicle 180 degrees within a confined space. Fire apparatus, particularly fire department pumpers, often need to turn around to complete an operation such as a reverse hose lay. Although turning fire apparatus around may not be difficult in adequate space, it becomes more complicated in narrow streets or intersections.

This exercise may be performed in an area that is at least 50 feet (15.25 m) wide and 100 feet (30.5 m) long (Figure 4.55). The apparatus begins in the center of one end of the test area. The driver then pulls forward, moves toward one side or the other, and begins the turning process. There is no limit to the number of direction changes that are required before the apparatus is turned 180 degrees and driven through the same opening it entered. A spotter may be used during the process of turning the vehicle around. Successful completion of this exercise means that the apparatus has been turned 180 degrees and driven through the original entrance point with no course markers being struck or without leaving the defined course.

If the apparatus is so small that it can complete a U-turn without stopping and backing in a course of the dimensions described here, make the course smaller so that backing is required.

structed back from the "garage door" opening. This bay should be 10 feet (3 m) longer than the vehicle. A straight line may be provided inside the bay and a traffic cone may be placed where the front left wheel is supposed to stop. Again, the local jurisdiction may choose to alter these dimensions based on local conditions and apparatus size. The test procedure has the driver/operator moving past the setback area (which is on the left), backing the apparatus, making a left turn in reverse through the setback area, and into the apparatus bay area. This exercise should then be repeated from the opposite direction. The driver/operator is considered to have successfully completed this exercise when able to back into the apparatus bay without having to stop and pull forward and without striking any obstructions or markers.

Serpentine Course

This exercise simulates maneuvering around parked and stopped vehicles and tight corners. In the serpentine exercise, at least three markers are placed an equal distance apart in a line (Figure 4.54). The markers should be between 30 and 38 feet (9 m and 12 m) apart, depending on the size of the apparatus being used. Adequate space must be provided on each side of the markers for the apparatus to move freely. The driver/operator is required to maneuver the vehicle first backward through the cones

Diminishing-Clearance

The diminishing-clearance exercise measures a driver/operator's ability to steer the apparatus in a straight line, to judge distances from wheel to object, and to stop at a finish line. The speed at which the apparatus is driven is optional, but it should be fast enough to require the driver/operator to exercise quick judgment. The course for this exercise is arranged by two rows of stanchions that form a lane 75 feet (23 m) long. The lane narrows from a width of 9 feet 6 inches (2.9 m) to a diminishing

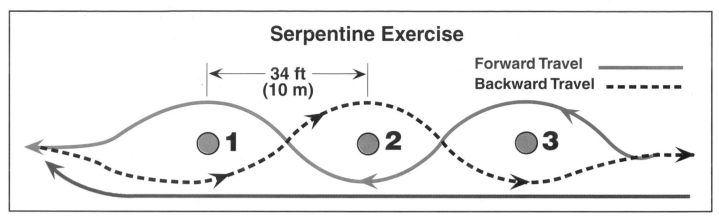

Figure 4.54 The serpentine exercise is intended to simulate maneuvering the apparatus in tight locations and around parked vehicles.

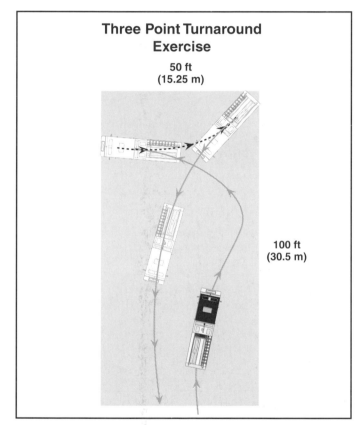

Figure 4.55 The turning around exercise simulates the need to change the direction of the apparatus at a fire scene in order to make a reverse hose lay to a water supply source.

clearance of 8 feet 2 inches (2.5 m) (Figure 4.56). The driver/operator must maneuver the apparatus through this lane without touching stanchions. At a point 50 feet (15 m) beyond the last stanchion, the driver/operator must stop with the front bumper within 6 inches (150 mm) of the finish line. Obviously, these dimensions need to be adjusted for larger vehicles such as Aircraft Rescue and Fire Fighting (ARFF) apparatus.

Road Tests

Prior to being certified to drive emergency apparatus, driver/operators should demonstrate their ability to operate the apparatus on public thoroughfares. Driver/operators should only be allowed to operate on public thoroughfares after they have demonstrated the ability to control the apparatus they are driving. Each department will want to develop an established route that driver/operator candidates should follow. This route should cover all of the usual driving conditions that can be expected within that jurisdiction. However, as a minimum, NFPA 1002 says that any road test that leads to certification should include at least the following elements:

• Four left and four right turns

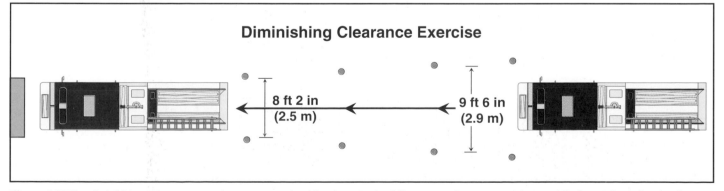

Figure 4.56 The diminishing clearance exercise measures the driver/operator's ability to steer the apparatus in a straight line and to judge the distance from the wheel to various objects.

- A straight section of urban business street or two-lane rural road at least one mile (1.6 km) in length

- One through intersection and two intersections where a stop must be made

- A railroad crossing

- One curve, either left or right

- A section of limited-access highway that includes a conventional on-ramp, off-ramp, and is long enough to allow for at least two lane changes

- A downgrade that is steep enough and long enough to require gear changing to maintain speed

- An upgrade that is steep enough and long enough to require gear changing to maintain speed

- One underpass, low-clearance bridge

During testing, the evaluation of a driver/operator candidate on a road test is very subjective. In general, they should be evaluated on their adherence to posted traffic requirements and departmental policies, as well as their ability to safely control the vehicle.

Summary of Good Driving Practices

A collision or vehicular failure caused by irresponsible driving has many repercussions and is inexcusable. Lives that could have been saved may be lost; property that should have been protected will be destroyed; firefighters on the apparatus, innocent bystanders, or other drivers may be injured or killed, leaving the department and the driver/operator open to lawsuits and civil prosecution for such things as negligent homicide or manslaughter; and the apparatus will be useless for an indefinite time, leaving citizens with less protection.

The following are some critical points to remember for safe operation and driving of fire apparatus:

- Remember that speed is less important than arriving safely at the destination.

- Slow down for intersections and stop when faced with a red light or stop sign. Anticipate the worst possible situation.

- Drive defensively. Be aware of everything that is happening or likely to happen 360 degrees around the apparatus.

- Expect that some motorists and pedestrians will neither hear nor see the apparatus warning devices.

- Be aware of the route's general road and traffic conditions. Adjust this expectation with the season, weather, day of the week, and time of day.

- Remember that icy, wet, or snow-packed roads increase braking distance.

- Do not grind the gears on manual transmission vehicles.

- Do not use the clutch pedal as a footrest.

- Do not exceed 10 mph (15 km/h) when leaving the station.

- Do not race the engine when the apparatus is standing still. It is unnecessary and abuses the engine.

- Always use low gear when starting from a standstill. Using second or third gear and slipping the clutch damages the clutch and causes unnecessary, rapid wear.

- Keep the apparatus under control at all times.

- Take *nothing* for granted.

Chapter 5

Positioning Apparatus

Job Performance Requirements

This chapter provides information that will assist the reader in meeting the following job performance requirements from NFPA 1002, *Standard on Fire Apparatus Driver/Operator Professional Qualifications*, 1998 edition.

3-2.1* Produce effective hand or master streams, given the sources specified in the following list, so that the pump is safely engaged, all pressure control and vehicle safety devices are set, the rated flow of the nozzle is achieved and maintained, and the apparatus is continuously monitored for potential problems.

- Internal tank
- Pressurized source
- Static source
- Transfer from internal tank to external source

(a) *Requisite Knowledge:* Hydraulic calculations for friction loss and flow using both written formulas and estimation methods, safe operation of the pump, problems related to small-diameter or dead-end mains, low-pressure and private water supply systems, hydrant cooling systems, and reliability of static sources.

(b) *Requisite Skills:* **The ability to position a fire department pumper to operate at a fire hydrant and at a static water source,** power transfer from vehicle engine to pump, draft, operate pumper pressure control systems, operate the volume/pressure transfer valve (multistage pumps only), operate auxiliary cooling systems, make the transition between internal and external water sources, and assemble hose lines, nozzles, valves, and appliances.

3-2.2 Pump a supply line of 2½ in. (65 mm) or larger, given a relay pumping evolution the length and size of the line and the desired flow and intake pressure, so that the proper pressure and flow are provided to the next pumper in the relay.

(a) *Requisite Knowledge:* Hydraulic calculations for friction loss and flow using both written formulas and estimation methods, safe operation of the pump, problems related to small-diameter or dead-end mains, low-pressure and private water supply systems, hydrant cooling systems, and reliability of static sources.

(b) *Requisite Skills:* **The ability to position a fire department pumper to operate at a fire hydrant and at a static water source,** power transfer from vehicle engine to pump, draft, operate pumper pressure control systems, operate the volume/pressure transfer valve (multistage pumps only), operate auxiliary cooling systems, make the transition between internal and external water sources, and assemble hose lines, nozzles, valves, and appliances.

6-1.4* **Operate a wildland fire apparatus, given a predetermined route off of a public way that incorporates the maneuvers and features specified in the following list that the driver/ operator is expected to encounter during normal operations, so that the vehicle is safely operated in compliance with all applicable departmental rules and regulations, the requirements of NFPA 1500,** *Standard on Fire Department Occupational Safety and Health Program,* **Section 4-2, and the design limitations of the vehicle.**

- **Loose or wet soil**
- **Steep grades (30 percent fore and aft)**
- **Limited sight distance**
- **Blind curve**
- **Vehicle clearance obstacles (height, width, undercarriage, angle of approach, angle of departure)**
- **Limited space for turnaround**
- **Side slopes (20 percent side to side)**

(a) *Requisite Knowledge:* The effects on vehicle control of braking reaction time, load factors, general steering reactions, speed, and centrifugal force; applicable laws and regulations; principles of skid avoidance, night driving, shifting, and gear patterns; negotiating intersections, railroad crossings, and bridges; weight and height limitations for both roads and bridges; identification and operation of automotive gauges; and proper operation limits.

(b) *Requisite Skills:* The ability to operate passenger restraint devices, maintain safe following distances, maintain control of the vehicle while accelerating, decelerating, and turning, maintain reasonable speed for road, weather, and traffic conditions, operate safely during nonemergency conditions, operate under adverse environmental or driving surface conditions, and use automotive gauges and controls.

6-2.1* Produce effective fire streams, utilizing the sources specified in the following list, so that the pump is safely engaged, all pressure-control and vehicle safety devices are set, the rated flow of the nozzle is achieved, and the apparatus is continuously monitored for potential problems.

- Water tank
- Pressurized source
- Static source

(a) *Requisite Knowledge:* Hydraulic calculations for friction loss and flow using both written formulas and estimation methods, safe operation of the pump, proper apparatus placement, personal safety considerations, problems related to small-diameter or dead-end mains and low-pressure and private water supply systems, hydrant cooling systems, and reliability of static sources.

(b) *Requisite Skills:* **The ability to position a wildland fire apparatus to operate at a fire hydrant and at a static water source, properly place apparatus for fire attack,** transfer power from vehicle engine to pump, draft, operate pumper pressure control systems, operate the volume/pressure transfer valve (multistage pumps only), operate auxiliary cooling systems, make the transition between internal and external water sources, and assemble hose lines, nozzles, valves, and appliances.

6-2.2 Pump a supply line, given a relay pumping evolution the length and size of the line and pumping flow and desired intake pressure, so that adequate intake pressures and flow are provided to the next pumper in the relay.

(a) *Requisite Knowledge:* Hydraulic calculations for friction loss and flow using both written formulas and estimation methods, safe operation of the pump, problems related to small-diameter or dead-end main and to low-pressure and private water supply systems, hydrant cooling systems, and reliability of static sources.

(b) *Requisite Skills:* **The ability to position a wildland apparatus to operate at a fire hydrant and at a static water source,** transfer power from vehicle engine to pump, draft, operate pumper pressure control systems, operate the volume/pressure transfer valve (multistage pumps only), operate auxiliary cooling systems, make the transition between internal and external water sources, and assemble hose lines, nozzles, valves, and appliances.

Emergency incidents require that apparatus and the personnel assigned to them work in harmony in order to perform effectively. For incident control to be achieved efficiently and safely, apparatus must be positioned so that its use is maximized. The driver/operator's ability to properly position apparatus requires training, practice, and ingenuity. The driver/operator must also be able to execute certain maneuvers when called for by Incident Commanders or by pre-incident plans. Each type of apparatus will be positioned according to its purpose and overall strategic objectives. It must also function in coordination with other apparatus working the incident.

This chapter discusses proper positioning of pumping apparatus based on various functions for which it may be used such as fire attack or water supply. The chapter also discusses position considerations for wildland and support fire apparatus. Special positioning considerations for all types of apparatus are covered in the final portion of the chapter.

It should be noted that positioning for two common types of fireground operations are not covered in this chapter. Positioning pumpers in relay pumping operations is covered in Chapter 13. Positioning pumpers and tenders (tankers) during water shuttle operations is covered in Chapter 14.

Positioning Fire Department Pumpers

As stated in Chapter 2, the primary function of a fire department pumper is to provide water for fire fighting operations. Depending on the situation, the pumper may provide water directly to fire streams for incident control or supply water to other pumpers or aerial apparatus that in turn are directly attacking the fire. The driver/operator must understand the important principles involved with each of these situations. The following sections cover the most common positioning considerations for pumpers used for fire attack and water supply operations.

Fire Attack Pumpers

There is no one set rule for positioning pumpers supplying attack lines on the fireground (Figure 5.1). This is because multiple factors must be reviewed when determining position for both first-due and late-arriving pumpers. The following sections contain some guide-

Figure 5.1 The attack pumper is positioned according to the fire conditions and the accessibility of the fire scene. *Courtesy of Ron Jeffers.*

Figure 5.2 The pumper is generally parked near the building's main entrance when no fire conditions are obvious. Firefighters may then exit the apparatus and investigate the incident.

lines that will assist driver/operators in determining effective placement. As in all fire situations, standard operating procedures and the judgment of the responsible officer or driver/operator should be the deciding factors when committing/positioning the apparatus.

Positioning for Fire Attack

Determining the proper position for the attack pumper begins with sizing up the incident. This is particularly crucial for the first apparatus arriving on the scene. As the apparatus approaches the scene, the driver/operator and company officer should observe the incident conditions to determine the best place to park the apparatus. Later-arriving apparatus will be directed to locations based on SOPs or orders from the Incident Commander.

If the apparatus arrives at a location where no fire conditions are evident ("nothing-showing" mode), it is generally advisable to park near the main entrance to the occupancy (Figure 5.2). This allows fire company per-

sonnel to enter the structure and investigate the situation. The driver/operator should remain with the vehicle and prepare to make connections to the water supply or sprinkler/standpipe fire department connection or pull attack hoselines if the need arises.

When fire conditions are evident upon approaching the scene, look for the best tactical position in which to place the apparatus. There are a variety of considerations that influence this decision. Some of the more important ones include the following:

- *Departmental standard operating procedures* — Many fire departments have established procedures for the placement of each initial apparatus. When this is the case, the driver/operator should make every effort to follow these procedures because it will affect the placement of later-arriving apparatus.

- *Rescue situations* — Rescue is always the first tactical priority at any fire incident. If there is an obvious rescue situation, such as people hanging out of upper-story windows, the apparatus should be parked in order to facilitate the timely deployment of ground ladders or the aerial device (if so equipped) to effect the rescue. For more information on deploying an aerial device, see the IFSTA **Fire Department Aerial Apparatus** manual.

- *Water supply* — If the incident is small enough to be handled with the water that is carried on the apparatus, water supply is not a major consideration in positioning the apparatus. If an external supply is required, this must be considered by the driver/operator and company officer before parking in the final position. If the apparatus is located in a position that is difficult for other apparatus to access, such as up a narrow lane or driveway, a supply hose should be laid into the scene as the apparatus moves into position

(Figure 5.3). If a fire hydrant is located close enough to the fire building to allow the attack pumper to connect to it and still be in a safe, effective operating position, this should also be considered (Figure 5.4).

- *Method of attack* — The method of fire attack has a major impact on apparatus positioning. If the incident can be handled with preconnected handlines, the apparatus must be positioned so that the nozzle reaches the area that contains the seat of the fire. If portable master streams are going to be used, the apparatus must be parked close enough for hoselines to effectively supply them. If the turret on the apparatus is going to be used, the apparatus must be in a position that allows the fire stream to reach its intended target (Figure 5.5).

- *Exposures* — If the fire has the potential to threaten exposures, the apparatus should be parked in a position that allows fire streams to deploy in order to protect those exposures (Figure 5.6). In some cases, it will be necessary to "write off" the original fire building in order to save severely threatened exposures. This is a tactical decision that relates back to the size-up process.

When considering exposures, it is important to remember that we bring our own potential exposure with us: the apparatus itself. Avoid parking the apparatus in a location that subjects it to high levels of radiant heat, falling embers, or other products of combustion (Figure 5.7). Any position that requires wetting the apparatus to prevent damage is not a good position (Figure 5.8).

- *Wind direction* — Whenever possible, attempt to park the apparatus upwind of the incident. This negates the need for the driver/operator to wear protective breathing apparatus while operating the vehicle. It also reduces the possibility of the apparatus becoming an exposure should fire conditions worsen. Lastly, if hazardous materials are involved in the fire, parking upwind will lessen the chance of contaminating the vehicle and personnel assigned to it (Figure 5.9). More information on positioning the apparatus at hazardous material incidents is contained later in this chapter.

- *Terrain* — There are numerous ways in which terrain affects apparatus placement. Given the choice, always choose a paved surface over an unpaved surface. This eliminates the chance of the apparatus getting stuck once the area becomes wet. In most cases, it is also desirable to be uphill from the incident whenever possible. There is less wear placed on the fire pump by supplying hoselines downhill than pumping uphill (this will be further explained in the hydraulic section

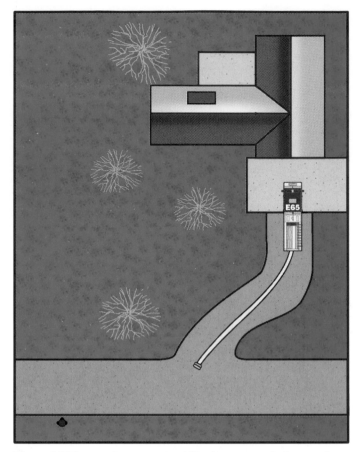

Figure 5.3 The attack pumper should lay its own supply line up a long, narrow driveway.

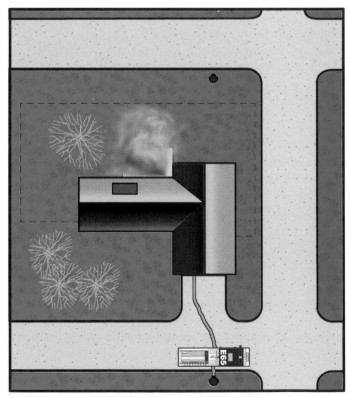

Figure 5.4 In ideal situations, the attack pumper may connect directly to a fire hydrant.

Figure 5.5 The turret stream must be able to reach the main body of fire.

Figure 5.7 The fireground has enough of its own challenges without firefighters bringing their own exposures. *Courtesy of Ron Jeffers.*

Figure 5.6 Exposure protection is a high priority at any fire scene. *Courtesy of Ron Jeffers.*

Figure 5.8 Avoid parking the apparatus so close to the fire that it requires wetting the truck to keep it cool. *Courtesy of Ron Jeffers.*

of this manual). Parking uphill from a hazardous material incident eliminates the chance of the product flowing underneath the truck. Always position uphill from a vehicle fire in case burning fuel begins leaking from the vehicle. The exception to the uphill rule is when positioning at a wildland fire. Wildland fires move uphill faster than on even terrain. Stay downhill from the main body of wildland fires if possible.

- *Relocation potential* — Always leave yourself a way out. Never position the apparatus in a location that does not allow an easy retreat should conditions warrant the need to move.

Figure 5.9 Park the apparatus to allow for an upwind attack from a hazardous materials incident. *Courtesy of Rich Mahaney.*

When laying a supply hose into the fire scene, make sure that the hose is laid to the side of the street if at all possible (Figure 5.10). This is particularly important with large diameter hoselines because once they are charged with water, they are difficult for later-arriving apparatus to drive over.

Another important consideration when determining a good position for attack pumpers is the condition of the fire building and the potential for a structural collapse. Buildings that have been subjected to extensive fire damage or buildings in poor condition before the fire may be subject to sudden collapse (Figure 5.11). For this reason, apparatus should be parked far enough away so that they are not in the collapse zone should one occur. The collapse zone should be at least equal to the height of the building (Figure 5.12). When possible, park at the corners of the building; this is generally considered the safest position should a collapse occur. Keep in mind that the corners of the building are also considered the optimum position for aerial apparatus. Make sure that aerial apparatus are able to get into the necessary position.

There are many indicators that a building may become unstable. Bulging walls; large cracks in the exterior; falling bricks, blocks, or mortar; and interior collapses are all signs that a serious exterior collapse may occur. Pre-incident planning aids in identifying buildings with a good potential for collapse. Buildings that are old and poorly maintained should be targeted. The presence of ornamental stars or large bolts with washers at various intervals on exterior walls indicate that reinforcement ties are present to hold otherwise unstable walls in place (Figure 5.13).

The intensity of the fire also dictates apparatus placement. Large, hot fires require the apparatus to be positioned farther away from the fire building. Consideration must also be given to the fire's potential growth. If the fire has the potential to grow or spread to exposures, the apparatus must be placed so that it is not trapped by the advancing fire.

Try not to park beneath power or other utility lines. This is particularly true if there is any chance that fire or weather conditions will cause the lines to fall.

Figure 5.11 Firefighters must always be alert to a potential collapse during fire fighting operations. *Courtesy of Harvey Eisner.*

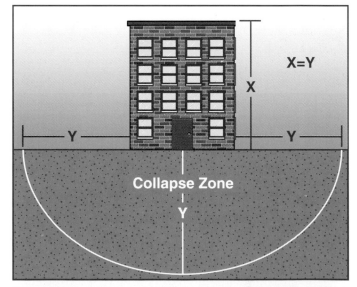

Figure 5.12 The collapse zone should equal the height of the building.

Figure 5.10 Lay the supply line to the side of the street.

Figure 5.13 Ornamental stars are an indicator of a building that is in a deteriorating state of condition.

Another consideration for placing apparatus is the debris that can fall from the fire building. This is of particular concern at high-rise fire incidents. Large pieces of glass and other debris may fall from many stories above street level (Figure 5.14). This can pose a serious hazard to personnel operating from the apparatus and to the apparatus itself. In these situations, the apparatus should be positioned away from the area in which debris is falling, and all personnel should be kept safe of the falling debris zone. If it is not possible to reposition, cover the apparatus with tarps or salvage covers to protect it from the falling debris.

Positioning to Support Aerial Apparatus

In most jurisdictions, pumping apparatus arrive on the scene prior to the arrival of the first aerial apparatus. Driver/operators of pumping apparatus must not only seek a good tactical location for the rig they are driving, but they must also keep in mind the needs of aerial apparatus that will soon arrive on the scene. Failure to leave a good position for the aerial apparatus can have serious negative consequences on the overall outcome of the incident.

In most cases, it is best to give the aerial apparatus the most optimum operating position and to locate the pumping apparatus a little farther away. The reason for

this is that the aerial device has a fixed length. If the apparatus is parked too far away from the building, there is nothing that can be done to correct the situation. Most pumpers have in excess of 1,000 feet (300 m) of hose. If they are parked out of the reach of their preconnected hoselines, they can always extend them with extra hose carried on the apparatus.

In order to facilitate the proper placement of pumpers and aerial apparatus, some departments choose to use the "inside/outside" method of apparatus placement. If the building is less than five stories tall, engine companies should park on the side of the street closest to the building and aerials park on the outside (Figure 5.15). The philosophy here is that the building is low enough to be reached by the aerial device even if it has to go over the closer engines. If the building is higher than five stories, the engines take the outside position and the aerials park next to the building (Figure 5.16). This allows for the

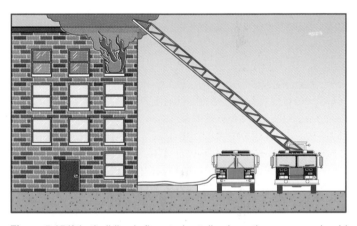

Figure 5.15 If the building is five stories tall or less, the pumpers should be parked at the inside position, with the aerial apparatus at the outside position.

Figure 5.14 High-rise fires often result in excessive and dangerous falling debris. *Courtesy of the City of Los Angeles Fire Department.*

Figure 5.16 If the building exceeds five stories, the aerial apparatus should be parked closest to the building.

aerials' maximum reach ability. The drivers of each type of apparatus can be trained where to park their vehicles depending on the height of the building.

Pumpers providing water for elevated stream operations should position as close to the aerial apparatus as possible (Figure 5.17). Friction and elevation loss are major considerations when supporting elevated streams, and pumpers must be aware of these losses. Pumpers equipped with their own elevated stream devices should position in the same manner as aerial apparatus providing fire suppression (Figure 5.18).

Supporting Fire Department Connections

Unlike some of the situations that were previously discussed, the positioning of pumpers to support sprinkler or standpipe fire department connections (FDC) is rather simple (Figure 5.19). Pumpers will generally position as close as possible to the sprinkler or standpipe FDC. This location should be established during pre-incident planning activities. Many jurisdictions have SOPs that require the first-due pumper to proceed directly to the FDC.

Most of the time, a fire hydrant is located very close to the FDC. This allows the pumper to connect to both the hydrant and the FDC with relative ease (Figure 5.20). On rare occasions, the location of the water supply is such that the pumper has to position at the source. This is obviously the case when support pumpers are using a draft source. In extreme cases where there is no water supply available near the sprinkler or fire department connection, it may be necessary to establish a relay to supply water.

There are situations when pumpers supporting sprinklers or standpipes must give priority to other apparatus. This is especially true when the spot normally occupied by the FDC pumper is required by an aerial apparatus. As previously stated, pumpers can easily operate at varying distances, but aerial apparatus require more precise locations. More information on operating the pumper at facilities with sprinklers and standpipes is contained in Chapter 8.

Figure 5.19 A typical fire department connection.

Figure 5.17 Pumpers may be required to supply water to aerial apparatus for elevated master streams. *Courtesy of Harvey Eisner.*

Figure 5.18 In some cases, the pump on the apparatus may be used to supply the elevated master stream. *Courtesy of Joel Woods.*

Figure 5.20 The pumper should connect to both the water supply and the fire department connection.

Water Source Supply Pumpers

Not all pumpers locate directly on the fire scene and pump into attack hoselines. In some situations, pumpers will locate at a distant water supply source and pump water to the apparatus at the fire scene (Figure 5.21). The following sections address important considerations for pumping apparatus that position at static and pressurized water supply sources.

Drafting Operations

Drafting operations are required when a pumper is going to be supplied from a static water supply source such as a pond, lake, stream, or cistern. Drafting pumpers may supply fireground apparatus directly or may serve as source pumpers for relay or water shuttle operations. These operations are common in rural areas but can just as easily be performed at urban incidents.

Fire departments should attempt to identify all suitable drafting locations in their response district and keep a record of them for future use. This information should be contained in map or plan books carried on the apparatus as well as retained by the dispatcher. Pre-incident planning information for specific occupancies or response areas may also contain this information. Preference should be given to drafting sites that are accessible from a paved surface and require a minimum length of suction hose or lift (Figure 5.22). Minimizing lift distances provides better discharge abilities and should be

of primary concern. In no case should the maximum lifts in Table 5.1 be exceeded if full pump capacity is to be used. Bridges, boat ramps, and large docks make for the best drafting locations.

Figure 5.21 Pumpers will often be required to supply water to the fire scene.

Figure 5.22 Try to use a drafting site that has a solid, paved approach.

Rated Capacity of Pump (gpm)	Intake Hose Diameter (inches)	Volume Discharged (gpm)										
		2 sections hose (20 feet)							3 sections hose (30 feet)			
		Feet of Lift										
		4	6	8	10	12	14	16	18	20	22	24
500	4	590	560	530	500	465	430	390	325	270	195	65
500	4½	660	630	595	560	520	480	430	370	310	225	70
750	4½	870	830	790	750	700	650	585	495	425	340	205
750	5	945	905	860	820	770	720	655	560	480	375	235
1,000	5	1,160	1,110	1,055	1,000	935	870	790	670	590	485	340
1,000	6	1,345	1,290	1,230	1,170	1,105	1,045	960	835	725	590	400
1,250	6	1,435	1,375	1,310	1,250	1,175	1,100	1,020	900	790	660	495
1,500	6	1,735	1,660	1,575	1,500	1,410	1,325	1,225	1,085	995	800	590
1,500	Dual 5	1,990	1,990	1,810	1,720	1,615	1,520	1,405	1,240	1,110	950	730
1,500	Dual 6	2,250	2,150	2,040	1,935	1,820	1,710	1,585	1,420	1,270	1,085	835

**Table 5.1 (U.S.)
Maximum Discharge at Various Lifts**

NOTE: Test conducted with net pump pressure of 150 psi, 1,000 feet altitude, 60°F water temperature, 28.94 inches Hg barometric pressure (poor weather).
Operation at lower than 150 psi net pump pressure will result in an increased discharge; at higher pressure, a decreased discharge.
This data based on a pumper with ability to discharge rated capacity when drafting at not more than a 10-foot lift. Many pumpers will exceed this performance and therefore will discharge greater quantities than shown at all lifts.
Source: American Insurance Association

Table 5.1 (metric)
Maximum Discharge at Various Lifts

Rated Capacity of Pump (L/min)	Intake Hose Diameter (mm)	Volume Discharged (L/min)										
		2 sections hose (6 m)							3 sections hose (9 m)			
							Meters of Lift					
		1.2	1.8	2.4	3	3.6	4.3	4.9	5.5	6.1	6.7	7.3
2 000	100	2 233	2 120	2 006	1 893	1 760	1 628	1 476	1 230	1 022	738	246
2 000	115	2 498	2 385	2 252	2 120	1 968	1 817	1 628	1 401	1 173	852	265
3 000	115	3 293	3 142	2 990	2 839	2 650	2 461	2 214	1 874	1 609	1 287	776
3 000	125	3 577	3 426	3 255	3 104	2 915	2 725	2 479	2 120	1 817	1 420	890
4 000	125	4 391	4 202	3 993	3 785	3 539	3 293	2 990	2 536	2 233	1 835	1 287
4 000	150	5 091	4 883	4 656	4 429	4 183	3 956	3 634	3 161	2 744	2 233	1 514
5 000	150	5 432	5 204	4 959	4 732	4 448	4 163	3 861	3 407	2 990	2 498	1 873
6 000	150	6 568	6 284	5 962	5 678	5 337	5 016	4 637	4 107	3 615	3 028	2 233
6 000	Dual 125	7 533	7 533	6 852	6 511	6 113	5 754	5 318	4 694	4 202	3 596	2 763
6 000	Dual 150	8 517	8 139	7 722	7 324	6 889	6 473	6 000	5 375	4 807	4 107	3 161

NOTE: Test conducted with net pump pressure of 1 050 kPa, 300 m altitude, 15.6°C water temperature, 735 mm Hg barometric pressure (poor weather).
Operation at lower than 1 050 kPa net pump pressure will result in an increased discharge; at higher pressure, a decreased discharge.
This data based on a pumper with ability to discharge rated capacity when drafting at not more than a 3 m lift. Many pumpers will exceed this performance and therefore will discharge greater quantities than shown at all lifts.
Source: American Insurance Association

Be wary of drafting from locations that are off paved surfaces. These surfaces may be unstable and cause the apparatus to sink into the ground. As well, the bank may slough off into the water, creating a serious tipping hazard for the apparatus.

When placing the suction hose directly into the static water source, the pumper should stop before reaching the source. First, connect the hard suction hose and strainer to the pumper. Then, drive the pumper into the final draft position (Figures 5.23a through c). This prevents firefighters from standing on unsure footing or in water currents while making connections. Attaching a rope to the end of the strainer before putting it in the water helps firefighters position it properly without having to enter the water (Figure 5.24).

The hard suction hose strainer should not rest on the bottom of the water source during drafting. The rope attached to the strainer may be tied off to the apparatus or a nearby object in order to hold the strainer off the bottom (Figure 5.25). Floats, such as a spare tire or plastic bucket, can be used to hold the strainer at an appropriate depth. The suction hose may also be laid on top of a ground ladder to keep it off the bottom (Figure 5.26). Some departments choose to use floating strainers that eliminate this problem (Figure 5.27).

Many rural jurisdictions choose to designate drafting sites within their jurisdiction and place dry hydrants at those locations (Figure 5.28). A dry hydrant consists of a suction hose connection on the shore and a length of pipe equipped with a strainer that extends into the water supply source (Figure 5.29). This allows suction hose from the pumper to be quickly connected to the water supply source when a drafting operation is needed. For more detailed information on operating a pumper from draft, refer to Chapter 11 of this manual.

Hydrant Operations

In most jurisdictions, the most common water supply source is a fire hydrant. There are a number of ways in which a fire department pumper may be connected to a fire hydrant. Each of these is detailed in the following sections.

Historically, hard suction hose has been used to connect a pumper to a fire hydrant. However, driver/operators should be aware that hard suction hose is designed to withstand the negative pressures associated with drafting operations. Most are *not* designed or intended to be used under positive-pressure conditions. There have been many instances of hard suction hose coupling failure or hose failure when connected to fire hydrants. For this reason, *IFSTA recommends that hard suction hose never be connected to a fire hydrant*.

Many jurisdictions have SOPs that require the driver/operator to place gated valves on the small

Figure 5.23a Stop the pumper a little short of the drafting site.

Figure 5.23b Connect the intake hose to the fire pump while you are on secure footing.

Figure 5.23c Once the intake hose and strainer are connected, the apparatus can be carefully moved into a good drafting position and the hose and strainer may be placed in the water.

Figure 5.24 The rope tied to the end of the intake hose assists in positioning the hose.

Figure 5.25 Tie off the rope to the apparatus or another solid object.

Figure 5.26 In some cases, it may be advantageous to lay the intake hose on a ground ladder.

Figure 5.27 Floating strainers keep the intake hose near the water's surface.

Figure 5.28 Jurisdictions that rely on a static water source often install dry hydrants at key locations.

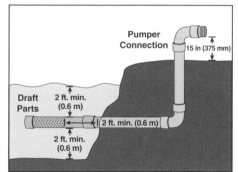

Figure 5.29 This diagram shows a good installation of a dry hydrant.

diameter discharges of the hydrant when making a connection to the large diameter discharge (Figure 5.30). This allows additional hoses to be connected to the hydrant later on without having to shut down the hydrant.

Large diameter intake hose connections. The preferred type of hose for connection to a fire hydrant is large diameter intake hose, sometimes referred to as a soft sleeve or soft suction hose (actually, soft suction is a misnomer). Intake hose sections are commonly 10 to 50 feet (3 m to 15 m) long. In order to properly position the

Figure 5.30 A gate valve placed on the small diameter discharge allows for another connection to be made later without interrupting the flow of water through the large diameter intake.

apparatus, the driver/operator must know the length of the intake hose on the apparatus. The driver/operator must, through practice, judge the proper distance from the hydrant. This distance is judged from the hydrant, rather than the curb line because most hydrants are located different distances from the curb. Whether the hydrant outlet faces the street or is parallel to the curb line is another regulating factor. If the front wheels of the apparatus are turned to a 45-degree angle, the driver/operator can easily adjust the distance to or from the hydrant by moving the unit forward or backward (Figure 5.31).

Side intake connections. The driver/operator must stop the pumper with the pump intake a few feet (meters) short of being in line with the hydrant outlet (Figure 5.32). Stopping short of the hydrant outlet permits the intake hose to slightly curve, preventing kinks that drastically restrict flow.

A good way to minimize kinks in soft sleeve hose is to put two full twists in the hose when making the connection between the hydrant and the pumper (Figure 5.33). These twists prevent the formation of kinks, yet do not affect the hose's ability to flow water. Practice with the

hose is necessary to become proficient in making twists and in determining how many are required to avoid kinks. Twists should not be put in the hose if either or both ends are equipped with Storz couplings. This could result in one of the coupling connections coming apart when the hose is charged.

Front and rear intake connections. Similar precautions and judgment must be used when positioning pumpers with front and rear pump intakes. The driver/operator must stop the pumper either a few feet (meters) short or a few feet (meters) beyond the hydrant to permit the hose to curve (Figure 5.34). Only practice with the individual pumper will develop proper positioning for intake hose connections. When using front or rear intake connections, the vehicle should be aimed or angled in the direction of the hydrant (Figure 5.35). This angle should be 45 degrees or less. However, when performing this maneuver, make sure that you do not block the entire street from other vehicles that may need to access the scene.

Connection to the 2½-inch (65 mm) hydrant outlets. When the maximum flow from a hydrant is not needed,

Figure 5.31 The front wheels of the apparatus should be at a 45 degree angle when positioning at a hydrant.

Figure 5.33 Unless Storz couplings are being used, two full twists should be made in the hose.

Figure 5.32 Stop the apparatus so that the intake is a few feet (meters) short of the hydrant.

Figure 5.34 Stop the apparatus so that the front intake is a few feet (meters) short of the hydrant.

or large diameter intake hose is not available, connection to the hydrant may be made with one or two 2½-inch (65 mm) outlets. This is done by connecting sections of 2½- or 3-inch (65 mm or 77 mm) hose to the pump (Figure 5.36). This type of operation is by far the easiest to set up. The smaller diameter hose tends to be in longer lengths than intake hose and allows maximum flexibility with regard to the location of the pumper. It is also light enough to be easily handled by one person. With the ease of handling, maneuvering time is decreased. This allows the pumper to connect and begin supplying water with a minimum of delay.

The main disadvantage of connecting to the 2½-inch (65 mm) outlet is that it limits the amount of water that can be supplied. This can be a serious problem if the water supply demand increases during the course of the incident (i.e. the fire gets bigger). This limited water supply is mainly due to the high amount of friction loss in the smaller hoselines. This can be minimized by using 3-inch (77 mm) hoselines instead of 2½-inch (65 mm) lines. A kink in the hoseline can also reduce the maximum flow appreciably (Figure 5.37). Removing kinks is one of the easiest ways to ensure the maximum possible flow.

The pump intakes to which the hoselines are connected are also important in terms of friction loss. The gated 2½-inch (65 mm) intake connections to the fire pump drastically limit total flow (Figure 5.38). The maximum flow through one of these 2½-inch (65 mm) intakes depends upon the way the piping is arranged; in general, 250 gpm (946 L/min) is an average capacity. A better arrangement is to bring the 2½- or 3-inch (65 mm or 77 mm) lines into the pump through the large intake connection (Figure 5.39). This can be done with a bell re-

Figure 5.36 In some cases, the pumper may use medium diameter hoselines to connect to the fire hydrant.

Figure 5.37 Kinks in the hoseline may severely reduce the flow into the fire pump.

Figure 5.35 The apparatus should be at a 45 degree angle when using the front intake.

Figure 5.38 Medium diameter supply hoses are usually connected to the 2½-inch (65 mm) gated intakes.

ducer or a suction siamese fitting that allows connection of more than one 2½- or 3-inch (65 mm or 77 mm) hose.

Multiple intake connections. Occasionally, a pumper will use both a large diameter intake (commonly referred to as the *steamer connection*) and smaller hoselines from one exceptionally strong hydrant. In these cases, the pumper position should be determined by the soft sleeve requirements because it is the shorter (and greater capacity) hose (Figure 5.40). The decision to add the additional smaller lines is one reason that many jurisdictions require driver/operators to place gate valves on the 2½-inch (65 mm) hydrant connections as discussed earlier.

Dual Pumping Operations

With dual pumping (often incorrectly referred to as tandem pumping; see following section), one strong hydrant may be used to supply two pumpers. This type of operation has several advantages, including better use of available water and shorter hose lays (particularly if the hydrant is close to the fire). Additional hoselines can be placed in operation more quickly, and apparatus may be grouped close together, allowing easier coordination. The method for making a dual pumping hookup is as follows:

Step 1: Pumper 1 connects to the hydrant steamer connection using a large intake hose. This pumper then pumps water through its lines to the fire (Figure 5.41).

Step 2: Pumper 2 is positioned intake-to-intake with Pumper 1. The hydrant is closed until the intake gauge of Pumper 1 reads 0 (about 5 psi [35 kPa]). The throttle of Pumper 1 will need adjusting. This makes the volumes of discharge and intake equal, so the cap of the unused intake can be removed (Figure 5.42). (**NOTE:** If Pumper 1 is equipped with a keystone or gate valve on its unused intake, the hydrant need not be turned down.)

Step 3: Pumper 2 is connected by intake hose to the unused large intake of Pumper 1.

Figure 5.41 The first pumper is hooked to the hydrant and begins discharging water. **Note:** The photos in this sequence show hard intake hose being used for this evolution. Hard intake hose should only be used if it is rated to handle positive pressures.

Figure 5.39 Some apparatus have a siamese attached to the large pump intake.

Figure 5.40 Maximum water supply may be achieved when multiple hoselines are attached between the hydrant and the fire pump.

Figure 5.42 The second pumper then may be connected to the large intake of the first pumper. Having a valve on the large intake makes this process easier.

Step 4: The hydrant is opened completely (Figure 5.43).

Step 5: Pumper 2 pumps water through its lines to the fire. Its supply is the water not being used by Pumper 1 that is passing through it (Figure 5.44).

Tandem Pumping Operations

Tandem pumping operations may be used when pressures higher than a single engine is capable of supplying are required. This sometimes occurs when the pumper is attempting to supply high-rise sprinkler or standpipe systems or long hose layouts. Tandem pumping operations are also commonly used when the attack pumper is only a short distance from a hydrant. The second pumper is placed directly on the hydrant to support supply lines to the attack unit (Figure 5.45). It is important to use caution when supplying hoselines with a tandem pumping operation because it is possible to supply greater pressure than the hose can withstand. Pressure supplied to the hose should not exceed the pressure at which the hose is annually tested by the department. Consult NFPA 1962, *Standard for the Care, Use, and Service Testing of Fire Hose Including Couplings and Nozzles,* for the test pressures recommended for the type of fire hose used by your fire department. Departments that routinely perform high-pressure tandem pumping operations may have hose designated for that function.

To perform tandem pumping, the two engines are positioned in basically the same manner as for dual pumping, although the pumpers may be as much as 300 feet (90 m) apart. In dual pumping, the pumpers are connected intake-to-intake; in tandem pumping, the pumper directly attached to the water supply source pumps water through its discharge outlet(s) into the intake(s) of the second engine. This enables the second engine to discharge water at a much higher pressure

Figure 5.44 Both pumpers are now being supplied by the same hydrant.

Figure 5.45 Tandem pumping is really a short form of relay pumping.

than a single engine could have supplied. The higher pressures result from the fact that the pumps are actually acting in series.

Positioning Wildland Fire Apparatus

Wildland fire apparatus are in unique situations for fireground positioning requirements. Because of the traveling nature of wildland fires, wildland fire apparatus are seldom positioned in the same spot for the duration of an incident. Wildland fire apparatus may reposition many times during the course of an incident. In many cases, they attack a fire while moving (Figure 5.46). Because of these facts, the guidelines for positioning wildland fire apparatus are perhaps a bit more flexible than those for structural fire apparatus. The two most common functions for wildland fire apparatus are providing structural protection and making a fire attack. The following sections highlight some of the more important considerations for both of these scenarios.

Structural Protection

The highest priority for most wildland fire fighting operations is the protection of structures that are exposed to the fire. This boundary between the wildland and struc-

Figure 5.43 If the first pumper did not have a valve on the large intake, the hydrant must be fully opened after the second pumper is connected to the first.

Figure 5.46 Pump and roll operations are common on wildland fires. *Courtesy of Monterey County (CA) Fire Training Officers Association.*

Figure 5.47 Structures that are closely surrounded by thick vegetation are target hazards.

Figure 5.48 Back apparatus down narrow driveways in the event a hasty retreat is required.

tural development is often referred to as the *wildland/urban interface.* Fires in the wildland/urban interface are among the most challenging for firefighters.

Many structures threatened by a wildland fire are not on wide, paved streets. They are often at the ends of long, narrow driveways opening from rural lanes. They are also often surrounded by dry, flammable vegetation (Figure 5.47). While getting into and out of a structure's location can be easy under normal conditions, it may be much more difficult if the driveway is completely obscured by smoke. For safety, engines should be backed in from the last known turnaround, and the crew should note the locations of landmarks along the way (Figure 5.48).

Once the apparatus reaches the structure it has been assigned to protect, it should be positioned so that it is safe and convenient from which to work. This can be accomplished by performing the following procedures:

- Park off the roadway to avoid blocking other fire apparatus or evacuating vehicles.

- Scrape away fuel, if necessary, to avoid parking in flammable vegetation.

- Park on the lee side of the structure to minimize exposure to heat and blowing fire embers.

- Park near (but not too close to) the structure so that hoselines can be kept short.

- Keep cab doors closed and windows rolled up to keep out burning material.

- Place the engine's air-conditioning system (if so equipped) in recirculation mode to avoid drawing in smoke from outside.

- Do not park next to or under hazards such as the following:
  — Power lines
  — Trees or snags
  — LPG tanks or other pressure vessels
  — Structures that might burn

Making a Fire Attack

Apparatus being used to make a wildland fire attack are expected to operate from a variety of positions during the course of attack (Figure 5.49). The driver/operator must constantly be aware of the fire's current location and direction of travel so that the apparatus and its crew are never placed in a position of peril. Because the apparatus is usually being operated close to the fire, the driver/operator's vision is often severely obscured by smoke. High brush and dense vegetation also tend to limit the driver/operator's ability to clearly see the surface on which he is operating. For these reasons, special precautions must be taken to ensure the safety of the apparatus and the crew assigned to it.

When the vehicle operates under conditions of reduced visibility because of smoke or darkness, it should be driven at an appropriately reduced speed. A spotter (scout) may be needed to walk ahead of the vehicle to

Figure 5.49 Apparatus may be parked in a variety of positions to attack a wildland fire.

Figure 5.50 A spotter should walk ahead of apparatus being driven in heavy smoke.

help locate and avoid obstacles such as logs, stumps, rocks, low-hanging limbs, ditches, and gullies (Figure 5.50). Spotters should be equipped with reliable handlights, wear highly visible clothing, and stay within the driver's field of view at all times.

When the apparatus is operated in a stationary position, such as when pumping from draft, it should be positioned for maximum protection from heat and flames. Natural or man-made firebreaks such as streams or roads can be used. Driver/operators should consider potential hazards, such as falling trees, rolling rocks, incoming air drops, and heavy equipment building control lines, when selecting a position for the apparatus. A short 1½- or 1¾-inch (38 mm or 45 mm) line should be deployed and charged for protection of the apparatus (Figure 5.51). Whenever the apparatus is parked, the wheels should always be chocked. The apparatus should be parked facing the exit direction.

On steep hillsides, loose or unstable ground can cause the apparatus to slide or overturn, especially if it has a relatively high center of gravity. Even on level terrain, a vehicle can become mired in soft ground, sand, or mud, leaving it vulnerable to being overrun by a fire (Figure 5.52). Do not drive apparatus across a bridge unless the bridge is known to be strong enough to support the vehicle's weight (Figure 5.53). Do not attempt to ford streams with apparatus that is not designed to do so.

Driving fire apparatus on the shoulders of railroad roadbeds can result in tire damage from the coarse, angular rock of which the roadbeds are made (Figure 5.54). In addition, this rock has relatively little cohesion, so apparatus may be in danger of sliding and/or rolling over on these steep inclines. Also, unless it can be confirmed that train traffic has been halted, this can be a vulnerable and dangerous position.

When apparatus is used in a mobile attack, hoselines should be kept as short as possible (Figure 5.55). This helps to keep the apparatus mobile because shorter hoselines are less likely to become looped around stumps

Figure 5.51 A handline should be deployed to protect the apparatus.

Figure 5.52 One pitfall of off-road operations is the potential for getting stuck in soft ground. *Courtesy of the National Interagency Fire Center.*

Figure 5.53 Respect the weight limitations for any bridges that are encountered.

Figure 5.54 Rail beds contain coarse rocks and other unstable driving conditions.

Figure 5.55 Keep hose lengths to a minimum on pump and roll operations.

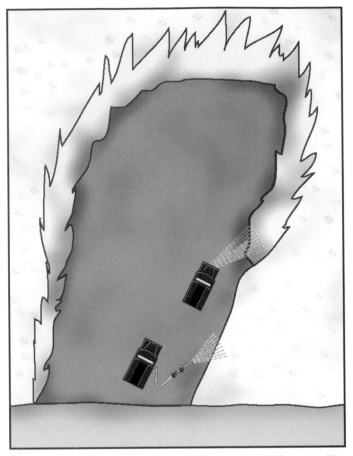

Figure 5.56 Tandem attacks are an effective method for controlling wildland fires.

Figure 5.57 Apparatus being driven in smoke should always have the headlights on. *Courtesy of the National Interagency Fire Center.*

or other objects and are easier to disentangle if they do get caught. Except when operating in the burned area, a small portion of water in the vehicle's tank should be reserved for protection of the apparatus and crew. When progressing along the fire's edge, the crew should make sure that the fire is out. To ensure complete extinguishment, engines can work in tandem or a single engine can work with a hand crew. The second engine or the hand crew mops up and patrols the line, making sure all fire is extinguished (Figure 5.56).

Experience using engines in the wildland environment has resulted in certain basic safety procedures being developed. These general engine-operation safety procedures are as follows:

- Engines should be parked in a safety zone and should not be left unattended at fires.

- Effective communication/coordination with the rest of the fireground organization is critical for safe and effective engine operations.

- Headlights should be on whenever the engine is running (Figure 5.57).

- Engines should be backed into one-way roads and driveways facing the escape route.

- All windows should remain rolled up to prevent burning embers from entering the cab of the vehicle.

- Engine and crews should draw back to the flanks rather than attempt a frontal attack if the fire is spreading rapidly upslope.

- Engine position should maximize protection from heat and fire. Take into consideration such hazards as overhead power lines, heavy fuel stands, and incoming air drops. Take advantage of natural breaks such as roads and orchards. Protect apparatus with 1- or 1½-inch (25 or 38 mm) lines.

- Apparatus should not be driven into unburned fuels higher than the bumper or running board without a spotter. Hidden stumps, logs, or other hazards can disable the vehicle. Spotters are also needed for nighttime driving when the terrain is not visible.

- Crews should use areas of burned fuel whenever possible (Figure 5.58). Apparatus attacking from the unburned side must leave sufficient clearance distances from the fire line to allow for loss of water and mechanical failure.

- Crews should be aware of fire conditions. Travel and position apparatus accordingly at all times.

- The location of operating crews should be considered when moving apparatus. Do not drive into smoke where crews may be operating. If apparatus must drive through smoke, sound the horn or siren intermittently, use warning and headlights, and drive slowly.

For additional information about operating fire apparatus at wildland fires, see IFSTA's **Fundamentals of Wildland Fire Fighting** manual.

Positioning Support Apparatus

A variety of apparatus other than pumping apparatus may be present on the fireground, depending on the resources available in any given jurisdiction. As mentioned earlier, aerial apparatus are common in many jurisdictions. Information on positioning aerial apparatus on the fire scene can be found in IFSTA's **Fire Department Aerial Apparatus** manual. In addition to pumping and aerial apparatus, the following types of apparatus may be present on the fireground:

- Rescue/squad apparatus
- Command vehicles
- Breathing air supply apparatus
- Emergency medical service (EMS) vehicles

Rescue/Squad Apparatus

In many jurisdictions, rescue companies, sometimes referred to as squads, are dispatched to fire incidents (Figure 5.59). Personnel assigned to them are commonly used as extra manpower on the fire scene or to perform truck company functions in the absence of an aerial

Figure 5.58 It is safest to operate in the burned (black) area.

Figure 5.59 Rescue companies may perform a variety of functions on the fireground. *Courtesy of Ron Jeffers.*

apparatus on the scene. Much of the equipment carried on rescue vehicles is similar to that carried on aerial apparatus. In addition, rescue vehicles may be equipped with large-scale electrical generation and lighting capabilities that make them useful during nighttime operations (Figure 5.60). Some rescue vehicles are also equipped to refill self-contained breathing apparatus (SCBA) cylinders.

In general, the positioning of rescue apparatus is not as critical as that of pumping and aerial apparatus. Rescue apparatus should be parked as close to the scene as possible, without blocking access to other apparatus. As well, the rescue apparatus should have a clear exit path from the scene in the event that it is needed at a second incident that might occur during the course of operations at the current scene. This is particularly crucial in small communities where only one rescue apparatus is available to cover the entire jurisdiction. If the rescue apparatus is going to be used for scene lighting or SCBA cylinder refilling, it needs to be parked strategically for those purposes.

Command Vehicles

Most fire departments have officers that ride in and use some sort of command staff vehicles. There is a wide variety of command vehicles, ranging from passenger sedans all the way up to large mobile command post vehicles (Figures 5.61 a through c). Volunteer fire officers commonly use their personal vehicles as command vehicles. Choosing a location for the command post/vehicle is fairly critical from an incident management standpoint. The location should be a logical one that is fairly conspicuous to all of the responders operating on the fire scene. Ideally, the command vehicle should be positioned on the corner of a building so that the Incident Commander is afforded a view of two sides of the building (Figure 5.62). Driveways, parking lots, yards, and cross streets all make good locations for command vehicles.

The general guidelines for positioning command vehicles are as follows:

- Provide for maximum visibility of the incident (attempt to have a clear view of two sides).

- Provide for maximum visibility of the area surrounding the incident.

- Place in a position that is easy to locate for other responders operating on the scene.

Figure 5.60 Rescue companies are commonly used to provide lighting for nighttime operations. This rescue vehicle has floodlights fixed to the end of its hydraulic crane to provide elevated lighting. *Courtesy of Joel Woods.*

Figure 5.62 The command post should afford a view of two sides of the incident if possible. *Courtesy of Bill Tompkins.*

Figure 5.61a Some incident commanders operate out of their response vehicles. *Courtesy of Ron Jeffers.*

Figure 5.61b Some jurisdictions utilize small command post vehicles for major operations.

Figure 5.61c An example of a large, custom command post vehicle.

- Position somewhere outside of the immediate danger zone.

- Avoid blocking the movement of other fire apparatus or interfering with incident operations.

- Display some type of light or sign that readily identifies the vehicle as the command post. This identification varies from jurisdiction to jurisdiction. Commonly used command post markers include pennants, flags, traffic cones, signs, banners, or flashing green lights (Figures 5.63 a through c).

Figure 5.63a A simple sign may mark the command post.

Figure 5.63b This command post is marked with a portable sign.

Figure 5.63c Some jurisdictions denote the command post using a flashing green light.

Breathing Air Supply Apparatus

Extended fire fighting and hazardous materials operations often require firefighters to expend a large quantity of SCBA cylinders throughout the course of the incident. Few departments carry enough spare cylinders on front-line fire apparatus to handle these extended operations. This requires sending special apparatus to the scene to refill the empty SCBA cylinders (Figure 5.64). Two types of equipment are used to refill SCBA cylinders on the scene: cascade systems and breathing air compressors.

Cascade systems are large breathing air cylinders that are connected together in banks. Cascade systems on mobile fire apparatus typically range from a bank of 4 to 12 large cylinders (Figure 5.65). Cascade systems allow air to be transferred from the large cylinders into the smaller SCBA cylinders. Cascade systems have a limited duration of use before they themselves must be refilled. This duration depends on the number and size of the cylinders in the system.

Figure 5.64 Mobile air units are sent to incidents that require the use and filling of many breathing apparatus cylinders.

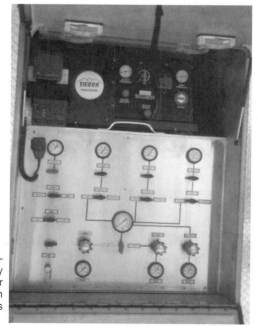

Figure 5.65 Cascade systems may be used to fill air cylinders when an air compressor is not available.

Breathing air compressors are engine-driven appliances that take in atmospheric air, purify it, and compress it (Figure 5.66). They continue to refill SCBA cylinders as long as their motors are running.

Smaller jurisdictions carry cascade systems or breathing air compressors on rescue apparatus or, in some cases, even aerial apparatus. Larger jurisdictions have special apparatus that are dedicated strictly to refilling/replacing SCBA cylinders. These mobile breathing air supply apparatus may carry large quantities of extra SCBA cylinders as well as equipment to refill expended cylinders. These apparatus may be equipped with large cascade systems, breathing air compressors, or both. They may also be equipped with long hose reels that allow cylinders to be refilled at a remote location such as inside a large building or on the upper floors of a high-rise structure.

The positioning of mobile air supply apparatus is much the same as that for rescue vehicles. The apparatus should be close enough to the scene so that the firefighters do not have to carry SCBA cylinders an extraordinary distance (Figure 5.67). These apparatus should not block scene access for other vehicles. If the hose reel is going to be used for remote filling, the apparatus needs to be positioned so that the hose can be appropriately deployed. Remember that apparatus using breathing air compressors to refill SCBA cylinders need to be parked upwind of the fire in clear air space. The breathing air compressors have filter sensors (interlocks) that prevent their use if the incoming air is contaminated.

Some jurisdictions have special SOPs for the location of breathing air supply apparatus. These SOPs may require the apparatus to be positioned near the command post or in the area where firefighter rehabilitation is conducted. Driver/operators need to know the SOPs for their department so that the apparatus is located appropriately.

Emergency Medical Service Vehicles

Emergency medical service vehicles commonly respond to fire and hazardous materials incidents to treat and transport injured civilians and to stand by in case an emergency responder needs medical assistance. In some jurisdictions, this has been a common practice for many years. In other jurisdictions, it has only become a standard practice since the adoption of NFPA 1500, *Standard on Fire Department Occupational Safety and Health Program*, which lists this practice as a requirement for emergency scene operations.

The two primary types of EMS vehicles that may respond to fire scenes are paramedic/quick responder units (nontransport) and ambulances (transport)

Figure 5.66 Some mobile air units are equipped with breathing air compressors.

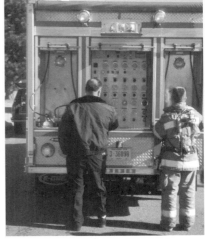

Figure 5.67 The mobile air supply apparatus should be close enough to the scene so that the firefighters do not have to carry SCBA cylinders an extraordinary distance.

(Figures 5.68 a and b). These may be basic or advanced life support vehicles. Paramedic/quick responder units are utility vehicles that carry emergency medical technicians (EMTs) or paramedics and the equipment they need to treat victims. These vehicles are not equipped to transport victims to the hospital. Ambulances carry the necessary equipment to both treat and transport victims.

As with other types of support vehicles, EMS vehicles should be parked close to the scene but do not block access for other fire and emergency vehicles (Figure 5.69). On incidents where there are victims requiring EMS intervention, the Incident Commander should establish a triage and treatment area. Obviously, most EMS vehicles will locate in that vicinity. Always park the EMS vehicle in a position that allows the vehicle to easily leave the scene to transport a victim should it be required.

On incidents where there are no immediate EMS situations to handle, the EMS vehicles and personnel will be in a standby mode. The most obvious location for EMS vehicles and staff would be in the area where firefighter rehabilitation is being conducted. If the incident is spread over a large area, there may be multiple EMS/rehab areas established.

Figure 5.68a EMS responder vehicles are not intended to transport victims.

Figure 5.68b A typical ambulance. *Courtesy of Joel Woods.*

Figure 5.69 EMS vehicles should not block the access of fire fighting vehicles.

Special Positioning Situations

A variety of other scenarios and conditions affect apparatus placement. The driver/operator must be familiar with these circumstances so that the apparatus is placed in a safe, yet effective location for each situation.

Staging

Often, apparatus placement at the scene of a fire or medical incident is limited by the order in which responding apparatus arrive. A late-arriving ladder truck may be blocked from a better position by earlier-arriving apparatus. Standard operating procedures governing apparatus placement is one way to prevent this type of situation from occurring. An apparatus staging procedure facilitates the orderly positioning of apparatus and allows the Incident Commander to fully utilize the potential of each unit and crew.

Through improvements in incident management strategies, an apparatus staging procedure in two levels has been developed that can be used for any multicompany response. Level I staging is applied to the initial response to a fire or medical incident involving more than one engine company. Level II staging is used in greater alarm situations where a large number of emergency vehicles are responding to an incident. Level II staging procedures must be initiated by the Incident Commander when requesting additional alarms or by a dispatcher when a large initial response is called for.

Level I staging is used on every emergency response when two companies performing like functions are dispatched (for example, two engine companies). The first-due engine company, truck company, rescue or squad company, and command officer proceed directly to the scene. Later-arriving units park or stage at least one block away from the scene in their direction of travel (Figure 5.70). The Incident Commander may order the staged units to lay additional supply lines, send personnel to the scene, or proceed to the scene and set up. Engine companies in jurisdictions that typically perform straight or forward hose lays should stage near hydrants or other sources of water. Staged apparatus should not allow their paths to the scene to become blocked.

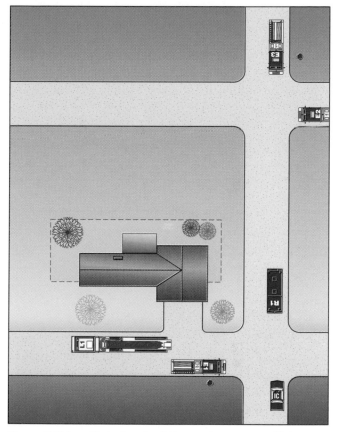
Figure 5.70 Level I staging should be used on every multiple-company response.

Level II staging is used when numerous emergency vehicles will be responding to an incident. Incidents that require mutual aid or that result in multiple alarms need Level II staging. When the additional units are requested by the Incident Commander, an apparatus staging area is designated (Figure 5.71). Companies are informed of the staging area location when they are dispatched and respond directly to that location. A parking lot or open field can serve as a staging area. Generally, the company officer of the first company to arrive at the staging area becomes the staging officer. On large-scale incidents, a chief officer may be assigned to the staging officer function. The staging officer should communicate available resources and resource needs to the Incident Commander. Company officers should report to the staging officer as they arrive and park. When the Incident Com-

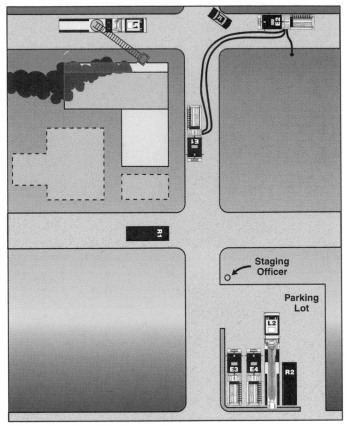

Figure 5.71 Level II staging is used on multiple-alarm incidents.

mander requires additional assistance, companies are summoned through the staging officer and sent to the scene. The staging area should be a secure area that is free of nonemergency traffic. The apparatus belonging to the staging officer should park near the entrance to the staging area and should leave its emergency lights flashing/rotating. All subsequent apparatus that arrive at the staging area should turn off their emergency lights when they park. This makes it easy for incoming companies to identify the location of the staging officer.

Operations on Highways

Some of the most dangerous scenarios faced by firefighters are operations on highways, interstates, turnpikes, and other busy roadways. The most common types of incidents on these thoroughfares are motor vehicle accidents and/or fires. The potential for multiple-injury accidents or accidents involving hazardous materials is also high. There are numerous challenges relative to apparatus placement, operational effectiveness, and responder safety when dealing with incidents on busy roadways.

Problems associated with simply accessing the scene can be a challenge to emergency responders. This is particularly true on limited-access highways and turnpikes. Apparatus may have to respond over long distances between exits to reach an incident. In some cases, apparatus will be required to travel a long distance before there is a turn-around that allows them the ability to get to the opposite side of the median if necessary. Apparatus should not be driven against the normal flow of traffic, unless the road has been closed by police units. Incidents occurring on bridges may require the use of aerial apparatus or ground ladders in order to reach the scene from below.

Water supply can be a problem on roadways that are in rural areas or even on limited-access highways in urban areas. Long hose lays or water operations may be needed to supply water to the incident scene. Hydrant placement on highways may be infrequent or may not exist. When responding to a highway incident, it is often desirable to have one pumper respond to the nearest overpass or underpass. This pumper can assist units on the highway in establishing a water supply if the source is off the highway. It may be necessary to stretch hoselines or use an aerial device from an overpass or underpass to get water to the level of the highway. Some highway systems are equipped with dry standpipe risers (Figure 5.72). These require one pumper that is off the highway to establish a water supply and pump into the standpipe inlet. Units on the highway can then connect to the standpipe discharge outlet and receive a steady flow of water.

The driver/operator should use prudence when responding to an incident on a highway or turnpike. A fire apparatus usually travels slower than the normal flow of traffic and the use of warning lights and sirens may create traffic conditions that actually slow the fire unit's response. The siren should not be used except to clear slow traffic. A minimum of warning lights should be used at

Figure 5.72 Some highway systems are equipped with dry standpipes. *Courtesy of Andy Mount.*

the scene to prevent blinding other drivers or distracting them, possibly leading to another accident. However, all warning lights should *never* be turned off because that might compromise the safety of firefighters working at the scene.

Cooperation between police and fire department personnel at highway incidents is essential. At least one lane next to the incident lane should be closed (Figure 5.73). Additional or all traffic lanes may have to be closed if the extra lane does not provide a safe barrier. Fire apparatus

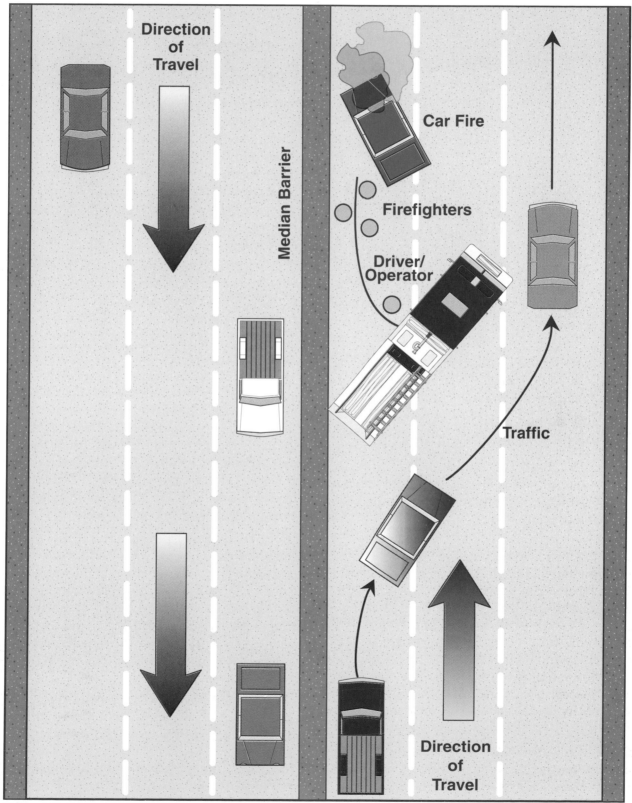

Figure 5.73 At least one lane next to the incident should be closed.

should be placed between the flow of traffic and the firefighters working on the incident to act as a shield. The apparatus should be parked on an angle so that the operator is protected from traffic by the tailboard. Front wheels should be turned away from the firefighters working highway incidents so that the apparatus will not be driven into them if struck from behind. Also consider parking additional apparatus 150 to 200 feet (45 m to 60 m) behind the shielding apparatus to act as an additional barrier between firefighters and the flow of traffic.

All crew members must use extreme caution when getting off the apparatus so that they are not struck by passing traffic. Similarly, driver/operators are extremely vulnerable to being struck by motorists if they step back beyond the protection offered by properly spotted apparatus. Departments that commonly respond to highway incidents tend to prefer top-mounted pump panels because they afford a greater degree of protection to the driver/operator (Figure 5.74).

Hazardous Materials Incidents

Hazardous materials incidents are increasingly common in today's fire service. Emergencies involving hazardous materials may occur in a transportation setting or a fixed facility. The possibility that a hazardous material may be involved in an incident should be considered by the company officer and driver/operator on virtually every fire or transportation incident. It is not possible to cover all of the information needed to train the driver/operator in hazardous materials identification in this manual. However, there are a few general considerations that driver/operators should keep in mind when responding to a potential hazardous materials emergency.

If you are the first-arriving apparatus, never drive directly into the scene without first attempting to identify the material that is involved. Failure to heed this advice could result in the apparatus becoming an ignition source for flammable gases or in contamination of the apparatus and its crew. Always stop well short of the incident scene until the nature of the hazard is understood. Do not park over manholes. Flammable materials flowing into the underground system could ignite and explode.

Try to obtain information on the wind speed and direction while en route to the scene. This may be obtained from the dispatcher or by observing the conditions as you respond. If at all possible, approach the incident and park from the upwind and uphill side.

Once on the scene, most jurisdictions use a series of control zones to organize the emergency scene. The zones prevent sightseers and other unauthorized persons from interfering with first responders, help regulate

Figure 5.74 Pumpers with top-mount pump panels are preferred on highway incidents.

movement of first responders within the zones, and minimize contamination. Control zones are not necessarily static and can be adjusted as the incident changes. Zones divide the levels of hazard of an incident, and what a zone is called generally depicts this level. The three most common terms for the hazardous materials zones are the hot zone, warm zone, and cold zone (Figure 5.75).

The *hot zone* (also called restricted zone, exclusion zone, or red zone) is an area surrounding the incident that has been contaminated by the released material. This area will be exposed to the gases, vapors, mists, dusts, or runoff of the material. The zone extends far enough to prevent people outside the zone from suffering ill effects from the released material.

The *warm zone* (also called the contamination reduction zone, limited-access zone, or yellow zone) is an area abutting the hot zone and extending to the cold zone. It is considered safe for workers to enter briefly without special protective clothing, unless assigned a task requiring increased protection. The warm zone is used to support workers in the hot zone and to decontaminate personnel and equipment exiting the hot zone. Decontamination usually takes place within a corridor (decon corridor) located in the warm zone.

The *cold zone* (also called the support zone or green zone) encompasses the warm zone and is used to carry out all other support functions of the incident. Workers in the cold zone are not required to wear personal protective clothing because the zone is considered safe. The command post, the staging area, and the triage/treatment area are located within the cold zone. Most of the time the driver/operator and his apparatus are positioned in the cold zone.

Avoid staging in the same location when responding to bomb threats and other potential terrorist incidents. It is possible that an explosive device could be placed in the staging location with the intent of harming emergency personnel. For more information on responding to hazardous materials emergencies, see IFSTA's **Hazardous Materials for First Responders** manual.

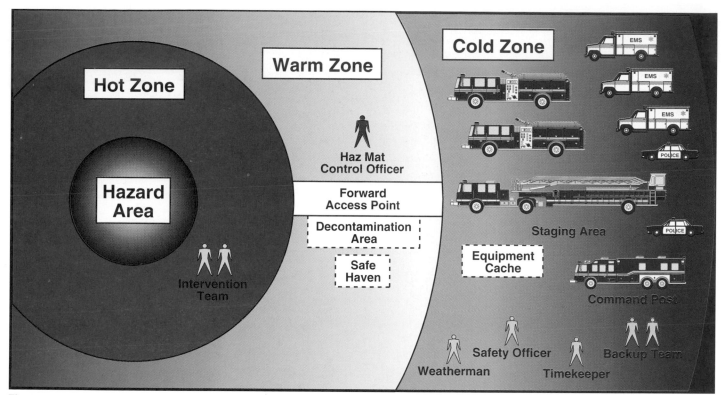

Figure 5.75 The standard zone system that is used for hazardous materials incidents.

Operating Near Railroads

Occasionally, emergency incidents occur in close proximity to railroads. Driver/operators should understand the hazards associated with operating near railroads and take any measures possible to minimize those hazards. It is not always possible to stop the flow of trains on the track during emergency operations. Always treat a railroad track as a potentially active line.

Never park the apparatus on the railroad tracks (Figure 5.76). As well, keep the apparatus far enough away from the tracks so that it will not be struck by a passing train. When possible, park the apparatus on the same side of the tracks as the incident. This negates the need to stretch hoselines across the tracks or for personnel to be traversing back and forth between each side.

If it becomes absolutely necessary to stretch attack or supply lines across a railroad track, attempt to confirm from the rail company that train traffic has been halted on that set of tracks. If this is not possible to confirm, attempt to run the hose beneath the rails or use aerial apparatus to run hose over the top of the area (Figure 5.77). Use caution when operating aerial apparatus in the vicinity of rail lines that operate from high-voltage, overhead electrical lines.

Emergency Medical Incidents

Many of the calls that fire departments respond to are emergency medical incidents. While apparatus position-

Figure 5.76 Never park the apparatus on the railroad tracks.

Figure 5.77 In some cases, hose may be run beneath railroad tracks so that neither train nor fireground operations are interrupted.

ing on these incidents is typically not only important from a tactical standpoint, it is important from a safety standpoint. It is important to allow the ambulance the best position for patient loading. Many firefighters have been injured or killed when struck by oncoming traffic while operating on emergency medical incidents.

When possible, park the apparatus off the street. This virtually eliminates any of the hazards associated with oncoming traffic. Attempt to locate in a driveway, parking lot, or yard. Make sure, however, that the surface is stable enough to support the weight of the fire apparatus.

If it is not possible to locate off the street, use the apparatus as a shield between the work area and oncoming traffic. Park larger apparatus (such as a pumper) between smaller apparatus (such as an ambulance) and the oncoming flow of traffic (Figure 5.78). In particular, guard the patient-loading area of the ambulance by shielding it with another vehicle. If possible, place traffic cones to direct oncoming traffic away from the apparatus (Figure 5.79).

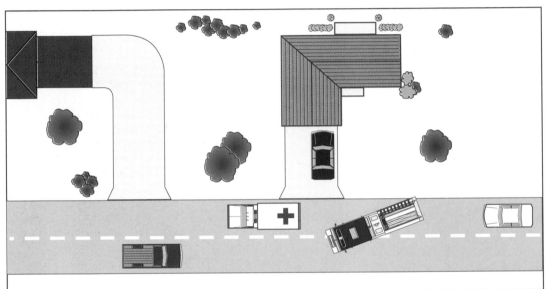

Figure 5.78 Use the pumper to protect the ambulance loading area.

Figure 5.79 Traffic cones may be used to redirect approaching vehicles.

Chapter 6

What is Water and Where Does it Come From?

This chapter provides information that will assist the reader in meeting the following job performance requirements from NFPA 1002, *Standard on Fire Apparatus Driver/Operator Professional Qualifications*, 1998 edition.

3-2.1* Produce effective hand or master streams, given the sources specified in the following list, so that the pump is safely engaged, all pressure control and vehicle safety devices are set, the rated flow of the nozzle is achieved and maintained, and the apparatus is continuously monitored for potential problems.

- Internal tank
- Pressurized source
- Static source
- Transfer from internal tank to external source

(a) *Requisite Knowledge:* Hydraulic calculations for friction loss and flow using both written formulas and estimation methods, safe operation of the pump, **problems related to small-diameter or dead-end mains, low-pressure and private water supply systems, hydrant cooling systems, and reliability of static sources**.

(b) *Requisite Skills:* The ability to position a fire department pumper to operate at a fire hydrant and at a static water source, power transfer from vehicle engine to pump, draft, operate pumper pressure control systems, operate the volume/pressure transfer valve (multistage pumps only), operate auxiliary cooling systems, make the transition between internal and external water sources, and assemble hose lines, nozzles, valves, and appliances.

3-2.2 Pump a supply line of 2½ in. (65 mm) or larger, given a relay pumping evolution the length and size of the line and the desired flow and intake pressure, so that the proper pressure and flow are provided to the next pumper in the relay.

(a) *Requisite Knowledge:* Hydraulic calculations for friction loss and flow using both written formulas and estimation methods, safe operation of the pump, **problems related to small-diameter or dead-end mains, low-pressure and private water supply systems, hydrant cooling systems, and reliability of static sources**.

(b) *Requisite Skills:* The ability to position a fire department pumper to operate at a fire hydrant and at a static water source, power transfer from vehicle engine to pump, draft, operate pumper pressure control systems, operate the volume/pressure transfer valve (multistage pumps only), operate auxiliary cooling systems, make the transition between internal and external water sources, and assemble hose lines, nozzles, valves, and appliances.

6-2.1* Produce effective fire streams, utilizing the sources specified in the following list, so that the pump is safely engaged, all pressure-control and vehicle safety devices are set, the rated flow of the nozzle is achieved, and the apparatus is continuously monitored for potential problems.

- Water tank
- Pressurized source
- Static source

(a) *Requisite Knowledge:* Hydraulic calculations for friction loss and flow using both written formulas and estimation methods, safe operation of the pump, proper apparatus placement, personal safety considerations, **problems related to small-diameter or dead-end mains and low-pressure and private water supply systems, hydrant cooling systems, and reliability of static sources**.

(b) *Requisite Skills:* The ability to position a wildland fire apparatus to operate at a fire hydrant and at a static water source, properly place apparatus for fire attack, transfer power from vehicle engine to pump, draft, operate pumper pressure control systems, operate the volume/pressure transfer valve (multistage pumps only), operate auxiliary cooling systems, make the transition between internal and external water sources, and assemble hose lines, nozzles, valves, and appliances.

6-2.2 Pump a supply line, given a relay pumping evolution the length and size of the line and pumping flow and desired intake pressure, so that adequate intake pressures and flow are provided to the next pumper in the relay.

For the past 30 or so years, the fire service has seen the introduction of a wide variety of extinguishing agents. Dry chemicals, dry powders, foam concentrates, and halogenated hydrocarbons are all examples of these "modern" extinguishing agents. Yet, despite these advances, plain water remains, by far, the most commonly used fire extinguishing agent in today's fire service. In order to become an effective pumping apparatus driver/operator, a firefighter must have a working knowledge of the chemistry of water that makes it an effective extinguishing agent and the physics associated with moving it. This chapter highlights those important principles.

Characteristics of Water

Water (H_2O) is a compound of hydrogen and oxygen formed when two parts hydrogen (H) combine with one part oxygen (O). Between 32°F and 212°F (0°C and 100°C), water exists in a liquid state (Figure 6.1). Below 32°F (0°C) (the freezing point of water), it converts to a solid state of matter called *ice*. Above 212°F (100°C) (the boiling point of water), it converts into a gas called *water vapor* or *steam*. Water cannot be seen in this vapor form. It only becomes visible as it rises away from the surface of the liquid water and begins to condense.

For all practical purposes, water is considered to be incompressible, and its weight varies at different temperatures. Water's density, or its weight per unit of volume, is measured in pounds per cubic foot (kg/L). Water is heaviest (approximately 62.4 lb/ft³ [1 kg/L]) close to its freezing point. Water is lightest (approximately 60 lb/ft³ [0.96 kg/L]) close to its boiling point. For fire protection purposes, ordinary fresh water is generally considered to weigh 62.5 lb/ft³ or 8.33 lb/gal (1 kg/L).

Extinguishing Properties of Water

Water has the ability to extinguish fire in several ways. The primary way water extinguishes fire is by *cooling*, or absorbing heat from the fire. Another way is by *smothering* (excluding oxygen). This works especially well on the surface of heavy flammable liquids. Smothering also occurs to some extent when water converts to steam in a confined space.

As an extinguishing agent, water is affected by two natural laws of physics: The Law of Specific Heat and The Law of Latent Heat of Vaporization. These laws are vitally important when considering the heat-absorbing ability of water. The amount of heat absorbed by water is also affected by the amount of surface area of the water exposed to the heat. Another important consideration is specific gravity.

The Law of Specific Heat

Specific heat is a measure of the heat-absorbing capacity of a substance. Water is not only noncombustible, but it can also absorb large amounts of heat. Amounts of heat transfer are measured in British thermal units (Btus) or in joules (J) (1 Btu = 1.055 kJ). A *Btu* is the amount of heat required to raise the temperature of 1 pound of water 1°F. The joule, also a unit of work, has taken the place of the calorie in the SI (International System of Units) heat measurement (1 calorie = 4.19 joules).

The specific heat of any substance is the ratio between the amount of heat needed to raise the temperature of a specified quantity of a material and the amount of heat needed to raise the temperature of an identical quantity of water by the same number of degrees. The specific heat of different substances varies. Table 6.1 shows some fire extinguishing agents and the specific heat comparison (by weight) with water.

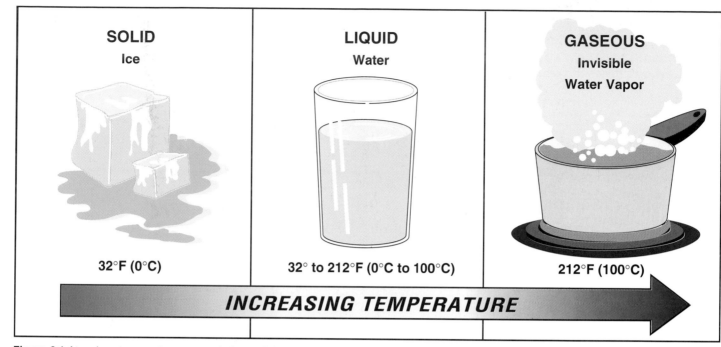

| SOLID | LIQUID | GASEOUS |
| Ice | Water | Invisible Water Vapor |

| 32°F (0°C) | 32° to 212°F (0°C to 100°C) | 212°F (100°C) |

INCREASING TEMPERATURE

Figure 6.1 At various temperatures, water is in a solid, liquid, or gaseous form.

Table 6.1	
Specific Heat of Extinguishing Agents	
Agent	**Specific Heat**
Water	1.00
Calcium chloride solution	0.70
Carbon dioxide (solid)	0.12
Carbon dioxide (gas)	0.19
Sodium bicarbonate	0.22

Figure 6.2 It takes five times as much heat to raise the temperature of 1 pound of water 1°F as is does an equal amount of carbon dioxide.

Using Table 6.1, divide the specific heat of water (1.00) by the specific heat of carbon dioxide gas (0.19). Note that it takes more than five times the amount of heat to raise the temperature of 1 pound of water 1°F than it takes to raise the temperature of the same amount of carbon dioxide gas (Figure 6.2). Another way of stating this is that water absorbs five times as much heat as does an equal amount of carbon dioxide. A comparison of the materials listed shows that water is clearly the best material for absorbing heat.

The Law of Latent Heat of Vaporization
The *latent heat of vaporization* is the quantity of heat absorbed by a substance when it changes from a liquid to a vapor. The temperature at which a liquid absorbs enough heat to change to vapor is known as its *boiling point.* At sea level, water begins to boil or vaporize at

212°F (100°C). Vaporization, however, does not completely occur the instant water reaches the boiling point. Each pound of water requires approximately 970 Btu (1 023 kJ) of additional heat to completely convert into steam (Figures 6.3 a and b).

The latent heat of vaporization is significant in fighting fire because the temperature of the water is not increased beyond 212°F during the absorption of the 970 Btu for every pound of water. Suppose we have 1 gallon (U.S.) of water at 60°F. Each gallon of water weighs 8.33 pounds. It requires 152 Btu (212 – 60 = 152) to raise each

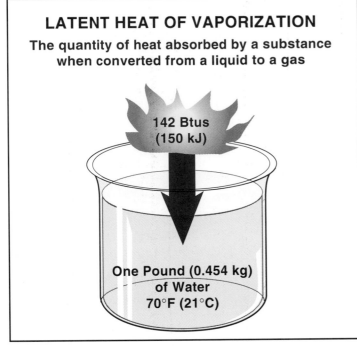

LATENT HEAT OF VAPORIZATION
The quantity of heat absorbed by a substance when converted from a liquid to a gas

142 Btus
(150 kJ)

One Pound (0.454 kg)
of Water
70°F (21°C)

Figure 6.3a To bring water from 70°F (21°C) to its boiling point (212°F [100°C]), 142 Btus (150 kJ) are needed.

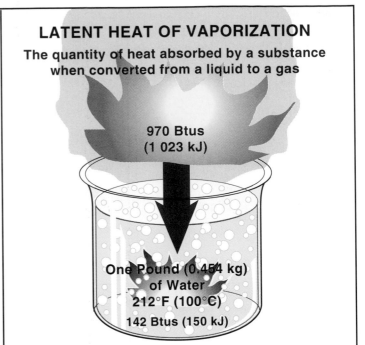

LATENT HEAT OF VAPORIZATION
The quantity of heat absorbed by a substance when converted from a liquid to a gas

970 Btus
(1 023 kJ)

One Pound (0.454 kg)
of Water
212°F (100°C)
142 Btus (150 kJ)

Figure 6.3b When water has reached its boiling point, 970 additional Btus (1 023 kJ) are required to turn it into steam. This high heat of vaporization makes water a good extinguishing agent.

pound of water to 212°F. Thus, 1 gallon of water absorbs 1,266 Btu (152 Btu/lb × 8.33 lb) getting to 212°F. Because the conversion to steam requires another 970 Btu per pound, an additional 8,080 Btu (970 Btu/pound × 8.33 lb) will be absorbed through this process. This means that 1 gallon of water will absorb 9,346 Btu (1,266 + 8,080) of heat if all the water is converted to steam. If water from a 100 gpm (400 L/min) fog nozzle (at 60°F) is projected into a highly heated area, it can absorb approximately 934,600 Btu of heat per minute if all of the water is converted to steam.

The amount of heat a combustible object can produce depends upon the material from which it is composed. The rate at which the object gives off heat depends upon such factors as its physical form, amount of surface exposed, and air or oxygen supply.

Surface Area of Water
The speed with which water absorbs heat increases in proportion to the water surface exposed to the heat. For example, if a 1-inch (25 mm) cube of ice is dropped into a glass of water, it will take quite awhile for the ice cube to absorb its capacity of heat (melt). This is because only 6 square inches (3 870 mm²) of the ice are exposed to the water (Figure 6.4). If the same cube of ice is divided into ⅛-inch (3 mm) cubes and these cubes are dropped into the water, 48 square inches (30 970 mm²) of the ice are exposed to the water. Although the smaller cubes equal the same mass of ice as the larger cube, the smaller cubes melt faster. That is why crushed ice melts in a drink faster

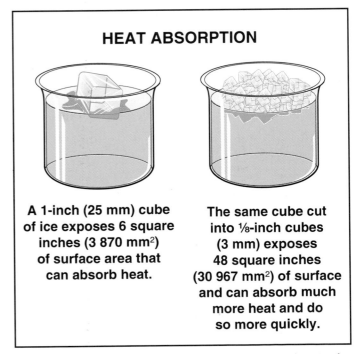

HEAT ABSORPTION

A 1-inch (25 mm) cube of ice exposes 6 square inches (3 870 mm²) of surface area that can absorb heat.

The same cube cut into ⅛-inch cubes (3 mm) exposes 48 square inches (30 967 mm²) of surface and can absorb much more heat and do so more quickly.

Figure 6.4 Dividing water into smaller particles increases its rate of heat absorption.

than ice cubes. This principle also applies to water in a liquid state. If water is divided into many drops, the rate of heat absorption increases hundreds of times.

Another characteristic of water that is sometimes an aid to fire fighting is its expansion capability when converted to steam. This expansion helps cool the fire area by driving heat and smoke from the area. The amount of expansion varies with the temperatures of the fire area. At 212°F (100°C), water expands approximately 1,700 times its original volume (Figure 6.5).

To illustrate steam expansion, consider a nozzle discharging 150 gallons (568 L) of water fog every minute

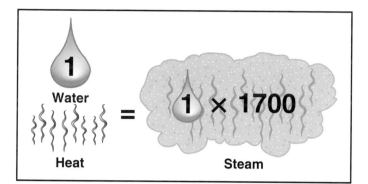

Figure 6.5 Water converted to steam expands 1,700 times. This expansion process absorbs heat and forces hot air and combustion gases out of a confined space.

into an area heated to approximately 500°F (260°C), causing the water fog to convert to steam (Figure 6.6). During 1 minute of operation, 20 cubic feet (0.57 m³) of water is discharged and vaporized. This 20 cubic feet (0.57 m³) of water expands to approximately 48,000 cubic feet (1 359 m³) of steam. This is enough steam to fill a room approximately 10 feet (3 m) high, 50 feet (15 m) wide, and 96 feet (29 m) long. In hotter atmospheres, steam expands to even greater volumes.

Steam expansion is not gradual, but rapid. If a room is already full of smoke and gases, the steam generated displaces these gases when adequate ventilation openings are provided (Figure 6.7). As the room cools, the steam condenses and allows the room to refill with cooler air. The use of a fog stream in a direct or combination fire attack requires that adequate ventilation be provided ahead of the hoseline. Otherwise, there is a high possibility of steam or even fire rolling back over and around the

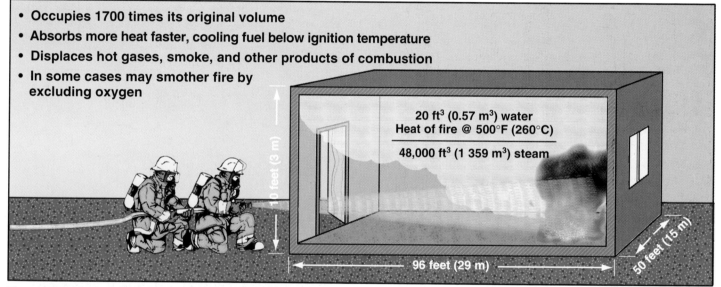

Figure 6.6 A nozzle discharging 150 gallons (568 L) of water fog for one minute generates enough steam to fill a room approximately 10 feet (3 m) high, 50 feet (15 m) wide, and 96 feet (29 m) long.

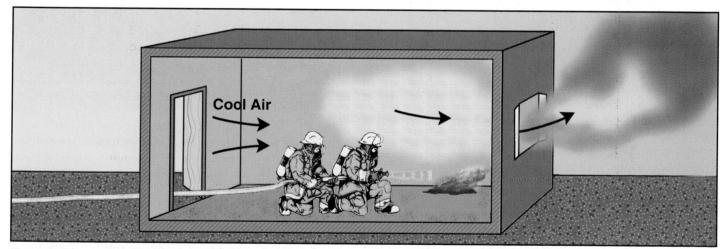

Figure 6.7 The steam generated displaces heat and fire gases when adequate ventilation openings are provided.

hose team, and the potential for burns or steam burns is great. There are some observable results of the proper application of a water fire stream into a room: Fire is extinguished or reduced in size, visibility may be maintained, and room temperature is reduced.

Water can also smother fire when it floats on liquids that are heavier than water, such as carbon disulfide. If the material is water soluble, such as alcohol, the smothering action is not likely to be effective. Water may also smother fire by forming an emulsion over the surface of certain combustible liquids. When a spray of water agitates the surface of these liquids, the agitation causes the water to be temporarily suspended in emulsion bubbles on the surface. The emulsion bubbles then smother the fire. This action can only take place when the combustible liquid has sufficient viscosity. *Viscosity* is the tendency of a liquid to possess internal resistance to flow. For example, water has low viscosity, while molasses has high viscosity. A heavy fuel oil, such as No. 6 grade, retains an emulsified surface longer than a lighter grade such as No. 2 home heating fuel or diesel fuel. The water in the emulsion absorbs heat from the oil adjacent to it, reduces the oil temperature, and decreases the amount of combustible vapors that are emitted.

Specific Gravity

The density of liquids in relation to water is known as *specific gravity*. Water is given a value of 1. Liquids with a specific gravity less than 1 are lighter than water and therefore float on water. Those with a specific gravity greater than 1 are heavier than water and sink to the bottom. If the other liquid also has a specific gravity of 1, it mixes evenly with water. Most flammable liquids have a specific gravity of less than 1. Therefore, if a firefighter confronted with a flammable liquid fire flows water on it improperly, the whole fire can just float away on the water and ignite everything in its path. The use of foam can control this situation because it floats on the surface of the flammable liquid and smothers the fire.

Advantages and Disadvantages of Water

Water has a number of characteristics that make it an excellent extinguishing agent:

- Water has a greater heat-absorbing capacity than other common extinguishing agents.
- A relatively large amount of heat is required to change water into steam. This means that more heat is absorbed from the fire.
- The greater the surface area of water exposed, the more rapidly heat is absorbed. The exposed surface area of water can be expanded by using fog streams or deflecting solid streams off objects.

- Water converted into steam occupies 1,700 times its original volume.
- Water is plentiful and readily available in most jurisdictions.

There are some disadvantages to using water as a fire extinguishing agent. The following disadvantages are due to some additional properties that water possesses:

- Water has a high surface tension and does not readily soak into dense materials. However, when wetting agents are mixed with water, the water's surface tension is reduced and its penetrating ability is increased.
- Water may be reactive with certain fuels such as combustible metals (Figure 6.8).
- Water has low levels of opacity and reflectivity that allow radiant heat to easily pass through it.
- Water freezes at 32°F (0°C), which is a problem in jurisdictions that frequently experience freezing atmospheric conditions. Water freezing poses a hazard to firefighters by coating equipment, roofs, ladders, and other surfaces. In addition, ice forming in and on equipment may cause it to malfunction.
- Water readily conducts electricity, which can be hazardous to firefighters working around energized electrical equipment.

Figure 6.8 Water reacts violently with burning combustible metals. *Courtesy of Linda Gheen.*

Water Pressure and Velocity

Pressure has a variety of meanings. Ordinarily, one thinks of pressure as force exerted on one substance by another. In this manual, however, *pressure* is defined as force per unit area. Pressure may be expressed in pounds per square foot (psf), pounds per square inch (psi) or kilopascals (kPa).

Pressure can easily be confused with force. *Force* is a simple measure of weight and is usually expressed in pounds or kilograms. This measurement is directly related to the force of gravity, which is the amount of attraction the earth has for all bodies. If several objects of the same size and weight are placed on a flat surface, they each exert an equal force on that surface.

For example, three square containers of equal size (1 × 1 × 1 foot [0.3 m by 0.3 m by 0.3 m]) containing 1 cubic foot (0.028 m³) of water and weighing 62.5 pounds (28 kg) each are placed next to each other (Figure 6.9). Each container exerts a force of about 62.5 psf (about 306 kg/m²) with a total of about 187.5 pounds (85 kg) of force over a 3-square-foot (0.3 m²) area.

If the containers are stacked on top of each other, the total force exerted — 187.5 pounds or 85 kg — remains the same, but the area of contact is reduced to 1 square foot (0.1 m²) (Figure 6.10). The pressure then becomes 187.5 psf (about 919 kg/m²).

To understand how force is determined, it is necessary to know the weight of water and the height that a column of water occupies. The weight of 1 cubic foot of water is approximately 62.5 pounds. Because 1 square foot contains 144 square inches, the weight of water in a 1-square-inch column of water 1 foot high equals 62.5 pounds divided by 144 square inches or 0.434 pounds. A 1-square-inch column of water 1 foot high therefore exerts a pressure at its base of 0.434 psi (Figure 6.11). The height required for a 1-square-inch column of water to produce 1 psi at its base equals 1 foot divided by 0.434 psi/ft or 2.304 feet; therefore, 2.304 feet of water column exerts a pressure of 1 psi at its base.

In metrics, a cube that is 0.1 m × 0.1 m × 0.1 m (a cubic decimeter) holds 1 liter of water. The weight of 1 liter of water is 1 kilogram. The cube of water exerts 1 kPa (1 kg) of pressure at the bottom of the cube. One cubic meter of water holds 1 000 liters of water and weighs 1 000 kg. Because the cubic meter of water is comprised of 100 columns of water, each 10 decimeters tall, each column exerts 10 kPa at its base (Figure 6.12).

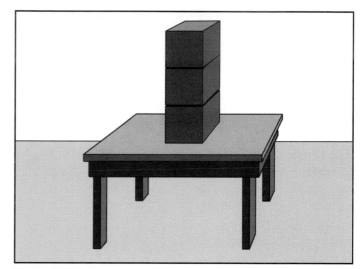

Figure 6.10 Each container measures 1 cubic foot (0.028 m³) and weighs 62.5 pounds (28 kg). If they are stacked, they exert a pressure of 187.5 pounds per square foot (919 kg/m²)

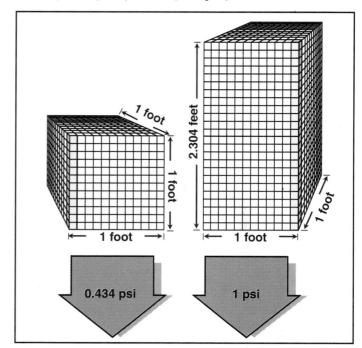

Figure 6.11 These examples clearly illustrate the relationship between height and pressure in the U.S. system of measurement.

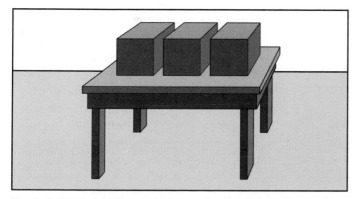

Figure 6.9 Each container measures 1 cubic foot (0.028 m³) and weighs 62.5 pounds (28 kg). Each container therefore exerts a pressure of 62.5 pounds per square foot (306 kg/m²).

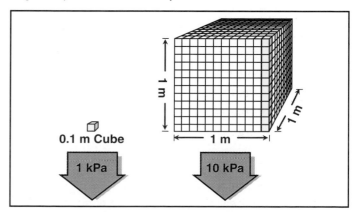

Figure 6.12 These examples clearly illustrate the relationship between height and pressure in the metric system of measurement.

Principles of Pressure

The speed with which a fluid travels through hose or pipe is developed by pressure upon that fluid. The speed at which this fluid travels is often referred to as *velocity*. It is important to identify the type of pressure because the word pressure in connection with fluids has a very broad meaning. There are six basic principles that determine the action of pressure upon fluids. It is very important that the driver/operator clearly understand these principles before attempting to understand the types of pressure.

First Principle

Fluid pressure is perpendicular to any surface on which it acts. This principle is illustrated by a vessel having flat sides and containing water (Figure 6.13). The pressure exerted by the weight of the water is perpendicular to the walls of the container. If this pressure is exerted in any other direction, as indicated by the slanting arrows, the water would start moving downward along the sides and rising in the center.

Second Principle

Fluid pressure at a point in a fluid at rest is the same intensity in all directions. To put it another way, fluid pressure at a point in a fluid at rest has no direction (Figure 6.14).

Third Principle

Pressure applied to a confined fluid from without is transmitted equally in all directions (Figure 6.15). This principle is illustrated by a hollow sphere to which a water pump is attached. A series of gauges is set into the sphere around its circumference. When the sphere is filled with water and pressure is applied by the pump, all gauges will register the same pressure. (This is true if they are on the same grade line with no change in elevation.)

Fourth Principle

The pressure of a liquid in an open vessel is proportional to its depth (Figure 6.16). This principle is illustrated by three vertical containers, each 1 square inch (645 mm²) in cross-sectional area. The depth of the water is 1 foot (0.3 m) in the first container, 2

feet (0.6 m) in the second, and 3 feet (0.9 m) in the third container. The pressure at the bottom of the second container is twice that of the first, and the pressure at the bottom of the third container is three times that of the first. Thus, the pressure of a liquid in an open container is proportional to its depth.

Fifth Principle

The pressure of a liquid in an open vessel is proportional to the density of the liquid. This principle is illustrated by two containers (Figure 6.17). One container holds mercury 1 inch (25 mm) deep, the other holds water 13.55 inches (344 mm) deep, yet the pressure at the bottom of each container is approximately the same. Thus, mercury is 13.55 times denser than water. Therefore, the pressure of a liquid in an open vessel is proportional to the density of the liquid.

Figure 6.13 The pressure exerted by the weight of the fluid is perpendicular to the walls of the container.

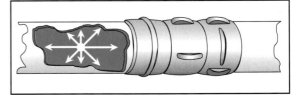

Figure 6.14 When a fluid is at rest, fluid intensity is the same in all directions.

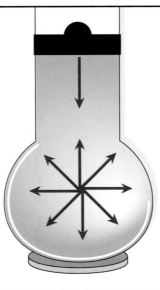

Figure 6.15 Pressure that is transmitted to a confined fluid from without is transmitted equally in all directions.

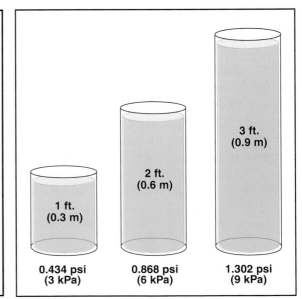

1 ft. (0.3 m)

2 ft. (0.6 m)

3 ft. (0.9 m)

0.434 psi (3 kPa)

0.868 psi (6 kPa)

1.302 psi (9 kPa)

Figure 6.16 The pressure of a fluid in an open vessel is proportional to its depth.

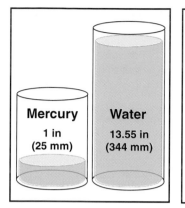

Figure 6.17 The pressure of a fluid in an open vessel is proportional to the density of the fluid.

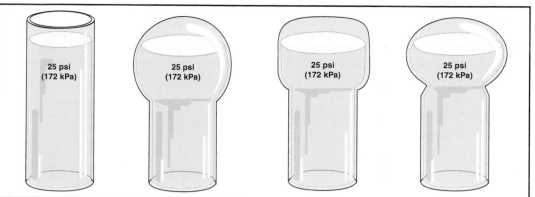

Figure 6.18 The pressure of a fluid on the bottom of a container is independent of the container's shape.

Sixth Principle

The pressure of a liquid on the bottom of a vessel is independent of the shape of the vessel. This principle is illustrated by showing water in several different shaped containers, each having the same cross-sectional area at the bottom and the same height (Figure 6.18). The pressure is the same in each container.

Types of Pressure

There are a number of terms used for different types of pressure that are encountered in water supply systems and in the fire service. The driver/operator should be acquainted with each of these terms so that he uses them in their proper context.

Atmospheric Pressure

The atmosphere surrounding the earth has depth and density and exerts pressure upon everything on earth. Atmospheric pressure is greatest at low altitudes and least at very high altitudes. At sea level, the atmosphere exerts a pressure of 14.7 psi (101 kPa), which is considered standard atmospheric pressure.

A common method of measuring atmospheric pressure is by comparing the weight of the atmosphere with the weight of a column of mercury: the greater the atmospheric pressure, the taller the column of mercury. A pressure of 1 psi (6.90 kPa) makes the column of mercury about 2.04 inches (52 mm) tall. At sea level, then, the column of mercury is 2.04 × 14.7, or 29.9 inches (759 mm) tall (Figure 6.19).

The readings of most pressure gauges are psi (or kPa) in addition to the existing atmospheric pressure. For example, a gauge reading 10 psi (70 kPa) at sea level is actually indicating 24.7 psi (170 kPa) (14.7 + 10 [100 + 70 kPa]). Engineers distinguish between such a gauge reading and actual atmospheric pressure by writing *psig*,

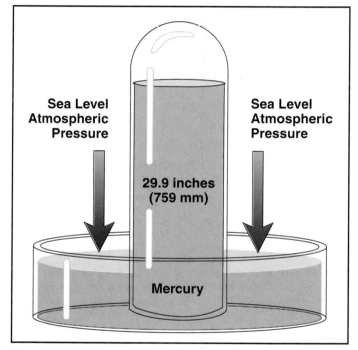

Figure 6.19 A pressure of 14.7 psi (101 kPa) causes the mercury column in this barometer to rise 29.9 inches (759 mm).

which means "pounds per square inch gauge." The notation for actual atmospheric pressure is *psia*, meaning "pounds per square inch absolute" (the psi above a perfect vacuum, absolute zero). Any pressure less than atmospheric pressure is called *vacuum*. Absolute zero pressure is called a *perfect vacuum*.

When a gauge reads -5 psig (-35 kPa), it is actually reading 5 psi (35 kPa) less than the existing atmospheric pressure (at sea level, 14.7 minus 5, or 9.7 psi [100 minus 35, or 65 kPa]). ***Throughout this manual, psi means psig.***

Head Pressure

Head in the fire service refers to the height of a water supply above the discharge orifice. In Figure 6.20, the water supply is 100 feet (30 m) above the hydrant discharge opening. This is referred to as 100 feet (30 m) of head. To convert head in feet (meters) to head pressure, divide the number of feet by 2.304 (for metric, divide the number of meters by 0.1). The result is the number of feet (meters) that 1 psi (7 kPa) raises a column of water. The water source in Figure 6.20 has a head pressure of 43.4 psi (300 kPa) (Table 6.2).

Static Pressure

The water flow definition of *static pressure* is stored potential energy available to force water through pipe, fittings, fire hose, and adapters. *Static* means at rest or without motion. Pressure on water may be produced by an elevated water supply, by atmospheric pressure, or by a pump. If the water is not moving, the pressure exerted is static. A truly static pressure is seldom found in municipal water systems because there is always some flow

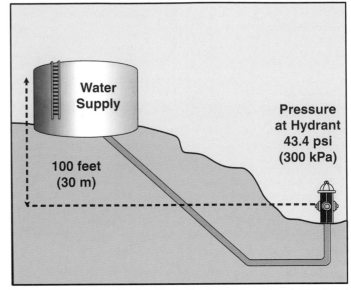

Figure 6.20 The head in this illustration is 100 feet (30 m). The head pressure is 43.4 psi (300 kPa).

Table 6.2
Head in Feet (Meters) and Head Pressure

U.S.				METRIC			
Feet of Head	Pounds per Square Inch	Pounds per Square Inch	Feet of Head	Meters of Head	kPa	kPa	Meters of Head
5	2.17	5	11.50	1	10	5	.5
10	4.33	10	23.00	2	20	10	1
15	6.50	15	34.60	3	30	15	1.5
20	8.66	20	46.20	4	40	20	2
25	10.83	25	57.70	5	50	25	2.5
30	12.99	30	69.30	6	60	30	3
35	15.16	35	80.80	7	70	35	3.5
40	17.32	40	92.30	8	80	40	4
50	21.65	50	115.40	9	90	50	5
60	26.09	60	138.50	10	100	60	6
70	30.30	70	161.60	15	150	70	7
80	34.60	80	184.70	20	200	80	8
90	39.00	90	207.80	25	250	90	9
100	43.30	100	230.90	30	300	100	10
120	52.00	120	277.00	40	400	200	20
140	60.60	140	323.20	50	500	300	30
160	69.20	160	369.40	60	600	400	40
200	86.60	180	415.60	70	700	500	50
300	129.90	200	461.70	80	800	600	60
400	173.20	250	577.20	90	900	700	70
500	216.50	275	643.00	100	1 000	800	80
600	259.80	300	692.70	200	2 000	900	90
800	346.40	350	808.10	300	3 000	1 000	100
1,000	433.00	500	1,154.50				

in the pipes due to normal domestic or industrial needs. Nevertheless, the pressure in a water system before water flows from a hydrant is considered static pressure (Figure 6.21).

Normal Operating Pressure

Normal operating pressure is that pressure found in a water distribution system during normal consumption demands. As soon as water starts to flow through a distribution system, static pressure no longer exists. The demands for water consumption fluctuate continuously, causing water flow to increase or decrease in the system. The difference between static pressure and normal operating pressure is the friction caused by water flowing through the various pipes, valves, and fittings in the system.

Residual Pressure

Residual pressure is that part of the total available pressure not used to overcome friction loss or gravity while forcing water through pipe, fittings, fire hose, and adapters. *Residual* means a remainder or that which is left. For example, during a fire flow test, residual represents the pressure left in a distribution system within the vicinity of one or more flowing hydrants. In a water distribution system, residual pressure varies according to the amount of water flowing from one or more hydrants, water consumption demands, and the size of the pipe. It is important to remember that residual pressure must be identified at the location where the reading is taken, not at the flow hydrant.

Flow Pressure (Velocity Pressure)

Flow pressure is that forward velocity pressure at a discharge opening while water is flowing (Figure 6.22). Because a stream of water emerging from a discharge opening is not encased within a tube, it exerts forward pressure but not sideways pressure. The forward velocity of flow pressure can be measured by using a pitot tube and gauge. If the size of the opening is known, a firefighter can use the measurement of flow pressure to calculate the quantity of water flowing in gpm or L/min.

Pressure Loss and Gain: Elevation and Altitude

Although the words elevation and altitude are often used interchangeably, the fire service makes a distinction between the two. *Elevation* refers to the center line of the pump or the bottom of a static water supply source above or below ground level. *Altitude* is the position of an object above or below sea level. Both are important in producing fire streams.

When a nozzle is above the pump, there is a pressure loss. When the nozzle is below the pump, there is a

Figure 6.21 The pressure in a water system before water flows from a hydrant is considered static pressure.

Figure 6.22 The flow pressure is usually measured by inserting a pitot tube and gauge in the stream flowing from an open hydrant discharge.

pressure gain. These losses and gains occur because of gravity. Both pressure loss and pressure gain are referred to as *elevation pressure*.

Altitude affects the production of fire streams because atmospheric pressure drops as height above sea level increases. This pressure drop is of little consequence to about 2,000 feet (600 m). Above this height, though, the lessened atmospheric pressure can be of concern. At high altitudes, fire department pumpers must work harder to produce the pressures required for effective fire streams. They must work harder because the less dense atmosphere reduces the pumper's effective lift when drafting. Above sea level, atmospheric pressure decreases approximately 0.5 psi (3.5 kPa) for every 1,000 feet (300 m).

Friction Loss

The concept of pressure loss due to friction in a hose or piping system play an important part in the remainder of this manual. The common term for pressure loss due to friction is simply friction loss. The fire service definition of *friction loss* is that part of the total pressure lost while forcing water through pipe, fittings, fire hose, and adapters. The driver/operator must have a clear understanding of the principles of friction loss in order to be able to effectively produce fire streams with a fire department pumper.

In a fire hose, friction loss is caused by the following:

- Movement of water molecules against each other
- Linings in fire hose
- Couplings
- Sharp bends
- Change in hose size or orifice by adapters
- Improper gasket size

Anything that affects movement of water may cause additional friction loss. Good quality fire hose has a smoother inner surface and causes less friction loss than lower quality hose. The friction loss in old hose may be as much as 50 percent greater than that in new hose.

The principles of friction loss in piping systems are the same as in fire hose. Friction loss is caused by the following:

- Movement of water molecules against each other
- Inside surface of the piping
- Pipe fittings
- Bends
- Control valves

The rougher the inner surface of the pipe (commonly referred to as the *coefficient of friction*), the more friction loss that occurs.

Friction loss can be measured by inserting in-line gauges in a hose or pipe. The difference in the residual pressures between gauges when water is flowing is the friction loss. The difference in pressure in a fire hose between a nozzle and a pumper is a good example of friction loss.

Principles of Friction Loss

There are four basic principles that govern friction loss in fire hose and pipes. These principles are discussed in the sections that follow.

First Principle

If all other conditions are the same, friction loss varies directly with the length of the hose or pipe. This prin-

ciple can be illustrated by one hose that is 100 feet (30 m) long and another hose that is 200 feet (60 m) long (Figure 6.23). A constant flow of 200 gpm (800 L/min) is maintained in each hose. The 100-foot (30 m) hose has a friction loss of 10 psi (70 kPa). The 200-foot (60 m) hose has twice as much friction loss, or 20 psi (140 kPa). (**NOTE:** You will learn how to calculate friction loss later in this manual.)

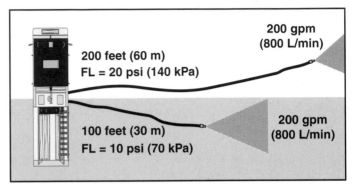

Figure 6.23 With all other variables the same, the friction loss varies directly with the length of the hose.

Second Principle

When hoses are the same size, friction loss varies approximately with the square of the increase in the velocity of the flow (Figure 6.24). This principle points out that friction loss develops much faster than the change in velocity. (Remember that velocity is proportional to flow.) For example, a length of 3-inch (77 mm) hose flowing 200 gpm (800 L/min) has a friction loss of 3.2 psi (22.4 kPa). As the flow doubles from 200 to 400 gpm (800 L/min to 1 600 L/min), the friction loss increases four times ($2^2 = 4$) to 12.8 psi (89.6 kPa). When the original flow is tripled from 200 to 600 gpm (800 L/min to 2 400 L/min), friction loss increases nine times ($3^2 = 9$) to 28.8 psi (201.6 kPa).

Third Principle

For the same discharge, friction loss varies inversely as the fifth power of the diameter of the hose. This principle readily proves the advantage of larger size hose and can be illustrated by one hose that is 2½ inches (65 mm) in diameter and another that is 3 inches (77 mm) in diameter. The friction loss in the 3-inch (77 mm) hose is:

$$\frac{(2\frac{1}{2})^5}{3^5} = \frac{98}{243} = 0.4 \text{ that of the } 2\frac{1}{2}\text{-inch hose}$$

$$\frac{(65)^5}{(77)^5} = \frac{1\ 160\ 290\ 625}{2\ 706\ 784\ 157} = 0.4 \text{ that of the 65 mm hose}$$

Fourth Principle

For a given flow velocity, friction loss is approximately the same, regardless of the pressure on the water. This principle explains why friction loss is the same when hoses or pipes at different pressures flow the same amount of water. For example, if 100 gpm (400 L/min) passes through a 3-inch (77 mm) hose within a certain time, the water must travel at a specified velocity (feet per second [meters per second]). For the same rate of flow to pass through a 1½-inch (38 mm) hose, the velocity must be greatly increased. Four 1½-inch (38 mm) hoses are needed to flow 100 gpm (400 L/min) at the same velocity required for a single 3-inch (77 mm) hose (Figure 6.25).

While pipe sizes are fixed, some brands of fire hose tend to expand to a larger inside diameter under higher pressures than other brands. Even though both brands of hose may be marketed at 1¾-inch (45 mm) hose, the one brand may expand to close to 2 inches (50 mm) when charged. Keep in mind that this tendency decreases the velocity and therefore decreases the friction loss.

Other Factors Affecting Friction Loss

One of the physical properties of water is that it is practically incompressible. This means that the same volume of water supplied into a fire hose under pressure at one end will be discharged at the other end. The size of hose determines the velocity for a

given volume of water. The smaller the hose, the greater the velocity needed to deliver the same volume.

Friction loss in a system increases as the length of hose or piping increases. Flow pressure will always be greatest near the supply source and lowest at the farthest point in the system. A condition existing in a practical water system and a fire hose layout is shown in Figure 6.26. An elevated tank is filled with water to a height of 150 feet (45 m). The pipe connections to the fire hydrant are at the bottom of the tank. From the hydrant, 300 feet (90 m) of 2½-inch (65 mm) hose is laid along the street with a valve on the end. Imagine a glass tube connected to the hose every 100 feet (30 m) standing upright 150 feet (45 m) to the same height as the elevated tank. With the valve closed, the water in all the tubes would stand at Line A, which is the same level as the water in the tank. This line indicates the static pressure.

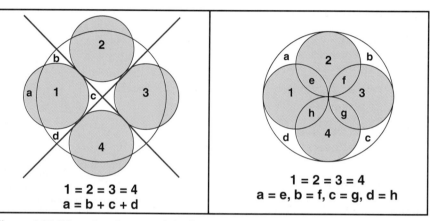

Figure 6.25 When the diameter of the hose is doubled, the area of the hose opening increases approximately four times.

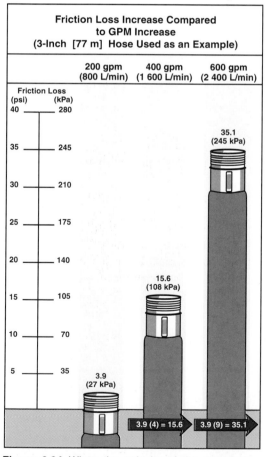

Figure 6.24 When the velocity of the stream is increased, the friction loss increases at an incrementally higher rate.

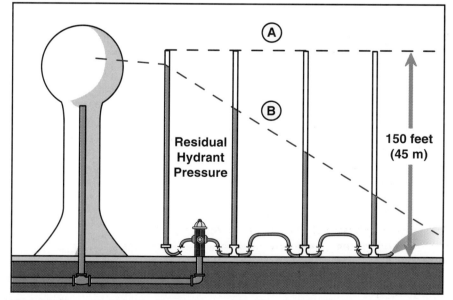

Figure 6.26 Line A indicates the static pressure. The distance between Lines A and B indicates the pressure loss due to friction. Note that there is no nozzle attached to the end of the hose.

When the valve on the nozzle end is opened, water flows moderately at a low pressure. If the opening is made directly at the hydrant, the flow will be much greater at a higher pressure. In other words, the flow pressure is not as great at the end of the hose as it is at the hydrant. Instead of the water being up to the level of the standpipe, it is up to Line B. The difference in the water level of the tubes indicates the pressure used to overcome the friction loss in the sections of hose between the tubes. The loss of pressure at each 100-foot (30 m) interval and the reduced discharge indicate the friction loss in the line.

An open fire hose produces a stream that normally has no use in fire fighting. Some type of nozzle is needed to shape the stream. When a closed nozzle with a 1-inch (25 mm) tip is added to the system, the water level in the glass tubes returns to Line A. When the nozzle is opened, the water level drops to Line C. Notice that Line C in Figure 6.27 is considerably higher than Line B in Figure 6.26 and that a fire stream with some reach has been produced.

If the 1-inch (25 mm) tip is replaced with a ¾-inch (19 mm) tip, the water level in the glass tubes rises even higher. The velocity increases, but the amount of flow decreases. By decreasing the amount of water flowing, a firefighter reduces the speed of the water in the hose; consequently, there is less friction loss.

Observe the height of the water in the first glass tubes in Figures 6.26 and 6.27. The water level in these tubes indicates a good supply of residual pressure left in the water main. Under these conditions, it is advisable to place a pumper in the line at the hydrant (Figure 6.28). Using a pump at this point provides additional force, thus increasing the pressure in the hose. This additional pressure makes it possible to produce effective fire streams. Using a pump also makes it possible to add hose, even to the extent of providing master streams.

It is important to remember that there are practical limits to the velocity or speed at which a stream can travel. If the velocity is increased beyond these limits, the friction becomes so great that the entire stream is agitated by resistance. This agitation causes a degree of turbulence called *critical velocity*. Beyond this point, it becomes necessary to parallel or siamese hoselines to increase the flow and reduce friction.

Reducing Friction Loss

Certain characteristics of hose layouts affect friction loss, including the following:

- Hose length
- Hose diameter
- Sharp bends (kinks) in the hose

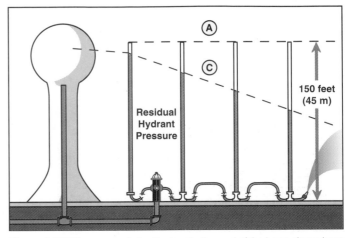

Figure 6.27 When a nozzle is attached to the end of the hose, the volume of water flowing is decreased. This results in less friction loss.

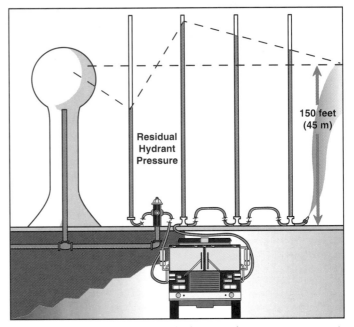

Figure 6.28 When a pumper is used to increase the water pressure, note the drop in the residual pressure and increase in the discharge pressure. The water in the first tube has been lowered due to the increased flow rate.

It is usually possible to minimize sharp bends or kinks in the hose by using proper hose handling techniques. To reduce friction loss due to hose length or diameter, it is necessary to reduce the length of the hose or increase its diameter. (This may not be possible during a fire situation.) Although the hose must be long enough to reach the needed location, any extra hose should be eliminated to reduce excess friction loss.

Because it is generally safe to increase the flow of water for fire fighting, it is usually acceptable to use larger diameter hose to reduce friction loss. However, hose size should not be increased so much that the hose becomes difficult to handle. Realistically, hose larger than 3 inches (77 mm) in diameter cannot be used for handlines.

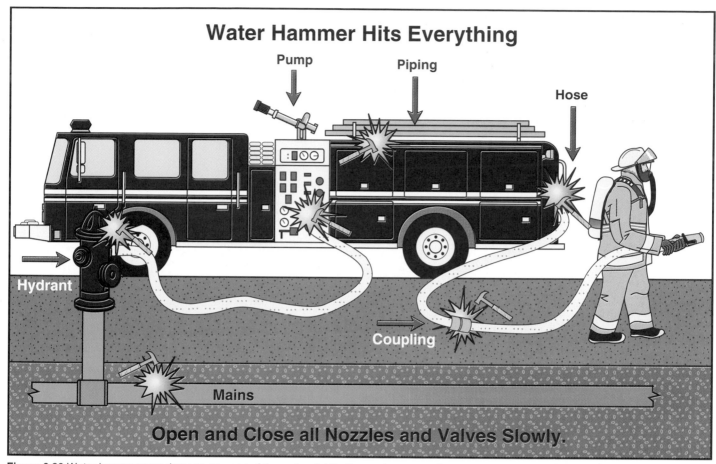

Figure 6.29 Water hammer may damage any part of the water distribution system or any fire apparatus connected to the system.

Water Hammer

Water moving through a pipe or hose has both weight and velocity. The weight of water increases as the pipe or hose size increases. Suddenly stopping water moving through a hose or pipe results in an energy surge being transmitted in the opposite direction, often at many times the original pressure. This surge is referred to as *water hammer*. Water hammer can damage the pump, appliances, hose, or the municipal water system itself (Figure 6.29). Always open and close nozzle controls, hydrants, valves, and hose clamps slowly to prevent water hammer. Equip apparatus inlets and remote outlets with pressure relief devices to prevent damage to equipment.

Principles of Municipal Water Supply Systems

Public and/or private water systems provide the methods for supplying water to populated areas. As the population increases in rural areas, rural communities seek to improve water distribution systems from reliable sources.

The water department may be a separate, city-operated utility or a regional or private water authority. Its principal function is to provide potable water. Water department officials should be considered the experts in water supply problems. The fire department must work with the water department in planning fire protection coverage. Water department officials should realize that fire departments are vitally concerned with water supply and work with them on water supply needs and the locations and types of fire hydrants.

The intricate working parts of a water system are many and varied. Basically, the system is composed of the following fundamental components, which are explored in the following subsections (Figure 6.30):

- Source of water supply
- Means of moving water
- Water processing or treatment facilities
- Water distribution system, including storage

Sources of Water Supply

The primary water supply can be obtained from either surface water or groundwater. Although most water systems are supplied from only one source, there are instances where both sources are used. Two examples of surface water supply are rivers and lakes (Figure 6.31). Groundwater supply can be water wells or water-producing springs.

Figure 6.31 A static water supply source requires some type of fire pump to take water into the system.

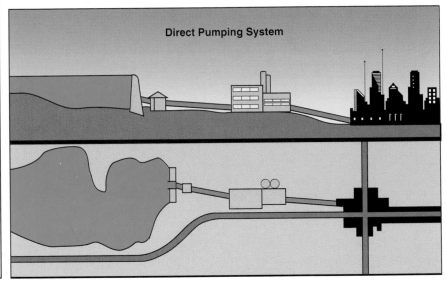

Direct Pumping System

Figure 6.32 A direct pumping system is used when the water source does not have sufficient elevation to create adequate pressure.

Figure 6.30 Every water system has four primary elements.

Sources

Means of Moving Water

Treatment Facilities

Distribution Systems

The amount of water that a community needs can be determined by an engineering estimate. This estimate is the total amount of water needed for domestic and industrial use and for fire fighting use. In cities, the domestic/industrial requirements far exceed that required for fire protection. In small towns, the requirements for fire protection may exceed other requirements.

Means of Moving Water

There are three methods of moving water in a system:

* Direct pumping system
* Gravity system
* Combination system

Direct Pumping System

Direct pumping systems use one or more pumps that take water from the primary source and discharge it through the filtration and treatment processes (Figure 6.32). From there, a series of pumps force the water into the distribution system. If purification of the water is not needed, the water can be pumped directly into the distribution system from the primary source. Failures in supply lines and pumps can usually be overcome by duplicating these units and providing a secondary power source.

Gravity System

A gravity system uses a primary water source located at a higher elevation than the distribution system (Figure 6.33). The gravity flow from the higher elevation provides the water pressure. This pressure is usually only sufficient when the primary water source is located at least several hundred feet (meters) higher than the highest point in the water distribution system. The most common examples include a mountain reservoir that supplies water to a city below or a system of elevated tanks in a city itself.

Combination System

Most communities use a combination of the direct pumping and gravity systems (Figure 6.34). In most

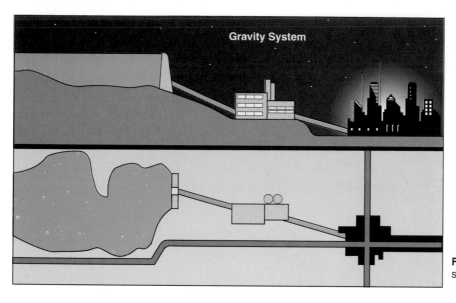

Figure 6.33 A gravity system is used where the water source is elevated.

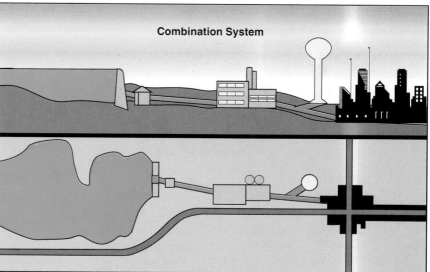

Figure 6.34 A combination of direct pumping and gravity is used to allow water storage during low demand. Later, this water can be used when consumption exceeds pump capacity.

cases, the gravity flow is supplied by elevated storage tanks. These tanks serve as emergency storage and provide adequate pressure through the use of gravity. When the system pressure is high during periods of low consumption, automatic valves open and allow the elevated storage tanks to fill. When the pressure drops during periods of heavy consumption, the storage containers provide extra water by feeding it back into the distribution system. Providing a good combination system involves reliable, duplicated equipment and proper-sized, strategically located storage containers.

The storage of water in elevated reservoirs can also ensure water supply when the system becomes otherwise inoperable. Storage should be sufficient to provide domestic and industrial demands plus the demands expected in fire fighting operations. Such storage should also be sufficient to permit making most repairs, alterations, or additions to the system. Location of the storage and the capacity of the mains leading from this storage are also important factors.

Many industries provide their own private systems, such as elevated storage tanks, that are available to the fire department. Water for fire protection may be available to some communities from storage systems, such as cisterns, that are considered a part of the distribution system. The fire department pumper removes the water from these sources by drafting (process of obtaining water from a static source into a pump that is above the source's level) and provides pressure by its pump. Drafting operations are detailed later in this manual.

Processing or Treatment Facilities

The treatment of water for the water supply system is a vital process (Figure 6.35). Water is treated to remove contaminants that may be detrimental to the health of those who use or drink it. Water may be treated by coagulation, sedimentation, filtration, or the addition of chemicals, bacteria, or other organisms. In addition to removing things from the water, some things may be added such as fluoride or oxygen.

Figure 6.35 In most cases, the water treatment facility is located adjacent to the supply source.

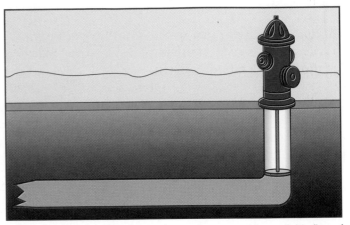

Figure 6.36 Dead-end hydrants do not always provide a reliable flow of water for fire fighting operations.

The fire department's main concern regarding treatment facilities is that a maintenance error, natural disaster, loss of power supply, or fire could disable the pumping station(s) or severely hamper the purification process. Any of these situations would drastically reduce the volume and pressure of water available for fire fighting operations. Another problem would be the inability of the treatment system to process water fast enough to meet the demand. In either case, fire officials must have a plan to deal with these potential shortfalls.

Water Distribution System

The distribution system of the overall water supply system is the part that receives the water from the pumping station and delivers it throughout the area served. The ability of a water system to deliver an adequate quantity of water relies upon the carrying capacity of the system's network of pipes. When water flows through pipes, its movement causes friction that results in a reduction of pressure. There is much less pressure loss in a water distribution system when fire hydrants are supplied from two or more directions. A fire hydrant that receives water from only one direction is known as a *dead-end hydrant* (Figure 6.36). When a fire hydrant receives water from two or more directions, it is said to have *circulating feed* or a *looped line* (Figure 6.37). A distribution system that provides circulating feed from several mains constitutes a *grid system*. A grid system should consist of the following components (Figure 6.38):

- *Primary feeders* — Large pipes (mains), with relatively widespread spacing, that convey large quantities of water to various points of the system for local distribution to the smaller mains

- *Secondary feeders* — Network of intermediate-sized pipes that reinforce the grid within the various loops of the primary feeder system and aid the concentration of the required fire flow at any point

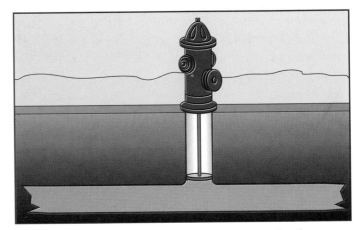

Figure 6.37 Looped hydrants receive water from two directions.

- *Distributors* — Grid arrangement of smaller mains serving individual fire hydrants and blocks of consumers

To ensure sufficient water, two or more primary feeders should run from the source of supply to the high-risk and industrial districts of the community by separate routes. Similarly, secondary feeders should be arranged in loops as far as possible to give two directions of supply to any point. This practice increases the capacity of the supply at any given point and ensures that a break in a feeder main will not completely cut off the supply.

In residential areas, the recommended size for fire hydrant supply mains is at least 6 inches (150 mm) in diameter. These should be closely gridded by 8-inch (200 mm) cross-connecting mains at intervals of not more than 600 feet (180 m). In the business and industrial districts, the minimum recommended size is an 8-inch (200 mm) main with cross-connecting mains every 600 feet (180 m). Twelve-inch (300 mm) mains may be used on principal streets and in long mains not cross-connected at frequent intervals. Water mains as large as 48 inches (1.2 m) may be found in major cities.

Water Main Valves

The function of a valve in a water distribution system is to provide a means for controlling the flow of water through the distribution piping. Valves should be located at frequent intervals in the grid system so that only small districts are cut off if it is necessary to stop the flow at specified points. Valves should be operated at least once a year to keep them in good condition. The actual need for valve operation in a water system rarely occurs, sometimes not for many years. Valve spacing should be such that only a minimum length of pipe is out of service at one time.

One of the most important factors in a water supply system is the water department's ability to promptly operate the valves during an emergency or breakdown of equipment. A well-run water utility has records of the locations of all valves. Valves should be inspected and operated on a regular basis. If each fire company is informed of the locations of valves in the distribution system, their condition and accessibility can be noted during fire hydrant inspections. The water department is then informed if any valves need attention.

Valves for water systems are broadly divided into *indicating* and *nonindicating* types. An indicating valve visually shows whether the gate or valve seat is open, closed, or partially closed. Valves in private fire protection systems are usually of the indicating type. Two common indicator valves are the post indicator valve (PIV) and the outside screw and yoke (OS&Y) valve. The *post indicator valve* is a hollow metal post that is attached to the valve housing (Figure 6.39). The valve stem inside this post has the words *OPEN* and *SHUT* printed on it so that the position of the valve is shown. These are commonly used on private water supply systems. The *OS&Y valve* has a yoke on the outside with a threaded stem that controls the gate's opening or closing (Figure 6.40). The threaded

Figure 6.39 The PIV shows the status of the valve through a small window

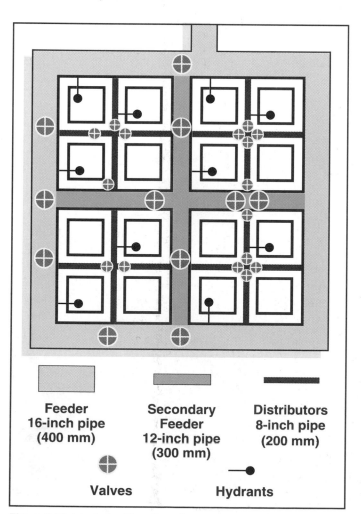

Feeder
16-inch pipe
(400 mm)

Secondary
Feeder
12-inch pipe
(300 mm)

Distributors
8-inch pipe
(200 mm)

Valves

Hydrants

Figure 6.38 A typical grid system of water supply pipes.

Figure 6.40 An OS&Y valve.

portion of the stem is out of the yoke when the valve is open and inside the yoke when the valve is closed. These are most commonly used on sprinkler systems but may be found in some water distribution system applications.

Nonindicating valves in a water distribution system are normally buried or installed in manholes (Figure 6.41). These are the most common types of valves used on most public water distribution systems. If a buried valve is properly installed, the valve can be operated aboveground through a valve box. A special socket wrench on the end of a reach rod operates the valve.

Control valves in water distribution systems may be either gate valves or butterfly valves. Both valves can be of the indicating or nonindicating type. Gate valves may be of the rising stem or the nonrising stem type. The rising stem type is similar to the OS&Y valve (Figure 6.42). On the nonrising stem type, the gate either rises or lowers to control the water flow as the valve nut is turned by the valve key (wrench). Nonrising-stem gate valves should be marked with a number indicating the number of turns necessary to completely close the valve. If a valve resists turning after fewer than the indicated number of turns, it usually means debris or other obstructions are in the valve. Butterfly valves are tight closing, and they usually have a rubber or a rubber-composition seat that is bonded to the valve body (Figure 6.43). The valve disk rotates 90

degrees from the fully open to the tight-shut position. The nonindicating butterfly type also requires a valve key. Its principle of operation provides satisfactory water control after long periods of inactivity.

The advantages of proper valve installation in a distribution system are readily apparent. If valves are installed according to established standards, it normally will be necessary to close off only one or perhaps two fire hydrants from service while a single break is being repaired. The advantage of proper valve installation is, however, reduced if all valves are not properly maintained and kept fully open. High friction loss is caused by valves that are only partially open. When valves are closed or partially closed, the condition may not be noticeable during ordinary domestic flows of water. As a result, the impairment will not be known until a fire occurs or until detailed inspections and fire flow tests are made. A fire department will experience difficulty in obtaining water in areas where there are closed or partially closed valves in the distribution system.

Water Pipes

Water pipe that is used underground is generally made of cast iron, ductile iron, asbestos cement, steel, plastic, or concrete. Whenever pipe is installed, it should be the proper type for the soil conditions and pres-

Figure 6.41 A typical underground water supply valve arrangement.

Figure 6.42 A rising stem gate valve.

Figure 6.43 Butterfly valves are common on private water supply systems.

sures to which it will be subjected. When water mains are installed in unstable or corrosive soils or in difficult access areas, steel or reinforced concrete pipe may be used to give the strength needed. Some locations that may require extra protection include areas beneath railroad tracks and highways, areas close to heavy industrial machinery, areas prone to earthquakes, or areas of rugged terrain.

The internal surface of the pipe, regardless of the material from which it is made, offers resistance to water flow. Some materials, however, have considerably less resistance to water flow than others. Personnel from the engineering division of the water department should determine the type of pipe best suited for the conditions at hand.

The amount of water able to flow through a pipe and the amount of friction loss created can also be affected by other factors. Frequently, friction loss is increased by encrustation of minerals on the interior surfaces of the pipe. Another problem is sedimentation that settles out of the water. Both of these conditions result in a restriction of the pipe size, increased friction loss, and a proportionate reduction in the amount of water that can be drawn from the system.

Water System Capacity

When engineers design a water distribution system, there are three basic rates of consumption that they consider in their design. These form an established base to which the fire flow requirements can be added during the design process. They also allow the engineers and fire protection people alike to determine the adequacy of the water distribution system. The driver/operator should be familiar with these terms because they encounter them often during water supply testing.

The *average daily consumption* (ADC) is the average of the total amount of water used in a water distribution system over the period of one year. The *maximum daily consumption* (MDC) is the maximum total amount of water that was used during any 24-hour interval within a 3-year period. Unusual situations, such as refilling a reservoir after cleaning, should not be considered in determining the maximum daily consumption. The *peak hourly consumption* (PHC) is the maximum amount of water used in any 1-hour interval over the course of a day.

The maximum daily consumption is normally about one and one-half times the average daily consumption. The peak hourly rate normally varies from two to four times the normal hourly rate. The effect these varying consumption rates have on the ability of the system to deliver required fire flows varies with the system design. Both maximum daily consumption and peak hourly consumption should be considered to ensure that the water supplies and pressure do not reach dangerously low levels during these periods. Adequate water must be available during these periods in the event that a fire occurs.

Private Water Supply Systems

In addition to the public water supply systems that service most communities, fire department personnel must also be familiar with the basic principles of any private water supply systems that may be found within their response jurisdiction. Private water supply systems are most commonly found on large commercial, industrial, or institutional properties. They may service one large building or a series of buildings on the complex. In general, the private water supply system exists for one of the three following purposes:

- To provide water strictly for fire protection purposes

- To provide water for sanitary and fire protection purposes

- To provide water for fire protection and manufacturing processes

The design of private water supply systems is typically similar to that of the municipal systems described earlier in this chapter. Most commonly, private water supply systems receive their water from a municipal water supply system. In some cases, the private system may have its own water supply source, independent of the municipal water distribution system (Figure 6.44).

In a few cases, a property may be served by two sources of water supply for fire protection: one from the municipal system and the other from a private source. In many cases, the private source of water for fire protection provides nonpotable (not for drinking) water. When this is the case, adequate measures must be taken to prevent contamination caused by the backflow of

Figure 6.44 Private water supply systems usually have their own pump house.

nonpotable water into the municipal water supply system. There are a variety of backflow prevention measures that can be employed to avoid this problem. Some jurisdictions do not allow the interconnection of potable and nonpotable water supply systems. This means that the protected property is required to maintain two completely separate systems.

Almost universally, private water supply systems maintain separate piping for fire protection and domestic/industrial services. This is in distinct contrast to most municipal water supply systems in which fire hydrants are connected to the same mains that supply water for domestic/industrial use. Separate systems are cost prohibitive for most municipal applications but are economically practical in many private applications. There are a number of advantages to having separate piping arrangements in a private water supply system, including the following:

- The property owner has control over the water supply source.

- Either of the systems (fire protection or domestic/industrial) are unaffected by service interruptions to the other system.

Keep in mind that private water supply systems that rely solely on the municipal water distribution system as their water supply source are subject to service interruptions in the event that the municipal system experiences a failure.

Fire department personnel must be familiar with the design and reliability of private water supply systems in their jurisdiction. Large, well-maintained systems may provide a reliable source of water for fire protection purposes. Small capacity, poorly maintained, or otherwise unreliable private water supply systems should not be relied upon to provide all the water necessary for adequate fire fighting operations. Historically, many significant fire losses can be traced, at least in part, to the failure of a private water supply system that was being used by municipal fire departments working the incident. Problems such as the discontinuation of electrical service to a property whose fire protection system is supplied by electrically driven fire pumps have resulted in disastrous losses.

If there is any question about the reliability of a private water supply system or of its ability to provide an adequate amount of water for a large-scale fire fighting operation, the fire department should make arrangements to augment the private water supply. This may be accomplished by relaying water from the municipal water supply system or by drafting from a reliable static water supply source close to the scene.

Chapter 7

Fire Hose Nozzles and Flow Rates

Job Performance Requirements

This chapter provides information that will assist the reader in meeting the following job performance requirements from NFPA 1002, *Standard on Fire Apparatus Driver/Operator Professional Qualifications*, 1998 edition.

3-2.1* Produce effective hand or master streams, given the sources specified in the following list, so that the pump is safely engaged, all pressure control and vehicle safety devices are set, the rated flow of the nozzle is achieved and maintained, and the apparatus is continuously monitored for potential problems.

- Internal tank
- Pressurized source
- Static source
- Transfer from internal tank to external source

(a) *Requisite Knowledge:* **Hydraulic calculations for friction loss and flow using both written formulas and estimation methods**, safe operation of the pump, problems related to small-diameter or dead-end mains, low-pressure and private water supply systems, hydrant cooling systems, and reliability of static sources.

(b) *Requisite Skills:* The ability to position a fire department pumper to operate at a fire hydrant and at a static water source, power transfer from vehicle engine to pump, draft, operate pumper pressure control systems, operate the volume/pressure transfer valve (multistage pumps only), operate auxiliary cooling systems, make the transition between internal and external water sources, and **assemble hose lines, nozzles, valves, and appliances**.

3-2.2 Pump a supply line of 2½ in. (65 mm) or larger, given a relay pumping evolution the length and size of the line and the desired flow and intake pressure, so that the proper pressure and flow are provided to the next pumper in the relay.

(a) *Requisite Knowledge:* **Hydraulic calculations for friction loss and flow using both written formulas and estimation methods**, safe operation of the pump, problems related to small-diameter or dead-end mains, low-pressure and private water supply systems, hydrant cooling systems, and reliability of static sources.

(b) *Requisite Skills:* The ability to position a fire department pumper to operate at a fire hydrant and at a static water source, power transfer from vehicle engine to pump, draft, operate pumper pressure control systems, operate the volume/pressure transfer valve (multistage pumps only), operate auxiliary cooling systems, make the transition between internal and external water sources, and **assemble hose lines, nozzles, valves, and appliances**.

6-2.1* Produce effective fire streams, utilizing the sources specified in the following list, so that the pump is safely engaged, all pressure-control and vehicle safety devices are set, the rated flow of the nozzle is achieved, and the apparatus is continuously monitored for potential problems.

- Water tank
- Pressurized source
- Static source

(a) *Requisite Knowledge:* **Hydraulic calculations for friction loss and flow using both written formulas and estimation methods**, safe operation of the pump, proper apparatus placement, personal safety considerations, problems related to small-diameter or dead-end mains and low-pressure and private water supply systems, hydrant cooling systems, and reliability of static sources.

(b) *Requisite Skills:* The ability to position a wildland fire apparatus to operate at a fire hydrant and at a static water source, properly place apparatus for fire attack, transfer power from vehicle engine to pump, draft, operate pumper pressure control systems, operate the volume/pressure transfer valve (multistage pumps only), operate auxiliary cooling systems, make the transition between internal and external water sources, and **assemble hose lines, nozzles, valves, and appliances**.

6-2.2 Pump a supply line, given a relay pumping evolution the length and size of the line and pumping flow and desired intake pressure, so that adequate intake pressures and flow are provided to the next pumper in the relay.

A *fire stream* can be defined as a stream of water or other extinguishing agent after it leaves a fire hose and nozzle until it reaches the desired point (Figure 7.1). During the time a stream of water or extinguishing agent passes through space, it is influenced by velocity, gravity, wind, and friction with the air. The condition of the stream when it leaves the nozzle is influenced by operating pressures, nozzle design, nozzle adjustment, and the condition of the nozzle orifice. The type of fire stream that is applied to a fire depends on the type of nozzle being used. Pumping apparatus driver/operators must be familiar with the different types of nozzles carried on their apparatus. Each type of nozzle has its own required flow rate and discharge pressure. The particular type of nozzle being used affects the hydraulic calculations that must be performed by the driver/operator. There also may be cases when the driver/operator is responsible for selecting an appropriate nozzle to perform a particular evolution. The driver/operator must understand the capabilities of each nozzle in order to make the correct choice.

This chapter provides the driver/operator with basic information on handline and master stream nozzles for attacking fires involving ordinary combustible materials with plain water or Class A foam. The appropriate pressures at which each nozzle should be operated are discussed in detail. The latter portion of the chapter discusses the methods for calculating the nozzle reaction from all types of nozzles. For more information on nozzles used for attacking flammable and combustible liquids fires with Class B fire fighting foams, see Chapter 15.

Figure 7.1 A fire stream must hit the seat of a fire to be effective.

nozzle is capable of delivering any other stream than the one for which it was designed. For example, it is not possible to obtain a solid stream from a fog stream nozzle. The best that can be achieved in this case is a pseudosolid stream, more correctly termed a straight stream. The sections that follow present the basics of solid stream nozzles and fog stream nozzles. Nozzles that produce broken streams are covered in the section on special purpose nozzles later in this chapter.

Solid Stream Nozzles

A *solid stream* is a fire stream produced from a fixed orifice, smoothbore nozzle (Figure 7.2). The solid stream nozzle is designed to produce a stream as compact as

Fire Hose Nozzles

The fire service utilizes three basic types of fire streams: solid, fog, and broken. Nozzles have been developed for each type of stream. It is important to emphasize that no possible with little shower or spray. A solid stream has the ability to reach areas that other streams might not reach. They may be used on handlines, portable or apparatus-mounted master streams, or elevated master streams.

Solid stream nozzles are designed so that the shape of the water in the nozzle is gradually reduced until it reaches a point a short distance from the outlet. At this point, the nozzle becomes a cylindrical bore whose length is from one to one and one-half times its diameter (Figure 7.3). The purpose of this short, truly cylindrical bore is to give the water its round shape before discharge. A smooth-finish waterway contributes to both the shape and reach of the stream. Alteration or damage to the nozzle can significantly alter stream shape and performance.

The velocity of the stream (nozzle pressure) and the size of the discharge opening determine the flow from a solid stream nozzle. When solid stream nozzles are used on handlines, they should be operated at 50 psi (350 kPa) nozzle pressure. A solid stream master stream device should be operated at 80 psi (560 kPa).

Determining the Flow From a Solid Stream Nozzle

In some cases, it may be necessary for a driver/operator to determine the amount of water that is being discharged from a solid stream nozzle. This may be required when performing friction loss calculations, determining pump discharge pressure, or when a solid stream nozzle is attached to a fire hydrant during water supply testing. The following formula may be used:

$$\textbf{GPM} = \textbf{29.7} \times \textbf{d}^2 \times \sqrt{\textbf{NP}}$$

Where: GPM = Discharge in gallons per minute

29.7 = A constant

d = Diameter of the orifice in inches

NP = Nozzle pressure in psi

$$\textbf{L/min} = \textbf{0.067} \times \textbf{d}^2 \times \sqrt{\textbf{NP}}$$

Where: L/min = Discharge in liters per minute

0.067 = A constant

d = Diameter of the orifice in millimeters

NP = Nozzle pressure in kPa

Using this formula, it is possible to determine water flow from any solid stream nozzle when the nozzle pressure and tip diameter are known. Table 7.1 lists various nozzle tip diameters. The following example illustrates the application of this formula.

Example: Determine the water flow from a 1-inch tip operating at 50 psi.

$$GPM = (29.7)(d)^2(\sqrt{NP})$$
$$GPM = (29.7)(1)^2(\sqrt{50})$$
$$GPM = (29.7)(1)(7.07)$$
$$GPM = 210$$

Example: Determine the flow of water from a 25 mm tip operating at 350 kPa.

$$L/min = (0.067)(d)^2(\sqrt{NP})$$
$$L/min = (0.067)(25)^2(\sqrt{350})$$
$$L/min = (0.067)(625)(18.7)$$
$$L/min = 780$$

Fog Stream Nozzles

A stream of water with a certain velocity remains in a solid mass, not losing continuity until it strikes an object, is overcome by gravity, or is changed by friction with the air. For a stream to be broken into finely divided particles, it must be driven against an obstruction with sufficient velocity to shatter the mass. The angle at which the stream of water is deflected from an obstruction determines the reduction in forward velocity of the stream, as well as the pattern or shape the stream assumes. Therefore, a wide-angle deflection produces a wide-angle fog, and a narrow-angle deflection produces a narrow-angle fog.

Figure 7.2 A typical solid stream nozzle.

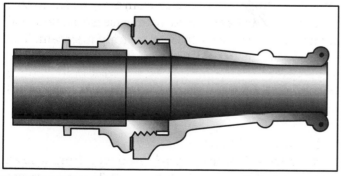

Figure 7.3 Most solid stream nozzles have a standard design for the shape of the waterway.

Table 7.1
Common Nozzle Diameters

U.S. (Inches)	Metric (mm) (Actual)	Metric (mm) (Known As)
½	12.700	13
⅝	15.880	16
¾	19.050	19
⅞	22.225	22
1	25.400	25
1⅛	28.575	29
1¼	31.750	32
1⅜	34.925	35
1½	38.100	38
1¾	44.450	45
2	50.800	50
2¼	57.150	57
2½	63.500	65
2¾	69.850	70
3	76.200	77

The following terms are important to know when discussing the mechanical principles of fog streams:

• *Periphery*—The line bounding a rounded surface; the outward boundary of an object distinguished from its internal regions

• *Deflection* — A turning or state of being turned; a turning from a straight line or given course; a bending; a deviation

• *Impinge*—To strike or dash about or against; clashing with a sharp collision; to come together with force

A fog stream may be produced by deflection at the periphery or by impinging jets of water or by a combination of these. Periphery-deflected streams are produced by deflecting water from the periphery of an inside circular stem in a periphery-deflected fog nozzle (Figure 7.4). This water is again deflected by the exterior barrel. The relative positions of the deflecting stem and the exterior barrel determine the shape of the fog stream.

The impinging stream nozzle drives several jets of water together at a set angle to break the water into finely divided particles (Figure 7.5). Impinging stream nozzles usually produce a wide-angle fog pattern, but a narrow pattern is possible. Impinging jets are also used on periphery-deflected stream nozzles.

The reach of the fog stream is directly dependent on the width of the stream, the size of the water droplets, and the amount of water flowing (Figure 7.6). When selecting

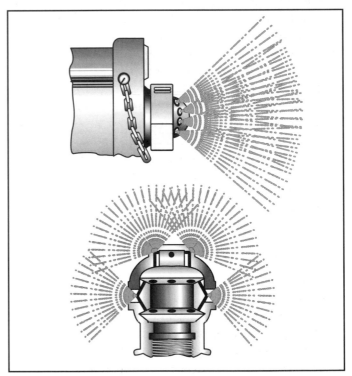

Figure 7.5 The impinging stream nozzle directs several jets of water together to form a fog pattern.

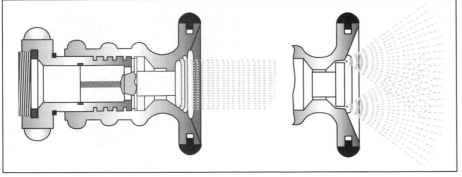

Figure 7.4 The relative positions of the deflection stem and the exterior barrel of a periphery-deflected fog nozzle determine the stream pattern.

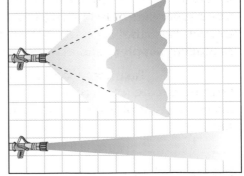

Figure 7.6 The reach of the fog stream is dependent on the width of the stream. Wider streams have less reach than narrow patterns.

a particular fog nozzle, you must be familiar with its capabilities to ensure that it has the desired reach capabilities for the size of stream desired.

When water is discharged at angles to the direct line of discharge, the reaction forces largely counterbalance each other, thus reducing nozzle reaction. This balancing of forces is the reason why fog patterns are easier to handle than solid or straight stream patterns.

A variety of fog nozzles are available. The purpose of this manual is to describe the general design and operation of various fog nozzles, not to recommend one type over another.

Constant Flow Nozzles

Constant flow nozzles are designed to flow a specific amount of water at a specific nozzle discharge pressure on all stream patterns. Most constant flow nozzles utilize a periphery-deflected stream. Most constant flow nozzles are also equipped with an adjustable pattern setting. These nozzles discharge the same volume of water regardless of the pattern setting (Figure 7.7). As the exterior barrel is rotated to change the fog stream pattern, the space between the deflecting stem and the internal throat remains the same. As a result, the same quantity of water passes through. Most constant flow nozzles are intended to be operated at a nozzle pressure of 100 psi (700 kPa), although some lower pressure nozzles (75 psi [535 kPa] nozzles are common) for special applications (such as high-rise fire fighting) are in existence.

Manually Adjustable Nozzles

A refinement of the constant flow nozzle is the adjustable gallonage nozzle. This nozzle has a number of constant flow settings, enabling the firefighter to select a flow rate that best suits the existing conditions. The nozzle supplies the selected flow at the rated nozzle discharge pressure (Figure 7.8). If the driver/operator is unable to supply the proper pressure, actual flow will differ from that indicated at the nozzle. Most of these nozzles are designed to supply the gallonage marked on each setting at a nozzle pressure of 100 psi (700 kPa). The driver/operator must know the flow at which the nozzle is set in order to properly supply the hoseline and nozzle.

CAUTION: Take care when adjusting flow settings. Nozzles that are set on a low flow may not provide the proper amount of water to sufficiently cool a burning fuel.

Automatic Nozzles

In the past, there were several types of variable flow fog nozzles used by the fire service. These included the rotary control nozzle and the ball-valve "mystery" nozzle.

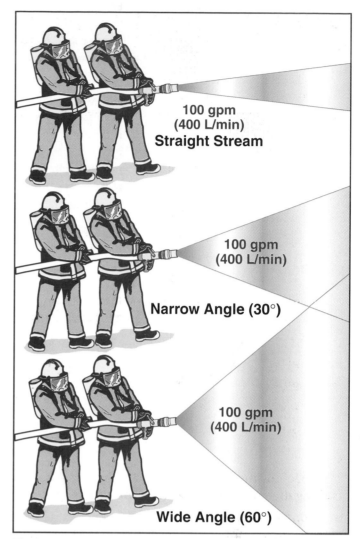

100 gpm (400 L/min) Straight Stream

100 gpm (400 L/min) Narrow Angle (30°)

100 gpm (400 L/min) Wide Angle (60°)

Figure 7.7 A constant gallonage nozzle delivers the same flow, regardless of the stream pattern setting.

Figure 7.8 With an adjustable flow nozzle, the selected flow remains the same whether a straight, narrow fog, or a wide fog stream pattern is used.

These nozzles had different flow rates depending on the fog pattern that was in use and the pressure being supplied to the nozzle. These older types of variable flow nozzles are rarely found in use in today's fire service.

The most common variable flow nozzles in use today are automatic nozzles. *Automatic nozzles*, also referred to as *constant pressure nozzles*, are basically variable flow nozzles with pattern-change capabilities and the ability to maintain the same nozzle pressure (Figure 7.9). If the gallonage supplied to the nozzle changes, the automatic nozzle maintains approximately the same nozzle pressure and pattern. This feature is made possible by a baffle that moves automatically, varying the spacing between the baffle and the throat (Figure 7.10).

A stream from an automatic nozzle can "look good," but may not be supplying sufficient water for extinguishment or protection. It should be the goal of the driver/operator to provide an acceptable flow of water at the discharge pressure for which the nozzle is designed. Most nozzles are designed for a 100 psi (700 kPa) discharge pressure. However, some nozzles may be designed for lower pressures such as 75 psi (535 kPa). These are commonly used in high-rise fire fighting. The acceptable flow of water varies with the hoseline size, nozzle design, and incident demand. More information on making these calculations is contained in Chapter 8.

CAUTION: Make sure that adequate pump discharge pressures are used to supply hoselines equipped with automatic nozzles. Nozzles receiving inadequate pressures may not provide the proper amount of water to sufficiently cool a burning fuel even though the stream "looks good."

An automatic nozzle serves as a pressure regulator — within its flow limits — for the pumper as lines are added or shut down. In this way, all available water may be used continuously if desired. If water supply is inadequate, the maximum volume is that which can be achieved without the pump "running away" (cavitating). The throttle is backed off slightly from the point where the pump starts to speed up or becomes erratic or when supply hose becomes soft and begins to collapse. This action provides the maximum flow possible until the water supply can be supplemented.

An automatic nozzle maintains a constant nozzle pressure of approximately 100 psi (700 kPa), no matter how much the pump discharge pressure is above this figure. As pump discharge pressure is increased, the nozzle automatically enlarges its effective opening size — within the range of the nozzle — to match the flow.

Figure 7.9 An automatic nozzle.

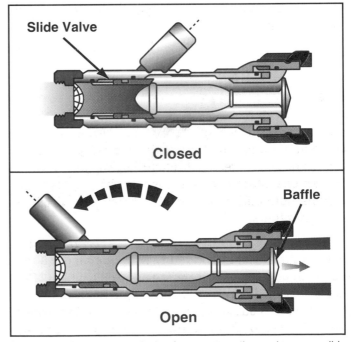

Figure 7.10 One common design for an automatic nozzle uses a slide barrel control inside the nozzle.

High-Pressure Fog Nozzles

High-pressure fog nozzles operate at pressures up to 800 psi (5 600 kPa) (Figure 7.11). They develop a fog stream with considerable forward velocity but deliver a relatively low volume of water. This nozzle delivers water in a very fast-moving, fine spray. High-pressure fog nozzles may use an impinging stream. High-pressure nozzles and lines are best used for fighting wildland fires. These lines are not recommended for structural fire fighting because they generally only flow in the neighborhood of 8 to 15 gpm (32 L/min to 60 L/min). These flows are insufficient to enable firefighters to safely and effectively attack structural fires.

Figure 7.11 High-pressure fog nozzles generally differ in appearance from other fire nozzles.

Selecting Nozzles

As mentioned at the beginning of the chapter, the driver/operator may be required to select the correct nozzle to perform a particular task. The following sections cover the types of nozzles most commonly carried on fire department pumping apparatus.

Handline Nozzles

Handline nozzles are designed to be placed on mobile attack lines that can be easily maneuvered by firefighters (Figure 7.12). They may be of the solid, fog, or broken stream type. Handline nozzles range in size from small booster line nozzles for ¾-inch (19 mm) booster hose, to large fog or solid stream nozzles that are designed to be placed on the end of a 3-inch (77 mm) hoseline. Generally, 350 gpm (1 400 L/min) is the maximum amount of water that can safely flow through a handline nozzle. Flows greater than that require pressures that make the hoselines difficult and dangerous for firefighters to handle.

Figure 7.12 Handline nozzles are used to attack smaller fires. *Courtesy of Bob Esposito.*

Master Stream Nozzles

The term *master stream* is applied to any fire stream that is too large to be controlled without mechanical aid. Master stream devices are powerful and generate a considerable amount of nozzle reaction force. It is extremely important, therefore, that firefighters take proper safety precautions. Master streams may be either solid or fog streams; both utilize a nozzle of sufficient size to deliver the higher flows.

The master stream is the "big gun" of the fire department. Solid master streams are usually operated at 80 psi (560 kPa) and fog master streams at 100 psi (700 kPa). Master stream flows are usually 350 gpm (1 400 L/min) or greater. Master streams are used when handlines would be ineffective, conditions are unsafe, or when manpower is limited (Figure 7.13). A handline stream may be inef-

Figure 7.13 Master streams are most commonly used on large, defensive fire attack operations.

fective if the fire is so large or intense that heat is generated faster than it can be absorbed by the stream. An extremely hot fire may also make it impossible for firefighters to approach close enough to use handlines. Master streams not only deliver large volumes of water but also have a greater reach than handheld streams.

Because master stream devices are used from fixed positions, most of them have some means for moving the stream in either a vertical or horizontal plane or both. To permit such adjustments, the water must pass through one or more sharp bends. On some larger master stream devices, there are two bends that form a loop in the shape of a ram's horns (Figure 7.14). Some other master stream devices have a single bent-pipe waterway (Figure 7.15).

The amount of friction loss varies from device to device. Although the friction loss in a master stream device may be assumed to be some fixed figure, each department must determine the friction loss in the devices it has available, either by flow test or manufacturer's documentation. Appendix B illustrates the procedure for determining friction loss in master stream devices.

There are four basic categories of master stream devices, whose characteristics are highlighted in the following sections:

- Monitor
- Deluge set
- Turret pipe
- Elevated master stream

Figure 7.14 The appearance of these devices is the basis for their ram's horn nozzle name. *Courtesy of Elkhart Brass Mfg. Co.*

Figure 7.15 Some master streams have a single, bent pipe waterway. *Courtesy of Elkhart Brass Mfg. Co.*

Monitor

Monitors are often incorrectly referred to as deluge sets, but they differ in one important way: with a monitor, the stream direction and angle can be changed while water is being discharged.

There are three basic types of monitors: fixed, combination, and portable (Figures 7.16 a through c). The fixed monitor is permanently mounted on the apparatus. The combination monitor is also mounted on the apparatus. It can be used there as a turret or removed and used as a portable monitor. The portable monitor can be carried to the location where it is needed.

Turret Pipe

A turret pipe is mounted on a fire apparatus deck and is connected directly to the pump by permanent pipe (Figure 7.17). The turret pipe is also sometimes called a *deck gun* or *deck pipe*. It is supplied by permanent piping from the pump.

Deluge Set

The deluge set consists of a short length of large diameter hose with a large nozzle or large playpipe supported at the discharge end by a tripod (Figure 7.18). There is a siamese connection at the supply end. The direction and angle of the stream cannot be changed while the deluge set is discharging water.

Elevated Master Stream

Elevated master streams are those large capacity nozzles that are designed to be placed on the end of a fire department aerial apparatus elevating device. They may be permanently attached to elevating platforms and prepiped aerial ladders or they may be detachable.

A *ladder pipe* is a master stream device used in conjunction with aerial ladders. Detachable ladder pipes are attached to the rungs of an aerial ladder and are supplied by fire hose (usually 3-inch [77 mm] hose) (Figure 7.19). These ladder pipes must be operated manually by a

Figure 7.16a This master stream device may not be removed from the apparatus for use.

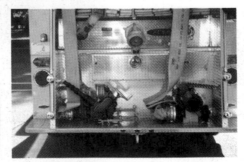

Figure 7.16b Some master stream devices are designed solely for portable use.

Figure 7.16c This master stream device may be operated from the apparatus or removed to a remote location.

Figure 7.17 A turret pipe.

Figure 7.19 Detachable ladder pipes are most common on light duty aerial ladders.

Figure 7.18 A deluge set does not allow the stream to be easily moved.

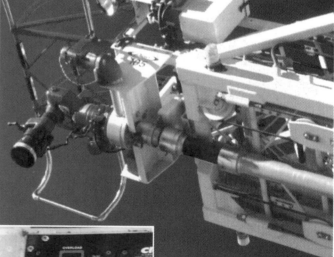

Figure 7.20 Most modern aerial ladders have piped waterways.

firefighter at the tip of the ladder or by using a rope from the ground. Movement of these streams is limited to vertical up and down motions. Horizontal movement of the nozzle would place dangerous stress on the aerial ladder.

Some aerial ladders and all aerial platform and elevated master stream devices have prepiped waterways instead of hose (Figure 7.20). Prepiped ladders generally have the ladder pipe attached to the end of the waterway, which is on the bottom of the ladder. These ladder pipes may be operated from the top either manually or by a power control switch located there. They can also usually be operated from the turntable or pump panel area by remote power controls (Figure 7.21). The power system to operate the master stream may be either electric, hydraulic, or pneumatic.

Master streams used with elevating platforms are basically similar to those with prepiped aerial ladders. The primary difference is that they are located on the aerial

Figure 7.21 Piped waterways are usually equipped with remote control nozzles that may be operated from the pump panel or turntable control pedestal.

platform and can be more easily maneuvered by firefighters at the tip of the aerial device. Some elevating platforms are equipped with two master streams on one platform (Figure 7.22). This gives added flexibility during large-scale fire operations.

Special Purpose Nozzles

Some fire situations (chimney fires, basement fires, fires in inaccessible spaces, and exposure problems) call for special stream nozzles. The degree to which such nozzles are needed depends on the particular fire situation or fuel involved. None of the following listed nozzles are required to be carried on fire department pumpers according to NFPA 1901, *Standard for Automotive Fire Apparatus*. Each department must decide which, if any, of these nozzles are required to handle fire situations that are anticipated within its jurisdiction.

All of the nozzles described in the following sections are broken stream nozzles. Broken stream nozzles differ from fog stream nozzles. Fog streams use deflection or impinging streams to create a fog pattern. Broken streams are created when water is forced through a series of small holes on the discharge end of the nozzle. In general, broken streams produce larger droplets of water than do fog streams. This tends to give them better reach and penetrating power. This is what makes them useful for special applications.

Cellar Nozzles

Cellar nozzles, also called *distributors*, are often used on basement fires (Figure 7.23). These nozzles can be lowered through holes cut in the floor or through some other suitable opening (Figure 7.24). To achieve best results, check for unseen obstructions that may lessen the nozzle's effectiveness. Cellar nozzles may not be equipped with shutoffs, so an in-line shutoff valve should be placed at a convenient location back from the nozzle. These nozzles may also be used to attack attic fires. The method of

Figure 7.23 A typical cellar nozzle. *Courtesy of Elkhart Brass Mfg. Co.*

Figure 7.22 Some aerial platforms are equipped with two master stream devices. *Courtesy of Joel Woods.*

operation is similar to fighting a cellar fire, except that the nozzle is pushed through a hole in the ceiling to attack the fire above.

Water Curtain Nozzles

A water curtain nozzle produces a fan-shaped stream designed to act as a water curtain between a fire and a combustible material (Figure 7.25). A water curtain may also be used to protect firefighters from heat. To be effective, the water curtain must cover a wide area and be reasonably heavy. A water curtain is only effective in absorbing convected heat from a fire. Radiated heat is transmitted through the water curtain. Radiated heat can only be reduced by placing water on the exposure. Thus,

Figure 7.24 The cellar nozzle is inserted through a hole that is made in the floor.

Figure 7.25 Water curtain nozzles were never proven to be overly effective and are rarely found in use today.

a water curtain between the fire and combustible material is not as effective as the same amount of water flowing over the surface of the combustible material. It is better, therefore, to direct fire streams onto exposed surfaces. This may be accomplished by angling the nozzle so that the water cascades down the side of the exposure being protected.

Piercing Nozzles

Piercing nozzles are commonly used to apply water to areas that are otherwise inaccessible to water streams. Piercing nozzles may also be used to deliver aqueous film forming foam (AFFF) to a confined area. The piercing nozzle is generally a 3- to 6-foot (1 m to 2 m) hollow steel rod 1½ inches (38 mm) in diameter (Figure 7.26). The discharge end of the piercing nozzle is usually a hardened steel point suitable for driving through concrete block or other types of wall or partition assemblies. Built into the point is an impinging jet nozzle generally capable of delivering about 100 gpm (400 L/min) of water. Opposite the pointed nozzle end of the piercing nozzle is the driving end. This end of the nozzle is driven with a sledgehammer to force the point through an obstruction (Figure 7.27).

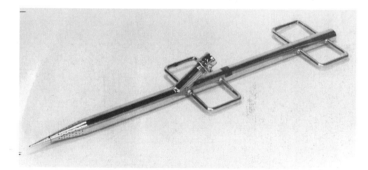

Figure 7.26 Piercing nozzles have a blunt end that is struck with a sledge to drive the pointed end through a barrier. *Courtesy of Superior Flamefighter Inc.*

Figure 7.27 At least two firefighters are needed to place a piercing nozzle in service.

Chimney Nozzles

Chimney nozzles have been developed to attack chimney flue fires. The chimney nozzle is designed to be placed on the end of a booster hose. The nozzle is a solid piece of brass or steel with numerous, very small impinging holes (Figure 7.28). At a nozzle pressure of 100 psi (700 kPa), a chimney nozzle generally produces only 1.5 to 3 gpm (6 L/min to 12 L/min) of water in a very fine, misty fog cone. The hose and nozzle are lowered down the entire length of the chimney and then quickly pulled out (Figure 7.29). The mist from the nozzle immediately turns to steam and chokes the flue fire as well as loosens the soot on the inside of the chimney. Because water is used in such small quantities and converts to steam so quickly, the flue liner is not damaged by sudden cooling. Because this process may damage booster hose, it is better to use an old section of hose on the end of the regular section of hose when using a chimney nozzle.

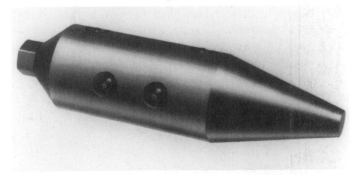

Figure 7.28 A typical chimney nozzle. *Courtesy of Jaffrey Fire Protection.*

Nozzle Pressure and Reaction

As water is discharged from a nozzle at a given pressure, a force pushes back on the firefighters handling the hoseline. This counterforce, known as *nozzle reaction*, clearly illustrates Newton's Third Law of Motion. This law states that for every action there is an equal and opposite reaction. Therefore, the greater the nozzle discharge pressure, the greater the resulting nozzle reaction. It is the resulting nozzle reaction to a given pressure that forces us to limit the amount of nozzle pressure that can be supplied to an attack line. Firefighters can be seriously injured and fire attacks greatly hampered by nozzles violently whipping around from excess nozzle reaction.

Tests have revealed that the practical working limits for velocity of fire streams are within 60 to 120 feet per second (18.3 m to 36.6 m per second). These limiting velocities are produced by nozzle pressures that range from 25 to 100 psi (175 kPa to 700 kPa). Nearly all fog nozzles are designed to operate at a nozzle pressure of 100 psi (700 kPa). This pressure is manageable and acceptable for both handlines and master streams.

Lower nozzle pressures must be used with solid stream nozzles due to the greater amount of nozzle reaction experienced with these nozzles. As a rule, 50 psi (350 kPa) should be used as the nozzle pressure for solid stream handlines. If greater reach and volume are needed, the nozzle pressure may be raised to 65 psi (455 kPa) without becoming unmanageable. Above this point, solid streams become increasingly difficult to handle.

Portable master stream devices equipped with solid stream nozzles are not to be operated above 80 psi (560

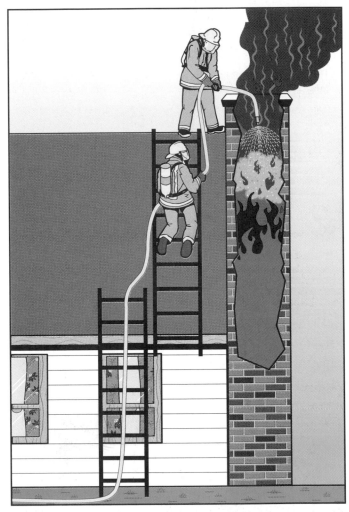

Figure 7.29 The chimney nozzle is lowered down the flue to extinguish the fire.

kPa) unless approved by the manufacturer of the device. Some manufacturers may also specify lower nozzle pressure, particularly on larger nozzle tip sizes. However, fixed master stream devices with solid stream nozzles may be operated at higher pressures (80 to 100 psi [560 kPa to 700 kPa]) as required. Solid stream nozzles used on aerial devices should be limited to a discharge pressure of 80 psi (560 kPa).

Determining the nozzle reaction from a given hose layout is certainly not something that a driver/operator would ever do on the fire scene. However, there are other instances in which these calculations can prove useful. The most common situation would be determining hose and nozzle configurations for preconnected attack lines that will be placed on the apparatus. The driver/operator can use the equations in the following sections to determine whether it is realistic for the given number of crew members to handle the hose lay that is being proposed.

Calculating Nozzle Reaction for Solid Stream Nozzles

In order to calculate the nozzle reaction from hoselines or appliances utilizing solid stream nozzles, use the following equation:

NR = $1.57 \times d^2 \times$ NP

Where: NR = Nozzle reaction in pounds

1.5 = A constant

d = Nozzle diameter in inches

NP = Nozzle pressure in psi

NR = $0.0015 \times d^2 \times$ NP

Where: NR = Nozzle reaction in newtons

0.0015 = A constant

d = Nozzle diameter in mm

NP = Nozzle pressure in kPa

The following example illustrates the application of this formula:

Example: Determine the nozzle reaction from a hoseline equipped with a 1¼-inch tip operating at a nozzle pressure of 50 psi.

$$NR = (1.57)(d)^2(NP)$$
$$NR = (1.57)(1.25)^2(50)$$
$$NR = (1.57)(1.56)(50)$$
$$NR = 122.5 \text{ pounds}$$

Example: Determine the nozzle reaction from a hoseline equipped with a 29 mm tip operating at 350 kPa.

$$NR = (0.0015)(d)^2(NP)$$
$$NR = (0.0015)(29)^2(350)$$
$$NR = (0.0015)(841)(350)$$
$$NR = 441.5 \text{ N}$$

When using the U.S. system of measurement, the rule-of-thumb formula $NR = Q/3$ may be used to approximate solid stream nozzle reaction on the fireground. In this case, Q equals the total flow of water in gpm through the nozzle.

Calculating Nozzle Reaction for Fog Stream Nozzles

In order to calculate the nozzle reaction from hoselines or appliances utilizing fog stream nozzles, use the following equation:

NR = $0.0505 \times Q \times \sqrt{NP}$

Where: NR = Nozzle reaction in pounds

0.0505 = A constant

Q = Total flow through the nozzle in gpm

NP = Nozzle pressure in psi

NR = $0.0156 \times Q \times \sqrt{NP}$

Where: NR = Nozzle reaction in newtons

0.0156 = A constant

Q = Total flow through the nozzle in L/min

NP = Nozzle pressure in kPa

(**NOTE:** The value of Q in the previous equation represents the total flow of water through the nozzle. This is not to be confused with the value of Q (Q = Flow/100) that will be used in the friction loss calculations in Chapter 8.)

The following example illustrates the application of this formula:

Example: Find the nozzle reaction of a hoseline with a fog nozzle flowing 200 gpm at 100 psi.

$$NR = (0.0505)(Q)(\sqrt{NP})$$
$$NR = (0.0505)(200)(\sqrt{100})$$
$$NR = (0.0505)(200)(10)$$
$$NR = 101 \text{ pounds}$$

Example: Find the nozzle reaction of a hoseline with a fog nozzle flowing 800 L/min at 700 kPa.

$$NR = (0.0156)(Q)(\sqrt{NP})$$
$$NR = (0.0156)(800)(\sqrt{700})$$
$$NR = (0.0156)(800)(26.44)$$
$$NR = 330 \text{ N}$$

When using the U.S. system of measurement, the rule-of-thumb formula $NR = Q/2$ may be used to approximate fog nozzle reaction on the fireground. In this case, Q equals the total flow of water in gpm through the nozzle.

Theoretical Pressure Calculations

Job Performance Requirements

This chapter provides information that will assist the reader in meeting the following job performance requirements from NFPA 1002, *Standard on Fire Apparatus Driver/Operator Professional Qualifications*, 1998 edition.

3-2.1* Produce effective hand or master streams, given the sources specified in the following list, so that the pump is safely engaged, all pressure control and vehicle safety devices are set, **the rated flow of the nozzle is achieved and maintained**, and the apparatus is continuously monitored for potential problems.

- Internal tank
- Pressurized source
- Static source
- Transfer from internal tank to external source

(a) *Requisite Knowledge:* **Hydraulic calculations for friction loss and flow using both written formulas and estimation methods**, safe operation of the pump, problems related to small-diameter or dead-end mains, low-pressure and private water supply systems, hydrant cooling systems, and reliability of static sources.

(b) *Requisite Skills:* The ability to position a fire department pumper to operate at a fire hydrant and at a static water source, power transfer from vehicle engine to pump, draft, operate pumper pressure control systems, operate the volume/pressure transfer valve (multistage pumps only), operate auxiliary cooling systems, make the transition between internal and external water sources, and assemble hose lines, nozzles, valves, and appliances.

3-2.2 Pump a supply line of 2½ in. (65 mm) or larger, given a relay pumping evolution the length and size of the line and the desired flow and intake pressure, so that the **proper pressure and flow are provided to the next pumper in the relay**.

(a) *Requisite Knowledge:* **Hydraulic calculations for friction loss and flow using both written formulas and estimation methods**, safe operation of the pump, problems related to small-diameter or dead-end mains, low-

pressure and private water supply systems, hydrant cooling systems, and reliability of static sources.

(b) *Requisite Skills:* The ability to position a fire department pumper to operate at a fire hydrant and at a static water source, power transfer from vehicle engine to pump, draft, operate pumper pressure control systems, operate the volume/pressure transfer valve (multistage pumps only), operate auxiliary cooling systems, make the transition between internal and external water sources, and assemble hose lines, nozzles, valves, and appliances.

3-2.4 Supply water to fire sprinkler and standpipe systems, given specific system information and a fire department pumper, so that water is supplied to the system at the proper volume and pressure.

(a) *Requisite Knowledge*: **Calculation of pump discharge pressure**; hose layouts; location of fire department connection; alternative supply procedures if fire department connection is not usable; operating principles of sprinkler systems as defined in NFPA 13, *Standard for the Installation of Sprinkler Systems*, NFPA 13D, *Standard for the Installation of Sprinkler Systems in One- and Two-Family Dwellings and Manufactured Homes*, and NFPA 13R, *Standard for the Installation of Sprinkler Systems in Residential Occupancies Up To and Including Four Stories in Height*; fire department operations in sprinklered occupancies as defined in NFPA 13E, *Guide for Fire Department Operations in Properties Protected by Sprinkler and Standpipe Systems*; and operating principles of standpipe systems as defined in NFPA 14, *Standard for the Installation of Standpipe and Hose Systems*.

6-2.1* Produce effective fire streams, utilizing the sources specified in the following list, so that the pump is safely engaged, all pressure-control and vehicle safety devices are set, the rated flow of the nozzle is achieved, and the apparatus is continuously monitored for potential problems.

- Water tank
- Pressurized source
- Static source

(a) *Requisite Knowledge:* **Hydraulic calculations for friction loss and flow using both written formulas and estimation methods**, safe operation of the pump, proper apparatus placement, personal safety considerations, problems related to small-diameter or dead-end mains and low-pressure and private water supply systems, hydrant cooling systems, and reliability of static sources.

(b) *Requisite Skills:* The ability to position a wildland fire apparatus to operate at a fire hydrant and at a static water source, properly place apparatus for fire attack, transfer power from vehicle engine to pump, draft, operate pumper pressure control systems, operate the volume/pressure transfer valve (multistage pumps only), operate auxiliary cooling systems, make the transition between internal and external water sources, and assemble hose lines, nozzles, valves, and appliances.

6-2.2 Pump a supply line, given a relay pumping evolution the length and size of the line and pumping flow and desired intake pressure, so that **adequate intake pressures and flow are provided to the next pumper in the relay**.

(a) *Requisite Knowledge:* **Hydraulic calculations for friction loss and flow using both written formulas and estimation methods**, safe operation of the pump, problems related to small-diameter or dead-end main and to low-pressure and private water supply systems, hydrant cooling systems, and reliability of static sources.

(b) *Requisite Skills:* The ability to position a wildland apparatus to operate at a fire hydrant and at a static water source, transfer power from vehicle engine to pump, draft, operate pumper pressure control systems, operate the volume/pressure transfer valve (multistage pumps only), operate auxiliary cooling systems, make the transition between internal and external water sources, and assemble hose lines, nozzles, valves, and appliances.

The prime objective of the driver/operator is to provide fire suppression crews with the water flow and pressure needed to achieve efficient fire control and/or extinguishment. To meet this objective, driver/operators must understand the theoretical aspects of fire stream development. They must then be able to convert these theories into practice during fireground operations in order to produce effective fire streams.

In this chapter, the driver/operator learns the theoretical methods of calculating pressure loss in a variety of hose lays and fireground situations. He learns to consider the effects of friction within the hose assembly, as well as the effect that changes in elevation have on supplying hoselines. Lastly, the driver/operator learns to use the figures derived from the pressure loss calculations to determine the pump discharge pressure that is required to adequately supply fire streams.

Driver/operators rarely, if ever, perform the calculations contained in this chapter while on the fireground. On the fireground, they are more likely to use the methods described in Chapter 9. However, it is important that the driver/operator be able to calculate theoretical friction loss for a variety of other reasons, including the following:

• It gives the driver/operator a better understanding of the basis for the fireground methods described in Chapter 9.

• It allows the driver/operator to predetermine accurate pump discharge pressures for preconnected hoselines and common hose lays used on the apparatus he is responsible for operating (Figure 8.1).

• It serves as a tool for pre-incident planning at properties that require hose deployment that is out of the ordinary for the fire department.

Throughout the following sections, there are a number of example problems designed to familiarize the driver/operator with the concepts used in performing hydraulic calculations. It is not the intent of any of the

Figure 8.1 Preconnected handlines are the most commonly used fire service hose lays. *Courtesy of Ron Jeffers.*

following example calculations to limit the driver/operator to one method of solution. The example problems serve as a guide for those unfamiliar with the calculations. The examples show the proper way to solve a particular problem. They are worked out step by step to show the reader the entire problem-solving process. The exercises enable the reader to practice solving problems using the format presented in the examples.

NOTE: In order to avoid the clutter and confusion of trying to switch back and forth between the U.S. and metric methods of performing these calculations, this chapter appears in two versions. This version contains the U.S. system of measurements. Immediately following this version of Chapter 8, you will find a second version of Chapter 8 in metrics.

Before getting into the main body of text of this chapter, it is important to review the common sizes of fire hose and solid stream nozzle tip sizes used in today's fire service. These figures are used throughout the remainder of this manual. For comparison purposes, both the U.S. and metric measurements are contained in Tables 8.1 and 8.2. All metrics contained in IFSTA publications follow a document titled, *"Training Guidelines for the Metric Conversion of Fire Departments in Canada."* The National Fire Protection Association does not follow these same guidelines. Thus, you will note some differences in what a particular size hose is called in metric terms in Table 8.1. Jurisdictions that use the NFPA hose sizes need to refer to this table when using the rest of this manual.

Table 8.2 Common Nozzle Diameters		
U.S. (Inches)	Metric (mm) (Actual)	Metric (mm) (Known As)
½	12.700	13
⅝	15.880	16
¾	19.050	19
⅞	22.225	22
1	25.400	25
1⅛	28.575	29
1¼	31.750	32
1⅜	34.925	35
1½	38.100	38
1¾	44.450	45
2	50.800	50
2¼	57.150	57
2½	63.500	65
2¾	69.850	70
3	76.200	77

Total Pressure Loss: Friction Loss and Elevation Pressure Loss

To produce effective fire streams, it is necessary to know the amount of friction loss in the fire hose and any pressure loss or gain due to elevation. As was discussed in Chapter 6, friction loss can be caused by a number of factors: hose conditions, coupling conditions, or kinks. The primary determinant, however, is the volume of water flowing per minute.

Calculation of friction loss must also take into account the length and diameter of the hoseline and any major hose appliances attached to the line. Because the amount of hose used between an engine and the nozzle is not always the same, the driver/operator must be capable of determining friction loss in a given length of hose. Elevation differences, such as hills, gullies, aerial devices, or multistoried buildings, create a pressure loss or gain known as *elevation pressure*. The development of elevation pressure occurs anytime there is a change in elevation between the nozzle and the pump (Figure 8.2).

Combined, friction loss and elevation pressure loss are referred to as *total pressure loss" (TPL)*. This is not to be confused with total pump discharge pressure, which also includes nozzle pressure. Total pump discharge pres-

Table 8.1 Common Hose Sizes		
U.S. (Inches)	Canada (mm)	NFPA (mm)
¾	20	20
1	25	25
1½	38	38
1¾	45	44
2	50	51
2½	65	65
3	77	76
3½	90	89
4	100	100
4½	115	113
5	125	125
6	150	150

Figure 8.2 One example of an elevation change is the use of an elevated master stream.

sure is discussed later in this chapter. Both friction loss and elevation pressure are expressed in pounds per square inch (psi).

Determining Friction Loss

There are two methods used to determine friction loss: actual tests and calculations. The most accurate of these methods determines friction loss through true field tests. Field tests involve using in-line gauges to measure friction loss at various flows through an actual hose layout. The calculation method involves the use of mathematical friction loss equations or field application methods. This chapter focuses on the use of mathematical equations to determine friction loss. Field applications are covered in Chapter 9.

It is important to emphasize that the only truly accurate method for determining pressure loss in any particular hose lay is by measuring the pressure at both ends of the hose and subtracting the difference; however, this is generally not practical in real-life situations. The calculation of pressure loss through the use of formulas or field applications is, at best, an inexact science. The figures derived from these calculation methods most likely will be different than if actual tests were performed. However, the figures that are obtained through the calculation method usually are close enough to the actual conditions that they can be relied upon to be safe for fireground operations.

In the past, the fire service used the formula $2Q^2 + Q$ as the basis for mathematical friction loss calculations. This formula was based on the average friction loss found in 2½-inch hose that was constructed in the 1930s (Figure 8.3). Conversion factors were needed to find friction loss in other sizes of hose. Furthermore, the formula did not take into consideration the length of the hoseline, and it was placed in a separate step.

Improvement in the technology of fire hose construction in the past 30 years or so made the old friction loss formula obsolete (Figure 8.4). The figures derived by that formula provided unrealistically high friction loss figures for today's hose. Thus, it was clear that a new friction loss formula needed to be developed. The new formula needed to take into account the size of the fire hose, the quantity of water flowing, and the length of the hose lay. These three factors gave rise to the following formula for computing friction loss:

EQUATION A
$FL = CQ^2L$

Where:

FL = Friction loss in psi

C = Friction loss coefficient (from Table 8.3)

Q = Flow rate in hundreds of gpm (flow/100)

L = Hose length in hundreds of feet (length/100)

Table 8.3 contains the commonly used friction loss coefficients for the various sizes of hose in use today. These coefficients are used by IFSTA and the NFPA. They are only approximations of the friction loss for each size hose. The actual coefficients for any particular piece of hose varies with the condition of the hose and the manufacturer. When using the coefficients provided in this manual, the results reflect a worst-case situation. In other words, the results are probably slightly higher than the actual friction loss. For departments that require more exact friction loss calculations, different coefficients should be obtained from the manufacturer of their fire hose or through actual calculation, which is covered in the next section of this chapter.

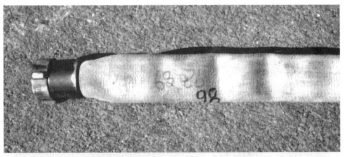

Figure 8.3 Older style hose had a woven outer jacket.

Figure 8.4 Most newer fire hose has a rubber or similar outer jacket.

Table 8.3
Friction Loss Coefficient — Single Hoselines

Hose Diameter (Inches)	Coefficient
¾ (booster)	1,100
1 (booster)	150
1¼ (booster)	80
1½	24
1¾ with 1½-inch couplings	15.5
2	8
2½	2
3 with 2½-inch couplings	0.8
3 with 3-inch couplings	0.677
3½	0.34
4	0.2
4½	0.1
5	0.08
6	0.05
Standpipes	
4	0.374
5	0.126
6	0.052

The steps for determining friction loss using Equation A are as follows:

Step 1: Obtain from Table 8.3 the friction loss coefficient for the hose being used.

Step 2: Determine the number of hundreds of gallons of water per minute flowing (Q) through the hose by using the equation Q = gpm/100.

Step 3: Determine the number of hundreds of feet of hose (L) by using the equation L = feet/100.

Step 4: Plug the numbers from Steps 1, 2, and 3 into Equation A to determine the total friction loss.

Example 1

If 300 gpm is flowing from a nozzle, what is the total pressure loss due to friction for 400 feet of 2½-inch hose?

C = 2 from Table 8.3

$$Q = \frac{gpm}{100} \qquad Q = \frac{300}{100} \qquad Q = 3$$

$$L = \frac{hose\ length}{100} \qquad L = \frac{400}{100} \qquad L = 4$$

$FL = CQ^2L \qquad FL = (2)(3)^2(4) \qquad FL = (2)(9)(4)$

FL = 72 psi total friction loss

Example 2

What is the total pressure loss due to frictic[on in] 4-inch hose when 750 gpm is flowing?

C = 0.2 from Table 8.3

$$Q = \frac{gpm}{100} \qquad Q = \frac{750}{100} \qquad Q = 7.5$$

$$L = \frac{hose\ length}{100} \qquad L = \frac{600}{100} \qquad L = 6$$

$FL = CQ^2L \qquad FL = (0.2)(7.5)^2(6) \qquad FL = (0.2)(56.25)(6)$

FL = 67.5 psi total friction loss

Example 3

What is the total pressure loss due to friction in 250 feet of 1¾-inch hose when 150 gpm is flowing?

C = 15.5 from Table 8.3

$$Q = \frac{gpm}{100} \qquad Q = \frac{150}{100} \qquad Q = 1.5$$

$$L = \frac{hose\ length}{100} \qquad L = \frac{250}{100} \qquad L = 2.5$$

$FL = CQ^2L \quad FL = (15.5)(1.5)^2(2.5) \quad FL = (15.5)(2.25)(2.5)$

FL = 87.2 psi total friction loss

Determining Your Own Friction Loss Coefficients

If you wish to calculate more accurate results for the fire hose that is carried on your apparatus, rather than use the results from the standard friction loss coefficients, it is recommended that you test your hose to determine the actual coefficients. If a department has the equipment available, the test procedure for determining friction loss for a particular hose is rather simple. Before performing these tests, however, several basic principles must be considered. In order to get results indicative of averages that can be expected on the fireground, it is necessary to use the same hose that would be used on the fireground. Conduct tests on hose that is in service not hose that is sitting on a storage rack or hose that has never been put into service (unless new hose is about to be placed into service).

When conducting tests on a particular type of hose, test only one type at a time. For example, if the department is testing 3-inch cotton/polyester, double-jacket hose, do not have any hose of different size or construction mixed in with test hose. If the department regularly mixes different kinds of hose, it may be very difficult to get a friction loss coefficient suitable for all situations. Different hose have varying amounts of friction loss due to differences in construction, fabrics, rubber liners, couplings, and wear.

Test results are only as accurate as the equipment used to measure the results. Thus, it is important that all measuring devices (pitot tubes, in-line gauges, flowmeters, and so on) be properly calibrated for optimum results. The following is a list of the equipment needed to conduct these tests:

- Pitot tube or flowmeter (Figures 8.5 a and b)
- Two in-line gauges, preferably calibrated in increments of 5 psi or less
- Hose to be tested
- Smoothbore nozzle if using pitot tube
- Any type nozzle if using flowmeter

The following is a step-by-step procedure for determining friction loss in any size hose:

Step 1: Lay out, on a level surface, the lengths of hose to be tested. Lay out 300 feet if the hose is in lengths of 50 feet or 400 feet if it is in lengths of 100 feet.

Step 2: Connect one end of the hoseline to a discharge on the pumper being used to carry out the tests. Connect a nozzle to the opposite end of the hoseline. If using a pitot tube to determine the nozzle pressure and corresponding flow of water, use a smoothbore nozzle. If a flowmeter is used to determine the flow, any nozzle is suitable (Figure 8.6).

Step 3: Insert gauge 1 in the hoseline at the connection between the first and second sections of hose away from the discharge. Make this connection 50 feet from the pumper if the hose being tested is in lengths of 50 feet. If 100-foot sections of hose are used, then gauge 1 should be 100 feet from the pumper.

Step 4: Insert gauge 2 at a distance of 200 feet from gauge 1, regardless of the length of the hose sections. Depending on the length of the hose sections, there should be either 50 or 100 feet between gauge 2 and the nozzle. If a portable flowmeter is being used, *DO NOT* insert it any-

Figure 8.5a A pitot tube and gauge are used to measure the velocity pressure of a stream of water.

Figure 8.5b Flowmeters provide a direct reading of the volume of water being discharged through an opening.

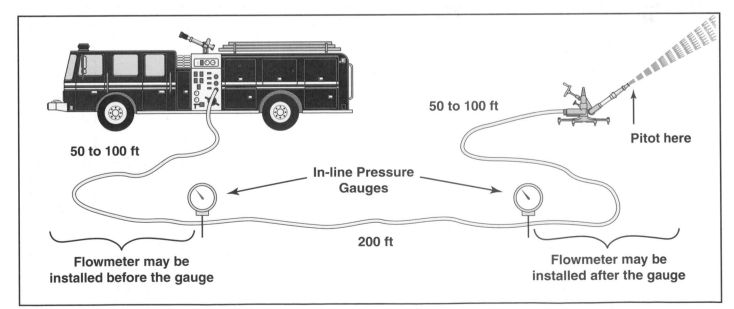

Figure 8.6 This layout can be used to determine the friction loss coefficient for any fire hose.

where between the gauges. It may be inserted in the hoseline anywhere but between the test gauges.

Step 5: When all appliances have been inserted into the hoseline, begin the tests. Supply water to the hoseline at a constant pump discharge pressure for the duration of each test run. Because three or four test runs should be made for each size hose, different pump discharge pressures may be used for different runs as long as they are not changed within the same test run. Use sufficient pump discharge pressure to produce a satisfactory fire stream from the nozzle.

Step 6: Once water is flowing, record the pump discharge pressure, the readings from gauge 1 and gauge 2, and the flowmeter or pitot gauge readings (whichever is being used) in the appropriate spaces of Figure 8.7.

Figure 8.7
Friction Loss Coefficient Determination Chart

Date: __ / __ / ____   Hose Size _____ Inches   Hose Construction _____

Person Conducting Tests _____

Column 1	Column 2	Column 3	Column 4	Column 5	Column 6	Column 7	Column 8	Column 9
Test Run No.	Pump Discharge Pressure psi	Pressure @ Gauge 1 psi	Pressure @ Gauge 2 psi	Nozzle Pressure* psi	Flow From Flowmeter or by Equation**	$\left(\dfrac{GPM}{100}\right)^2$ or $\left(\dfrac{Col.\ 6}{100}\right)^2$	Friction Loss per 100 feet or $\dfrac{(Col.\ 3-Col.\ 4)}{2}$	C $\left(\dfrac{Col.\ 8}{Col.\ 7}\right)$
1								
2								
3								
4								

Total of all Col.9 Answers

Average C = ⎯⎯⎯⎯⎯

No. of Tests Conducted

Average C =

* Not necessary if flowmeter is used
**GPM = 29.7 d² √NP

Figure 8.7 This chart may be used to determine the friction loss coefficient for fire hose.

Step 7: Complete Figure 8.7 as instructed on the table. This provides you with the friction loss coefficient for your particular hose.

Figure 8.8 represents the test results from a fire department that chose to determine its own coefficient for the 2½-inch hose used within the department. The hose was tested at three different pump discharge pressures. By completing Figure 8.8, the department determined that the actual coefficient for its 2½-inch hose was 1.6. This represents a 20 percent decrease from the standard coefficient for 2½-inch hose used in this manual (C=2).

Appliance Pressure Loss

Fireground operations often require the use of hoseline appliances. These appliances include reducers, increasers, gates, wyes, manifolds, aerial apparatus, and

Figure 8.8
Friction Loss Coefficient Determination Chart

Date: ___/___/___   Hose Size _____ Inches   Hose Construction _____

Person Conducting Tests _____

Column 1 Test Run No.	Column 2 Pump Discharge Pressure psi	Column 3 Pressure @ Gauge 1 psi	Column 4 Pressure @ Gauge 2 psi	Column 5 Nozzle Pressure* psi	Column 6 Flow From Flowmeter or by Equation**	Column 7 $\left(\frac{GPM}{100}\right)^2$ or $\left(\frac{Col.\ 6}{100}\right)^2$	Column 8 Friction Loss per 100 feet or $\left(\frac{Col.\ 3 - Col.\ 4}{0.2}\right)$	Column 9 C $\left(\frac{Col.\ 8}{Col.\ 7}\right)$
1	130	122	107	—	220	$\left(\frac{220}{100}\right)^2=$ 4.84	7.5	1.56
2	150	143	123	—	250	$\left(\frac{250}{100}\right)^2=$ 6.25	10	1.6
3	170	157	131	—	280	$\left(\frac{280}{100}\right)^2=$ 7.84	13	1.64
4	—	—	—	—	—	—	—	—

Average C = $\dfrac{4.8}{3}$

Total of all Col.9 Answers

No. of Tests Conducted

Average C = 1.6

* Not necessary if flowmeter is used
**GPM = 29.7 d² √NP

Figure 8.8 The results of each test are recorded on the chart.

standpipe systems (Figures 8.9 a through d). The friction loss created varies with each type of appliance. Appliance friction loss is insignificant in cases where the total flow through the appliance is less than 350 gpm, and for the purpose of this manual, it is not included in calculations of flows less than that. Friction loss in these appliances varies with the rated capacity of the device and the flow. Generally, it is safe to assume a loss of 25 psi or greater when flowing at the rated capacity. For siameses and wyes, the friction loss varies with the size of the device and the flow. **For this text, we will assume a 0 psi loss for flows less than 350 gpm and a 10 psi loss for *each appliance* (other than master stream devices) in a hose assembly when flowing 350 gpm or more.** Friction loss caused by handline nozzles are not considered in the calculations in this manual, as it is generally insignificant in the overall pressure loss in a hose assembly. **For this manual, we will assume a friction loss of 25 psi in all master stream appliances, regardless of the flow.**

As with fire hose, the only sure way to determine the exact friction loss of each appliance is for individual fire departments to conduct their own friction loss tests. A method of determining this friction loss is found in Appendix B.

Figure 8.9c A siamese.

Figure 8.9a A reducer.

Figure 8.9b A gated wye.

Figure 8.9d Aerial devices create friction loss.

Determining Elevation Pressure

Fireground operations often require the use of hoselines at varying elevations. Elevation pressure, which is created by elevation differences between the nozzle and the pump, must be considered when determining total pressure loss.

Water exerts a pressure of 0.434 psi per foot of elevation. When a nozzle is operating at an elevation higher than the apparatus, this pressure is exerted back against the pump (Figure 8.10). To compensate for this pressure "loss," elevation pressure must be added to friction loss to determine total pressure loss. Operating a nozzle lower than the pump results in pressure pushing against the nozzle (Figure 8.11). This "gain" in pressure is compensated for by subtracting the elevation pressure from the total friction loss.

In order to simplify elevation pressure calculations on the fireground, the following formula may be used:

EQUATION B
EP = 0.5H

Where:

EP = Elevation pressure in psi

0.5 = A constant

H = Height in feet

It is generally easier to determine elevation pressure in a multistoried building by another method. By counting the number of stories of elevation, the following equation may be used.

EQUATION C
EP = 5 psi x (number of stories -1)

The determination of elevation pressure is shown in the following examples.

Example 4

Calculate the total pressure loss due to elevation pressure for a hoseline operating at the top of a 100-foot hill (Figure 8.12).

EP = 0.5H

EP = (0.5)(100)

EP = 50 psi

Example 5

A hoseline operating on a ninth-floor structure fire is connected to the building's standpipe system (Figure 8.13). What is the total pressure loss due to elevation at the base of the standpipe system?

EP = 5 psi x (number of stories −1)

EP = 5 psi x (9 −1)

EP = (5)(8)

EP = 40 psi

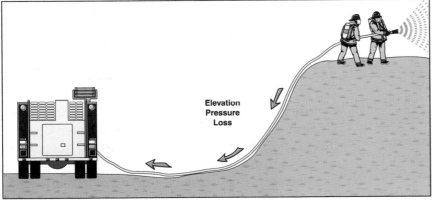

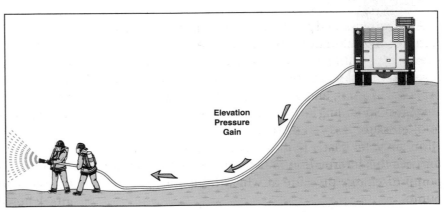

Figure 8.10 In this scenario, an elevation pressure loss would be noted at the pump.

Figure 8.11 In this scenario, an elevation pressure gain would be noted at the pump.

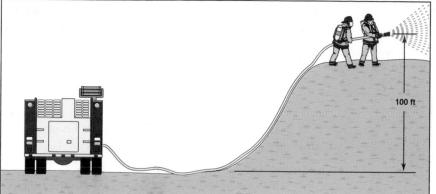

Figure 8.12 Example 4.

100 ft

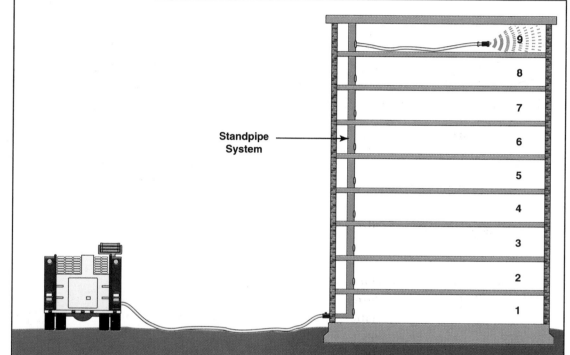

Standpipe
System

Figure 8.13 Example 5.

Hose Layout Applications

Hose layouts include single hoselines, multiple hoselines, wyed or manifold hoselines, and siamese hoselines. In each type of layout, friction loss is affected by such factors as hose diameter and whether the hose layouts are of equal or unequal length.

As was mentioned previously, the combination of friction loss and elevation pressure is referred to as total pressure loss (TPL). Pressure changes are possible due to hose friction loss, appliance friction loss (when flows exceed 350 gpm), and any pressure loss or gain due to elevation. By adding all the affecting pressure losses, the total pressure loss can be determined for any hose lay. Ultimately, (and later in this chapter) the driver/operator uses this information to determine the appropriate pump discharge pressure at which to operate the fire pump.

Hose layouts can be divided into two basic categories: simple hose layouts and complex hose layouts. The principles used in determining the total pressure loss in each category are essentially the same. The complexity of the friction loss calculation determines whether the layout is simple or complicated.

Simple Hose Layouts

Simple hose layouts include single hoselines, equal length multiple hoselines, equal wyed hoselines, and equal length siamesed hoselines. These more frequently used hose layouts generally do not present great difficulty in determining total pressure loss.

Single Hoseline

The most commonly used hose lay is the single-hoseline layout (Figure 8.14). This hose lay, whether used as an attack line or supply line, presents the simplest friction

Figure 8.14 Smaller incidents are handled with a single attack line. *Courtesy of Ron Jeffers.*

loss calculations. The following examples demonstrate how to determine the total pressure loss in a single-hoseline layout.

Example 6

A pumper is supplying a 300-foot hoseline with 125 gpm flowing. The hoseline is composed of 200 feet of 2½-inch hose reduced to 100 feet of 1½-inch hose (Figure 8.15). What is the pressure loss due to friction in the hose assembly?

2½-inch Hose

C = 2 from Table 8.3

$Q = \dfrac{gpm}{100}$ $Q = \dfrac{125}{100}$ $Q = 1.25$

$L = \dfrac{feet}{100}$ $L = \dfrac{200}{100}$ $L = 2$

$FL = CQ^2L$ $FL = (2)(1.25)^2(2)$ $FL = (2)(1.5625)(2)$

FL = 6.25 psi

1½-inch Hose

C = 24 from Table 8.3

$Q = \dfrac{gpm}{100}$ $Q = \dfrac{125}{100}$ $Q = 1.25$

$L = \dfrac{feet}{100}$ $L = \dfrac{100}{100}$ $L = 1$

$FL = CQ^2L$ $FL = (24)(1.25)^2(1)$ $FL = (24)(1.5625)(1)$

FL = 37.5 psi

Total Pressure Loss
TPL = 37.5 + 6.25 = **43.75 psi pressure loss in the hose assembly**

Example 7

A fire is discovered on the third floor of a structure. An arriving engine company proceeds to the second floor of the structure and connects 150 feet of 1¾-inch hose to the standpipe outlet. Determine the total pressure loss due to friction and elevation pressure at the standpipe system fire department connection when 175 gpm is flowing. (Disregard friction loss in the standpipe system.)

1¾-inch Hose

C = 15.5 from Table 8.3

$Q = \dfrac{gpm}{100}$ $Q = \dfrac{175}{100}$ $Q = 1.75$

$L = \dfrac{feet}{100}$ $L = \dfrac{150}{100}$ $L = 1.5$

$FL = CQ^2L$ $FL = (15.5)(1.75)^2(1.5)$
$FL = (15.5)(3.0625)(1.5)$

FL =71 psi

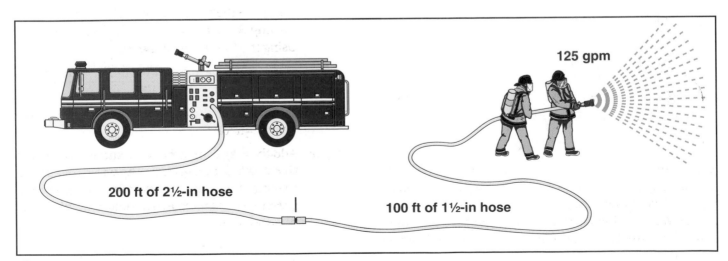

125 gpm

200 ft of 2½-in hose

100 ft of 1½-in hose

Figure 8.15 Example 6.

Elevation Pressure

$\quad$ EP = 5 psi x (number of stories − 1)

$\quad$ EP = 5 psi x (3 − 1)

$\quad$ EP = (5)(2)

$\quad$ EP = 10 psi

Total Pressure Loss

$\quad$ TPL = 71 + 10 = **81 psi pressure loss in the hose assembly**

Multiple Hoselines (Equal Length)

Fireground operations may involve the use of more than one hoseline from a pumper (Figure 8.16). These hoselines, whether of equal or varying diameters, are often the same length.

When determining the friction loss in equal length multiple lines whose diameters are the same, it is only necessary to perform calculations for one line. This is because each of the other hoselines will have approximately the same friction loss. Conversely, when the diameters of the hoselines vary, friction loss calculations must be made for each hoseline. The pump discharge pressure is then set for the highest pressure. The valve on the hose requiring the lesser amount of pressure is partially closed to reduce the pressure from that discharge. The following example shows how to determine total pressure loss when multiple hoselines are used.

Example 8

Three 2½-inch hoselines, each 400 feet long, are laid with 1-inch tips. Each line has a nozzle pressure of 50 psi. What is the total pressure loss due to friction in each hoseline?

$\quad$ GPM = 29.7 $d^2\sqrt{NP}$

$\quad$ GPM = (29.7) (1)^2($\sqrt{50}$)

$\quad$ GPM = 210 gpm

$\quad$ C = 2 from Table 8.3

$\quad Q = \dfrac{gpm}{100} \qquad Q = \dfrac{210}{100} \qquad Q = 2.1$

$\quad L = \dfrac{feet}{100} \qquad L = \dfrac{400}{100} \qquad L = 4$

$\quad$ FL = CQ^{2}L $\qquad$ FL = (2)(2.1)^2(4) $\qquad$ FL = (2)(4.41)(4)

$\quad$ **FL = 35.3 psi in each hoseline**

Wyed Hoselines (Equal Length)

A common hose assembly involves the use of one supply line, usually a 2½-, 3-, or 4-inch hose, wyed into two or more smaller attack lines. These attack lines may range in size from 1½ inches to 2½ inches and are generally equal in length. It is important that these wyed lines be

Figure 8.16 The pump operator may be required to supply multiple hoselines at a large fire incident. *Courtesy of Ron Jeffers.*

the same length and diameter in order to avoid two different nozzle pressures and an exceptionally difficult friction loss problem.

Determining friction loss in wyed hoselines of equal length and diameter is not a complex problem. When the nozzle pressure, hose length, and diameter are the same on both lines, an equal split of the total water flowing occurs at the wye appliance. This enables only one of the wyed hoselines to be considered when computing the total pressure lost. Use the following steps when calculating the friction loss in an equal length wyed hoseline assembly.

Step 1: $\quad$ Compute the number of hundreds of gpm flowing in each wyed hoseline by using the equation:

$$Q = \frac{\text{flow rate (gpm)}}{100}$$

Step 2: $\quad$ Determine the friction loss in one of the wyed attack lines using the equation:

$$FL = CQ^2L$$

Step 3: $\quad$ Compute the total number of hundreds of gpm flowing through the supply line to the wye by using the following equation:

$$Q_{Total} = \frac{(\text{gpm in attack line 1}) + (\text{gpm in attack line 2})}{100}$$

Step 4: $\quad$ Determine the friction loss in the supply line using Equation A, FL = (C)(Q$_{Total}$)^2(L)

Step 5: $\quad$ Add the friction loss from the supply line, one of the attack lines, 10 psi for the wye appliance (if the total flow exceeds 350 gpm), and elevation pressure (if applicable) to determine the total pressure loss.

The following examples illustrate how to compute the total pressure loss in an equal length wyed hose assembly.

Example 9

Determine the pressure lost due to friction in a hose assembly in which two 1½-inch hoselines, each 100 feet long and flowing 95 gpm, are wyed off 400 feet of 2½-inch hose (Figure 8.17).

1½-inch Hose

C = 24 from Table 8.3

$$Q = \frac{gpm}{100} \qquad Q = \frac{95}{100} \qquad Q = 0.95$$

$$L = \frac{feet}{100} \qquad L = \frac{100}{100} \qquad L = 1$$

$FL = CQ^2L \qquad FL = (24)(0.95)^2(1) \qquad FL = (24)(0.9025)(1)$

FL = 21.7 psi in each 1½-inch hoseline

2½-inch Hose

C = 2 from Table 8.3

$$Q_{Total} = \frac{(gpm \text{ in attack line 1}) + (gpm \text{ in attack line 2})}{100}$$

$$Q_{Total} = \frac{95 + 95}{100}$$

$$Q_{Total} = \frac{190}{100}$$

$$Q_{Total} = 1.9$$

$$L = \frac{feet}{100} \qquad L = \frac{400}{100} \qquad L = 4$$

$FL = (C)(Q_{Total})^2L \qquad FL = (2)(1.9)^2(4) \qquad FL = (2)(3.61)(4)$

FL = 28.9 psi in the 2½-inch hoseline

Total Pressure Loss

TPL = 21.7 + 28.9 = **50.6 psi total pressure loss in this hose assembly**

NOTE: The total flow through this hose assembly was less than 350 gpm, so it was not necessary to consider the pressure loss in the wye.

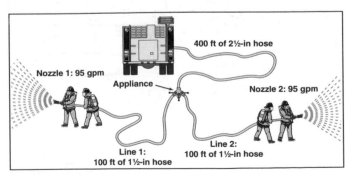

Figure 8.17 Example 9.

Example 10

Determine the pressure lost due to friction in a hose assembly in which two 2½-inch hoselines, each 200 feet long and flowing 250 gpm, are wyed off 500 feet of 4-inch hose.

2½-inch Hose

C = 2 from Table 8.3

$$Q = \frac{gpm}{100} \qquad Q = \frac{250}{100} \qquad Q = 2.5$$

$$L = \frac{feet}{100} \qquad L = \frac{200}{100} \qquad L = 2$$

$FL = CQ^2L \qquad FL = (2)(2.5)^2(2) \qquad FL = (2)(6.25)(2)$

FL = 25 psi in each 2½-inch hoseline

4-inch Hose

C = 0.2 from Table 8.3

$$Q_{Total} = \frac{(gpm \text{ in attack line 1}) + (gpm \text{ in attack line 2})}{100}$$

$$Q_{Total} = \frac{250 + 250}{100}$$

$$Q_{Total} = \frac{500}{100}$$

$$Q_{Total} = 5$$

$$L = \frac{feet}{100} \qquad L = \frac{500}{100} \qquad L = 5$$

$FL = (C)(Q_{Total})^2L \qquad FL = (0.2)(5)^2(5) \qquad FL = (0.2)(25)(5)$

FL = 25 psi in the 4-inch hoseline

Total Pressure Loss

TPL = 25 + 25 + 10 = **60 psi total pressure loss in this hose assembly**

NOTE: The total flow through this hose assembly exceeded 350 gpm, so it was necessary to add 10 psi for the pressure loss in the wye.

Siamesed Hoselines (Equal Length)

When the flow rate is increased through hose, additional pressure is needed to overcome friction in the hose. When conditions require great volumes of water, or when hose lays are long, the friction loss in the hose is also greater. To keep friction loss within reasonable limits, firefighters may lay two or more parallel hoselines and siamese them together at a point close to the fire. When two hoselines of equal lengths are siamesed to supply a fire stream, friction loss is approximately 25 percent of that of a single hoseline at the same nozzle pressure. When three hoselines of equal length are siamesed, the friction loss is approximately 10 percent of that of a single line if equal nozzle pressure is maintained.

There are a number of ways to determine the amount of friction loss in siamesed hoselines. The easiest method is to use Equation A, which we used for determining the amount of pressure lost due to friction in single hoselines. When calculating friction loss in siamesed lines, however, it is necessary to use a different set of coefficients (C) than for single hoselines. These coefficients are found in Table 8.4. The following is a step-by-step procedure for determining friction loss in siamesed lines:

Step 1: Compute the total number of hundreds of gpm flowing by using the equation:

$$Q = \frac{\text{gpm flowing}}{100}$$

Step 2: Determine the friction loss in the attack line using Equation A ($FL = CQ^2L$). Use Table 8.3 for the coefficient in this step.

Step 3: Determine the amount of friction loss in the siamesed lines using Equation A ($FL=CQ^2L$). Use Table 8.4 to obtain the coefficient in this step.

Step 4: Add the friction loss from the siamesed lines, attack line, 10 psi for the siamese appliance (if flow is greater than 350 gpm), and elevation pressure (if applicable) to determine the total pressure loss.

Table 8.4
Friction Loss Coefficients — Siamesed Lines of Equal Length

Number of Hoses and Their Diameter (inches)	Coefficient
Two 2½"	0.5
Three 2½"	0.22
Two 3" with 2½" couplings	0.2
One 3" with 2½" couplings, one 2½"	0.3
One 3" with 3" couplings, one 2½"	0.27
Two 2½", one 3" with 2½" couplings	0.16
Two 3" with 2½" couplings, one 2½"	0.12

Example 11

Determine the pressure lost due to friction in a hose assembly in which two 3-inch hoses with 2½-inch couplings, each 1,000 feet long, are used to supply a siamese to which 300 feet of 2½-inch hose is attached (Figure 8.18). The solid stream nozzle on the 2½-inch hose has 1¼-inch tip with a nozzle pressure of 50 psi.

GPM = 29.7 d²√NP

GPM = (29.7) (1.25)²(√50)

GPM = 328 gpm

2½-inch Attack Line

C = 2 from Table 8.3

$Q = \dfrac{\text{gpm}}{100}$ $Q = \dfrac{328}{100}$ Q = 3.28

$L = \dfrac{\text{feet}}{100}$ $L = \dfrac{300}{100}$ L = 3

FL = CQ²L   FL = (2)(3.28)²(3)   FL = (2)(10.76)(3)

FL = 65 psi loss in the 2½-inch hoseline

Siamesed Lines

C = 0.2 from Table 8.4

$Q = \dfrac{\text{gpm}}{100}$ $Q = \dfrac{328}{100}$ Q = 3.28

$L = \dfrac{\text{feet}}{100}$ $L = \dfrac{1000}{100}$ L = 10

FL = CQ²L   FL = (0.2)(3.28)²(10)   FL = (0.2)(10.76)(10)

FL = 21.5 psi in the siamesed lines

Total Pressure Loss

TPL = 65 + 21.5 = **86.5 psi total pressure loss in this hose assembly**

NOTE: The total flow through this hose assembly was less than 350 gpm, so it was not necessary to consider the pressure loss in the siamese.

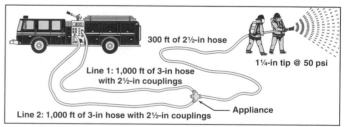

300 ft of 2½-in hose

1¼-in tip @ 50 psi

Line 1: 1,000 ft of 3-in hose with 2½-in couplings

Line 2: 1,000 ft of 3-in hose with 2½-in couplings

Appliance

Figure 8.18 Example 11.

Example 12

Determine the total pressure loss in a hose assembly in which two 2½-inch hoses, each 750 feet long, are being used to supply a siamese to which 200 feet of 2½-inch hose is attached. The nozzle on the attack line is flowing 300 gpm. The nozzle is located 30 feet uphill of the siamese.

2½-inch Attack Line

C = 2 from Table 8.3

$Q = \dfrac{gpm}{100}$ $Q = \dfrac{300}{100}$ $Q = 3$

$L = \dfrac{feet}{100}$ $L = \dfrac{200}{100}$ $L = 2$

$FL = CQ^2L$ $FL = (2)(3)^2(2)$ $FL = (2)(9)(2)$ $FL = 36$ psi

$EP = 0.5H$

$EP = (0.5)(30)$

$EP = 15$ psi

$TPL_{Attack\ Line} = FL + EP$ $TPL_{Attack\ Line} = 36 + 15$

$TPL_{Attack\ Line} = 51$ psi

Siamesed Lines

C = 0.5 from Table 8.4

$Q = \dfrac{gpm}{100}$ $Q = \dfrac{300}{100}$ $Q = 3$

$L = \dfrac{feet}{100}$ $L = \dfrac{750}{100}$ $L = 7.5$

$FL = CQ^2L$ $FL = (0.5)(3)^2(7.5)$ $FL = (0.5)(9)(7.5)$

FL = 33.8 psi in the siamesed lines

Total Pressure Loss

TPL = 51 + 33.8 = **84.8 psi total pressure loss in this hose assembly**

Complex Hose Layouts

Fireground operations may involve the use of hose layouts that challenge the mathematical ability of the driver/operator. These hose lays, which include standpipe operations, unequal length multiple and wyed hoselines, manifold hoselines, and master streams, require the driver/operator to make additional calculations to determine the total pressure loss.

Standpipe Operations

In most cases, fire departments have predetermined pressures that the driver/operator is expected to pump into the fire department connection (FDC) of a standpipe system. These pressures are contained in the department's SOPs, in the pre-incident plan for that particular property, or on a faceplate adjacent to the FDC

(Figures 8.19 a through c). In order to be able to determine the required pressure for the standpipe system, it is necessary to determine the total pressure loss. The following example shows how to accomplish this. Treat the FDC like any other hose appliance. If the flow in the system exceeds 350 gpm, add 10 psi of friction loss for the FDC.

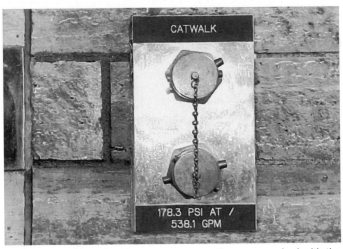

Figure 8.19a Some fire department connections are marked with the required supply pressure on them. *Courtesy of the City of Phoenix Department of Development Services.*

Figure 8.19b In some cases, the pumping instructions are located on a sign adjacent to a fire department connection. *Courtesy of Mount Prospect (IL) Fire Department.*

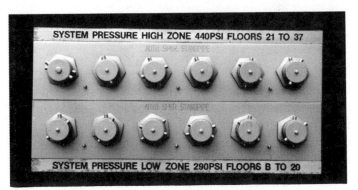

Figure 8.19c In this example, each zone of the system has a different required FDC inlet pressure. *Courtesy of Bil Murphy.*

Example 13

A fire is discovered on the fifth floor of a structure. The arriving engine company proceeds to the fourth floor of the occupancy and connects three 50-foot sections of 2½-inch hose to the standpipe outlet. The hose is equipped with a solid stream nozzle that has a 1¼-inch tip. The standpipe is 6 inches in diameter. What is the total pressure loss due to friction and elevation pressure in the standpipe system and attached hose?

$$GPM = 29.7 \ d^2 \sqrt{NP}$$

$$GPM = (29.7) \ (1.25)^2 (\sqrt{50})$$

$$GPM = 328 \ gpm$$

Friction Loss in Standpipe

Based on 10 feet per floor, assume that 30 feet of standpipe is used.

C = 0.052 from Table 8.3

$$Q = \frac{gpm}{100} \qquad Q = \frac{328}{100} \qquad Q = 3.28$$

$$L = \frac{feet}{100} \qquad L = \frac{30}{100} \qquad L = 0.3$$

$$FL = CQ^2L \qquad FL = (0.052)(3.28)^2(0.3)$$
$$FL = (0.052)(10.76)(0.3)$$

FL = 0.17 psi loss in the standpipe

2½-inch Attack Line

C = 2 from Table 8.3

$$Q = \frac{gpm}{100} \qquad Q = \frac{328}{100} \qquad Q = 3.28$$

$$L = \frac{feet}{100} \qquad L = \frac{150}{100} \qquad L = 1.5$$

$$FL = CQ^2L \qquad FL = (2)(3.28)^2(1.5) \qquad FL = (2)(10.76)(1.5)$$

FL = 32.3 psi loss in the 2½-inch hoseline

Elevation Pressure Loss

EP = (5)(No. of Stories – 1)

EP = (5)(5 – 1)

EP = (5)(4)

EP = 20 psi

Total Pressure Loss

TPL = 0.17 + 32.3 + 20 = **52.47 psi total pressure loss in this standpipe and hose assembly**

One of the purposes of the preceding example was to demonstrate that friction loss within hard piping is minimal. In the example, the friction loss in the standpipe was less than one-fifth of a psi. Therefore, it is not usually necessary to calculate this friction loss because it has little effect on the overall problem facing the driver/operator. The fact that hard piping, or connections, pose minimal friction loss concerns is important to remember for the remainder of this manual. It is this principle that allows us to use 2½-inch couplings on 3-inch hose with minimal flow restrictions when compared to 3-inch hose with 3-inch couplings.

Multiple Hoselines (Unequal Length)

Occasionally, a situation may arise where multiple hoselines of equal or unequal diameter are not the same length. This can result from the addition of a new hoseline to a pumper or the addition of hose lengths to an existing line. When unequal length hoselines are used, the amount of friction loss varies in each line. For this reason, friction loss must be calculated in each hoseline.

Example 14

Two 2½-inch hoselines, one 500 feet long and the other 300 feet long, are equipped with 250 gpm fog nozzles (Figure 8.20). Find the total pressure loss due to friction in each line.

Line 1

C = 2 from Table 8.3

$$Q = \frac{gpm}{100} \qquad Q = \frac{250}{100} \qquad Q = 2.5$$

$$L = \frac{feet}{100} \qquad L = \frac{500}{100} \qquad L = 5$$

$$FL = CQ^2L \qquad FL = (2)(2.5)^2(5) \qquad FL = (2)(6.25)(5)$$

FL = 62.5 psi loss in Line 1

Line 2

C = 2 from Table 8.3

$$Q = \frac{gpm}{100} \qquad Q = \frac{250}{100} \qquad Q = 2.5$$

$$L = \frac{feet}{100} \qquad L = \frac{300}{100} \qquad L = 3$$

$$FL = CQ^2L \qquad FL = (2)(2.5)^2(3) \qquad FL = (2)(6.25)(3)$$

FL = 37.5 psi loss in Line 2

Total Pressure Loss

The total pressure loss in the system is based on the highest loss of the two lines, which in this case would be Line 1 at **62.5 psi**.

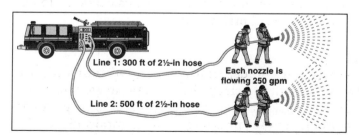

Figure 8.20 Example 14.

Example 15

A pumper is supplying four lines. Two lines are 200 feet of 1¾-inch hose equipped with a ¾-inch tip operating at 50 psi. The other two lines are 150 feet of 1½-inch hose equipped with a ⅝-inch tip operating at 50 psi. Determine the total pressure loss due to friction in each hoseline (Figure 8.21).

Lines 1 & 2

GPM = 29.7 d²√NP

GPM = (29.7) (0.625)²(√50)

GPM = 82 gpm

C = 24 from Table 8.3

$Q = \dfrac{gpm}{100}$ $Q = \dfrac{82}{100}$ Q = 0.82

$L = \dfrac{feet}{100}$ $L = \dfrac{150}{100}$ L = 1.5

FL = CQ²L   FL = (24)(0.82)²(1.5)

FL = (24)(0.6724)(1.5)

FL = 24 psi loss in each 1½-inch hoseline

Lines 3 & 4

GPM = 29.7 d²√NP

GPM = (29.7) (0.75)²(√50)

GPM = 118 gpm

C = 15.5 from Table 8.3

$Q = \dfrac{gpm}{100}$ $Q = \dfrac{118}{100}$ Q = 1.18

$L = \dfrac{feet}{100}$ $L = \dfrac{200}{100}$ L = 2

FL = CQ²L   FL = (15.5)(1.18)²(2)

FL = (15.5)(1.3924)(2)

FL = 43 psi loss in each 1½-inch hoseline

Total Pressure Loss

The total pressure loss in the system is based on the highest loss of the two sets of lines, which in this case would be Lines 3 and 4 at **43 psi.**

Wyed Hoselines (Unequal Length) and Manifold Hoselines

The addition of hose lengths to an existing wyed hoseline assembly may result in unequal length attack lines. Generally, the principles used to determine friction loss in equal length wyed hoselines also apply to unequal length wyed hoselines. Because the length of the attack lines is different, the split of the total water flowing is not equal at the wye appliance. For this reason, it is necessary to determine friction for each of the unequal length wyed lines.

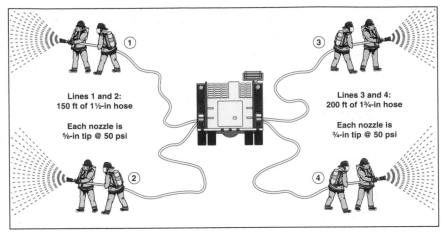

Figure 8.21 Example 15.

Lines 1 and 2:
150 ft of 1½-in hose

Each nozzle is
⅝-in tip @ 50 psi

Lines 3 and 4:
200 ft of 1¾-in hose

Each nozzle is
¾-in tip @ 50 psi

Some fireground operations involve the use of a water thief or manifold appliance (Figures 8.22 a and b). Generally, a hose assembly involving a manifold consists of one large diameter hoseline supplying several smaller attack lines. These attack lines, as with any other hose assembly, may be equal or unequal in length and diameter. Keep in mind that when hose lengths are unequal in length and/or diameter, the total pressure loss in the system is based on the highest pressure loss in any of the lines. In a real-life situation, the hoselines requiring less than the maximum pressure are gated down at the manifold. Gating down the lines at the manifold so that the proper pressure is achieved in each line is, at best, a guessing game, unless the manifold is equipped with a pressure gauge on each discharge.

As with a wyed hoseline, friction loss in a manifold assembly is not overly difficult to determine. The following steps can be used to calculate friction loss in unequal length wyed or manifold hoselines.

Step 1: Compute the number of hundreds of gpm flowing in each of the wyed hoselines by using the following equation:

$$Q = \frac{\text{discharge gpm}}{100}$$

Step 2: Determine the friction loss in each of the wyed lines using Equation A:

FL = CQ²L

Figure 8.22a A water thief.

Figure 8.22b An LDH manifold.

Step 3: Compute the total number of hundreds of gpm flowing in the supply line to the wye or manifold by adding the sum of the flows in the attack lines and dividing by 100.

Step 4: Determine the friction loss in the supply line using the equation:

$$FL = (C)(Q_{Total})^2(L)$$

Step 5: Add the friction loss from the supply line, the wye or manifold appliance (if total flow is greater than 350 gpm), elevation loss, and the wyed line with the greatest amount of friction loss to determine the total pressure loss.

The following examples illustrate how to compute the total pressure loss in an unequal wyed hose assembly and a manifold hose assembly.

Example 16

Determine the total pressure loss due to friction and elevation pressure for a hoseline assembly in which one 400 foot, 3-inch hose with 2½-inch couplings is supplying two attack lines (Figure 8.23). The first attack line consists of 200 feet of 1¾-inch hose that is stretched through the front door of a structure. The first attack line is flowing 150 gpm. The second attack line is 150 feet of 1½-inch hose that is carried up a ground ladder and stretched through a second floor window. The second line is flowing 95 gpm.

Attack Line 1 (1¾-inch)

C = 15.5 from Table 8.3

$Q = \dfrac{gpm}{100}$ $\quad Q = \dfrac{150}{100}$ $\quad Q = 1.5$

$L = \dfrac{feet}{100}$ $\quad L = \dfrac{200}{100}$ $\quad L = 2$

$FL = CQ^2L$ $\quad FL = (15.5)(1.5)^2(2)$ $\quad FL = (15.5)(2.25)(2)$

FL = 69.8 psi loss in Attack Line 1

Attack Line 2 (1½-inch)

C = 24 from Table 8.3

$Q = \dfrac{gpm}{100}$ $\quad Q = \dfrac{95}{100}$ $\quad Q = 0.95$

$L = \dfrac{feet}{100}$ $\quad L = \dfrac{150}{100}$ $\quad L = 1.5$

$FL = CQ^2L$ $\quad FL = (24)(0.95)^2(1.5)$ $\quad FL = (24)(0.9025)(1.5)$

FL = 32.5 psi

EP = (5)(No. of Stories −1)

EP = (5)(2 −1)

EP = (5)(1)

EP = 5 psi

TPL = 32.5 + 5 = **37.5 psi loss in Attack Line 2**

Supply Line (3-inch with 2½-inch couplings)

C = 0.8 from Table 8.3

$Q_{Total} = \dfrac{(Attack\ Line\ 1\ gpm) + (Attack\ Line\ 2\ gpm)}{100}$

$Q_{Total} = \dfrac{150 + 95}{100}$ $\quad Q_{Total} = \dfrac{245}{100}$

$Q_{Total} = 2.45$

$L = \dfrac{feet}{100}$ $\quad L = \dfrac{400}{100}$ $\quad L = 4$

$FL = (C)(Q_{Total})^2(L)$ $\quad FL = (0.8)(2.45)^2(4)$

$FL = (0.8)(6.0025)(4)$

FL = 19.2 psi in the supply line

Total Pressure Loss

The total pressure loss in the system is based on the highest loss of the two attack lines, which in this case would be Line 1, and the friction loss in the supply line. Because the flow rate was less than 350 gpm, no appliance loss is required.

TPL = 69. 8 + 19.2 = **89 psi loss is this hose assembly**

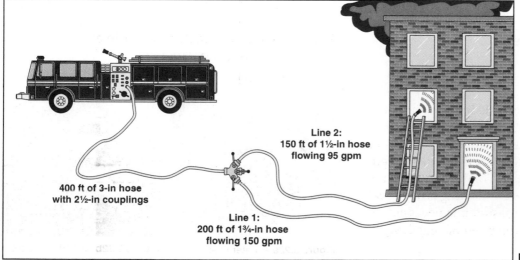

Line 2:
150 ft of 1½-in hose
flowing 95 gpm

400 ft of 3-in hose
with 2½-in couplings

Line 1:
200 ft of 1¾-in hose
flowing 150 gpm

Figure 8.23 Example 16.

Example 17

Determine the total pressure loss due to friction for a hoseline assembly in which one 500-foot, 3-inch hose with 3-inch couplings is supplying three attack lines that are attached to a water thief. The first attack line consists of 150 feet of 1½-inch hose flowing 125 gpm. The second attack line is 100 feet of 1½-inch hose flowing 95 gpm. The third attack line is 150 feet of 2½-inch hose flowing 225 gpm.

Attack Line 1 (1½-inch)

C = 24 from Table 8.3

$$Q = \frac{gpm}{100} \quad Q = \frac{125}{100} \quad Q = 1.25$$

$$L = \frac{feet}{100} \quad L = \frac{150}{100} \quad L = 1.5$$

$$FL = CQ^2L \quad FL = (24)(1.25)^2(1.5)$$

$$FL = (24)(1.5625)(1.5)$$

FL = 56.3 psi loss in Attack Line 1

Attack Line 2 (1½-inch)

C = 24 from Table 8.3

$$Q = \frac{gpm}{100} \quad Q = \frac{95}{100} \quad Q = 0.95$$

$$L = \frac{feet}{100} \quad L = \frac{100}{100} \quad L = 1$$

$$FL = CQ^2L \quad FL = (24)(0.95)^2(1) \quad FL = (24)(0.9025)(1)$$

FL = 21.7 psi loss in Attack Line 2

Attack Line 3 (2½-inch)

C = 2 from Table 8.3

$$Q = \frac{gpm}{100} \quad Q = \frac{225}{100} \quad Q = 2.25$$

$$L = \frac{feet}{100} \quad L = \frac{150}{100} \quad L = 1.5$$

$$FL = CQ^2L \quad FL = (2)(2.25)^2(1.5) \quad FL = (2)(5.0625)(1.5)$$

FL = 15.2 psi loss in Attack Line 3

Supply Line (3-inch with 3-inch couplings)

C = 0.677 from Table 8.3

$$Q_{Total} = \frac{(Sum\ of\ Attack\ Lines\ 1,\ 2,\ \&\ 3\ gpm)}{100}$$

$$Q_{Total} = \frac{125 + 95 + 225}{100} \quad Q_{Total} = \frac{445}{100}$$

$$Q_{Total} = 4.45$$

$$L = \frac{feet}{100} \quad L = \frac{500}{100} \quad L = 5$$

$$FL = (C)(Q_{Total})^2(L) \quad FL = (0.677)(4.45)^2(5)$$

$$FL = (0.677)(19.8025)(5)$$

FL = 67 psi in the supply line

Total Pressure Loss

The total pressure loss in the system is based on the highest loss of the three attack lines, which in this case would be Line 1, and the friction loss in the supply line. Because the total flow rate exceeds 350 gpm, it is necessary to add the 10 psi appliance loss.

TPL = 56.3 + 67 + 10 = **133.3 psi loss in this hose assembly**

Example 18

Determine the total pressure loss due to friction for a hoseline assembly in which one 700-foot, 5-inch hose is supplying three hoselines that are attached to a large diameter hose manifold (Figure 8.24). Two of the hoselines are 150 feet of 3-inch hose with 2½-inch couplings that are supplying a portable master stream device. The master stream is discharging 80 psi through a 1½-inch tip. Add 25 psi for the appliance loss in the master stream device. The third hoseline is 150 feet of 2½-inch hose flowing 275 gpm.

Master Stream

$$GPM = 29.7\ d^2\sqrt{NP}$$

$$GPM = (29.7)\ (1.5)^2(\sqrt{80})$$

$$GPM = 598\ gpm$$

C = 0.2 from Table 8.4

$$Q = \frac{gpm}{100} \quad Q = \frac{598}{100} \quad Q = 5.98$$

$$L = \frac{feet}{100} \quad L = \frac{150}{100} \quad L = 1.5$$

$$FL = CQ^2L \quad FL = (0.2)(5.98)^2(1.5)$$

$$FL = (0.2)(35.7604)(1.5)$$

FL = 10.7 psi

TPL = 10.7 + 25 = **35.7 psi loss in the lines supplying the master stream**

2½-inch Handline

C = 2 from Table 8.3

$$Q = \frac{gpm}{100} \quad Q = \frac{275}{100} \quad Q = 2.75$$

$$L = \frac{feet}{100} \quad L = \frac{150}{100} \quad L = 1.5$$

$$FL = CQ^2L \quad FL = (2)(2.75)^2(1.5) \quad FL = (2)(7.5625)(1.5)$$

FL = 22.7 psi loss in the 2½-inch handline

Supply Line (5-inch)

C = 0.08 from Table 8.3

$$Q_{Total} = \frac{(Master\ Stream\ gpm) + (Hand\ Line\ gpm)}{100}$$

$$Q_{Total} = \frac{598 + 275}{100} \qquad Q_{Total} = \frac{873}{100}$$

$$Q_{Total} = 8.73$$

$$L = \frac{feet}{100} \qquad L = \frac{700}{100} \qquad L = 7$$

$$FL = (C)(Q_{Total})^2(L) \qquad FL = (0.08)(8.73)^2(7)$$

$$FL = (0.08)(76.2129)(7)$$

FL = 42.7 psi pressure loss in the supply line

Total Pressure Loss

The total pressure loss in the system is based on the highest loss of the three attack lines, which in this case are the lines supplying the master stream, and the friction loss in the supply line. Because the total flow rate exceeds 350 gpm, it is necessary to add the 10 psi appliance loss.

TPL = 35.7 + 42.7 + 10 = **88.4 psi loss in this hose assembly**

Master Streams

The principles upon which master streams are developed are essentially the same as for other fire streams. Master streams, however, require a greater volume of water than do handlines. Multiple hoselines, siamesed hoselines, or large diameter single hoselines are generally used as supply lines for these large volumes of water. If a master stream requires a water flow greater than the capacity of a single pumper, multiple pumpers may be used to supply the master stream appliance. As stated earlier in this manual, add a 25 psi pressure loss to all calculations involving master stream devices.

The hose lays used to supply master streams are essentially the same as those used for other fire streams. For this reason, the concepts used in determining friction loss are also the same. The only difference occurs if unequal length or diameter hoselines are used to supply a master stream appliance. If this situation is present, use an average of the hose lengths for ease of calculation. To obtain an average, add the length of each hoseline and then divide by the number of hoselines being used. Use the coefficients for wyed hoselines contained in Table 8.4 with the total flow through the nozzle to complete the friction loss calculations. In some cases, multiple lines of equal diameter but unequal length are used, and no coefficient for that combination exists in Table 8.4. In that case, average the length of the lines and assume an equal amount of water is going through each hose.

In this manual, the aerial devices with piped waterways are treated in the same manner as master stream appliances: using a friction loss of 25 psi to include the intake, internal piping, and nozzle. Elevation pressure loss is calculated separately. For exact figures, consult the aerial manufacturer for specific friction loss data, or perform field tests to derive data. If a traditional detachable ladder pipe and hose assembly are used, the friction loss within the siamese (if used), hose, and ladder pipe must all be accounted for. The following examples show how to calculate the total pressure loss in a master stream hose layout.

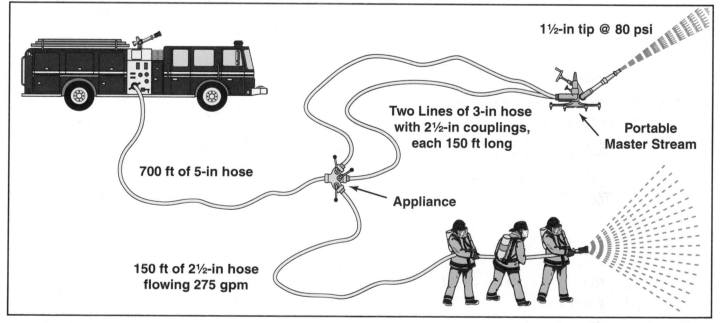

Figure 8.24 Example 18.

Example 19

Determine the total pressure loss in a hose assembly when a 3,000 gpm master stream device is supplied by three 5-inch hoselines (Figures 8.25 a and b). Two of the hoselines are 900 feet in length, and one is 1,200 feet. Assume 1,000 gpm is flowing through each hoseline.

Friction Loss in the 5-inch Hoselines

C = 0.08 from Table 8.3

$$Q = \frac{gpm}{100} \quad Q = \frac{1,000}{100} \quad Q = 10$$

$$\text{Average Length } (L_{Ave}) = \frac{(\text{Sum of Hose Lengths})}{3}$$

$$L_{Ave} = \frac{(900 + 900 + 1,200)}{3} \quad L_{Ave} = 1,000$$

$$L = \frac{L_{Ave}}{100} \quad L = \frac{1,000}{100} \quad L = 10$$

FL = CQ²L   FL = (0.08)(10)²(10)   FL = (0.08)(100)(10)

FL = 80 psi loss in each 5-inch hose

Total Pressure Loss

TPL = 80 + 25 = **105 psi pressure loss in this hose assembly**

Figure 8.25a Some extremely high-flow master stream devices may be fed by multiple large diameter hoselines.

Example 20

A pumper is supplying 450 feet of 4½-inch hose that is feeding an aerial device with a piped waterway. The aerial device is elevated 65 feet and is discharging 900 gpm (Figure 8.26). Determine the total pressure loss in this hose assembly.

4½-inch Hose

C = 0.1 from Table 8.3

$$Q = \frac{gpm}{100} \quad Q = \frac{900}{100} \quad Q = 9$$

$$L = \frac{feet}{100} \quad L = \frac{450}{100} \quad L = 4.5$$

FL = CQ²L   FL = (0.1)(9)²(4.5)   FL = (0.1)(81)(4.5)

FL = 36.5 psi

Elevation Pressure

EP = (0.5)(H)

EP = (0.5)(65)

EP = 32.5 psi

Total Pressure Loss

TPL = 36.5 + 32.5 + 25 = **94 psi pressure loss in this hose assembly**

Example 21

Determine the total pressure loss in the hose assembly when a fire department pumper is supplying two 3-inch hoselines with 2½-inch couplings, each 300 feet long. These hoselines are connected to a siamese appliance that is in turn supplying 100 feet of 3½-inch hose attached to a detachable ladder pipe. The ladder pipe is elevated 70 feet and is discharging through a 1⅝-inch diameter solid stream nozzle at 80 psi.

Master Stream

GPM = 29.7 d²√NP

GPM = (29.7) (1.625)²(√80)

GPM = 701 gpm

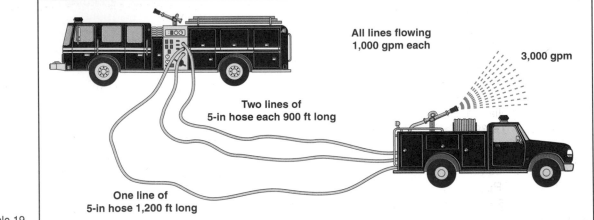

All lines flowing 1,000 gpm each

3,000 gpm

Two lines of 5-in hose each 900 ft long

One line of 5-in hose 1,200 ft long

Figure 8.25b Example 19.

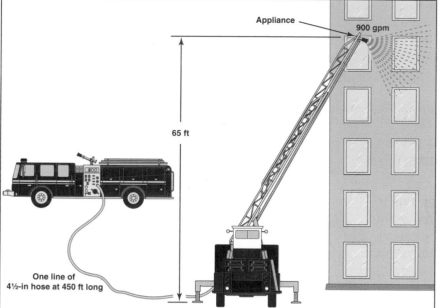

Figure 8.26 Example 20.

Appliance

900 gpm

65 ft

One line of
4½-in hose at 450 ft long

3½-inch Hose and Master Stream

C = 0.34 from Table 8.3

$Q = \dfrac{gpm}{100}$ $Q = \dfrac{701}{100}$ $Q = 7.01$

$L = \dfrac{feet}{100}$ $L = \dfrac{100}{100}$ $L = 1$

$FL = CQ^2L$ $FL = (0.34)(7.01)^2(1)$

$FL = (0.34)(49.1401)(1)$

FL = 16.7 psi

EP = (0.5)(H)

EP = (0.5)(70)

EP = 35 psi loss

$TPL_{3½} = FL + EP + Appliance\ Loss$
$TPL_{3½}$ = 16.7 + 35 + 25 = 76.7 psi loss in the 3½-inch hose and the master stream

3-inch Hoselines and Siamese

C = 0.2 from Table 8.4

$Q = \dfrac{gpm}{100}$ $Q = \dfrac{701}{100}$ $Q = 7.01$

$L = \dfrac{feet}{100}$ $L = \dfrac{300}{100}$ $L = 3$

$FL = CQ^2L$ $FL = (0.2)(7.01)^2(3)$ $FL = (0.2)(49.1401)(3)$

FL = 29.5 psi

$TPL_3 = FL + Appliance\ Loss$
TPL_3 = 29.5 + 10 = 39.5 psi loss in the dual 3-inch lines and siamese appliance

Total Pressure Loss
$TPL = TPL_{3½} + TPL_3$
TPL = 76.7 + 39.5 = **116.2 psi pressure loss in this hose assembly**

Determining Pump Discharge Pressure

Numerous skills possessed by the driver/operator are reflected during the development of fire streams. These skills enable the driver/operator to provide fire suppression crews with a fire stream at the proper pressure. In order to fully utilize these skills, the driver/operator must understand pump discharge pressure and its relation to nozzle pressure.

In order to deliver the necessary water flow to the fire location, the pump discharge pressure at the apparatus must be enough to overcome the sum of all pressure losses. These pressure losses, combined with the required nozzle pressure, are used to determine the pump discharge pressure. Pump discharge pressure (PDP) can be calculated using the following equation:

EQUATION D
PDP = NP + TPL

   Where:

   PDP = Pump discharge pressure in psi

   NP = Nozzle pressure in psi

   TPL = Total pressure loss in psi (appliance, friction, and elevation losses)

Fire apparatus supplying multiple hoselines, wyed hoselines, or manifold hoselines may be required to provide different pump discharge pressures for each attack line. Because this is not possible, you must use another method to compensate for these individual pressure requirements. Set the pump discharge pressure for the hoseline with the greatest pressure demand. For multiple hoselines, gate back the remaining hoselines at the discharge outlets. For wyed or manifold hoselines, gate back the hoselines at the appliance. Gate back the hoselines until the desired pressure on each line is obtained. As discussed earlier in this chapter, use the following nozzle pressures to ensure safety and efficiency:

- Solid stream nozzle (handline) — 50 psi
- Solid stream nozzle (master stream) — 80 psi
- Fog nozzle (all types) — 100 psi

Pressure losses for elevated master streams and turret pipes vary depending on the manufacturer of the device. As discussed earlier in this chapter, for the sake of calculation, we consider each of these to have a pressure loss of 25 psi.

Later in this manual, you learn that virtually all fire pumps in use today are centrifugal pumps. These pumps are able to take advantage of incoming water pressure into the pump. Thus, if a pumper is required to discharge 150 psi, and it has an intake pressure of 50 psi coming into the pump, the pump only needs to "create" an additional 100 psi to meet the demand. This concept is called *net pump discharge pressure* (NPDP).

The following examples show how to calculate pump discharge pressure using Equation D.

Example 22

A pumper is supplying 500 feet of 2½-inch hose that is flowing 300 gpm through a fog stream nozzle. Determine the pump discharge pressure required to supply the hoseline.

C = 2 from Table 8.3

$Q = \dfrac{gpm}{100}$ $Q = \dfrac{300}{100}$ $Q = 3$

$L = \dfrac{feet}{100}$ $L = \dfrac{500}{100}$ $L = 5$

$FL = CQ^2L$ $FL = (2)(3)^2(5)$ $FL = (2)(9)(5)$

FL = 90 psi

Pump Discharge Pressure

PDP = NP + TPL   PDP = 100 + 90   **PDP = 190 psi**

This means that the pumper will have to be discharging 190 psi in order for the nozzle to be receiving the appropriate amount of pressure. If the pumper had an incoming pressure of 50 psi, the pump would only need to generate 140 psi to make up the difference.

Example 23

Two 2½-inch hoselines, one 300 feet long and the other 500 feet long, are each equipped with 250 gpm fog nozzles. Determine the pump discharge pressure required to supply these hoselines (Figure 8.27).

Hoseline 1

C = 2 from Table 8.3

$Q = \dfrac{gpm}{100}$ $Q = \dfrac{250}{100}$ $Q = 2.5$

$L = \dfrac{feet}{100}$ $L = \dfrac{300}{100}$ $L = 3$

$FL = CQ^2L$ $FL = (2)(2.5)^2(3)$ $FL = (2)(6.25)(3)$

FL = 37.5 psi

Hoseline 2

C = 2 from Table 8.3

$Q = \dfrac{gpm}{100}$ $Q = \dfrac{250}{100}$ $Q = 2.5$

$L = \dfrac{feet}{100}$ $L = \dfrac{500}{100}$ $L = 5$

$FL = CQ^2L$ $FL = (2)(2.5)^2(5)$ $FL = (2)(6.25)(5)$

FL = 62.5 psi

Pump Discharge Pressure

Use the hose with the highest pressure loss to determine the PDP:

PDP = NP + TPL   PDP = 100 + 62.5   **PDP = 162.5 psi**

Set the pump discharge pressure for the hoseline with the higher pressure demand (162.5 psi). Gate back the discharge outlet on the remaining hoseline until the desired hoseline pressure is obtained. (**NOTE:** Where flowmeters are used, gate back to desired flow.)

Example 24

Determine the pump discharge pressure for a fire department pumper that is using two 3-inch hoselines with 2½-inch couplings to supply an elevated master device with fixed piping 200 feet away. The elevated master stream is discharging 1,000 gpm through a fog nozzle that is elevated 60 feet.

C = 0.2 from Table 8.4

$Q = \dfrac{gpm}{100}$ $Q = \dfrac{1,000}{100}$ $Q = 10$

$L = \dfrac{feet}{100}$ $L = \dfrac{200}{100}$ $L = 2$

$FL = CQ^2L$ $FL = (0.2)(10)^2(2)$ $FL = (0.2)(100)(2)$

FL = 40 psi

EP = (0.5)(H)

EP = (0.5)(60)

EP = 30 psi

TPL = FL + EP + Appliance Loss

TPL = 40 + 30 + 25 = **95 psi pressure loss in this hose assembly.**

Pump Discharge Pressure

PDP = NP + TPL   PDP = 100 + 95   **PDP = 195 psi**

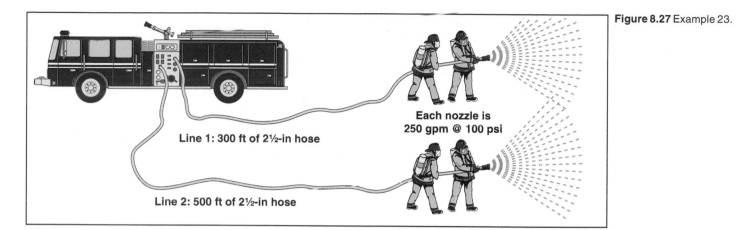

Figure 8.27 Example 23.

Line 1: 300 ft of 2½-in hose

Line 2: 500 ft of 2½-in hose

Each nozzle is
250 gpm @ 100 psi

Example 25

Determine the pump discharge pressure in a hoseline assembly in which one 300-foot, 3-inch hose with 3-inch couplings is supplying three attack lines that are attached to a water thief. The first attack line consists of 100 feet of 1½-inch hose flowing 125 gpm. The second attack line is 200 feet of 1½-inch hose flowing 95 gpm. The third attack line is 150 feet of 2½-inch hose flowing 250 gpm. All nozzles are fog streams.

Attack Line 1 (1½-inch)

C = 24 from Table 8.3

$Q = \dfrac{gpm}{100}$ $Q = \dfrac{125}{100}$ $Q = 1.25$

$L = \dfrac{feet}{100}$ $L = \dfrac{100}{100}$ $L = 1$

FL = CQ²L   FL = (24)(1.25)²(1)   FL = (24)(1.5625)(1)

FL = 37.5 psi loss in Attack Line 1

Attack Line 2 (1½-inch)

C = 24 from Table 8.3

$Q = \dfrac{gpm}{100}$ $Q = \dfrac{95}{100}$ $Q = 0.95$

$L = \dfrac{feet}{100}$ $L = \dfrac{200}{100}$ $L = 2$

FL = CQ²L   FL = (24)(0.95)²(2)   FL = (24)(0.9025)(2)

FL = 43.4 psi loss in Attack Line 2

Attack Line 3 (2½-inch)

C = 2 from Table 8.3

$Q = \dfrac{gpm}{100}$ $Q = \dfrac{250}{100}$ $Q = 2.5$

$L = \dfrac{feet}{100}$ $L = \dfrac{150}{100}$ $L = 1.5$

FL = CQ²L   FL = (2)(2.5)²(1.5)   FL = (2)(6.25)(1.5)

FL = 18.8 psi loss in Attack Line 3

Supply Line (3-inch with 3-inch couplings)

C = 0.677 from Table 8.3

$Q_{Total} = \dfrac{(\text{Sum of Attack Lines 1, 2, \& 3 gpm})}{100}$

$Q_{Total} = \dfrac{125 + 95 + 250}{100}$ $Q_{Total} = \dfrac{470}{100}$

$Q_{Total} = 4.7$

$L = \dfrac{feet}{100}$ $L = \dfrac{300}{100}$ $L = 3$

FL = (C)(Q$_{Total}$)²(L)   FL = (0.677)(4.7)²(3)

FL = (0.677)(22.09)(3)

FL = 44.9 psi in the supply line

The total pressure loss in the system is based on the highest loss of the three attack lines, which in this case would be Line 2, and the friction loss in the supply line. Because the total flow rate exceeds 350 gpm, it is necessary to add the 10 psi appliance loss.

TPL = FL$_{Attack}$ + FL$_{Supply}$ + Appliance Loss

TPL = 43.4 + 44.9 + 10 = 98.3 psi loss in this hose assembly

Pump Discharge Pressure

PDP = NP + TPL   PDP = 100 + 98.3   **PDP = 198.3 psi**

Set the pump discharge pressure for the hoseline with the higher pressure demand (198.3 psi). Gate back the discharge outlet on the water thief for the other hoseline until the desired hoseline pressures are obtained.

Example 26

Determine the pump discharge pressure in the hose assembly when a fire department pumper is supplying two 3-inch hoselines with 2½-inch couplings, each 200 feet long. These hoses are connected to a siamese appliance that is in turn supplying 100 feet of 3-inch hose with 3-inch couplings attached to a detachable ladder pipe. The ladder pipe is elevated 60 feet and is discharging through a 1½-inch diameter solid stream nozzle at 80 psi.

Master Stream

$$GPM = 29.7\ d^2\sqrt{NP}$$

$$GPM = (29.7)(1.5)^2(\sqrt{80})$$

$$GPM = 598\ gpm$$

3-inch Hose With 3-inch Couplings and Master Stream

C = 0.677 from Table 8.3

$$Q = \frac{gpm}{100} \qquad Q = \frac{598}{100} \qquad Q = 5.98$$

$$L = \frac{feet}{100} \qquad L = \frac{100}{100} \qquad L = 1$$

$$FL = CQ^2L \qquad FL = (0.677)(5.98)^2(1)$$

$$FL = (0.677)(35.7604)(1)$$

FL = 24.2 psi

$$EP = (0.5)(H)$$

$$EP = (0.5)(60)$$

$$EP = 30\ psi\ loss$$

$$TPL_{3\,w/\,3} = FL + EP + Appliance\ Loss$$

$TPL_{3\,w/\,3} = 24.2 + 30 + 25 =$ **79.2 psi loss in the 3-inch hose with 3-inch couplings and the master stream**

3-inch Hoselines and Siamese

C = 0.2 from Table 8.4

$$Q = \frac{gpm}{100} \qquad Q = \frac{598}{100} \qquad Q = 5.98$$

$$L = \frac{feet}{100} \qquad L = \frac{200}{100} \qquad L = 2$$

$$FL = CQ^2L \qquad FL = (0.2)(5.98)^2(2)$$

$$FL = (0.2)(35.7604)(2)$$

FL = 14.3 psi

$$TPL_{Dual\,3s} = FL + Appliance\ Loss$$

$TPL_{Dual\,3s} = 14.3 + 10 =$ **24.3 psi loss in the dual 3-inch lines and siamese appliance**

$$TPL_{Assembly} = TPL_{3\,w/\,3} + TPL_{Dual\,3s}$$

$TPL_{Assembly} = 79.2 + 24.3 =$ **103.5 psi pressure loss in this hose assembly**

Pump Discharge Pressure

$$PDP = NP + TPL_{Assembly} \qquad PDP = 80 + 103.5$$

PDP = 183.5 psi

Determining Net Pump Discharge Pressure

Net pump discharge pressure takes into account all factors that contribute to the amount of work the pump must do to produce a fire stream.

When a pumper is being supplied by a hydrant or a supply line from another pumper, the net pump discharge pressure is the difference between the pump discharge pressure and the incoming pressure from the hydrant. If the PDP, for instance, is 150 psi and the intake gauge reads 50 psi, the net pump discharge pressure is 100 psi. This is shown by the following formula:

EQUATION E
$NPDP_{PPS}$ = PDP – Intake reading

Where:

$NPDP_{PPS}$ = Net pump discharge pressure from a positive pressure source

PDP = Pump discharge pressure

Note that this equation does not apply to situations where the pumper is operating at a draft. These situations are covered in more detail in Chapter 12 of this manual.

The application of Equation E is shown in the following example:

Example 27

A pumper operating from a hydrant is discharging water at 170 psi. The incoming pressure from the hydrant registers 20 psi on the intake gauge. Determine the net pump discharge pressure.

$$NPDP_{PPS} = PDP - Intake\ reading$$

$$NPDP_{PPS} = 170\ psi - 20\ psi$$

$NPDP_{PPS}$ = 150 psi

Theoretical Pressure Calculations

Job Performance Requirements

This chapter provides information that will assist the reader in meeting the following job performance requirements from NFPA 1002, *Standard on Fire Apparatus Driver/Operator Professional Qualifications*, 1998 edition.

3-2.1* Produce effective hand or master streams, given the sources specified in the following list, so that the pump is safely engaged, all pressure control and vehicle safety devices are set, **the rated flow of the nozzle is achieved and maintained**, and the apparatus is continuously monitored for potential problems.

* Internal tank

* Pressurized source

* Static source

* Transfer from internal tank to external source

(a) *Requisite Knowledge:* **Hydraulic calculations for friction loss and flow using both written formulas and estimation methods**, safe operation of the pump, problems related to small-diameter or dead-end mains, low-pressure and private water supply systems, hydrant cooling systems, and reliability of static sources.

(b) *Requisite Skills:* The ability to position a fire department pumper to operate at a fire hydrant and at a static water source, power transfer from vehicle engine to pump, draft, operate pumper pressure control systems, operate the volume/pressure transfer valve (multistage pumps only), operate auxiliary cooling systems, make the transition between internal and external water sources, and assemble hose lines, nozzles, valves, and appliances.

3-2.2 Pump a supply line of 2½ in. (65 mm) or larger, given a relay pumping evolution the length and size of the line and the desired flow and intake pressure, so that the **proper pressure and flow are provided to the next pumper in the relay**.

(a) *Requisite Knowledge:* **Hydraulic calculations for friction loss and flow using both written formulas and estimation methods**, safe operation of the pump, problems related to small-diameter or dead-end mains, low-

pressure and private water supply systems, hydrant cooling systems, and reliability of static sources.

(b) *Requisite Skills:* The ability to position a fire department pumper to operate at a fire hydrant and at a static water source, power transfer from vehicle engine to pump, draft, operate pumper pressure control systems, operate the volume/pressure transfer valve (multistage pumps only), operate auxiliary cooling systems, make the transition between internal and external water sources, and assemble hose lines, nozzles, valves, and appliances.

3-2.4 Supply water to fire sprinkler and standpipe systems, given specific system information and a fire department pumper, so that water is supplied to the system at the proper volume and pressure.

(a) *Requisite Knowledge*: **Calculation of pump discharge pressure**; hose layouts; location of fire department connection; alternative supply procedures if fire department connection is not usable; operating principles of sprinkler systems as defined in NFPA 13, *Standard for the Installation of Sprinkler Systems*, NFPA 13D, *Standard for the Installation of Sprinkler Systems in One- and Two-Family Dwellings and Manufactured Homes*, and NFPA 13R, *Standard for the Installation of Sprinkler Systems in Residential Occupancies Up To and Including Four Stories in Height*; fire department operations in sprinklered occupancies as defined in NFPA 13E, *Guide for Fire Department Operations in Properties Protected by Sprinkler and Standpipe Systems*; and operating principles of standpipe systems as defined in NFPA 14, *Standard for the Installation of Standpipe and Hose Systems*.

6-2.1* Produce effective fire streams, utilizing the sources specified in the following list, so that the pump is safely engaged, all pressure-control and vehicle safety devices are set, the rated flow of the nozzle is achieved, and the apparatus is continuously monitored for potential problems.

* Water tank

* Pressurized source

* Static source

The prime objective of the driver/operator is to provide fire suppression crews with the water flow and pressure needed to achieve efficient fire control and/or extinguishment. To meet this objective, driver/operators must understand the theoretical aspects of fire stream development. They must then be able to convert these theories into practice during fireground operations.

In this chapter, the driver/operator learns the theoretical methods of calculating pressure loss in a variety of hose lays and fireground situations. He learns to consider the effects of friction within the hose assembly, as well as the effect that changes in elevation have on supplying hoselines. Lastly, the driver/operator learns to use the figures derived from the pressure loss calculations to determine the pump discharge pressure required to adequately supply fire streams.

Driver/operators rarely, if ever, perform the calculations contained in this chapter while on the fireground. On the fireground, they are more likely to use the methods described in Chapter 9. However, it is important that the driver/operator be able to calculate theoretical friction loss for a variety of reasons, including the following:

- It gives the driver/operator a better understanding of the basis for the fireground methods described in Chapter 9.

- It allows the driver/operator to predetermine accurate pump discharge pressures for preconnected hoselines and common hose lays used on the apparatus he is responsible for operating (Figure 8.1).

- It serves as a tool for pre-incident planning at properties that require hose deployment that is out of the ordinary for the fire department.

Throughout the following sections, there are a number of example problems designed to familiarize the driver/operator with the concepts used in performing hydraulic calculations. It is not the intent of any of the

Figure 8.1 Preconnected handlines are the most commonly used fire service hose lays. *Courtesy of Ron Jeffers.*

following example calculations to limit the driver/operator to one method of solution. The example problems serve as a guide for those unfamiliar with the calculations. The examples show the proper way to solve a particular problem. They are worked out step by step to show the reader the entire problem-solving process. The exercises enable the reader to practice solving problems using the format presented in the examples.

NOTE: In order to avoid the clutter and confusion of trying to switch back and forth between the U.S. and metric methods of performing these calcuations, this chapter appears in two versions. This version contains the metric system of measurements. Immediately preceding this version of Chapter 8, you will find a separate version of Chapter 8 in the U.S. system of measurement.

Before getting into the main body of text of this chapter, it is important to review the common sizes of fire hose and solid stream nozzle tip sizes used in today's fire service. These figures are used throughout the remainder of this manual. For comparison purposes, both the U.S. and metric measurements are contained in Tables 8.1 and 8.2. All metrics contained in IFSTA publications follow a document titled, *"Training Guidelines for the Metric Conversion of Fire Departments in Canada."* The National Fire Protection Association does not follow these same guidelines. Thus, you will note some differences in what a particular size hose is called in metric terms in Table 8.1. Jurisdictions that use the NFPA hose sizes need to refer to this table when using the rest of this manual.

Table 8.2
Common Nozzle Diameters

U.S. (Inches)	Metric (mm) (Actual)	Metric (mm) (Known As)
½	12.700	13
⅝	15.880	16
¾	19.050	19
⅞	22.225	22
1	25.400	25
1⅛	28.575	29
1¼	31.750	32
1⅜	34.925	35
1½	38.100	38
1¾	44.450	45
2	50.800	50
2¼	57.150	57
2½	63.500	65
2¾	69.850	70
3	76.200	77

Table 8.1
Common Hose Sizes

U.S. (Inches)	Canada (mm)	NFPA (mm)
¾	20	20
1	25	25
1½	38	38
1¾	45	44
2	50	51
2½	65	65
3	77	76
3½	90	89
4	100	100
4½	115	113
5	125	125
6	150	150

Total Pressure Loss: Friction Loss and Elevation Pressure Loss

To produce effective fire streams, it is necessary to know the amount of friction loss in the fire hose and any pressure loss or gain due to elevation. As was discussed in Chapter 6, friction loss can be caused by a number of factors: hose conditions, coupling conditions, or kinks. The primary determinant, however, is the volume of water flowing per minute.

Calculation of friction loss must also take into account the length and diameter of the hoseline and any major hose appliances attached to the line. Because the amount of hose used between an engine and the nozzle is not always the same, the driver/operator must be capable of determining friction loss in a given length of hose. Elevation differences, such as hills, gullies, aerial devices, or multistoried buildings, create a pressure loss or gain known as *elevation pressure*. The development of elevation pressure occurs anytime there is a change in elevation between the nozzle and the pump (Figure 8.2).

Combined friction loss and elevation pressure loss are referred to as *total pressure loss (TPL)*. This is not to be confused with total pump discharge pressure, which also includes nozzle pressure. Total pump discharge pres-

Figure 8.2 One example of an elevation change is the use of an elevated master stream.

sure is discussed later in this chapter. Both friction loss and elevation pressure are expressed in kilopascals (kPa).

Determining Friction Loss

There are two methods used to determine friction loss: actual tests and calculations. The most accurate of these methods determines friction loss through true field tests. Field tests involve using in-line gauges to measure friction loss at various flows through an actual hose layout. The calculation method involves the use of mathematical friction loss equations or field application methods. This chapter focuses on the use of mathematical equations to determine friction loss. Field applications are covered in Chapter 9.

It is important to emphasize that the only truly accurate method for determining pressure loss in any particular hose lay is by measuring the pressure at both ends of the hose and subtracting the difference; however, this is generally not practical in real-life situations. The calculation of pressure loss through the use of formulas or field applications is, at best, an inexact science. The figures derived from these calculation methods are likely to be different than if actual tests were performed. However, the figures that are obtained through the calculation method usually are close enough to the actual conditions that they can be relied upon to be safe for fireground operations.

In the past, the fire service used the formula $2Q^2 + Q$ as the basis for mathematical friction loss calculations. This formula was based on the average friction loss found in 65 mm hose that was constructed in the 1930s (Figure 8.3). Conversion factors were needed to find friction loss in other sizes of hose. Furthermore, the formula did not take into consideration the length of the hoseline, and it was placed in a separate step.

Improvement in the technology of fire hose construction in the past 30 years or so made the old friction loss formula obsolete (Figure 8.4). The figures derived by that formula provided unrealistically high friction loss figures for today's hose. Thus, it was clear that a new friction loss formula needed to be developed. The new formula needed to take into account the size of the fire hose, the quantity of water flowing, and the length of the hose lay. These three factors gave rise to the following formula for computing friction loss:

EQUATION A
$FL = CQ^2L$

Where:

FL = Friction loss in kPa

C = Friction loss coefficient (from Table 8.3)

Q = Flow rate in hundreds of liters (flow/100)

L = Hose length in hundreds of meters (length/100)

Table 8.3 contains the commonly used friction loss coefficients for the various sizes of hose in use today. These coefficients were derived from the U.S. coefficients used by IFSTA and the NFPA. They are only approximations of the friction loss for each size hose. The actual coefficients for any particular piece of hose vary with the condition of the hose and the manufacturer. When using the coefficients provided in this manual, the results reflect a worst-case situation. In other words, the results are probably slightly higher than the actual friction loss. For departments that require more exact friction loss calculations, different coefficients should be obtained from the manufacturer of their fire hose or through actual calculation, which is covered in the next section of this chapter.

Figure 8.3 Older style hose had a woven outer jacket.

Figure 8.4 Most newer fire hose has a rubber or similar outer jacket.

Table 8.3
Friction Loss Coefficient – Single Hoselines

Hose Diameter and Type (mm)	Coefficient (C)
20 mm booster	1 741
25 mm booster	238
32 mm booster	127
38 mm	38
45 mm	24.6
50 mm	12.7
65 mm	3.17
70 mm with 77 mm couplings	2.36
77 mm with 65 mm couplings	1.27
77 mm with 77 mm couplings	1.06
90 mm	0.53
100 mm	0.305
115 mm	0.167
125 mm	0.138
150 mm	0.083
Standpipes	
100 mm	0.600
125 mm	0.202
150 mm	0.083

The steps for determining friction loss using Equation A are as follows:

Step 1: Obtain from Table 8.3 the friction loss coefficient for the hose being used.

Step 2: Determine the number of hundreds of liters of water per minute flowing (Q) through the hose by using the equation Q = liters/100.

Step 3: Determine the number of hundreds of meters of hose (L) by using the equation L = meters/100.

Step 4: Plug the numbers from Steps 1, 2, and 3 into Equation A to determine the total friction loss.

Example 1

If 1 200 L/min is flowing from a nozzle, what is the total pressure loss due to friction for 120 meters of 65 mm hose?

C = 3.17 from Table 8.3

$$Q = \frac{liters}{100} \quad Q = \frac{1\ 200}{100} \quad Q = 12$$

$$L = \frac{hose\ length}{100} \quad L = \frac{120}{100} \quad L = 1.2$$

$FL = CQ^2L \quad FL = (3.17)(12)^2(1.2) \quad FL = (3.17)(144)(1.2)$

FL = 548 kPa total friction loss

Example 2

What is the total pressure loss due to friction in 180 meters of 100 mm hose when 3 000 L/min is flowing?

C = 0.305 from Table 8.3

$$Q = \frac{liters}{100} \quad Q = \frac{3\ 000}{100} \quad Q = 30$$

$$L = \frac{hose\ length}{100} \quad L = \frac{180}{100} \quad L = 1.8$$

$FL = CQ^2L \quad FL = (0.305)(30)^2(1.8) \quad FL = (0.305)(900)(1.8)$

FL = 494 kPa total friction loss

Example 3

What is the total pressure loss due to friction in 75 meters of 45 mm hose when 600 L/min is flowing?

C = 24.6 from Table 8.3

$$Q = \frac{liters}{100} \quad Q = \frac{600}{100} \quad Q = 6$$

$$L = \frac{hose\ length}{100} \quad L = \frac{75}{100} \quad L = 0.75$$

$FL = CQ^2L \quad FL = (24.6)(6)^2(0.75) \quad FL = (24.6)(36)(0.75)$

FL = 664 kPa total friction loss

Determining Your Own Friction Loss Coefficients

If you wish to calculate more accurate results for the fire hose that is carried on your apparatus rather than use the results from the standard friction loss coefficients, it is recommended that you test your hose to determine the actual coefficients. If a department has the equipment available, the test procedure for determining friction loss for a particular hose is rather simple. Before performing these tests, however, several basic principles must be considered. In order to get results indicative of averages that can be expected on the fireground, it is necessary to use the same hose that would be used on the fireground. Furthermore, conduct tests on hose in service. Do not use hose sitting on a storage rack or hose that has never been put into service (unless new hose is about to be placed into service).

When conducting tests on a particular type of hose, test only one type at a time. For example, if the department is testing 77 mm cotton/polyester, double-jacket hose, do not have any hose of different size or construction mixed in with test hose. If the department regularly mixes different kinds of hose, it may be very difficult to get a friction loss coefficient suitable for all situations. Different hose have varying amounts of friction loss due to differences in construction, fabrics, rubber liners, couplings, and wear.

Test results are only as accurate as the equipment used to measure the results. Thus, it is important that all measuring devices (pitot tubes, in-line gauges, flowmeters, and so on) be properly calibrated for optimum results. The following is a list of the equipment needed to conduct these tests:

- Pitot tube or flowmeter (Figures 8.5 a and b)
- Two in-line gauges, preferably calibrated in increments of 50 kPa or less
- Hose to be tested
- Smoothbore nozzle if using pitot tube
- Any type nozzle if using flowmeter

The following is a step-by-step procedure for determining friction loss in any size hose:

Step 1: Lay out, on a level surface, lengths of the hose to be tested. Lay out 90 meters if the hose is in lengths of 15 meters; 120 meters if it is in lengths of 30 meters.

Step 2: Connect one end of the hoseline to a discharge on the pumper being used to carry out the tests. Connect a nozzle to the opposite end of the hoseline. If using a pitot tube to determine the nozzle pressure and corresponding flow of water, use a smoothbore nozzle. If a flowmeter is used to determine the flow, any nozzle is suitable (Figure 8.6).

Step 3: Insert gauge 1 in the hoseline at the connection between the first and second sections of hose away from the discharge. This connection should be 15 meters from the pumper if the hose being tested is in lengths of 15 meters. If 30-meter sections of hose are used, then gauge 1 should be 30 meters from the pumper.

Step 4: Insert gauge 2 at a distance of 60 meters from gauge 1, regardless of the length of the hose sections. Depending on the length of the hose sections, there should be either 15 meters or 30 meters between gauge 2 and the nozzle. If a portable flowmeter is being used, *DO NOT* in-

Figure 8.5a A pitot tube and gauge are used to measure the velocity pressure of a stream of water.

Figure 8.5b Flowmeters provide a direct reading of the volume of water being discharged through an opening.

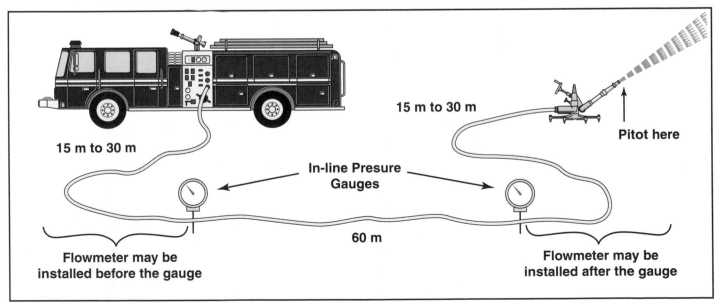

Figure 8.6 This layout can be used to determine the friction loss coefficient for any fire hose.

sert it anywhere between the gauges. It may be inserted in the hoseline anywhere but between the test gauges.

Step 5: When all appliances have been inserted into the hoseline, begin the tests. Supply water to the hoseline at a constant pump discharge pressure for the duration of each test run. Because three or four test runs should be made for each size hose, different pump discharge pressures may be used for different runs as long as they are not changed within the same test run. The pump discharge pressure should be sufficient to produce a satisfactory fire stream from the nozzle.

Step 6: Once water is flowing, record the pump discharge pressure, the readings from gauge 1 and gauge 2, and the flowmeter or pitot gauge readings (whichever is being used) in the appropriate spaces of Figure 8.7.

Figure 8.7
Friction Loss Coefficient Determination Chart

Date: ___ / ___ / ___   Hose Size _____ Millimeters   Hose Construction _____

Person Conducting Tests _____

Column 1	Column 2	Column 3	Column 4	Column 5	Column 6	Column 7	Column 8	Column 9
Test Run No.	Pump Discharge Pressure kPa	Pressure @ Gauge 1 kPa	Pressure @ Gauge 2 kPa	Nozzle Pressure* kPa	Flow From Flowmeter or by Equation**	$\left(\dfrac{\text{L/min}}{100}\right)^2$ or $\left(\dfrac{\text{Col. 6}}{100}\right)^2$	Friction Loss per 100 meters or $\left(\dfrac{\text{Col. 3-Col. 4}}{0.6}\right)$	C $\left(\dfrac{\text{Col. 8}}{\text{Col. 7}}\right)$
1								
2								
3								
4								

Total of all Col.9 Answers

Average C = ————————

No. of Tests Conducted

Average C =

* Not necessary if flowmeter is used
**L/min = 0.067 d² √NP

Figure 8.7 This chart may be used to determine the friction loss coefficient for fire hose.

Step 7: Complete Figure 8.7 as instructed on the table. This provides you with the friction loss coefficient for your particular hose.

Figure 8.8 represents the test results from a fire department that chose to determine its own coefficient for the 65 mm hose used within the department. The hose was tested at three different pump discharge pressures. By completing Figure 8.8, the department was able to deter-mine that the actual coefficient that should be used for its 65 mm hose is 2.34. This represents a 26 percent decrease from the standard coefficient for 65 mm hose used in this manual (C = 3.17).

Appliance Pressure Loss

Fireground operations often require the use of hoseline appliances. These appliances include reducers, increasers, gates, wyes, manifolds, aerial apparatus, and

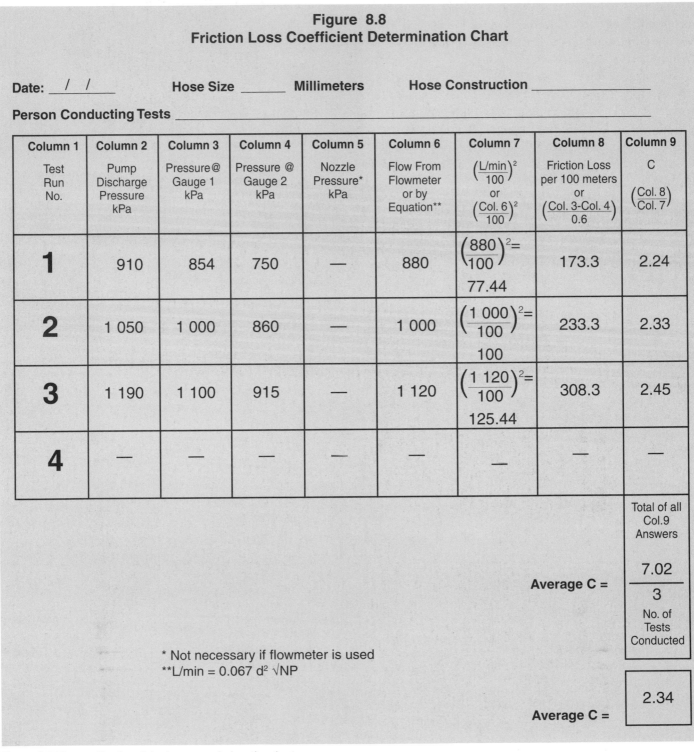

Figure 8.8
Friction Loss Coefficient Determination Chart

Date: ___/___/___ Hose Size _____ Millimeters Hose Construction _____

Person Conducting Tests _____

Column 1 Test Run No.	Column 2 Pump Discharge Pressure kPa	Column 3 Pressure @ Gauge 1 kPa	Column 4 Pressure @ Gauge 2 kPa	Column 5 Nozzle Pressure* kPa	Column 6 Flow From Flowmeter or by Equation**	Column 7 $\left(\frac{L/min}{100}\right)^2$ or $\left(\frac{Col.\ 6}{100}\right)^2$	Column 8 Friction Loss per 100 meters or $\left(\frac{Col.\ 3 - Col.\ 4}{0.6}\right)$	Column 9 C $\left(\frac{Col.\ 8}{Col.\ 7}\right)$
1	910	854	750	—	880	$\left(\frac{880}{100}\right)^2=$ 77.44	173.3	2.24
2	1 050	1 000	860	—	1 000	$\left(\frac{1\ 000}{100}\right)^2=$ 100	233.3	2.33
3	1 190	1 100	915	—	1 120	$\left(\frac{1\ 120}{100}\right)^2=$ 125.44	308.3	2.45
4	—	—	—	—	—	—	—	—

Total of all Col.9 Answers

Average C = $\dfrac{7.02}{3}$

No. of Tests Conducted

* Not necessary if flowmeter is used
**L/min = 0.067 d² √NP

Average C = 2.34

Figure 8.8 The results of each test are recorded on the chart.

standpipe systems (Figures 8.9 a through d). The amount of friction loss created varies with each type of appliance. Appliance friction loss is insignificant in cases where the total flow through the appliance is less than 1 400 L/min, and for the purpose of this manual, it is not included in calculations of flows less than that. Friction loss in these appliances varies with the rated capacity of the device and the flow. Generally, it is safe to assume a loss of 175 kPa or greater when flowing at the rated capacity. For siameses and wyes, the friction loss varies with the size of the device and the flow. **For this text, we will assume a 0 kPa loss for flows less than 1 400 L/min and a 70 kPa loss for** *each appliance* **(other than master stream devices) in a hose assembly when flowing 1 400 L/min or more.** Friction loss caused by handline nozzles is not considered in the calculations in this manual because it is generally insignificant in the overall pressure loss in a hose assembly. **For this manual, we will assume a friction loss of 175 kPa in all master stream appliances, regardless of the flow.**

As with fire hose, the only sure way to determine the exact friction loss of each appliance is for individual fire departments to conduct their own friction loss tests. A method of determining this friction loss is found in Appendix B.

Figure 8.9c A siamese.

Figure 8.9a A reducer.

Figure 8.9b A gated wye.

Figure 8.9d Aerial devices create friction loss.

Determining Elevation Pressure

Fireground operations often require the use of hoselines at varying elevations. Elevation pressure, which is created by elevation differences between the nozzle and the pump, must be considered when determining total pressure loss.

Water exerts a pressure of 10 kPa per meter of elevation. When a nozzle is operating at an elevation higher than the apparatus, this pressure is exerted back against the pump (Figure 8.10). To compensate for this pressure "loss," elevation pressure must be added to friction loss to determine total pressure loss. Operating a nozzle lower than the pump results in pressure pushing against the nozzle (Figure 8.11). This "gain" in pressure is compensated for by subtracting the elevation pressure from the total friction loss.

In order to simplify elevation pressure calculations on the fireground, the following formula may be used:

EQUATION B
EP = 10H
Where:

  EP = Elevation pressure in kPa

  10 = A constant

  H = Height in meters

It is generally easier to determine elevation pressure in a multistoried building by another method. By counting the number of stories of elevation, the following equation may be used.

EQUATION C
EP = 35 kPa x (number of stories –1)

The determination of elevation pressure is shown in the following examples.

Example 4
Calculate the total pressure loss due to elevation pressure for a hoseline operating at the top of a 30 meter hill (Figure 8.12).

  EP = 10H

  EP = (10)(30)

  EP = 300 kPa

Example 5
A hoseline operating on a ninth-floor structure fire is connected to the building's standpipe system (Figure 8.13). What is the total pressure loss due to elevation at the base of the standpipe system?

  EP = 35 kPa x (number of stories –1)

  EP = 35 kPa x (9 –1)

  EP = (35)(8)

  EP = 280 kPa

Figure 8.10 In this scenario, an elevation pressure loss would be noted at the pump.

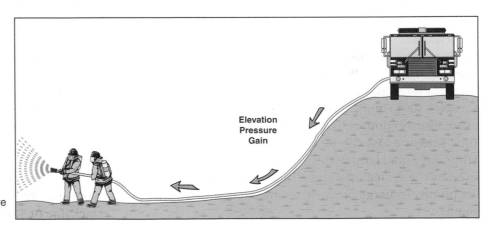

Figure 8.11 In this scenario, an elevation pressure gain would be noted at the pump.

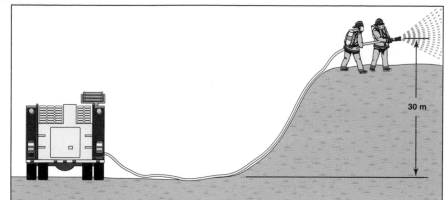

Figure 8.12 Example 4.

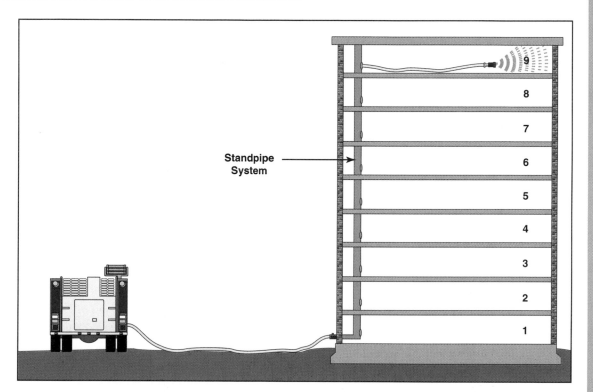

Standpipe System

Figure 8.13 Example 5.

Hose Layout Applications

Hose layouts include single hoselines, multiple hoselines, wyed or manifold hoselines, and siamesed hoselines. In each type of layout, friction loss is affected by such factors as hose diameter and whether the hose layouts are of equal or unequal length.

As was mentioned previously, the combination of friction loss and elevation pressure is referred to as total pressure loss (TPL). Pressure changes are possible due to hose friction loss, appliance friction loss (when flows exceed 1 400 L/min), and any pressure loss or gain due to elevation. By adding all the affecting pressure losses, the total pressure loss can be determined for any hose lay. Ultimately (and later in this chapter) this information is used by the driver/operator to determine the appropriate pump discharge pressure at which to operate the fire pump.

Hose layouts can be divided into two basic categories: simple hose layouts and complex hose layouts. The principles used in determining the total pressure loss in each category are essentially the same. The complexity of the friction loss calculation determines whether the layout is simple or complicated.

Simple Hose Layouts

Simple hose layouts include single hoselines, equal length multiple hoselines, equal wyed hoselines, and equal length siamesed hoselines. These more frequently used hose layouts generally do not present great difficulty in determining total pressure loss.

Single Hoseline

The most commonly used hose lay is the single-hoseline layout (Figure 8.14). This hose lay, whether used as an attack line or supply line, presents the simplest friction

Figure 8.14 Smaller incidents are handled with a single attack line. *Courtesy of Ron Jeffers.*

loss calculations. The following examples demonstrate how to determine the total pressure loss in a single-hoseline layout.

Example 6

A pumper is supplying a 90 m hoseline with 500 L/min flowing. The hoseline is composed of 60 meters of 65 mm hose reduced to 30 meters of 38 mm hose (Figure 8.15). What is the pressure loss due to friction in the hose assembly?

65 mm Hose

 C = 3.17 from Table 8.3

$$Q = \frac{liters}{100} \quad Q = \frac{500}{100} \quad Q = 5$$

$$L = \frac{meters}{100} \quad L = \frac{60}{100} \quad L = 0.6$$

FL = CQ²L FL = (3.17)(5)²(0.6) FL = (3.17)(25)(0.6)

FL = 47.6 kPa

38 mm Hose

 C = 38 from Table 8.3

$$Q = \frac{liters}{100} \quad Q = \frac{500}{100} \quad Q = 5$$

$$L = \frac{meters}{100} \quad L = \frac{30}{100} \quad L = 0.3$$

FL = CQ²L FL = (38)(5)²(0.3) FL = (38)(25)(0.3)

FL = 285 kPa

Total Pressure Loss
TPL = 47.6 + 285 = **332.6 kPa pressure loss in the hose assembly**

Example 7

A fire is discovered on the third floor of a structure. An arriving engine company proceeds to the second floor of the structure and connects 45 m of 45 mm hose to the standpipe outlet. Determine the total pressure loss due to friction and elevation pressure at the standpipe system fire department connection when 700 L/min is flowing. (Disregard friction loss in the standpipe system.)

45 mm Hose

 C = 24.6 from Table 8.3

$$Q = \frac{liters}{100} \quad Q = \frac{700}{100} \quad Q = 7$$

$$L = \frac{meters}{100} \quad L = \frac{45}{100} \quad L = 0.45$$

FL = CQ²L FL = (24.6)(7)²(0.45) FL = (24.6)(49)(0.45)

FL = 542.4 kPa

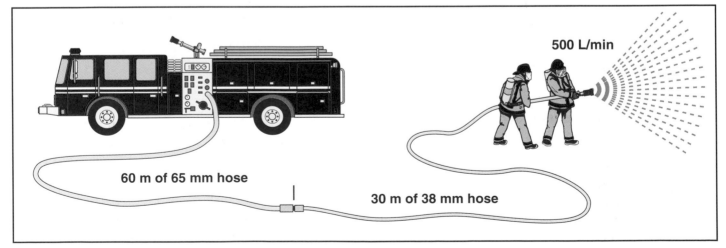

500 L/min

60 m of 65 mm hose

30 m of 38 mm hose

Figure 8.15 Example 6.

Elevation Pressure

EP = 35 kPa x (number of stories – 1)

EP = 35 kPa x (3 – 1)

EP = (35)(2)

EP = 70 kPa

Total Pressure Loss

TPL = 542.4 + 70 = **612.4 kpa pressure loss in the hose assembly**

Multiple Hoselines (Equal Length)

Fireground operations may involve the use of more than one hoseline from a pumper (Figure 8.16). These hoselines, whether of equal or varying diameters, are often the same length.

When determining the friction loss in equal length multiple lines whose diameters are the same, it is only necessary to perform calculations for one line. This is because each of the other hoselines have approximately the same friction loss. Conversely, when the diameters of the hoselines vary, friction loss calculations must be made for each hoseline. The pump discharge pressure is then set for the highest pressure. The valve on the hose requiring the lesser amount of pressure is partially closed to reduce the pressure from that discharge. The following example shows how to determine total pressure loss when multiple hoselines are used.

Example 8

Three 65 mm hoselines, each 120 meters long, are laid with 25 mm tips. Each line has a nozzle pressure of 350 kPa. What is the total pressure loss due to friction in each hoseline?

L/MIN = $0.067d^2\sqrt{NP}$

L/MIN = $(0.067)(25)^2(\sqrt{350})$

L/MIN = 783 L/min

C = 3.17 from Table 8.3

$Q = \dfrac{liters}{100}$ $Q = \dfrac{783}{100}$ $Q = 7.83$

$L = \dfrac{meters}{100}$ $L = \dfrac{120}{100}$ $L = 1.2$

FL=CQ^2L FL=$(3.17)(7.83)^2(1.2)$ FL=$(3.17)(61.3)(1.2)$

FL = 233.2 kPa in each hoseline

Wyed Hoselines (Equal Length)

A common hose assembly involves the use of one supply line, usually a 65 mm, 77 mm, or 100 mm hose, wyed into two or more smaller attack lines. These attack lines may range in size from 38 mm to 65 mm and are generally equal in length. It is important that these wyed lines be

Figure 8.16 The pump operator may be required to supply multiple hoselines at a large fire incident. *Courtesy of Ron Jeffers.*

the same length and diameter in order to avoid two different nozzle pressures and an exceptionally difficult friction loss problem.

Determining friction loss in wyed hoselines of equal length and diameter is not a complex problem. When the nozzle pressure, hose length, and diameter are the same on both lines, an equal split of the total water flowing occurs at the wye appliance. This enables only one of the wyed hoselines to be considered when computing the total pressure loss. The following steps may be used when calculating the friction loss in an equal length wyed hoseline assembly.

Step 1: Compute the number of hundreds of L/min flowing in each wyed hoseline by using the equation:

$$Q = \frac{\text{flow rate (L/min)}}{100}$$

Step 2: Determine the friction loss in one of the wyed attack lines using the equation:

$$FL = CQ^2L$$

Step 3: Compute the total number of hundreds of L/min flowing through the supply line to the wye by using the following equation:

$$Q_{Total} = \frac{(\text{L/min in attack line 1}) + (\text{L/min in attack line 2})}{100}$$

Step 4: Determine the friction loss in the supply line using Equation A, FL = $(C)(Q_{Total})^2(L)$

Step 5: Add the friction loss from the supply line, one of the attack lines, 70 kPa for the wye appliance (if the total flow exceeds 1 400 L/min), and elevation pressure (if applicable) to determine the total pressure loss.

The following examples illustrate how to compute the total pressure loss in an equal length wyed hose assembly.

Example 9

Determine the pressure loss due to friction in a hose assembly in which two 38 mm hoselines, each 30 meters long and flowing 400 L/min, are wyed off of 120 meters of 65 mm hose (Figure 8.17).

38 mm Hose

C = 38 from Table 8.3

$$Q = \frac{liters}{100} \quad Q = \frac{400}{100} \quad Q = 4$$

$$L = \frac{meters}{100} \quad L = \frac{30}{100} \quad L = 0.3$$

FL = CQ²L   FL = (38)(4)²(0.3)   FL = (38)(16)(0.3)

FL = 182.4 kPa in each 38 mm hoseline

65 mm Hose

C = 3.17 from Table 8.3

$$Q_{Total} = \frac{(L/min\ in\ attack\ line\ 1) + (L/min\ in\ attack\ line\ 2)}{100}$$

$$Q_{Total} = \frac{400 + 400}{100} \quad Q_{Total} = \frac{800}{100}$$

$$Q_{Total} = 8$$

$$L = \frac{meters}{100} \quad L = \frac{120}{100} \quad L = 1.2$$

FL = (C)(Q_{total})²(L)   FL = (3.17)(8)²(1.2)

FL = (3.17)(64)(1.2)

FL = 243.5 kPa in the 65 mm hoseline

Total Pressure Loss

TPL = 182.4 + 243.5 = **425.9 kPa total pressure loss in this hose assembly**

NOTE: The total flow through this hose assembly was less than 1 400 L/min, so it was not necessary to consider the pressure loss in the wye.

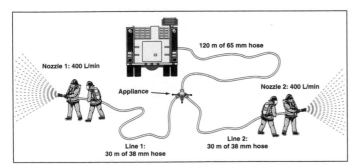

Figure 8.17 Example 9.

Example 10

Determine the pressure loss due to friction in a hose assembly in which two 65 mm hoselines, each 60 meters long and flowing 1 000 L/min, are wyed off of 150 meters of 100 mm hose.

65 mm Hose

C = 3.17 from Table 8.3

$$Q = \frac{liters}{100} \quad Q = \frac{1\ 000}{100} \quad Q = 10$$

$$L = \frac{meters}{100} \quad L = \frac{60}{100} \quad L = 0.6$$

FL = CQ²L   FL = (3.17)(10)²(0.6)   FL = (3.17)(100)(0.6)

FL = 190 kPa in each 65 mm hoseline

100 mm Hose

C = 0.305 from Table 8.3

$$Q_{Total} = \frac{(L/min\ in\ attack\ line\ 1) + (L/min\ in\ attack\ line\ 2)}{100}$$

$$Q_{Total} = \frac{1\ 000 + 1\ 000}{100}$$

$$Q_{Total} = \frac{2\ 000}{100} \quad Q_{Total} = 20$$

$$L = \frac{meters}{100} \quad L = \frac{150}{100} \quad L = 1.5$$

FL = (C)(Q_{total})²(L)   FL = (0.305)(20)²(1.5)

FL = (0.305)(400)(1.5)

FL = 183 kPa in the 100 mm hoseline

Total Pressure Loss

TPL = 190 + 183 + 70 = **443 kPa total pressure loss in this hose assembly**

NOTE: The total flow through this hose assembly exceeded 1 400 L/min, so it was necessary to add 70 kPa for the pressure loss in the wye.

Siamesed Hoselines (Equal Length)

When the flow rate is increased through hose, additional pressure is needed to overcome friction in the hose. When conditions require great volumes of water, or when hose lays are long, the friction loss in the hose is also greater. To keep friction loss within reasonable limits, firefighters may lay two or more parallel hoselines and siamese them together at a point close to the fire. When two hoselines of equal length are siamesed to supply a fire stream, friction loss is approximately 25 percent of that of a single hoseline at the same nozzle pressure. When three hoselines of equal length are siamesed, the friction loss is approximately 10 percent of that of a single line if equal nozzle pressure is maintained.

There are a number of ways to determine the amount of friction loss in siamesed hoselines. The easiest method is to use Equation A, just as was done for determining the amount of pressure loss due to friction in single hoselines. When calculating friction loss in siamesed lines, however, it is necessary to use a different set of coefficients (C) than for single hoselines. These coefficients are found in Table 8.4. The following is a step-by-step procedure for determining friction loss in siamesed lines:

Step 1: Compute the total number of hundreds of L/min flowing by using the equation:
$$Q = \frac{\text{L/min flowing}}{100}$$

Step 2: Determine the friction loss in the attack line using Equation A (FL = CQ²L). Use Table 8.3 for the coefficient in this step.

Step 3: Determine the amount of friction loss in the siamesed lines using Equation A (FL = CQ²L). Use Table 8.4 to obtain the coefficient in this step.

Step 4: Add the friction loss from the siamesed lines, attack line, 70 kPa for the siamese appliance (if flow is greater than 1 400 L/min), and elevation pressure (if applicable) to determine the total pressure loss.

Table 8.4
Friction Loss Coefficients — Siamesed Lines of Equal Length

Hose Diameter and Type (mm)	Coefficient C
Two 65 mm	0.789
Three 65 mm	0.347
Two 77 mm with 65 mm couplings	0.316
Two 77 mm with 77 mm couplings	0.268
One 77 mm with 65 mm couplings, one 65 mm	0.473
One 77 mm with 77 mm couplings, one 65 mm	0.426
Two 65 mm lines, one 77 mm with 65 mm couplings	0.253
Two 77 mm with 65 mm couplings, one 65 mm	0.189

Example 11

Determine the pressure loss due to friction in a hose assembly in which two 77 mm hoses with 65 mm couplings, each 300 meters long, are being used to supply a siamese to which 100 meters of 65 mm hose is attached (Figure 8.18). The solid stream nozzle on the 65 mm hose has a 32 mm tip with a nozzle pressure of 350 kPa.

L/min = 0.067 d²√NP

L/min = (0.067) (32)²(√350)

L/min = 1 284 L/min

65 mm Attack Line

C = 3.17 from Table 8.3

$Q = \dfrac{\text{L/min}}{100}$ $\quad Q = \dfrac{1\ 284}{100} \quad$ Q = 12.84

$L = \dfrac{\text{meters}}{100}$ $\quad L = \dfrac{100}{100} \quad$ L = 1

FL = CQ²L FL = (3.17)(12.84)²(1) FL = (3.17)(164.9)(1)

FL = 523 kPa loss in the 65 mm hoseline

Siamesed Lines

C = 0.316 from Table 8.4

$Q = \dfrac{\text{L/min}}{100}$ $\quad Q = \dfrac{1\ 284}{100} \quad$ Q = 12.84

$L = \dfrac{\text{meters}}{100}$ $\quad L = \dfrac{300}{100} \quad$ L = 3

FL = CQ²L FL = (0.316)(12.84)²(3)

FL = (0.316)(164.9)(3)

FL = 156.3 kPa in the siamesed lines

Total Pressure Loss

TPL = 523 + 156.3 = **679.3 kPa total pressure loss in this hose assembly**

NOTE: The total flow through this hose assembly was less than 1 400 L/min, so it was not necessary to consider the pressure loss in the siamese.

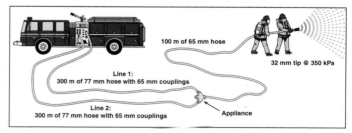

Figure 8.18 Example 11.

Example 12

Determine the total pressure loss in a hose assembly in which two 65 mm hoses, each 225 meters long, are being used to supply a siamese to which 60 meters of 65 mm hose is attached. The nozzle on the attack line is flowing 1 200 L/min. The nozzle is located 10 meters uphill of the siamese.

65 mm Attack Line

C = 3.17 from Table 8.3

$Q = \dfrac{L/min}{100}$ $Q = \dfrac{1\ 200}{100}$ $Q = 12$

$L = \dfrac{meters}{100}$ $L = \dfrac{60}{100}$ $L = 0.6$

$FL = CQ^2L$ $FL = (3.17)(12)^2(0.6)$

$FL = (3.17)(144)(0.6)$ $FL = 274$ kPa

$EP = 10H$

$EP = (10)(10)$

$EP = 100$ kPa

$TPL_{Attack\ Line} = FL + EP$ $TPL_{Attack\ Line} = 274 + 100$

$TPL_{Attack\ Line}$ = 374 kPa

Siamesed Lines

C = 0.789 from Table 8.4

$Q = \dfrac{L/min}{100}$ $Q = \dfrac{1\ 200}{100}$ $Q = 12$

$L = \dfrac{meters}{100}$ $L = \dfrac{225}{100}$ $L = 2.25$

$FL = CQ^2L$ $FL = (0.789)(12)^2(2.25)$

$FL = (0.789)(144)(2.25)$

FL = 255.6 kPa in the siamesed lines

Total Pressure Loss

TPL = 374 + 255.6 = **629.6 kPa total pressure loss in this hose assembly**

Complex Hose Layouts

Fireground operations may involve the use of hose layouts that challenge the mathematical ability of the driver/operator. These hose lays, which include standpipe operations, unequal length multiple and wyed hoselines, manifold hoselines, and master streams, require the driver/operator to make additional calculations to determine the total pressure loss.

Standpipe Operations

In most cases, fire departments have predetermined pressures that the driver/operator is expected to pump into the fire department connection (FDC) of a standpipe system. These pressures are contained in the department's SOPs, in the pre-incident plan for that particular property, or on a faceplate adjacent to the FDC

(Figures 8.19 a through c). In order to be able to determine the required pressure for the standpipe system, it is necessary to determine the total pressure loss. The following example shows how to accomplish this. Treat the FDC like any other hose appliance. If the flow in the system exceeds 1 400 L/min, add 70 kPa of friction loss for the FDC.

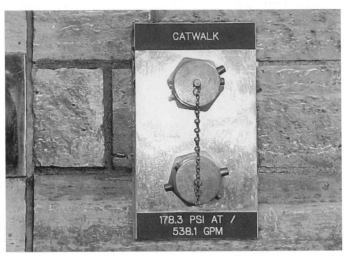

Figure 8.19a Some fire department connections are marked with the required supply pressure on them. *Courtesy of the City of Phoenix Department of Development Services.*

Figure 8.19b In some cases, the pumping instructions are located on a sign adjacent to a fire department connection. *Courtesy of Mount Prospect (IL) Fire Department.*

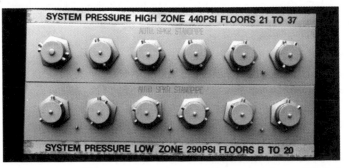

Figure 8.19c In this example, each zone of the system has a different required FDC inlet pressure. *Courtesy of Bil Murphy.*

Example 13

A fire is discovered on the fifth floor of a structure. The arriving engine company proceeds to the fourth floor of the occupancy and connects three 15 meter sections of 65 mm hose to the standpipe outlet. The hose is equipped with a solid stream nozzle that has a 32 mm tip. The standpipe is 150 mm in diameter. What is the total pressure loss due to friction and elevation pressure in the standpipe system and attached hose?

$$L/min = 0.067\ d^2\sqrt{NP}$$
$$L/min = (0.067)\ (32)^2(\sqrt{350})$$
$$L/min = 1\ 284\ L/min$$

Friction Loss in Standpipe

Based on 3 meters per floor, assume that 9 meters of standpipe is used.

C = 0.083 from Table 8.3

$$Q = \frac{L/min}{100} \quad Q = \frac{1\ 284}{100} \quad Q = 12.84$$

$$L = \frac{meters}{100} \quad L = \frac{9}{100} \quad L = 0.09$$

$$FL = CQ^2L \quad FL = (0.083)(12.84)^2(0.09)$$
$$FL = (0.083)(164.9)(0.09)$$

FL =1.23 kPa loss in the standpipe

65 mm Attack Line

C = 3.17 from Table 8.3

$$Q = \frac{L/min}{100} \quad Q = \frac{1\ 284}{100} \quad Q = 12.84$$

$$L = \frac{meters}{100} \quad L = \frac{45}{100} \quad L = 0.45$$

$$FL = CQ^2L \quad FL = (3.17)(12.84)^2(0.45)$$
$$FL = (3.17)(164.9)(0.45)$$

FL = 235.2 kPa loss in the 65 mm hoseline

Elevation Pressure Loss

$$EP = (35)(No.\ of\ Stories\ -1)$$
$$EP = (35)(5\ -1)$$
$$EP = (35)(4)$$
$$EP = 140\ kPa$$

Total Pressure Loss

TPL = 1.23 + 235.2 + 140 = **376.5 kPa total pressure loss in this standpipe and hose assembly**

One of the purposes of the preceding example was to demonstrate that friction loss within hard piping is minimal. In the example, the friction loss in the standpipe was less than 2 kPa. Therefore, it is not usually necessary to calculate this friction loss because it has little effect on the overall problem facing the driver/operator. The fact that hard piping, or connections, pose minimal friction loss concerns is important to remember for the remainder of this manual. It is this principle that allows us to use 65 mm couplings on 77 mm hose with minimal flow restrictions when compared to 77 mm hose with 77 mm couplings.

Multiple Hoselines (Unequal Length)

Occasionally, a situation may arise where multiple hoselines of equal or unequal diameter are not the same length. This can result from the addition of a new hoseline to a pumper or the addition of hose lengths to an existing line. When unequal length hoselines are used, the amount of friction loss varies in each line. For this reason, friction loss must be calculated in each hoseline.

Example 14

Two 65 mm hoselines, one 150 meters long and the other 100 meters long, are equipped with 1 000 L/min fog nozzles (Figure 8.20). Find the total pressure loss due to friction in each line.

Line 1

C = 3.17 from Table 8.3

$$Q = \frac{L/min}{100} \quad Q = \frac{1\ 000}{100} \quad Q = 10$$

$$L = \frac{meters}{100} \quad L = \frac{150}{100} \quad L = 1.5$$

$$FL = CQ^2L \quad FL = (3.17)(10)^2(1.5) \quad FL = (3.17)(100)(1.5)$$

FL = 475.5 kPa loss in Line 1

Line 2

C = 3.17 from Table 8.3

$$Q = \frac{L/min}{100} \quad Q = \frac{1\ 000}{100} \quad Q = 10$$

$$L = \frac{meters}{100} \quad L = \frac{100}{100} \quad L = 1$$

$$FL = CQ^2L \quad FL = (3.17)(10)^2(1) \quad FL = (3.17)(100)(1)$$

FL = 317 kPa loss in Line 2

Total Pressure Loss

The total pressure loss in the system is based on the highest loss of the two lines, which in this case would be Line 1 at **475.5 kPa.**

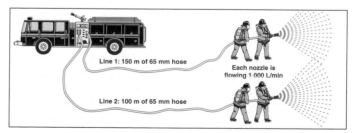

Figure 8.20 Example 14.

Example 15

A pumper is supplying four lines. Two lines are 60 meters of 45 mm hose equipped with a 19 mm tip operating at 350 kPa. The other two lines are 45 meters of 38 mm hose equipped with a 16 mm tip operating at 350 kPa. Determine the total pressure loss due to friction in each hoseline (Figure 8.21).

Lines 1 & 2

L/min = 0.067 d²√NP

L/min = (0.067) (16)²(√350)

L/min = 320.9 L/min

C = 38 from Table 8.3

$Q = \dfrac{L/min}{100}$ $Q = \dfrac{320.9}{100}$ $Q = 3.209$

$L = \dfrac{meters}{100}$ $L = \dfrac{45}{100}$ $L = 0.45$

FL=CQ²L FL=(38)(3.209)²(0.45) FL=(38)(10.3)(0.45)

FL = 176.1 kPa loss in each 38 mm hoseline

Lines 3 & 4

L/min = 0.067 d²√NP

L/min = (0.067) (19)²(√350)

L/min = 452.5 L/min

C = 24.6 from Table 8.3

$Q = \dfrac{L/min}{100}$ $Q = \dfrac{452.5}{100}$ $Q = 4.525$

$L = \dfrac{meters}{100}$ $L = \dfrac{60}{100}$ $L = 0.6$

FL = CQ²L   FL = (24.6)(4.525)²(0.6)

FL = (24.6)(20.48)(0.6)

FL = 302.2 kPa loss in each 45 mm hoseline

Total Pressure Loss

The total pressure loss in the system is based on the highest loss of the two sets of lines, which in this case would be Lines 3 and 4 at **302.2 kPa**.

Wyed Hoselines (Unequal Length) and Manifold Hoselines

The addition of hose lengths to an existing wyed hoseline assembly may result in unequal length attack lines. Generally, the principles used to determine friction loss in equal length wyed hoselines also apply to unequal length wyed hoselines. Because the lengths of the attack lines are different, the split of the total water flowing is not equal at the wye appliance. For this reason, it is necessary to determine friction loss for each of the unequal length wyed lines.

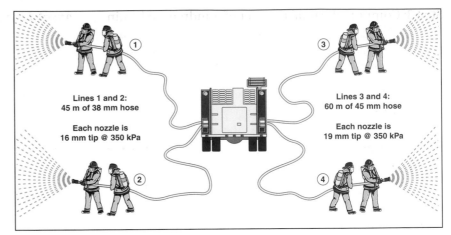

Figure 8.21 Example 15.

Some fireground operations involve the use of a water thief or manifold appliance (Figures 8.22 a and b). Generally, a hose assembly involving a manifold consists of one large diameter hoseline supplying several smaller attack lines. These attack lines, as with any other hose assembly, may be equal or unequal in length and diameter. Keep in mind that when hose lengths are unequal in length and/or diameter, the total pressure loss in the system is based on the highest pressure loss in any of the lines. In a real-life situation, the hoselines requiring less than the maximum pressure are gated down at the manifold. Gating down the lines at the manifold so that the proper pressure is achieved in each line is, at best, a guessing game, unless the manifold is equipped with a pressure gauge on each discharge.

As with a wyed hoseline, friction loss in a manifold assembly is not overly difficult to determine. The following steps can be used to calculate friction loss in unequal length wyed or manifold hoselines.

Step 1: Compute the number of hundreds of L/min flowing in each of the wyed hoselines by using the following equation:

$$Q = \frac{discharge\ L/min}{100}$$

Step 2: Determine the friction loss in each of the wyed lines using Equation A:

$$FL = CQ^2L$$

Figure 8.22a A water thief.

Figure 8.22b An LDH manifold.

Step 3: Compute the total number of hundreds of L/min flowing in the supply line to the wye or manifold by adding the sum of the flows in the attack lines and dividing by 100.

Step 4: Determine the friction loss in the supply line using the equation: $FL = (C)(Q_{total})^2(L)$

Step 5: Add the friction loss from the supply line, the wye or manifold appliance (if total flow is greater than 1 400 L/min), elevation loss, and the wyed line with the greatest amount of friction loss to determine the total pressure loss.

The following examples illustrate how to compute the total pressure loss in an unequal wyed hose assembly and a manifold hose assembly.

Example 16

Determine the total pressure loss due to friction and elevation pressure for a hoseline assembly in which one 120 meter, 77 mm hose with 65 mm couplings is supplying two attack lines (Figure 8.23). The first attack line consists of 60 meters of 45 mm hose that is stretched through the front door of a structure. The first line is flowing 600 L/min. The second attack line is 45 meters of 38 mm hose that is carried up a ground ladder and stretched through a second floor window. The second line is flowing 400 L/min.

Attack Line 1 (45 mm)

C = 24.6 from Table 8.3

$Q = \dfrac{\text{L/min}}{100}$ $Q = \dfrac{600}{100}$ $Q = 6$

$L = \dfrac{\text{meters}}{100}$ $L = \dfrac{60}{100}$ $L = 0.6$

$FL = CQ^2L$ $FL = (24.6)(6)^2(0.6)$ $FL = (24.6)(36)(0.6)$

FL = 531.4 kPa loss in Attack Line 1

Attack Line 2 (38 mm)

C = 38 from Table 8.3

$Q = \dfrac{\text{L/min}}{100}$ $Q = \dfrac{400}{100}$ $Q = 4$

$L = \dfrac{\text{meters}}{100}$ $L = \dfrac{45}{100}$ $L = 0.45$

$FL = CQ^2L$ $FL = (38)(4)^2(0.45)$ $FL = (38)(16)(0.45)$

FL = 273.6 kPa

EP = (35)(No. of Stories −1)

EP = (35)(2 −1)

EP = (35)(1)

EP = 35 kPa

TPL = 273.6 + 35 = **308.6 kPa loss in Attack Line 2**

Supply Line (77 mm with 65 mm couplings)

C = 1.27 from Table 8.3

$Q_{Total} = \dfrac{(\text{Attack Line 1 L/min}) + (\text{Attack Line 2 L/min})}{100}$

$Q_{Total} = \dfrac{600 + 400}{100}$ $Q_{Total} = \dfrac{1\,000}{100}$

$Q_{Total} = 10$

$L = \dfrac{\text{meters}}{100}$ $L = \dfrac{120}{100}$ $L = 1.2$

$FL = (C)(Q_{total})^2(L)$ $FL = (1.27)(10)^2(1.2)$

$FL = (1.27)(100)(1.2)$

FL = 152.4 kPa in the supply line

Total Pressure Loss

The total pressure loss in the system is based on the highest loss of the two attack lines, which in this case would be Line 1, and the friction loss in the supply line. Because the flow rate was less than 1 400 L/min, no appliance loss is required.

TPL = 531.4 + 152.4 = **683.8 kPa loss is this hose assembly**

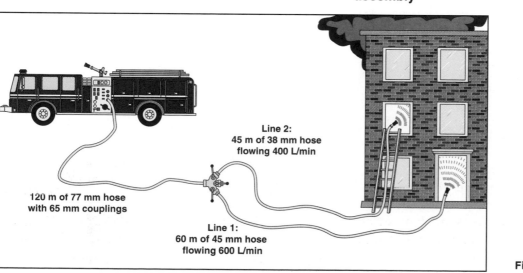

Line 2:
45 m of 38 mm hose
flowing 400 L/min

120 m of 77 mm hose
with 65 mm couplings

Line 1:
60 m of 45 mm hose
flowing 600 L/min

Figure 8.23 Example 16.

Example 17

Determine the total pressure loss due to friction for a hoseline assembly in which one 150 meter, 77 mm hose with 77 mm couplings is supplying three attack lines that are attached to a water thief. The first attack line consists of 45 meters of 38 mm hose flowing 500 L/min. The second attack line is 30 meters of 38 mm hose flowing 400 L/min. The third attack line is 45 meters of 65 mm hose flowing 900 L/min.

Attack Line 1 (38 mm)

C = 38 from Table 8.3

$$Q = \frac{L/min}{100} \quad Q = \frac{500}{100} \quad Q = 5$$

$$L = \frac{meters}{100} \quad L = \frac{45}{100} \quad L = 0.45$$

FL = CQ²L   FL = (38)(5)²(0.45)   FL = (38)(25)(0.45)

FL = 427.5 kPa loss in Attack Line 1

Attack Line 2 (38 mm)

C = 38 from Table 8.3

$$Q = \frac{L/min}{100} \quad Q = \frac{400}{100} \quad Q = 4$$

$$L = \frac{meters}{100} \quad L = \frac{30}{100} \quad L = 0.3$$

FL = CQ²L   FL = (38)(4)²(0.3)   FL = (38)(16)(0.3)

FL = 182.4 kPa loss in Attack Line 2

Attack Line 3 (65 mm)

C = 3.17 from Table 8.3

$$Q = \frac{L/min}{100} \quad Q = \frac{900}{100} \quad Q = 9$$

$$L = \frac{meters}{100} \quad L = \frac{45}{100} \quad L = 0.45$$

FL = CQ²L   FL = (3.17)(9)²(0.45)   FL = (3.17)(81)(0.45)

FL = 115.5 kPa loss in Attack Line 3

Supply Line (77 mm with 77 mm couplings)

C = 1.06 from Table 8.3

$$Q_{Total} = \frac{(Sum\ of\ Attack\ Lines\ 1,\ 2,\ \&\ 3\ L/min)}{100}$$

$$Q_{Total} = \frac{500 + 400 + 900}{100} \quad Q_{Total} = \frac{1\ 800}{100}$$

$$Q_{Total} = 18$$

$$L = \frac{meters}{100} \quad L = \frac{150}{100} \quad L = 1.5$$

FL = (C)(Q$_{total}$)²(L)   FL = (1.06)(18)²(1.5)

FL = (1.06)(324)(1.5)

FL = 515.2 kPa in the supply line

Total Pressure Loss

The total pressure loss in the system is based on the highest loss of the three attack lines, which in this case would be Line 1, and the friction loss in the supply line. Because the total flow rate exceeds 1 400 L/min, it is necessary to add the 70 kPa appliance loss.

TPL = 427.5 + 515.2 + 70 = **1 012.7 kPa loss in this hose assembly**

Example 18

Determine the total pressure loss due to friction for a hoseline assembly in which one 210 meter, 125 mm hose is supplying three hoselines that are attached to a large diameter hose manifold (Figure 8.24). Two of the hoselines are 45 meters of 77 mm hose with 65 mm couplings that are supplying a portable master stream device. The master stream is discharging 560 kPa through a 38 mm tip. Add 175 kPa for the appliance loss in the master stream device. The third hoseline is 45 meters of 65 mm hose flowing 1 100 L/min.

Master Stream

L/min = 0.067 d²√NP

L/min = (0.067) (38)²(√560)

L/min = 2 290 L/min

C = 0.316 from Table 8.4

$$Q = \frac{L/min}{100} \quad Q = \frac{2\ 290}{100} \quad Q = 22.9$$

$$L = \frac{meters}{100} \quad L = \frac{45}{100} \quad L = 0.45$$

FL = CQ²L   FL = (0.316)(22.9)²(0.45)

FL = (0.316)(524.4)(0.45)

FL = 74.6 kPa

TPL = 74.6 + 175 = **249.6 kPa loss in the lines supplying the master stream**

65 mm Handline

C = 3.17 from Table 8.3

$$Q = \frac{L/min}{100} \quad Q = \frac{1\ 100}{100} \quad Q = 11$$

$$L = \frac{meters}{100} \quad L = \frac{45}{100} \quad L = 0.45$$

FL = CQ²L   FL = (3.17)(11)²(0.45)   FL = (3.17)(121)(0.45)

FL = 172.6 kPa loss in the 65 mm handline

Supply Line (125 mm)

C = 0.138 from Table 8.3

$$Q_{Total} = \frac{(Master\ Stream\ L/min) + (Hand\ Line\ L/min)}{100}$$

$$Q_{Total} = \frac{2\ 290 + 1\ 100}{100} \quad Q_{Total} = \frac{3\ 390}{100}$$

$$Q_{Total} = 33.9$$

$$L = \frac{meters}{100} \quad L = \frac{210}{100} \quad L = 2.1$$

FL = (C)(Q$_{Total}$)^2(L)   FL = (0.138)(33.9)^2(2.1)

FL = (0.138)(1 149.2)(2.1)

FL = 333 kPa pressure loss in the supply line

Total Pressure Loss

The total pressure loss in the system is based on the highest loss of the three attack lines, which in this case would be the lines supplying the master stream, and the friction loss in the supply line. Because the total flow rate exceeds 1 400 L/min, it is necessary to add the 70 kPa appliance loss.

TPL = 249.6 + 333 + 70 = **652.6 kPa loss in this hose assembly**

Master Streams

The principles upon which master streams are developed are essentially the same as for other fire streams. Master streams, however, require a greater volume of water than do handlines. Multiple hoselines, siamesed hoselines, or large-diameter single hoselines are generally used as supply lines for these large volumes of water. If a master stream requires a water flow greater than the capacity of a single pumper, multiple pumpers may be used to supply the master stream appliance. As stated earlier in this manual, add a 175 kPa pressure loss to all calculations involving master streams devices.

The hose lays used to supply master streams are essentially the same as those used for other fire streams. For this reason, the concepts used in determining friction loss are also the same. The only difference occurs if unequal length or diameter hoselines are used to supply a master stream appliance. If this situation is present, use an average of the hose lengths for ease of calculation. To obtain an average, add the length of each hoseline and then divide by the number of hoselines being used. Use the coefficients for wyed hoselines contained in Table 8.4 may then be used with the total flow through the nozzle to complete the friction loss calculations. In some cases, multiple lines of equal diameter but unequal length are being used, and no coefficient for that combination exists in Table 8.4. In that case, average the length of the lines and assume an equal amount of water is going through each hose.

In this manual, the aerial devices with piped waterways are treated in the same manner as master stream appliances: using a friction loss of 175 kPa to include the intake, internal piping, and nozzle. Elevation pressure loss is calculated separately. For exact figures, consult the aerial manufacturer for specific friction loss data or perform field tests to derive data. If a traditional detachable ladder pipe and hose assembly is used, the friction loss within the siamese (if used), hose, and ladder pipe must all be accounted for. The following examples show how to calculate the total pressure loss in a master stream hose layout.

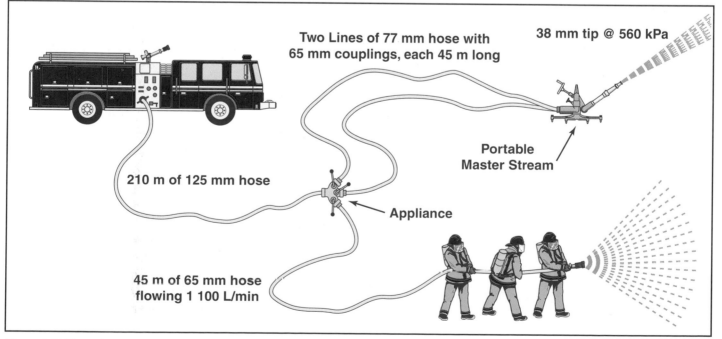

Two Lines of 77 mm hose with 65 mm couplings, each 45 m long

38 mm tip @ 560 kPa

210 m of 125 mm hose

Portable Master Stream

Appliance

45 m of 65 mm hose flowing 1 100 L/min

Figure 8.24 Example 18.

Example 19

Determine the total pressure loss in a hose assembly when a 12 000 L/min master stream device is being supplied by three 125 mm hoselines (Figures 8.25 a and b). Two of the hoseliness are 300 meters in length and one is 400 meters. Assume 4 000 L/min is flowing through each hoseline.

Friction Loss in the 125 mm Hoselines

C = 0.138 from Table 8.3

$$Q = \frac{L/min}{100} \quad Q = \frac{4\ 000}{100} \quad Q = 40$$

$$\text{Average Length } (L_{Ave}) = \frac{(\text{Sum of Hose Lengths})}{3}$$

$$L_{Ave} = \frac{(300 + 300 + 400)}{3} \quad L_{Ave} = 333$$

$$L = \frac{L_{Ave}}{100} \quad L = \frac{333}{100} \quad L = 3.33$$

$$FL = CQ^2L \quad FL = (0.138)(40)^2(3.33)$$

$$FL = (0.138)(1\ 600)(3.33)$$

FL = 735 kPa loss in each 125 mm hose

Total Pressure Loss
TPL = 735 + 175 = **910 kPa pressure loss in this hose assembly**

Figure 8.25a Some extremely high-flow master stream devices may be fed by multiple large diameter hoselines.

Example 20

A pumper is supplying 135 meters of 115 mm hose that is feeding an aerial device with a piped waterway. The aerial device is elevated 20 meters and is discharging 3 600 L/min (Figure 8.26). Determine the total pressure loss in this hose assembly.

115 mm Hose

C = 0.167 from Table 8.3

$$Q = \frac{L/min}{100} \quad Q = \frac{3\ 600}{100} \quad Q = 36$$

$$L = \frac{meters}{100} \quad L = \frac{135}{100} \quad L = 1.35$$

$$FL = CQ^2L \quad FL = (0.167)(36)^2(1.35)$$

$$FL = (0.167)(1\ 296)(1.35)$$

$$FL = 292.2 \text{ kPa}$$

Elevation Pressure

$$EP = (10)(H)$$

$$EP = (10)(20)$$

$$EP = 200 \text{ kPa}$$

Total Pressure Loss
TPL = 292.2 + 200 + 175 = **667.2 kPa pressure loss in this hose assembly**

Example 21

Determine the total pressure loss in the hose assembly when a fire department pumper is supplying two 77 mm hoselines with 65 mm couplings, each 100 meters long. These hoselines are connected to a siamese appliance that is in turn supplying 30 meters of 90 mm hose attached to a detachable ladder pipe. The ladder pipe is elevated 15 meters and is discharging through a 45 mm diameter solid stream nozzle at 560 kPa.

Master Stream

$$L/min = 0.067 \ d^2\sqrt{NP}$$

$$L/min = (0.067) \ (45)^2(\sqrt{560})$$

$$L/min = 3\ 211 \text{ L/min}$$

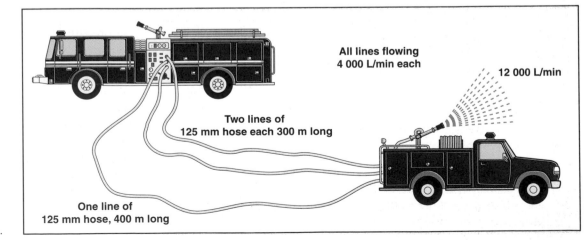

All lines flowing 4 000 L/min each

12 000 L/min

Two lines of 125 mm hose each 300 m long

One line of 125 mm hose, 400 m long

Figure 8.25b Example 19.

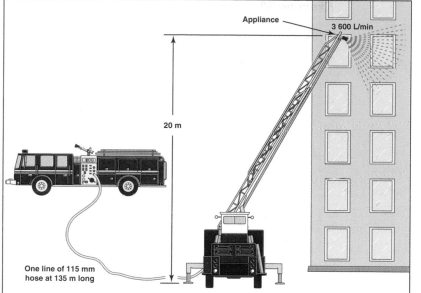

Appliance
3 600 L/min

20 m

One line of 115 mm
hose at 135 m long

Figure 8.26 Example 20.

90 mm Hose and Master Stream

C = 0.53 from Table 8.3

$Q = \dfrac{L/min}{100}$ $Q = \dfrac{3\ 211}{100}$ $Q = 32.11$

$L = \dfrac{meters}{100}$ $L = \dfrac{30}{100}$ $L = 0.3$

$FL = CQ^2L$ $FL = (0.53)(32.11)^2(0.3)$

$FL = (0.53)(1\ 031)(0.3)$

FL = 164 kPa

$EP = (10)(H)$

$EP = (10)(15)$

$EP = 150$ kPa loss

$TPL_{90} = FL + EP +$ Appliance Loss

TPL_{90} = 164 + 150 + 175 = 489 kPa loss in the 90 mm hose and the master stream

77 mm Hoselines and Siamese

C = 0.316 from Table 8.4

$Q = \dfrac{L/min}{100}$ $Q = \dfrac{3\ 211}{100}$ $Q = 32.11$

$L = \dfrac{meters}{100}$ $L = \dfrac{100}{100}$ $L = 1$

$FL = CQ^2L$ $FL = (0.316)(32.11)^2(1)$ $FL = (0.316)(1\ 031)(1)$

FL = 326 kPa

$TPL_{77} = FL +$ Appliance Loss

TPL_{77} = 326 + 70 = 396 kPa loss in the dual 77 mm lines and siamese appliance

Total Pressure Loss

$TPL = TPL_{90} + TPL_{77}$

TPL = 489 + 396 = 885 kPa pressure loss in this hose assembly

Determining Pump Discharge Pressure

Numerous skills possessed by the driver/operator are reflected during the development of fire streams. These skills enable the driver/operator to provide fire suppression crews with a fire stream at the proper pressure. To fully utilize these skills, the driver/operator must understand pump discharge pressure and its relation to nozzle pressure.

In order to deliver the necessary water flow to the fire location, the pump discharge pressure at the apparatus must be enough to overcome the sum of all pressure losses. These pressure losses, combined with the required nozzle pressure, are used to determine the pump discharge pressure (PDP). PDP can be calculated using the following equation:

EQUATION D
PDP = NP + TPL

Where:

PDP = Pump discharge pressure in kPa

NP = Nozzle pressure in kPa

TPL = Total pressure loss in kPa (appliance, friction, and elevation losses)

Fire apparatus supplying multiple hoselines, wyed hoselines, or manifold hoselines may be required to provide different pump discharge pressures for each attack line. Because this is not possible, another method must be used to compensate for these individual pressure requirements. Set the pump discharge pressure for the hoseline with the greatest pressure demand. For multiple hoselines, gate back the remaining hoselines at the discharge outlets. For wyed or manifold hoselines, gate back the hoselines at the appliance. Gate back the hoselines until the desired pressure on each line is obtained. As discussed earlier in this chapter, use the following nozzle pressures to ensure safety and efficiency:

- Solid stream nozzle (handline) — 350 kPa
- Solid stream nozzle (master stream) — 560 kPa
- Fog nozzle (all types) — 700 kPa

Pressure losses for elevated master streams and turret pipes vary depending on the manufacturer of the device. As discussed earlier in this chapter, for the sake of calculation, we consider each of these to have a pressure loss of 175 kPa.

The following examples show how to calculate pump discharge pressure using Equation D.

Example 22

A pumper is supplying 150 meters of 65 mm hose that is flowing 1 200 L/min through a fog stream nozzle. Determine the pump discharge pressure required to supply the hoseline.

C = 3.17 from Table 8.3

$$Q = \frac{L/min}{100} \quad Q = \frac{1\,200}{100} \quad Q = 12$$

$$L = \frac{meters}{100} \quad L = \frac{150}{100} \quad L = 1.5$$

FL = CQ²L   FL = (3.17)(12)²(1.5)   FL = (3.17)(144)(1.5)
FL = 685 kPa

Pump Discharge Pressure
PDP = NP + TPL   PDP = 700 + 685   **PDP = 1 385 kPa**

This means that the pumper must be discharging 1 385 kPa in order for the nozzle to be receiving the appropriate amount of pressure. If the pumper had an incoming pressure of 350 kPa, the pump would only need to generate 1 035 kPa to make up the difference.

Example 23

Two 65 mm hoselines, one 100 meters long and the other 150 meters long, are each equipped with 1 000 L/min fog nozzles. Determine the pump discharge pressure required to supply these hoselines (Figure 8.27).

Hoseline 1

C = 3.17 from Table 8.3

$$Q = \frac{L/min}{100} \quad Q = \frac{1\,000}{100} \quad Q = 10$$

$$L = \frac{meters}{100} \quad L = \frac{100}{100} \quad L = 1$$

FL = CQ²L   FL = (3.17)(10)²(1)   FL = (3.17)(100)(1)
FL = 317 kPa

Hoseline 2

C = 3.17 from Table 8.3

$$Q = \frac{L/min}{100} \quad Q = \frac{1\,000}{100} \quad Q = 10$$

$$L = \frac{meters}{100} \quad L = \frac{150}{100} \quad L = 1.5$$

FL = CQ²L   FL = (3.17)(10)²(1.5)   FL = (3.17)(10)(1.5)
FL = 475.5 kPa

Pump Discharge Pressure
Use the hose with the highest pressure loss to determine the PDP:

PDP = NP + TPL   PDP = 700 + 475.5   **PDP = 1 175.5 kPa**

Set the pump discharge pressure for the hoseline with the higher pressure demand (1 175.5 kPa). Gate back the discharge outlet on the remaining hoseline until the desired hoseline pressure is obtained. (**NOTE**: Where flowmeters are used, gate back to desired flow.)

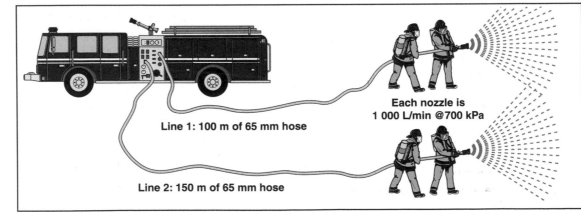

Each nozzle is
1 000 L/min @700 kPa

Line 1: 100 m of 65 mm hose

Line 2: 150 m of 65 mm hose

Figure 8.27 Example 23.

Example 24

Determine the pump discharge pressure for a fire department pumper that is using two 77 mm hoselines with 65 mm couplings to supply an elevated master stream device with fixed piping 60 meters away. The elevated master stream is discharging 4 000 L/min through a fog nozzle that is elevated 20 meters.

C = 0.316 from Table 8.4

$Q = \dfrac{L/min}{100}$ $Q = \dfrac{4\ 000}{100}$ $Q = 40$

$L = \dfrac{meters}{100}$ $L = \dfrac{60}{100}$ $L = 0.6$

$FL = CQ^2L$ $FL = (0.316)(40)^2(0.6)$

$FL = (0.316)(1\ 600)(0.6)$

FL = 303.4 kPa

EP = (10)(H)

EP = (10)(20)

EP = 200 kPa

TPL = FL + EP + Appliance Loss

TPL = 303.4 + 200 + 175 = 678.4 kPa pressure loss in this hose assembly.

Pump Discharge Pressure
PDP = NP + TPL PDP = 700 + 678.4 **PDP = 1 378.4 kPa**

Example 25

Determine the pump discharge pressure in a hoseline assembly in which one 90 meter, 77 mm hose with 77 mm couplings is supplying three attack lines that are attached to a water thief. The first attack line consists of 30 meters of 38 mm hose flowing 500 L/min. The second attack line is 60 meters of 38 mm hose flowing 400 L/min. The third attack line is 45 meters of 65 mm hose flowing 1 000 L/min. All nozzles are fog streams.

Attack Line 1 (38 mm)
C = 38 from Table 8.3

$Q = \dfrac{L/min}{100}$ $Q = \dfrac{500}{100}$ $Q = 5$

$L = \dfrac{meters}{100}$ $L = \dfrac{30}{100}$ $L = 0.3$

$FL = CQ^2L$ $FL = (38)(5)^2(0.3)$ $FL = (38)(25)(0.3)$

FL = 285 kPa loss in Attack Line 1

Attack Line 2 (38 mm)
C = 38 from Table 8.3

$Q = \dfrac{L/min}{100}$ $Q = \dfrac{400}{100}$ $Q = 4$

$L = \dfrac{meters}{100}$ $L = \dfrac{60}{100}$ $L = 0.6$

$FL = CQ^2L$ $FL = (38)(4)^2(0.6)$ $FL = (38)(16)(0.6)$

FL = 365 kPa loss in Attack Line 2

Attack Line 3 (65 mm)
C = 3.17 from Table 8.3

$Q = \dfrac{L/min}{100}$ $Q = \dfrac{1\ 000}{100}$ $Q = 10$

$L = \dfrac{meters}{100}$ $L = \dfrac{45}{100}$ $L = 0.45$

$FL = CQ^2L$ $FL = (3.17)(10)^2(0.45)$ $FL = (3.17)(100)(0.45)$

FL = 142.7 kPa loss in Attack Line 3

Supply Line (77 mm with 77 mm couplings)
C = 1.06 from Table 8.3

$Q_{Total} = \dfrac{(Sum\ of\ Attack\ Lines\ 1,\ 2,\ \&\ 3\ L/min)}{100}$

$Q_{Total} = \dfrac{500 + 400 + 1\ 000}{100}$ $Q_{Total} = \dfrac{1\ 900}{100}$

$Q_{Total} = 19$

$L = \dfrac{meters}{100}$ $L = \dfrac{90}{100}$ $L = 0.9$

$FL = (C)(Q_{Total})^2(L)$ $FL = (1.06)(19)^2(0.9)$

$FL = (1.06)(361)(0.9)$

FL = 344.4 kPa in the supply line

The total pressure loss in the system is based on the highest loss of the three attack lines, which in this case would be Line 2, and the friction loss in the supply line. Because the total flow rate exceeds 1 400 L/min, it is necessary to add the 70 kPa appliance loss.

TPL = FL$_{Attack}$ + FL$_{Supply}$ + Appliance Loss

TPL = 365 + 344.4 + 70 = 779.4 kPa loss in this hose assembly

Pump Discharge Pressure
PDP = NP + TPL PDP = 700 + 779.4 **PDP = 1 479.4 kPa**

Set the pump discharge pressure for the hoseline with the higher pressure demand (1 479.4 kPa). Gate back the discharge outlet on the water thief for the other hoseline until the desired hoseline pressures are obtained.

Example 26

Determine the pump discharge pressure in the hose assembly when a fire department pumper is supplying two 77 mm hoselines with 65 mm couplings, each 60 meters long. These hoses are connected to a siamese appliance that is in turn supplying 30 meters of 77 mm hose with 77 mm couplings attached to a detachable ladder pipe. The ladder pipe is elevated 20 meters and is discharging through a 38 mm diameter solid stream nozzle at 560 kPa.

Master Stream

L/min = 0.067 d²√NP

L/min = (0.067) (38)²(√560)

L/min = 2 290 L/min

77 mm Hose With 77 mm Couplings and Master Stream

C = 1.06 from Table 8.3

$$Q = \frac{L/min}{100} \quad Q = \frac{2\,290}{100} \quad Q = 22.9$$

$$L = \frac{meters}{100} \quad L = \frac{30}{100} \quad L = 0.3$$

$$FL = CQ^2L \quad FL = (1.06)(22.9)^2(0.3)$$

$$FL = (1.06)(524.4)(0.3)$$

FL = 166.8 kPa

$$EP = (10)(H)$$

$$EP = (10)(20)$$

EP = 200 kPa loss

$$TPL_{77\ w/\ 77} = FL + EP + Appliance\ Loss$$

$$TPL_{77\ w/\ 77} = 166.8 + 200 + 175 = \textbf{541.8 kPa loss in the 77}$$
mm hose with 77 mm couplings and the master stream

77 mm Hoselines and Siamese

C = 0.268 from Table 8.4

$$Q = \frac{L/min}{100} \quad Q = \frac{2\,290}{100} \quad Q = 22.9$$

$$L = \frac{meters}{100} \quad L = \frac{60}{100} \quad L = 0.6$$

$$FL = CQ^2L \quad FL = (0.268)(22.9)^2(0.6)$$

$$FL = (0.268)(524.4)(0.6)$$

FL = 84.3 kPa

$$TPL_{Dual\ 77s} = FL + Appliance\ Loss$$

$$TPL_{Dual\ 77s} = 84.3 + 70 = \textbf{154.3 kPa loss in the dual 77 mm}$$
lines and siamese appliance

$$TPL_{Assembly} = TPL_{77\ w/\ 77} + TPL_{Dual\ 77s}$$

$$TPL_{Assembly} = 541.8 + 154.3 = \textbf{696.1 kPa pressure loss in}$$
this hose assembly.

Pump Discharge Pressure

$$PDP = NP + TPL_{Assembly} \quad PDP = 560 + 696.1$$

PDP = 1 256.1 kPa

Determining Net Pump Discharge Pressure

Net pump discharge pressure takes into account all factors that contribute to the amount of work the pump must do to produce a fire stream.

When a pumper is being supplied by a hydrant or a supply line from another pumper, the net pump discharge pressure is the difference between the pump discharge pressure and the incoming pressure from the hydrant. If the PDP, for instance, is 1 000 kPa and the intake gauge reads 350 kPa, the net pump discharge pressure is 650 kPa. This can be shown by the following formula:

EQUATION E
$NPDP_{PPS}$ = PDP – Intake reading

Where:

$NPDP_{PPS}$ = Net pump discharge pressure from a positive pressure source

PDP = Pump discharge pressure

Note that this equation does not apply to situations where the pumper is operating at a draft. These situations are covered in more detail in Chapter 12 of this manual.

The application of Equation E is shown in the following example:

Example 27

A pumper operating from a hydrant is discharging water at 1 700 kPa. The incoming pressure from the hydrant registers 150 kPa on the intake gauge. Determine the net pump discharge pressure.

$$NPDP_{PPS} = PDP - Intake\ reading$$

$$NPDP_{PPS} = 1\,700\ kPa - 150\ kPa$$

$NPDP_{PPS}$ = 1 550 kPa

Chapter 9

Fireground Hydraulic Calculations

Job Performance Requirements

This chapter provides information that will assist the reader in meeting the following job performance requirements from NFPA 1002, *Standard on Fire Apparatus Driver/Operator Professional Qualifications*, 1998 edition. Particular portions of the job performance requirements (JPRs) that are met in this chapter are noted in bold text.

3-2.1* Produce effective hand or master streams, given the sources specified in the following list, so that the pump is safely engaged, all pressure control and vehicle safety devices are set, **the rated flow of the nozzle is achieved and maintained**, and the apparatus is continuously monitored for potential problems.

- Internal tank
- Pressurized source
- Static source
- Transfer from internal tank to external source

(a) *Requisite Knowledge:* **Hydraulic calculations for friction loss and flow using both written formulas and estimation methods**, safe operation of the pump, problems related to small-diameter or dead-end mains, low-pressure and private water supply systems, hydrant cooling systems, and reliability of static sources.

(b) *Requisite Skills:* The ability to position a fire department pumper to operate at a fire hydrant and at a static water source, power transfer from vehicle engine to pump, draft, operate pumper pressure control systems, operate the volume/pressure transfer valve (multistage pumps only), operate auxiliary cooling systems, make the transition between internal and external water sources, and assemble hose lines, nozzles, valves, and appliances.

3-2.2 Pump a supply line of 2½ in. (65 mm) or larger, given a relay pumping evolution the length and size of the line and the desired flow and intake pressure, so that the **proper pressure and flow are provided to the next pumper in the relay**.

(a) *Requisite Knowledge:* **Hydraulic calculations for friction loss and flow using both written formulas and estimation methods**, safe operation of the pump, prob-

lems related to small-diameter or dead-end mains, low-pressure and private water supply systems, hydrant cooling systems, and reliability of static sources.

(b) *Requisite Skills:* The ability to position a fire department pumper to operate at a fire hydrant and at a static water source, power transfer from vehicle engine to pump, draft, operate pumper pressure control systems, operate the volume/pressure transfer valve (multistage pumps only), operate auxiliary cooling systems, make the transition between internal and external water sources, and assemble hose lines, nozzles, valves, and appliances.

6-2.1* Produce effective fire streams, utilizing the sources specified in the following list, so that the pump is safely engaged, all pressure-control and vehicle safety devices are set, the rated flow of the nozzle is achieved, and the apparatus is continuously monitored for potential problems.

- Water tank
- Pressurized source
- Static source

(a) *Requisite Knowledge:* **Hydraulic calculations for friction loss and flow using both written formulas and estimation methods**, safe operation of the pump, proper apparatus placement, personal safety considerations, problems related to small-diameter or dead-end mains and low-pressure and private water supply systems, hydrant cooling systems, and reliability of static sources.

(b) *Requisite Skills:* The ability to position a wildland fire apparatus to operate at a fire hydrant and at a static water source, properly place apparatus for fire attack, transfer power from vehicle engine to pump, draft, operate pumper pressure control systems, operate the volume/pressure transfer valve (multistage pumps only), operate auxiliary cooling systems, make the transition between internal and external water sources, and assemble hose lines, nozzles, valves, and appliances.

The formulas that were presented and the calculations that were performed in Chapters 7 and 8 provide the fire apparatus driver/operator with useful background information into the hydraulics of fire fighting. However, the sense of urgency and the excitement present on the emergency scene seldom allow the driver/operator to perform these types of calculations in the field. On the fireground, the driver/operator commonly relies on one or more of the following methods for determining pressure loss and required pump discharge pressure:

- Flowmeters
- Hydraulic calculators
- Pump charts
- Hand method
- Condensed "Q" formula
- GPM flowing

Flowmeters

The ultimate purpose of all fireground hydraulic calculations is to discharge an appropriate amount of water from the nozzle(s) being used to attack a fire. In Chapter 8, as well as in much of this chapter, the calculation of pressure loss in the hose assembly has been used as the basis for determining a proper pump discharge pressure to supply to the hoseline(s). However, modern mechanical technology provides the driver/operator with an alternative to this process. Flowmeters reduce the amount of pressure calculations required of the driver/operator.

Rather than providing a readout of the pressure going through a discharge, flowmeters provide the water flow in gallons per minute (liters per minute) (Figure 9.1). The number displayed on the flowmeter requires no further calculation because it reflects how much water is moving through the discharge valve and consequently the nozzle. This quantity of water only diminishes before it reaches the nozzle if there is a leak or break in the hoseline.

Flowmeters are particularly advantageous when supplying hoselines or master stream devices equipped with automatic nozzles. As discussed in Chapter 6, nozzle pressure is automatically maintained at a predetermined level in automatic nozzles. This condition may lead driver/operators to supply insufficient discharge pressures. These pressures may result in what appears to be an acceptable fire stream, but in reality it has a very low flow in gallons (liters) per minute. If the discharge is equipped with a flowmeter, it is easier for the driver/operator to make sure that an adequate flow rate is being achieved. Flowmeters can make it possible for driver/operators to pump (within the limits of the pump) the correct volume of water to nozzles without having to know the length of hoseline, the amount of friction loss, or whether the nozzles are above or below the pump. This relieves the driver/operator from having to make calculations that are, at best, only close approximations of the amount of water reaching the nozzle.

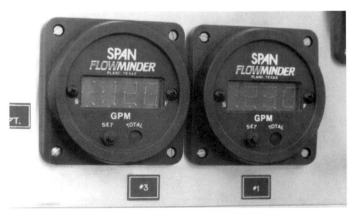

Figure 9.1 Flowmeters eliminate much of the guesswork for the driver/operator. *Courtesy of Class 1/Span Instruments.*

NFPA 1901, *Standard for Automotive Fire Apparatus*, allows flowmeters to be used instead of pressure gauges on all discharges 1½ to 3 inches (38 mm to 77 mm) in diameter. Discharges that are 3½ inches (90 mm) or larger may be equipped with flowmeters, but they must also have an accompanying pressure gauge. The flowmeter must provide a readout in increments no larger than 10 gpm (38 L/min) (Figure 9.2).

Types of Flowmeters

All flowmeters are designed to "read" water flow, but all flowmeters are not alike. Several designs are available, some of which are more reliable than others. The following two basic types of flowmeter sensors are commonly used in the fire service:

- Paddlewheel
- Spring probe

The first type of flowmeter used in fire apparatus was the paddlewheel type. The paddlewheel is mounted in the top of a straight section of pipe in such a manner that very little of the device extends into the waterway (Figure 9.3). This placement reduces the problems of impeded flow and damage by debris. Because the paddlewheel is located at the top of the pipe, sediment does not deposit on the paddlewheel. As water moves by the paddlewheel, a sensor measures the speed at which it is spinning and translates that information into a flow measurement.

In recent years, the spring probe flowmeter has gained increasing use in the fire service. This flowmeter uses a stainless steel spring probe to sense water movement in the discharge piping (Figure 9.4). The greater the flow of water through the piping, the more the spring probe is forced to bend. This sends a corresponding electrical charge to the digital display unit. Because the spring probe is the only moving part of the system, these devices are relatively maintenance free.

When properly calibrated and in good working condition, flowmeters should be accurate to a tolerance of ±3 percent. This means that the readout should not be more than 3 gallons (12 L) high or low for every 100 gpm (400 L/min) flowing.

Each discharge equipped with a flowmeter has a digital readout display mounted within 6 inches (150 mm) of the valve control for that discharge. In addition, some apparatus are equipped with a central flowmeter monitoring device that allows the driver/operator a number of options. Some of the information that the central monitor may provide includes the following:

- The flow through any particular discharge at that time
- The total amount of water being flowed through the pump at that time
- The total amount of water that has been flowed through the pump for the duration of that incident

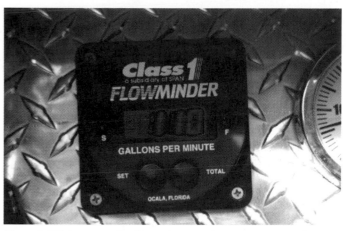

Figure 9.2 Flowmeters should provide a readout in increments of at least 10 gpm (40 L\min).

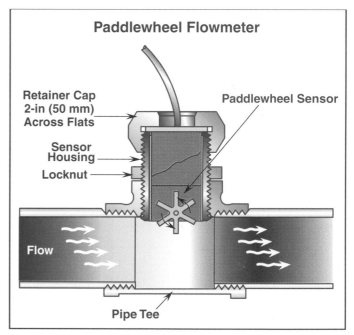

Figure 9.3 The earliest fire service flowmeters utilized a paddlewheel design. *Courtesy of Fire Research Corporation.*

Figure 9.4 The spring-probe flowmeter has become increasingly popular in recent years. *Courtesy of Class1/Span Instruments.*

Flowmeter Applications

There are a number of applications in which the use of a flowmeter can be helpful to the driver/operator. The following sections highlight some of these.

Diagnosing Waterflow Problems

The flowmeter can be used as a diagnostic tool to identify waterflow problems. If the flow does not increase when the driver/operator increases pressure, several problems are likely. Some examples of these might include that the hose is kinked or a midline valve (such as a gated wye) may be partially closed. If a firefighter communicates that water volume at the nozzle has suddenly diminished but there is no reduction in the flowmeter reading, it can be assumed that a hose has burst.

Relay Pumping

Use of a flowmeter during relay pumping makes it possible to feed a supply line without having to know the number of gallons (liters) flowing from the pumper receiving the water. This is done by monitoring the master discharge gauge and the flowmeter as the throttle is increased during the setup stage of the operation. As engine speed (rpm) increases so does the discharge and the gpm (L/min) reading from the flowmeter. Increase the engine speed until the flowmeter reading no longer increases. This sets the pump at the correct discharge pressure to supply an adequate flow to the receiving pumper. It also provides the driver/operator with a reading of the water volume being used by the receiving pumper. Although watching the flowmeter is helpful, also watch the master intake pressure gauge. Do not allow the incoming pressure to drop too much below 20 psi (140 kPa). More information on relay pumping can be found in Chapter 13 of this manual.

Standpipe Operations

When pumping to standpipes, it is difficult to determine where hoselines and nozzles are being placed in a multistory building. Pressure losses due to elevation must be accounted for when calculating discharge pressures. This problem is compounded when automatic nozzles are used because the nozzles compensate for inadequate nozzle pressures by providing a stream that may not be of adequate volume to control the fire.

When a flowmeter is used, the problem can be solved by determining the number and type of nozzles connected to the standpipe, adding their maximum rated flows, and then pumping the volume of water that matches this figure. An example is when three hose packs are taken to three different floors. One 2½-inch (65 mm) hose pack is equipped with a 300-gpm (1 200 L/min) fog nozzle, a second pack with a 1⅛-inch (29 mm) straight tip nozzle (desired flow = 250 gpm [1 000 L/min]), and a third 1¾-inch (45 mm) hose pack equipped with an automatic nozzle (desired flow = 150 gpm [600 L/min]). The total maximum flow for the three nozzles is 700 gpm (2 800 L/min). The operator increases the flow on the discharges supplying the fire department connection until that flow is achieved.

It is important that the driver/operator be in communication with the firefighters on the nozzles to ensure that nozzle pressures (and nozzle reactions) are correct. This is important because nozzles placed several floors apart may receive pressures somewhat greater or lesser than optimum. This is a problem no matter which method is used, and communication is the best way to make adjustments to correct the problem.

It is also important that the driver/operator realize that once the hoseline is charged, there is no flow through the system until a nozzle is opened. Thus, the driver/operator, through experience and training, must be able to set the pump for a discharge pressure that is relatively close to that which is required when the nozzle is flowing. Once the nozzle is fully opened and flowing, the driver/operator can adjust the discharge pressure until the appropriate amount of water is flowing.

Hydraulic Calculators

Hydraulic calculators enable the driver/operator to determine the pump discharge pressure required to supply a hose layout without having to perform tedious mental hydraulic calculations. There are three types of hydraulic calculators: manual, mechanical, and electronic. Manual and mechanical calculators operate by moving a slide or dial in which the water flow, size of hose, and length of the hose lay are indicated (Figure 9.5). By lining up each of these components properly, the driver/operator can then read the required pump discharge pressure. These are most commonly supplied to fire departments by hose and nozzle manufacturers. Contact your local fire equipment dealer for more information.

Figure 9.5 Hose and appliance manufacturers commonly supply their customers with mechanical friction loss calculators. *Courtesy of Akron Brass Company.*

Some apparatus may be equipped with electronic hydraulic calculators. These are specially programmed devices that allow the driver/operator to input the known information: water flow, size of hose, length of hose lay, and any elevation changes. Through preprogrammed formulas from either the factory or other available software, the calculator computes the required pump discharge pressure. These calculators may be portable or mounted near the pump panel (Figure 9.6). Inexpensive electronic programmable calculators can be preprogrammed for fireground calculations and carried on apparatus.

Pump Charts

Pump charts are used by some fire departments to reduce the need for calculations on the emergency scene. Pump charts contain the required pump discharge pressures for various hose lays and assemblies used within that jurisdiction. Pump charts may be placed on laminated sheets carried on the apparatus or on plates that are affixed to the pump panel. Fire departments may choose to develop their own pump charts or use those that are supplied by fire hose or nozzle manufacturers.

To develop and use the pump chart, it is important to understand the column headings. The nozzle heading should include only those nozzles and devices used by the department developing the chart. Also included are other hose layout applications such as sprinkler system support or relay pumping operations. The gpm (L/min) column indicates the flow being provided to that nozzle or layout. The NP column indicates the nozzle pressure being produced. The columns headed 100, 200 (30, 60) and so on indicate the number of feet (meters) of hose being used to supply a given nozzle or layout.

To use the chart, locate the nozzle column and the nozzle or layout being used (Table 9.1). Follow that line across to the vertical column headed by the number of feet (meters) of hose in that layout. The figure in the block where the two columns intersect is the required pump discharge pressure.

The first step in developing a pump chart is to identify all nozzles, devices, and layouts used by the department and enter them in the nozzle column. Then, enter gpm (L/min) flowing and the nozzle pressure desired for each item in the appropriate columns. If, for instance, a master stream device may be supplied by either 2½- or 3-inch (65 mm or 77 mm) hose, make a separate entry for each layout in the nozzle column, as indicated in the sample chart.

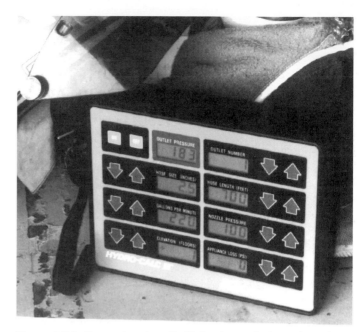

Figure 9.6 In the past, automatic friction loss calculators have been available. The companies that make this type of calculator tend to come and go; however, the devices remain in service in some jurisdictions. The company that produced this device is no longer in business.

The second step is to calculate the required pump discharge pressures for each of the listed layouts (using the formulas and tables given earlier in this manual or figures derived by field testing). Observe the following rules when making these calculations:

- Be certain to include friction loss in master stream appliances flowing in excess of 350 gpm (1 400 L/min).

- For wyed hoselines, the length of layout numbers indicate the number of feet (meters) of hose between the pumper and the wye. (If used as a preassembled unit, the attack hose leading from the wye is constant and therefore is not a factor in determining the length of the layout.)

- When a master stream nozzle may be supplied by a different number or size of hoselines, indicate these on the chart.

- Round pump discharge pressure to the nearest 5 psi (35 kPa).

- Do not list pump discharge pressures that exceed the test pressure used for the size of hose concerned.

- When calculating pump discharge pressures for relay operations, provide established departmental residual pressures at the intake of the pumper being supplied. This residual pressure may be indicated on the chart as a nozzle pressure.

The reliability of the chart depends entirely on the accuracy of the calculations. The following example shows how to use a pump chart.

Table 9.1
Pump Chart

Nozzles	GPM	NP	FL	50'	100'	150'	200'	250'	300'	400'	500'	600'	700'	800'	900'	1000'	Max'
Booster Lines	10	100	17	109	117	126	134	143	151	168	186	203	220	237	–	–	850'
	12	100	23	111	123	134	146	157	168	191	214	237	–	–	–	–	650'
	20	100	53	127	153	180	206	233	–	–	–	–	–	–	–	–	250'
	22	100	62	131	162	194	225	–	–	–	–	–	–	–	–	–	200'
	23	100	67	134	167	201	235	–	–	–	–	–	–	–	–	–	200'
	30	100	108	154	208	–	–	–	–	–	–	–	–	–	–	–	100'
1½" Fog Lines	125	100	53	127	153	180	207	233	–	–	–	–	–	–	–	–	250'
	100	100	35	117	135	152	169	187	204	239	–	–	–	–	–	–	400'
	95	100	31	116	131	147	163	179	194	226	–	–	–	–	–	–	450'
	60	100	13	107	113	120	126	133	139	152	166	179	192	205	218	231	1100'
	30	100	4	102	104	106	107	109	111	115	118	122	126	129	133	137	4050'

1½" Foam Lines — Attach 50' of 1½" hose to the pump then attach the eductor: set the eductor for 3% for hydrocarbon fires or 6% for polar solvent fires; switch the eductor from water to foam; then attach 100' of 1½" hose and a nozzle; set the nozzle for 60 gpm; insert the eductor tube into the foam bucket when ready for foam application; engine pressure should be 200 psi; do not increase or reduce engine pressure or change flow setting on the nozzle; do not shut nozzle off when the foam eductor tube is still in the foam bucket. A 3% setting will use one can of foam every 2 minutes. A 6% setting will use one can of foam every minute.

Nozzles	GPM	NP	FL	50'	100'	150'	200'	250'	300'	400'	500'	600'	700'	800'	900'	1000'	Max'
2½" Fog Lines	250	100	15	108	115	123	130	138	145	160	175	190	205	220	235	250	1000'
	200	100	10	105	110	115	120	125	130	140	150	160	170	180	190	200	1500'
	150	100	6	103	106	109	112	115	118	124	130	136	142	148	154	160	2500'
	120	100	4	102	104	106	108	110	112	116	120	124	129	133	137	141	3650'
Wyed Lines 1½" 150'	2x30	106	13	113	119	126	132	139	145	158	171	184	197	210	223	236	1100'
250'	2x30	109	13	116	122	129	135	142	148	161	174	187	200	213	226	239	1050'
150'	2x60	120	49	145	169	194	218	243	–	–	–	–	–	–	–	–	250'
Wyed Lines 2½" (150' 1½" Hose) Figure feet of 2½"	2x125	180	15	186	194	201	209	216	224	231	–	–	–	–	–	–	400'
	2x100	152	10	157	162	167	172	177	182	192	202	212	222	232	242	–	950'
	2x95	147	9	152	156	161	165	170	174	183	193	202	211	220	229	238	1100'
	2x60	120	4	122	124	126	128	130	132	136	140	144	149	153	157	161	3150'
	2x30	106	1	106	107	107	108	108	109	109	110	111	112	113	114	116	13100'
Relay "250" 1-2½"	250	20	15	28	35	43	50	58	65	80	95	110	125	140	155	170	1500'
2-2½"	250	20	4	22	24	26	28	30	32	37	41	45	49	53	57	62	5500'
Relay "375" 1-2½"	375	20	32	36	52	68	84	100	116	148	179	211	243	–	–	–	700'
2-2½"	375	20	9	24	29	33	38	42	47	55	64	73	82	91	100	109	2550'
Relay "500" 1-2½"	500	20	55	48	75	103	130	158	185	240	–	–	–	–	–	–	400'
2-2½"	500	20	15	28	35	43	51	58	66	81	96	112	127	142	157	173	1500'
Relay "750" 1-2½"	750	20	120	80	140	200	–	–	–	–	–	–	–	–	–	–	150'
2-2½"	750	20	33	37	53	70	87	103	120	153	187	220	–	–	–	–	650'
Relay "1000" 1-2½"	1000	20	210	125	230	–	–	–	–	–	–	–	–	–	–	–	100'
2-2½"	1000	20	58	48	78	107	137	166	195	–	–	–	–	–	–	–	350'
Relay "1250" 1-2½"	1250	20	325	183	–	–	–	–	–	–	–	–	–	–	–	–	50'
2-2½"	1250	20	90	65	110	155	200	246	–	–	–	–	–	–	–	–	250'

Sprinkler Systems _____ 150 psi _____

Standpipe Systems	GPM	NP	FL	50'	100'	150'	200'	250'	300'	400'	500'	600'	700'	800'	900'	1000'	Max'
1-1½" Fog	125	180	1	220	221	221	222	222	224	225	226	227	228	230	231	232	2500'
2-1½" Fog	250	180	4	222	224	226	228	230	232	236	240	244	248	–	–	–	750'
1-1½" & 1-2½"	375	180	9	225	229	234	238	243	247	–	–	–	–	–	–	–	300'
1-2½" Fog	250	123	4	165	167	169	171	173	175	179	183	187	191	195	199	203	2150'
2-2½" Fog	500	123	15	171	178	186	193	201	208	223	238	–	–	–	–	–	550'

Elevation Loss/Gain _____ ½ psi per foot _____

Maximum Engine Pressure 250 psi.

Sample Pump Chart. *Courtesy of Verdi Fire Department.*

Example 1

Using Table 9.2a, determine the pump discharge pressure for a master stream device that is equipped with a 1¾-inch solid stream nozzle tip and is supplied by three 2½-inch hoselines, each 600 feet long.

Solution: Locate the master stream device with a 1¾-inch nozzle tip in the left column of the chart. Choose the one that shows the device being supplied by three 2½-inch lines. Follow this line to the column for a 600-foot hose lay. You should land on the figure 189 psi. This is the required pump discharge pressure.

Example 2

Using Table 9.2b, determine the pump discharge pressure for a master stream device that is equipped with a 45 mm solid stream nozzle tip and is supplied by three 65 mm hoselines, each 180 meters long.

Solution: Locate the master stream device with a 45 mm nozzle tip in the left column of the chart. Choose the one that shows the device being supplied by three 65 mm lines. Follow this line to the column for a 180 meter hose lay. You should land on the figure 1 317 kPa. This is the required pump discharge pressure.

Appendixes C and D contain extensive pump chart information for various sizes of hose and hose lays.

Hand Method

One method used by many driver/operators over the years for determining friction loss in 2½-inch hose is the hand or "counting fingers" method. Starting with the thumb of the left hand, as illustrated in Figure 9.7, each finger is numbered at the base in terms of hundreds of gallons per minute. Returning to the thumb, and again moving from left to right, the tip of each finger is given a successive even number, beginning with two. Because nozzle capacities vary in gpm, the nearest half-hundred can be used with slight variations. The numbers 3, 5, 7, and 9 can be used for flows of 150, 250, 350, and 450 gpm, respectively. These half-hundred figures can be assigned to the spaces between the fingers. The friction loss for 100 feet of 2½-inch hose at a desired flow is determined by selecting the finger to which the desired flow has been assigned, and multiplying the number at the tip of the finger by the first digit at the base of the finger. Thus, the friction loss for a flow of 500 gpm can be found by using the numbers assigned to the little finger, or (5)(10) = 50

Table 9.2a (US) Pump Chart

Length of Lay in Feet

Nozzles	GPM	NP	100	200	300	400	500	600	700	800	900	1,000	1,100	1,200
Booster 1"	23	100		115										
Preconnect 1¾"	150	100		170										
Wyed Line:														
200' of 1½"on 2½" skid	190	100	150	157	164	172	179	186	193	201	208	215	222	229
2½" Fog	250	100	112	125	138	150	162	175	187	200	212	225	237	250
Master Stream:														
1¾" (Two 2½" Lines)	800	80	137	169	201	233								
1¾" (Three 2½" Lines)	800	80	119	133	147	161	175	189	203					
1¾" (Two 3" Lines)	800	80	118	131	144	157	170	183	196	209				
Relay:														
One 3" Line	250	20	25	30	35	40	45	50	55	60	65	70	75	80
One 3" Line	500	20	40	60	80	100	120	140	160	180	200			
Two 3" Lines	750	20	31	43	54	65	76	88	99	110	121	133	144	155
Two 3" Lines	1,000	20	100	180										
Sprinklers	Maintain 150 psi													
Elevation	Add 1/2 psi per foot or 5 psi per floor													

NOTE: All pressures rounded to the nearest whole number for this table only. Computations rounded to the nearest 5 psi are acceptable.

Table 9.2b (Metric)
Pump Chart

Length of Lay in Meters

Nozzles	L/min	NP	30	60	90	120	150	180	210	240	270	300	330	360
Booster 25 mm	90	700		815										
Preconnect 45 mm	570	700		1 180										
Wyed Line:														
60 m of 38 mm on														
65 mm skid	760	700	1 085	1 140	1 195	1 250	1 305	1 360	1 415	1 470	1 525	1 580	1 635	
65 mm Fog	1 000	700	795	890	985	1 080	1 175	1 270	1 365	1 460	1 555	1 650	1 745	
Master Stream:														
45 mm (Two 65 mm Lines)	3 200	560	957	1 179	1 401	1 623	1 845	2 067						
45 mm (Three 65 mm Lines)	3 200	560	832	929	1 026	1 123	1 220	1 317	1 414	1 511	1 608	1 705	1 802	1 899
45 mm (Two 77 mm Lines)	3 200	560	824	913	1 002	1 091	1 180	1 269	1 358	1 447	1 536	1 625	1 714	1 803
Relay:														
One 77 mm Line	1 000	140	178	216	254	292	330	368	406	444	482	520	558	596
One 77 mm Line	2 000	140	292	444	596	748	900	1 052	1 204	1 356	1 508	1 660	1 812	1 964
Two 77 mm Lines	3 000	140	225	310	395	480	565	650	735	820	905	990	1 075	1 160
Two 77 mm Lines	4 000	140	292	444	596	748	900	1 052						
Sprinklers	Maintain 1 050 kPa													
Elevation	Add 10 kPa per meter or 35 kPa per floor													

NOTE: All pressures rounded to nearest whole number for this table only. Computations rounded to the nearest 25 kPa are acceptable.

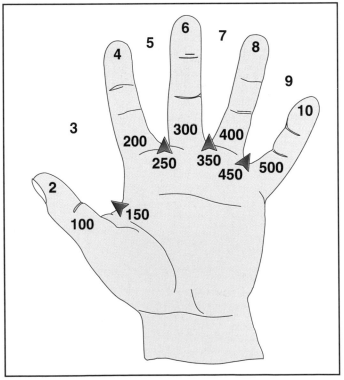

Figure 9.7 The counting fingers method for 2½-inch hose. *Courtesy of Maryland Fire and Rescue Institute.*

psi friction loss per 100 feet of 2 ½-inch hose. Likewise, the friction loss for a flow of 200 gpm is found using the numbers assigned to the index finger. Thus, (2)(4) = 8 psi friction loss per 100 feet of 2½-inch hose.

The answers provided by this method give a reasonable estimate of the friction loss that can be expected in that hoseline. If more accurate figures are required, one of the other methods previously discussed in this manual needs to be employed. (**NOTE:** This method has no conversion for the metric system of measurement.)

Example 3
Using the hand method, determine the total pressure loss due to friction in 400 feet of 2½-inch hose when a fog nozzle is flowing 250 gpm at 100 psi.

(2.5)(5) = 12.5 psi loss/100 feet

(12.5)(4) = **50 psi loss in the hose assembly**

Example 4
Using the hand method, what is the total pressure loss due to friction in a 2½-inch hoseline 300 feet long with 350 gpm flowing through the line?

(3.5)(7) = 24.5 psi loss/100 feet

(24.5)(3) = **73.5 psi loss in the hose assembly**

Figure 9.8 shows how the hand method may also be applied to 1¾-inch hose. In this case, you can calculate the friction loss in 100 feet of 1¾-inch hose by going to the finger that corresponds to the flow you are using and multiplying the number at the tip of the finger by the number at the base of the same finger. For example, if you are going to flow 150 gpm, multiply the number at the tip of your middle finger (3) with the number at the base. Thus the friction loss in 100 feet of 1¾-inch hose flowing 150 gpm is 3 times 9, or 27 psi.

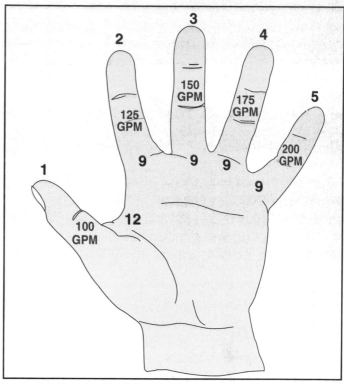

Figure 9.8 The counting fingers method for 1¾-inch hose.

Condensed "Q" Formula

The Condensed "Q" formula has been developed for fireground operations in which friction loss can be determined for 3-, 4-, and 5-inch hose. This method is shown in Equations F, G, and H. (**NOTE:** These equations may not be used with metric measurements.)

EQUATION F (3-inch hose)

FL per 100 feet = Q^2

Where:

FL = Friction loss in 100 feet of 3-inch hose

Q = Number of hundreds of gpm

Equation F can be used for 3-inch hose with either 2½-inch or 3-inch couplings. This formula is not as accurate as using $FL = CQ^2L$, but it is faster and sufficiently accurate for fireground operations.

NOTE: The amount of friction loss calculated using this formula will be 20 percent greater than if the same situation is calculated using $FL = CQ^2L$. This could mean as high as a 50 psi difference in a 1,000-foot, 3-inch hose lay.

The following example illustrates the use of Equation F.

Example 5

What will be the total pressure loss due to friction when 200 gpm is being discharged from a nozzle that is attached to 100 feet of 3-inch hose?

$$Q = \frac{discharge\ (gpm)}{100} \qquad FL = Q^2$$

$$Q = \frac{200}{100} \qquad FL = (2)^2$$

$$Q = 2 \qquad \textbf{FL = 4 psi per 100 feet of 3-inch hose}$$

EQUATION G (4-inch hose)

FL per 100 feet = $\dfrac{Q^2}{5}$

Where:

FL = Friction loss in 100 feet of 4-inch hose

Q = Number of hundreds of gallons per minute

The following example illustrates the use of Equation G.

Example 6

What will be the total pressure loss in 500 feet of 4-inch hose flowing 1,000 gpm?

$$Q = \frac{discharge\ (gpm)}{100} \qquad FL = \frac{Q^2}{5}$$

$$Q = \frac{1,000}{100} \qquad FL = \frac{(10)^2}{5}$$

$$Q = 10 \qquad FL = \frac{100}{5}$$

FL = 20 psi per 100 feet of 4-inch hose

Total Pressure Loss = 20 x 5

Total Pressure Loss = 100 psi

EQUATION H (5-inch hose)

$$FL \text{ per } 100 \text{ feet of hose} = \frac{Q^2}{15}$$

Where:

FL = Friction loss in 100 feet of 5-inch hose

Q = Number of hundreds of gallons per minute

The following example illustrates the use of Equation H.

Example 7

What will be the total pressure loss due to friction for 1,000 gpm flowing through 800 feet of 5-inch hose?

$$Q = \frac{\text{discharge (gpm)}}{100} \qquad FL = \frac{Q^2}{15}$$

$$Q = \frac{1,000}{100} \qquad FL = \frac{(10)^2}{15}$$

$$Q = 10 \qquad FL = \frac{100}{15}$$

FL = 6.6 psi per 100 feet of 5-inch hose

Total Pressure Loss = 6.6 x 8

Total Pressure Loss = 52.8 psi

GPM Flowing

The gpm flowing method permits friction loss to be calculated from the gpm flow and is applicable to both solid and fog streams. Another desirable feature of this method is that it can be used for hose sizes other than 2½-inch. By studying this method, the driver/operator can easily apply it to smaller sizes of hose. (**NOTE:** This method does not work for the metric system of measurement.)

In Table 9.3, note that a line separates the flow in gpm between 150 and 170. This separation is necessary because this method does not apply to 2½-inch hose with flows below 160 gpm. Flows less than 160 gpm are also seldom encountered through 2½-inch hose.

A driver/operator needs to know only the flow in gpm from a nozzle at a specified pressure. Then, by subtracting 10 from the first two numbers of the gpm flow, a sufficiently accurate friction loss figure per 100 feet of 2½-inch hose is obtained. A study of Table 9.3 further reveals that friction loss in 2½-inch hose increases 1 psi for every 10 gpm increase in flow.

Table 9.3
GPM Flowing #1

GPM Flowing 2½" Hose		Friction Loss per 100 ft. 2½" Hose
100		3
110-120		4
130-140		5
150		6
160-170		7
180		8
190	Subtract "10"	9
200	from the first	10
210	two numbers	11
220	of gpm	12
230		13
240		14
250		15
260		16
270		17
280		18
290		19
300		20

Example 8

What will be the total pressure loss due to friction in 400 feet of 2½-inch hose when using a 250 gpm fog nozzle at 100 psi?

Flow = 250 gpm

FL — From Table 9.3, 15 psi per 100 feet of 2½-inch hose

Friction Loss in 400 feet of 2½-inch hose

FL = (4)(15)

FL = 60 psi for the 400 feet of 2½-inch hose

In Table 9.4, this method is applied to 1½-inch hose. Table 9.4 is divided into three sections. The upper section represents flows from 50 to 75 gpm. This section is designed to show that friction loss in 1½-inch hose is the same as it is in 2½-inch hose with four times as much water flowing. For example, suppose the driver/operator needs to know the friction loss in 1½-inch hose with 70 gpm flowing. First, multiply the flow times four (4 x 70 = 280). Using Table 9.4, subtract 10 from 28. The result is 18 psi friction loss per 100 feet in 1½-inch hose.

Table 9.4
GPM Flowing #2

GPM Flowing 1½" Hose	GPM Flowing 2½" Hose	Friction Loss per 100 ft. 1½" or 2½" Hose
50	200	10
55	220	12
60 Same as 4	240 For increase of	14
65 times as	260 20 gpm flowing	16
70 much water	280 FL increases	18
flowing in	2 psi	
75 2½-inch	300	20
80	320	23
85	340	26
90	360 For increase of	29
95	380 20 gpm flowing FL increases	32
100	400 3 psi	35
105	420	39
110	440	43
115	460 For increase of 20 gpm flowing	47
120	480 FL increases	51
125	500 4 psi	55

The middle section of Table 9.4 represents flows from 80 to 100 gpm. Four times the flow through 1½-inch hose is shown in the 2½-inch column. The friction loss in 1½-inch hose, however, simply increases by 3 psi for every 20 gpm increase in flow through 2½-inch hose.

The lower section of Table 9.4 represents flows from 105 to 125 gpm. As in the top two sections, four times the flow through 1½-inch hose is shown for 2½-inch hose. The friction loss for these pressures increases 4 psi for every 20 gpm increase in flow through 2½-inch hose. The following example demonstrates how to use the gpm flowing method.

Example 9
What is the total pressure loss due to friction in a 200-foot, 1½-inch preconnect flowing 100 gpm?

Flow = 100 gpm

FL — From Table 9.4

FL = 35 psi per 100 feet of 1½-inch hose

Friction loss in 200 feet of 1½-inch hose

FL = (2)(35)

FL = 70 psi per 200 feet of 1½-inch hose

Chapter 10

Fire Pump Theory

This chapter provides information that will assist the reader in meeting the following job performance requirements from NFPA 1002, *Standard on Fire Apparatus Driver/Operator Professional Qualifications*, 1998 edition. Particular portions of the job performance requirements (JPRs) that are met in this chapter are noted in bold text.

3-2.1* Produce effective hand or master streams, given the sources specified in the following list, so that the pump is safely engaged, all **pressure control and vehicle safety devices** are set, the rated flow of the nozzle is achieved and maintained, and the apparatus is continuously monitored for potential problems.
* Internal tank
* Pressurized source
* Static source
* Transfer from internal tank to external source

(a) *Requisite Knowledge:* Hydraulic calculations for friction loss and flow using both written formulas and estimation methods, **safe operation of the pump**, problems related to small-diameter or dead-end mains, low-pressure and private water supply systems, hydrant cooling systems, and reliability of static sources.

(b) *Requisite Skills:* The ability to position a fire department pumper to operate at a fire hydrant and at a static water source, power transfer from vehicle engine to pump, draft, operate pumper pressure control systems, operate the volume/pressure transfer valve (multistage pumps only), operate auxiliary cooling systems, make the transition between internal and external water sources, and assemble hose lines, nozzles, valves, and appliances.

3-2.2 Pump a supply line of 2½ in. (65 mm) or larger, given a relay pumping evolution the length and size of the line and the desired flow and intake pressure, so that the proper pressure and flow are provided to the next pumper in the relay.

(a) *Requisite Knowledge:* Hydraulic calculations for friction loss and flow using both written formulas and estimation

methods, **safe operation of the pump**, problems related to small-diameter or dead-end mains, low-pressure and private water supply systems, hydrant cooling systems, and reliability of static sources.

(b) *Requisite Skills:* The ability to position a fire department pumper to operate at a fire hydrant and at a static water source, power transfer from vehicle engine to pump, draft, operate pumper pressure control systems, operate the volume/pressure transfer valve (multistage pumps only), operate auxiliary cooling systems, make the transition between internal and external water sources, and assemble hose lines, nozzles, valves, and appliances.

6-2.1* Produce effective fire streams, utilizing the sources specified in the following list, so that the pump is safely engaged, all **pressure-control and vehicle safety devices** are set, the rated flow of the nozzle is achieved, and the apparatus is continuously monitored for potential problems.
* Water tank
* Pressurized source
* Static source

(a) *Requisite Knowledge:* Hydraulic calculations for friction loss and flow using both written formulas and estimation methods, **safe operation of the pump**, proper apparatus placement, personal safety considerations, problems related to small-diameter or dead-end mains and low-pressure and private water supply systems, hydrant cooling systems, and reliability of static sources.

(b) *Requisite Skills:* The ability to position a wildland fire apparatus to operate at a fire hydrant and at a static water source, properly place apparatus for fire attack, transfer power from vehicle engine to pump, draft, operate pumper pressure control systems, operate the volume/pressure transfer valve (multistage pumps only), operate auxiliary cooling systems, make the transition between internal and external water sources, and assemble hose lines, nozzles, valves, and appliances.

High-pressure water systems in industrial or specialized applications are capable of providing a sufficient volume of water at suitable pressures for fire fighting operations. A few well-developed municipal systems may also have this capability, but the majority of water systems are unable to maintain adequate pressure in the hydrant system for effective fire fighting. In order to produce effective fire streams, it is necessary to increase existing pressures with fire pumps (Figure 10.1). Fire pumps are also necessary to provide the pressures needed to supply attack lines from the water tank mounted on the apparatus or from other static water supply sources such as portable tanks, lakes, streams, ponds, and rivers.

The earliest pumps used in the fire service were hand operated. To supply water under pressure, the firefighter pumped a handle that operated a piston in a cylinder (Figure 10.2). This action forced the water out of the pump with enough velocity to push it through the hose or nozzle. These hand-operated pumps were soon followed by rotary pumps. Rotary pumps contained a hand crank, which, when operated, caused a gear to rotate, again forcing the water out of the pump at a workable pressure. Both of these pumps are known as *positive displacement pumps* because a positive action takes place — all water and air are forced out of the pump body with each operating cycle. Modern apparatus still have some form of positive pressure pump connected to the main fire pump in order to allow the apparatus to use static water supply sources.

The modern fire department pumper is equipped with a centrifugal pump as its major pump. The *centrifugal pump* does not use positive action to force water from the pump; rather, it depends on the velocity of the water produced by centrifugal force to provide the necessary pump discharge pressure for effective operation. This chapter provides the driver/operator with the basic concepts surrounding the various types of fire pumps and their operation. Also included is information on the various components that make up the overall apparatus pumping system. Specific information on the actual operation of fire pumps is covered in Chapter 11. Foam extinguishing systems may be an integral part of the fire pump and apparatus. For more information on foam systems, see Chapter 13.

Figure 10.1 In order to produce effective fire streams, it is necessary to increase existing pressures with fire pumps.

Figure 10.2 Some of the earliest fire pumpers were hand-operated piston pumps.

Positive Displacement Pumps

The positive displacement pump has largely been replaced by the centrifugal pump for use as the main fire pump on modern fire apparatus. However, positive displacement pumps are still a necessary part of the overall pumping system on modern fire apparatus because unlike centrifugal pumps, they can pump air. For this reason, positive displacement pumps are used as priming pumps to get the water into the centrifugal pump during drafting operations. The basic principle of the positive displacement pump is a hydraulic law that stems from the near incompressibility of water: *When pressure is applied to a confined liquid, the same outward pressure is transmitted within the liquid, outward and equally in all directions.* In other words, by removing the trapped air in the pump, the water is forced into the pump casing by the atmospheric pressure. There are two basic types of positive displacement pumps: piston and rotary.

Piston Pumps

Piston pumps contain a piston that moves back and forth inside a cylinder. The pressure developed by this action causes intake and discharge valves to operate automatically and provides for the movement of the water through the pump. As the piston is driven forward, the air within the cylinder is compressed, creating a higher pressure inside the pump than the atmospheric pressure in the discharge manifold. This pressure causes the discharge valve to open and the air to escape through the discharge lines (Figure 10.3). This action continues until the piston completes its travel on the forward stroke and stops. At that point, pressures equalize and the discharge valve closes. As the piston begins the return stroke, the area in the cylinder behind the piston increases and the pressure decreases, creating a partial vacuum. At this time, the intake valve opens, allowing some of the air from the suction hose to enter the pump (Figure 10.4).

As the air from the suction hose is evacuated and enters the cylinder, the pressure within the hose and the intake area of the pump is reduced. Atmospheric pressure forces the water to rise within the hose until the piston completes its travel and the intake valve closes. As the forward stroke is repeated, the air is again forced out of the discharge. On the return stroke, more of the air in the intake section is removed and the column of water in the suction hose is raised. This action is repeated until all the air has been removed and the intake stroke results in water being introduced into the cylinder. The pump is now considered to be primed, and further strokes cause water to be forced into the discharge instead of air (Figure 10.5).

The forward stroke causes water to be discharged, and the return stroke causes the pump to fill with water again.

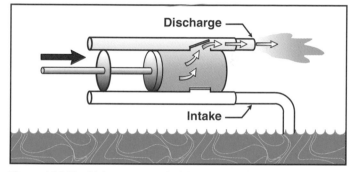

Figure 10.3 The higher pressure inside the pump causes the discharge valve to open, allowing air to escape through the discharge lines.

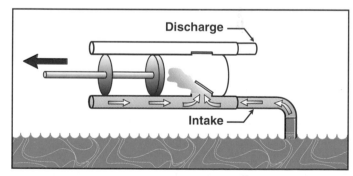

Figure 10.4 The partial vacuum created as the piston begins the return stroke causes the intake valve to open. This allows air from the suction hose to enter the pump.

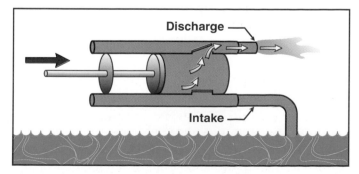

Figure 10.5 Once all the air has been evacuated, only water is pushed through the pump.

This is known as a *single-acting piston pump*. Obviously, this would not produce a usable fire stream because no water would be flowing during the return stroke, and the discharge would be a series of surges of water followed by an equal length of time with no water. A more constant stream can be produced by adding two additional valves. This is known as a *double-acting piston pump* because it both receives and discharges water on each stroke of the piston (Figure 10.6). Even with the double-acting pump, the output is a series of pressure surges with two periods of no flow. These occur when the piston ends its travel in each direction.

Because the pump cylinder contains a definite amount of water, just that amount is delivered by each stroke of the piston. The output capacity of the pump is determined by the size of the cylinder and the speed of the piston travel. There is a practical limit to the speed that a pump can be operated, so the capacity is usually determined by the size of the cylinder. In practice, it is more practical to build multicylinder pumps than one large single-cylinder pump. Multiple cylinders are more flexible and efficient because some of them can be disengaged when the pump's full capacity is not needed. Multicylinder pumps also provide a more uniform discharge because the cylinders are arranged to reach their peak flows at different parts of the cycle.

Older, large-capacity piston pumps are equipped with a pressure dome or air chamber on the discharge to even out the pulses (Figure 10.7). During peak pressures, water is forced into the air chamber, compressing the air in the chamber until the air pressure trapped in the dome is equal to the water pressure being applied. When the pressure in the pump discharge drops, the air pressure stored in the chamber forces the water into the line, thus counteracting the drop in pressure from the pump.

The piston pump has not been used as the major fire pump in pumpers for many years. In addition to the problems of the pulsating fire stream, it is very susceptible to wear. The efficiency of the piston pump depends upon a close tolerance between the piston and the walls of the cylinder. As the piston wears, water can escape from the discharge side of the pump to the intake, and the capacity of the pump is reduced accordingly. Wear also takes place on the walls of the cylinder. Lastly, pumping contaminated water can quickly damage the pump. For all these reasons, the piston pump has been phased out of the fire service as a high-capacity pump.

Although piston pumps are no longer used as high-capacity pumps, some are still in service for high-pressure stream fire fighting. These multicylinder, PTO-driven pumps provide pressures up to 1,000 psi (7 000 kPa) for

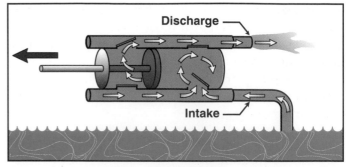

Figure 10.6 A double-action piston pump pushes water on both the forward and return strokes.

Figure 10.7 The pressure dome on this Ahrens-Fox pump helped to minimize the surge felt on hoselines.

high-pressure fog lines. The most common application for high-pressure pumps is in wildland fire fighting. This application also requires a dependable relief valve because pressures would quickly build to dangerous levels if the discharge flow were interrupted.

Rotary Pumps

Rotary-type pumps are the simplest of all fire apparatus pumps from the standpoint of design. Older fire apparatus used these pumps extensively as the major pump, but in recent years, their use has been confined to small capacity booster-type pumps and priming pumps. Most of the rotary-type pumps in use today are either of the rotary gear or rotary vane construction. They are driven by either a small electric motor or through a clutch that extends off the apparatus drive shaft.

Rotary Gear Pumps

The *rotary gear pump* consists of two gears that rotate in a tightly meshed pattern inside a watertight case. The gears are constructed so that they contact each other and are in close proximity to the case (Figure 10.8). With this arrangement, watertight and airtight pockets are formed by the gears within the case as they turn from the intake to the outlet. As each gear tooth reaches the discharge chamber, the air or water contained in that pocket is forced out of the pump. As the tooth returns to the intake side of the pump, the gears are meshed tightly enough to prevent the water or air that has been discharged from returning to the intake.

The total amount of water that can be pumped by a rotary gear pump depends upon the size of the pockets in the gears and the speed of rotation. The rotary gear pump is a positive displacement pump because each pocket in the gears contains a definite amount of water. As well, water is forced out of the pump each time the gears turn with a positive action. If the pump is trying to move more water than the discharge lines in use can take away, pressure builds up. An adequate pressure-relief device must be provided to handle any excess pressure that may result.

Like the piston pump, the rotary gear pump is very susceptible to damage from normal wear and from pumping water containing sand and other debris. To prevent damage to the casings, most gear pumps use bronze or another soft metal in the gears. These gears may be easily replaced if necessary. A strong alloy, such as cast iron, is used for the pump casing.

Some rotary gear pumps have power delivered to one gear, which then drives the other gear(s). In this case, the drive gear is usually made of steel with inserts of bronze to provide the strength needed to handle the torque that will be developed. Other designs use steel pilot gears mounted outside the case connected to the shaft that drives the pump gears.

Rotary Vane Pumps

The *rotary vane pump* is constructed with movable elements that automatically compensate for wear and maintain a tighter fit with closer clearances as the pump is used (Figure 10.9). In this type of pump, the rotor is mounted off-center inside the housing. The distance between the rotor and the housing is much greater at the intake than it is at the discharge. The vanes are free to move within the slot where they are mounted. As the rotor turns, the vanes are forced against the housing by centrifugal force. When the surface of the vane that is in contact with the casing becomes worn, centrifugal force causes it to extend further, thus automatically maintaining a tight fit. This self-adjusting feature makes the rotary vane pump much more efficient at pumping air than a standard rotary gear pump.

As the rotor turns, air is trapped between the rotor and the casing in the pockets formed by adjacent vanes. As the vanes turn, this pocket becomes smaller, which compresses the air and causes pressure to build up. This pocket becomes even smaller as the vanes progress toward the discharge opening. At this point, the pressure reaches its maximum level, forcing the trapped air out of

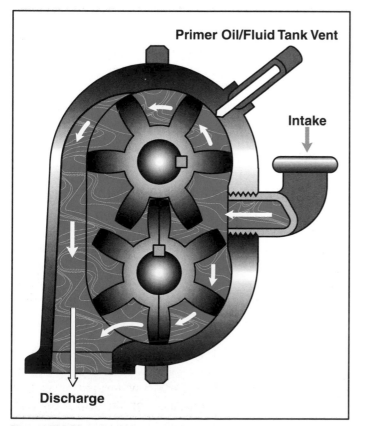

Figure 10.8 The basic design of a rotary gear pump.

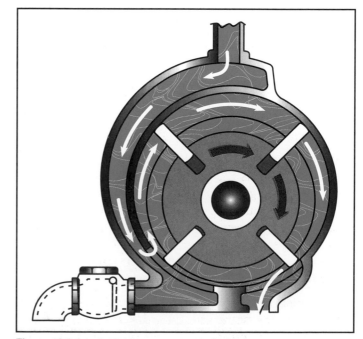

Figure 10.9 A typical rotary vane pump design.

the pump. The air or water is prevented from returning to the intake by the close spacing of the rotor at that point. As in the rotary gear pump, the air being evacuated from the intake side causes a reduced pressure (similar to a vacuum), and water is forced into the pump by atmospheric pressure until the pump fills with water. At this point, the pump is primed and forces water out of the discharge in the same manner as air was forced out.

Centrifugal Pumps

Nearly all modern fire apparatus utilize the centrifugal pump as their major pump (Figure 10.10). The *centrifugal pump* is classified as a nonpositive displacement pump because it does not pump a definite amount of water with each revolution. Rather, it imparts velocity to the water and converts it to pressure within the pump itself. This gives the pump a flexibility and versatility that has made it popular with the fire service. It has also virtually eliminated the positive displacement pump as a major fire pump in fire apparatus.

Principles of Operation and Construction of Centrifugal Pumps

In theory, the operation of a centrifugal pump is based on the principle that a rapidly revolving disk tends to throw water introduced at its center toward the outer edge of the disk. The faster the disk is turned, the farther the water is thrown, or the more velocity the water has. If the water is contained at the edge of the disk, the water at the center of the container begins to move outward. The velocity created by the spinning disk is converted to pressure by confining the water within the container. The water is limited in its movement by the walls of the container and moves upward in the path of least resistance. This shows that pressure has been created on the water. The height to which it rises, or the extent to which it overcomes the force of gravity, depends upon the speed of rotation.

Figure 10.10 The centrifugal fire pump is standard in the fire service today.

Fundamentally, the centrifugal pump consists of two parts: an impeller and a casing. The *impeller* (disk) transmits energy in the form of velocity to the water. The *casing* (container) collects the water and confines it in order to convert the velocity to pressure. Then, the casing directs the water to the discharge of the pump (Figure 10.11).

The impeller rotates very rapidly within the casing, generally from 2,000 to 4,000 rpm. Water is introduced from the intake into the eye of the impeller (Figure 10.12). To some extent, the volume capability of the pump is

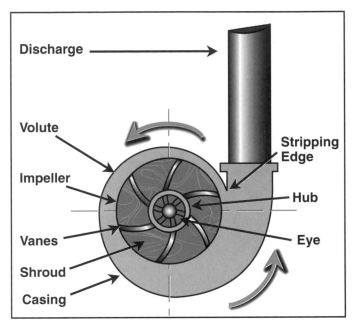
Figure 10.11 The major parts of a centrifugal fire pump.

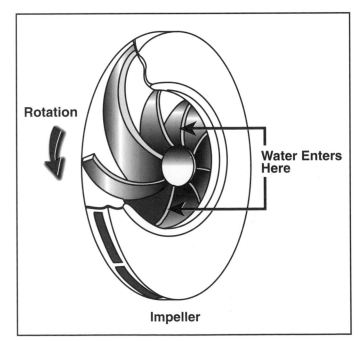
Figure 10.12 The water enters through the eye of the impeller.

dependent on the size of the eye. The larger the eye, the greater the flow capacity. As the water comes into contact with the vanes of the impeller, it is thrown by centrifugal force to the outside of the impeller (Figure 10.13). This water is confined in its travel by the shrouds of the impeller, which increase the velocity for a given speed of rotation. The impeller is mounted off-center in the casing. This placement creates a water passage that gradually increases in cross-sectional area as it nears the discharge outlet of the pump. This section of the pump is known as the *volute*. The increasing size of the volute is necessary because the amount of water passing through the volute increases as it approaches the discharge outlet. The gradually increasing size of the waterway reduces the velocity of the water, thus enabling the pressure to build up proportionately. There are three main factors that influence a centrifugal fire pump's discharge pressure:

• Amount of water being discharged

• Speed at which the impeller is turning

• Pressure of water when it enters the pump from a pressurized source (hydrant, relay, etc.).

If the discharge outlet is large enough in diameter to allow the water to escape as it is thrown from the impeller and collected in the volute, this pressure buildup is very small. If the discharge outlet is closed off, a very high pressure buildup results. With all other factors remaining constant, the amount of output pressure that a pump may develop is directly dependent upon the volume of water it is discharging. Simply stated, the greater the volume of water being flowed, the lower the discharge pressure.

Because the discharge pressure in the centrifugal pump results from the velocity and the amount of water that is moving, the speed of the impeller is an important factor in determining the pressure to be developed. The greater the speed of the impeller, the greater the pressure developed. This increase is approximately equal to the square of the change in impeller speed. For example, doubling the speed of the impeller results in four times as much pressure if all other factors remain constant.

A third factor that influences the discharge pressure is the intake pressure. By its nature, there is no positive mechanical blockage between the intake and the discharge outlet of a centrifugal pump. Isolation of discharge pressure from intake pressure in the impeller occurs entirely by the velocity of the moving water. Water flows through a centrifugal pump even if the impeller is not turning. When water is supplied to the eye of the impeller under pressure, it moves through the impeller by itself. Any movement of the impeller increases both the velocity of the water and the corresponding pressure buildup in the volute. Because the incoming pressure adds directly to the pressure being developed by the pump, incoming pressure changes are reflected in the discharge pressure (Figure 10.14).

The centrifugal pump depends on the velocity of the water to move water through the pump. It is unable to pump air and is not self-priming. In order for a centrifugal pump to draft water, some type of external primer must be supplied to remove the air and allow atmospheric pressure to force the water into the pump. One of the two types of rotary pumps described in the positive displacement pump section is typically used for this purpose. There are two basic types of centrifugal pumps used by the fire service: single-stage and two-stage.

Single-Stage Centrifugal Fire Pumps

Many centrifugal pumps used in the fire service are constructed with a single impeller and are referred to as

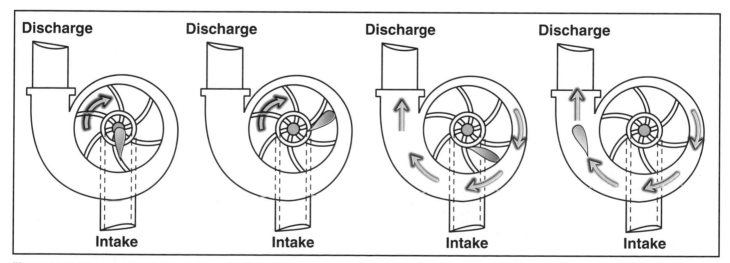

Figure 10.13 These schematics trace the path of water flow through the centrifugal fire pump.

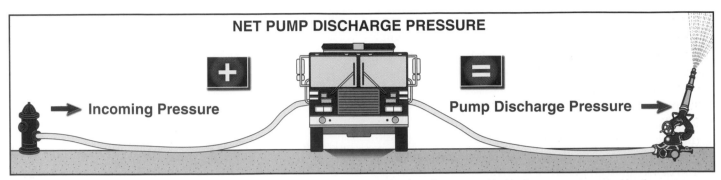

Figure 10.14 The net pump discharge pressure is the sum of the pressure coming into the pump added to the pressure created by the pump.

single-stage centrifugal pumps (Figure 10.15). Front-mount pumps, power take-off, separate engine driven, and midship transfer pumps use a single intake impeller and a simple casing to provide capacities up to 2,000 gpm (8 000 L/min) (Figure 10.16).

High-capacity pumps require large impellers with waterways that present a minimum of opposition to the movement of water. A law of physics states that for every action there is an equal and opposite reaction. The water in motion in the pump creates stress on the pump, its bearings, other moving parts, and the casings mounted to the truck frame.

To minimize the lateral thrust of large quantities of water entering the eye of the impeller, a double suction impeller was designed. The *double suction impeller* takes water in from both sides; the reaction being equal and opposite cancels the lateral thrust. It also provides a larger waterway for the movement of water through the impeller. Because the impeller turns at a very high rate of speed, a radial thrust is developed as the water is delivered to the discharge outlet. Stripping edges in the opposed discharge volutes divert the water 180 degrees apart. Water being removed at two places and traveling in opposite directions cancels the radial thrust. This design provides a hydraulically balanced pump, which lessens stress on the pump and chassis, and helps to lengthen the useful life of the pump and the apparatus.

Two-Stage Centrifugal Fire Pumps

The *two-stage centrifugal pump* has two impellers mounted within a single housing. The two impellers are usually mounted on a single shaft driven by a single drivetrain (Figure 10.17). Generally, the two impellers are identical and have the same capacity. What gives the two-stage pump its versatility and efficiency is its capability of connecting these two stages in series for maximum pressure or in parallel for maximum volume by use of a transfer valve.

Pumping in the Volume (Parallel) Position. When the pump is in the volume position, each of the impellers

Figure 10.15 A single-stage pump impeller. *Courtesy of Hale Fire Pump Company.*

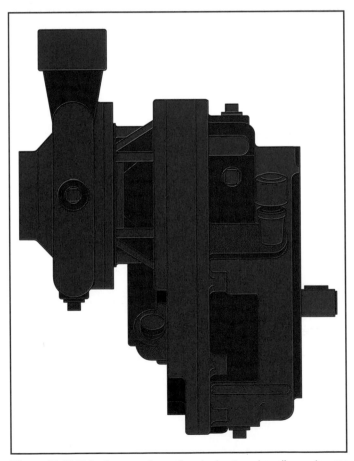

Figure 10.16 This diagram shows the single-stage impeller and pump casing. *Courtesy of Waterous Company.*

takes water from a source and delivers it to the discharge (Figure 10.18). Each of the impellers is capable of delivering its rated pressure while flowing 50 percent of the rated capacity; therefore, the total amount of water the pump can deliver is equal to the sum of each of the stages. If the pump is rated at 1,000 gpm (4 000 L/min) at 150 psi (1 000 kPa), each of the impellers supplies 500 gpm (2 000 L/min) to the pump discharge manifold (Figure 10.19). There the two streams combine, so the total amount available to the discharges is 1,000 gpm (4 000 L/min) at a net pump pressure of 150 psi (1 000 kPa). Changing the transfer valve to the series, or pressure, position greatly increases the maximum pressure attainable; however, increasing the pressure results in a corresponding reduction in volume. The driver/operator must remember that the pump receives its maximum flow rating at 150 psi (1 000 kPa). Increasing the discharge pressure of the pump above this figure results in a decreased volume of water being discharged.

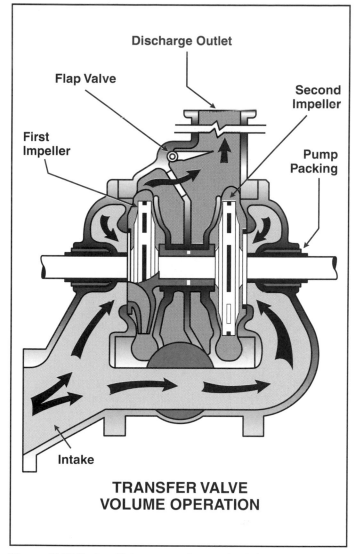

**TRANSFER VALVE
VOLUME OPERATION**

Figure 10.17 The impellers of a two-stage fire pump as they appear mounted on the impeller shaft. *Courtesy of Hale Fire Pump Company.*

Figure 10.18 During *Volume* operation, each of the impellers takes water from the source and delivers it to the discharge.

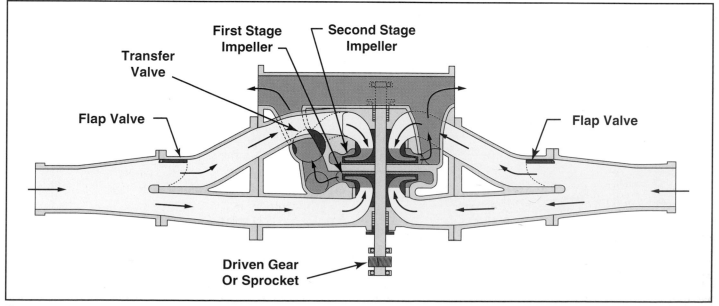

Figure 10.19 The route of water during *Volume* pumping operations. *Courtesy of Waterous Company.*

Pumping in the Pressure (Series) Position. When the transfer valve is in the pressure position, all the water from the intake manifold is directed into the eye of the first impeller (Figure 10.20). Depending on the pump manufacturer, the first stage increases the pressure and discharges 50 to 70 percent of the volume capacity through the transfer valve and into the eye of the second impeller. The second impeller increases the pressure and delivers the water at the higher pressure into the pump discharge port (Figure 10.21). At this point, the pressure is much higher than in the parallel (volume) position because the same stream of water has passed through two impellers, with each adding to the pressure. With only one impeller delivering water to the pump discharge port instead of two, as in the parallel (volume) position, the total volume of water is limited to the amount that one impeller can supply.

Each fire pump manufacturer has recommendations for when the transfer valve on their pump should be in the volume or pressure positions (Figure 10.22). The process of switching between pressure and volume is sometimes referred to as *changeover*. Historically, the fire service has always taught the rule of thumb that the transfer valve stays in the pressure position until it is necessary to supply more than one-half the rated volume capacity of the pump. However, advances in design and efficiency now allow most pump manufacturers to specify that the pump may remain in the pressure system until it is necessary to flow more than two-thirds of the rated volume capacity. At lower flow rates, operating in the series (pressure) position reduces the load and the required rpm of the engine. The driver/operator should consult the owner's manual for the specific pump being operated to obtain information on its recommended flow rate at which the transfer should occur.

When the transfer valve is operated on a two-stage pump, sudden changes in pressure occur as the water

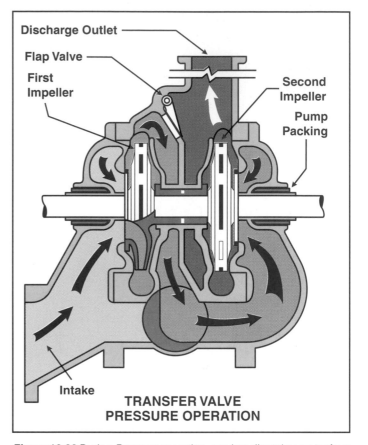

Figure 10.20 During *Pressure* operation, one impeller takes water from the supply source, adds pressure to it, and then discharges it into the second impeller. Here additional pressure is supplied to the water before it is discharged from the pump.

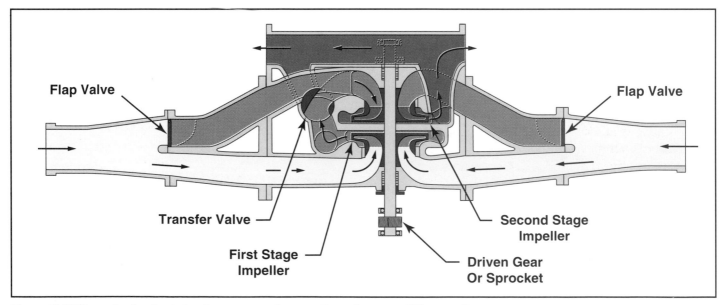

Figure 10.21 The route of water during *Pressure* pumping operations. *Courtesy of Waterous Company.*

Figure 10.22 The transfer valve is used to switch between the *Volume* and *Pressure* operations.

changes its direction of flow. For example, switching from volume to pressure results in an immediate doubling of the previous discharge pressure. This pressure can cause damage to hoselines and fire pumps as well as injury to firefighters on the hoselines if it takes place suddenly with excessive pressure on the pump. The maximum net pump discharge pressure at which the transfer valve should be operated varies depending on the manufacturer of the pump and the pump's age. However, in most cases, this recommended maximum pressure will not exceed 75 psi (535 kPa). There may be a slight interruption to fireground operations when this transfer occurs. The amount of time for the change to take place should be kept at a minimum. It is important that the changeover be coordinated with attack crews so that lines are not shut down at times that crews are in a precarious position.

The operator should attempt to anticipate the requirements that will be placed on the pumper as the fire fighting operation progresses and have the pump in the proper position. If there is any question as to the proper operation of the transfer valve, it is better to be in parallel (volume) than in series (pressure). While the parallel (volume) position may make it difficult to attain the desired pressure, it can supply 100 percent of the rated volume capacity at 150 psi (1 000 kPa). The series (pressure) mode of operation may make it impossible to maintain the necessary volume of water to the attack lines.

Operation of the transfer valve is performed manually on many two-stage pumps, particularly older ones. If this is the case, a built-in safeguard makes it physically impossible to accomplish the transfer while the pump is operating at high pressures. Newer pumps utilize a power-operated transfer valve. This transfer valve can be activated by using electricity, air pressure, vacuum from the

engine intake manifold (gasoline engines only), or even by using the water pressure itself to accomplish the transfer.

Many power-operated transfer valves operate at pressures as high as 200 psi (1 380 kPa). These pressures can present extreme danger to personnel and equipment, so special care must be used with these types of controls. If available, the power control may have some type of manual override to allow the transfer to be operated should the power equipment fail. It is important for all operators to be familiar with this manual override and to practice with it frequently.

The clapper (check) valves are essential in a two-stage pump (Figure 10.23). If they should stick open or closed, or get debris caught in them, the pump will not operate properly in the pressure (series) position. When the transfer valve is operated, the clapper valve allows the water to escape back into the intake, and it just churns through the pump instead of building up pressure. The operation of this valve can be inspected by removing the strainer from all big intake openings, reaching into the pump with some type of rod, and ensuring that the valve swings freely. In some cases, it may be necessary to hold the clapper valve open with the rod to ensure that the pump can be properly flushed.

Some manufacturers have used as many as four impellers connected in series to develop pressures up to 1,000 psi (6 900 kPa) for high-pressure fog fire fighting. Another approach is to supply a single-stage, high-pressure centrifugal pump mounted outboard on a conventional two-stage pump with a separate drive system. When pressures higher than 250 psi (1 725 kPa) are needed, the separate third stage can be engaged and used to increase the pressure from the second stage to a much higher level. These types of pumps are most commonly found in cities that have a substantial amount of high-rise structures. The increased pressure capabilities allow fire department pumpers to supply sprinkler and standpipe systems in tall buildings. Pumpers that are designed to supply high pressures must be equipped with fire hose that is rated and tested for these pressures.

Pump Wear Rings and Packing

Although there is no positive mechanical isolation of the discharge area of the pump from the intake in the impeller, the velocity of the water moving through the impeller prevents most water in the discharge from escaping back into the intake. The pressure in the volute is much higher than that in the intake side of the pump at the eye of the impeller, so a very close tolerance must be maintained between the pump casing and the hub of the impeller.

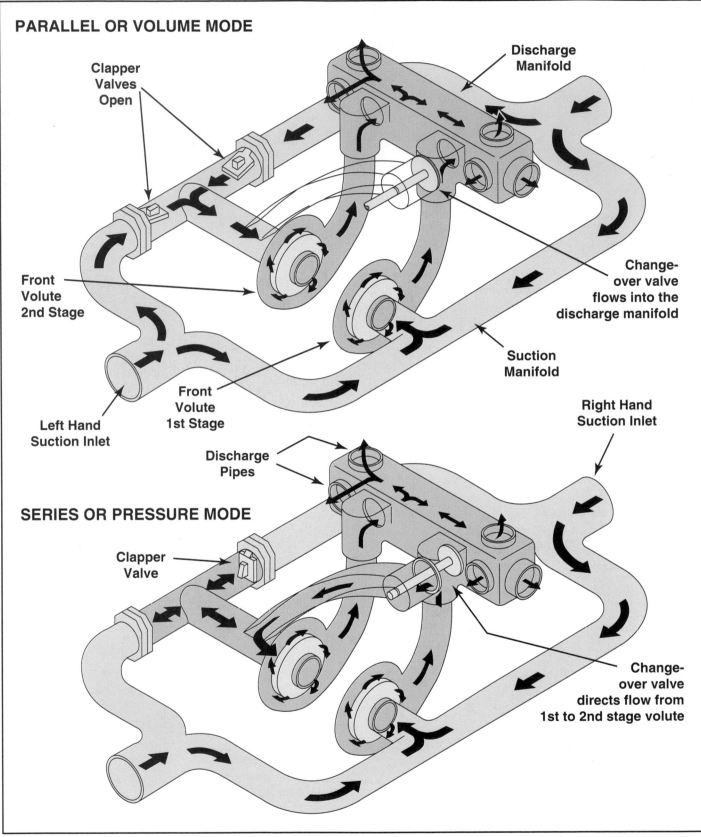

PARALLEL OR VOLUME MODE

Clapper
Valves
Open

Discharge
Manifold

Front
Volute
2nd Stage

Change-
over valve
flows into the
discharge manifold

Left Hand
Suction Inlet

Front
Volute
1st Stage

Suction
Manifold

Right Hand
Suction Inlet

Discharge
Pipes

SERIES OR PRESSURE MODE

Clapper
Valve

Change-
over valve
directs flow from
1st to 2nd stage volute

Figure 10.23 The position of the clapper valves is dependent on which mode of pumping is being used.

This opening is usually limited to .01 inch (0.25 mm) or less. Any increase in the opening lessens the pump's effectiveness.

Impurities, sediment, and dirt are present in every water supply. As these pass through the pump, they cause wear when they come in contact with the impeller, which is turning nearly 4,000 rpm when the pump approaches its capacity. This process is greatly accelerated when it is necessary to pump water with a high sand content. The sand particles passing between the impeller and the pump casing act like sandpaper in wearing down the metal surfaces. As the gap increases, greater amounts of water are allowed to slip back into the intake and thus are not available at the discharge. Eventually, the pump is no longer able to supply its rated capacity. The first indication that wear is becoming a problem is when increased engine rpm is required to pump the rated capacity in pump tests.

To restore the capacity of the pump without replacing the pump itself, replaceable wear rings or clearance rings are provided in the pump casing to maintain the desired spacing between the hub of the impeller and the casing (Figure 10.24). If the hub of the impeller is also worn down, it is possible to install smaller wear rings to compensate for the smaller size and maintain the proper clearance.

The centrifugal pump differs from the positive displacement pump in that little or no harm results from shutting off all discharges for short periods of time. When discharges are shut off, the energy being supplied to the impeller is dissipated in the form of heat as the water within the pump is allowed to churn. If this situation is allowed to continue for extended periods of time, the water in the pump becomes quite hot and the metal parts tend to expand. Keep in mind that heating occurs proportionately faster as the pump speed is increased. If the wear rings and the impeller expand too much, they may come in contact with each other and the friction of the two surfaces rubbing together may cause even more heat. In extreme cases, the wear rings may seize, causing serious pump damage. The best insurance against this happening is to ensure that some water is moving through the pump at all times.

It is best for the driver/operator to not put the pump in a position where it might overheat. The driver/operator can check the pump temperature by placing a hand on the direct pump intake pipe of the pump (Figure 10.25). If it is warm to the touch, open a discharge or circulator valve. If no water is expected to be discharged for an extended period of time, the pump should be disengaged until it is needed.

The impellers are fastened to a shaft that connects to a gear box (Figure 10.26). The gear box transfers the necessary energy to spin the impellers at a very high rate of speed. At the point where the shaft passes through the pump casing, a semi-tight seal must be maintained to prevent air leaks that could interfere with a drafting

Figure 10.24 Wear rings, which are replaceable, are vital in maintaining the proper spacing between the hub of the impeller and the casing.

Figure 10.25 In some cases, the pump operator can detect pump heating by feeling the intake piping.

Figure 10.26 The fire pump is turned by a gear box or, as in this case, a chain drive. *Courtesy of Waterous Company.*

operation. Packing rings are used to make this seal in most fire pumps (Figure 10.27). The most common type of packing is a material made of rope fibers impregnated with graphite or lead. The material is pushed into a stuffing box by a packing gland driven by a packing adjustment mechanism.

As packing rings wear with the operation of the shaft, the packing gland can be tightened and the leak controlled. Where the packing rings come into contact with the shaft, heat is developed by friction against the packing. To overcome this, a lantern ring (spacer) is supplied with the packing to provide cooling and lubrication. A small amount of water leaks out around the packing and prevents excessive heat buildup. If the packing is too tight, water is not allowed to flow between the packing and the shaft, and excessive heat buildup results. A scored shaft prevents even new packing material from making a good seal.

If the packing is too loose, as indicated by an excessive amount of water leaking from the pump during operation, air leaks adversely affect the pump's ability to draft. The adjustment of this packing must be made carefully and according to the manufacturer's instructions. In general, however, when the pump is operating under pressure, water may drip from the packing glands but not run in a steady stream.

The packing only receives the needed water for lubrication if the pump is full and operating under pressure.

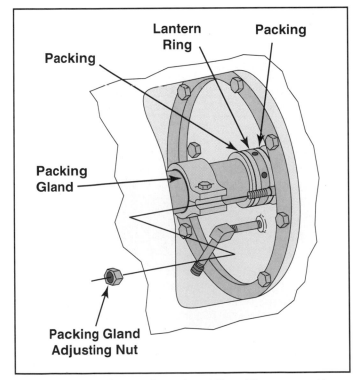

Figure 10.27 This diagram shows the position of the pump packing.

If the pump is operated dry for any length of time, it damages the shaft. This weakens the shaft and may cause it to fail in a future use.

Some fire departments make a practice of keeping the pump drained between fire calls. This is particularly common in cold weather climates. If the pump is not used for extended periods of time, the packing may dry out and excessive leakage results. After a period of dryness, adjustment to the packing should not be made until the pump is operating under pressure and the packing has had a chance to seal properly.

If the pump is equipped with mechanical seals instead of packing, they will not drip and will not require adjustment. Again, it is important that the pump be run periodically to lubricate the seals. It should be noted that mechanical seals are especially problematic if they freeze. Freezing may cause damage that necessitates immediate repair. This repair process is not simple. The entire pump and drive assembly may need to be disassembled in order to repair and replace the seals.

Pump Mounting and Drive Arrangements

There are many types of mounting arrangements and pump drive systems to consider when building a fire department pumper. Cost, appearance, space required, ease of maintenance, and tradition all enter into the decision, but the most important consideration is the use that the system receives. Each system has certain characteristics that make it more or less adaptable to a particular fire department's needs.

Auxiliary Engine-Driven Pumps

Auxiliary engine-driven pumps are those pumps that are powered by a gasoline or diesel engine independent of an engine used to drive the vehicle. Some manufacturers have models that are powered by special fuels, such as jet fuel, for military or other special applications. Although used on some structural pumpers, the most common applications for auxiliary engine-driven pumps are:

- Airport rescue and fire fighting (ARFF) vehicles
- Wildland fire apparatus (Figure 10.28)
- Mobile water supply apparatus
- Trailer-mounted fire pumps
- Portable fire pumps

Auxiliary engine-driven pumps offer the maximum amount of flexibility. With a separate engine, the pump can be mounted anywhere on the apparatus. The pump pressure is independent of the apparatus drive system, which makes it ideal for pump and roll operations. (Pump and roll refers to pumping water while the apparatus is in motion.) Auxiliary engine-driven pumps are used for ARFF, wildland, or mobile water supply apparatus. The pumping capacity of these units is generally 400 gpm (1 600 L/min) or less. In many cases, these pumps are part of a skid-mount assembly that includes a small water tank, booster reel, and hose trays for small-diameter attack lines. Fire departments purchase these skid-mount assemblies and mount them in the rear of a pickup truck to make a small attack fire apparatus.

Auxiliary engine-driven pumps used on ARFF apparatus and for trailer-mounted applications tend to be large capacity pumps of up to 4,000 gpm (16 000 L/min) or more. They are powered by full-sized diesel engines capable of up to 500 horsepower or more. Trailer-mounted pumps are commonly used at industrial and fire training facilities (Figure 10.29).

Power Take-Off Drive

In this arrangement, the pump is driven by a driveshaft that is connected to the power take-off (PTO) on the chassis transmission (Figure 10.30). PTO-driven fire pumps are most commonly used on initial attack, wild-

land, and mobile water supply apparatus. However, in later years, they have become increasingly popular on structural pumpers.

Proper mounting of these units is essential for dependable and smooth operation. The pump gear case must be mounted in a location that allows for a minimum of angles in the driveshaft. At the same time, it must not extend far enough below the chassis to be readily damaged as the unit travels on or off the road. Some type of skid plate can be used to provide protection if the pump extends too far below the chassis.

The PTO unit is powered by an idler gear in the truck transmission. The speed of the shaft is independent of the gear in which the road transmission is operating when the pump is in use, but it is under the control of the clutch. When the operator disengages the clutch to stop momentarily or to change gears, the pump also stops turning. The PTO pump does permit "pump and roll" operation, but it is not as effective as the separate engine unit.

The pressure being developed is determined by the speed of the engine; therefore, the pressure changes when the driver changes the vehicle speed. If the unit is

Figure 10.28 Auxiliary engine pumps are commonly used on small wildland fire apparatus.

Figure 10.29 Trailer pumps have auxiliary engines.

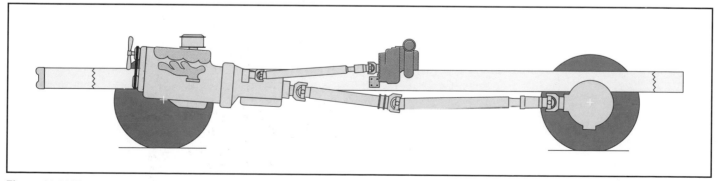

Figure 10.30 The basic layout of a PTO pumping system.

designed for pump and roll operation, a pressure gauge should be mounted inside the cab in full view of the driver, and the vehicle should be driven by the pressure gauge instead of the speedometer while hoselines are in operation. However, the driver/operator must still take caution not to drive the apparatus too fast for conditions.

The conventional power take-off unit limits the capacity of the pump to about 500 gpm (2 000 L/min). This is because the PTO unit is mounted on the side of the transmission, and the maximum strain this housing can take limits the available horsepower to approximately 35. As well, the pump manufacturers set limitations for rpms supplied to the pump drive.

In recent years, some manufacturers are providing "full torque" power take-offs that permit the installation of pumps as large as 1,250 gpm (5 000 L/min). This type of PTO is especially common on some of the modern automatic transmissions where the flywheel of the engine drives the PTO unit. This provides enough torque to drive larger pumps that traditional PTOs were incapable of powering.

Front-Mount Pumps

On some apparatus, the front bumper is extended and a pump is mounted between the bumper and the grill (Figure 10.31). This unit is driven through a gear box and a clutch connected by a universal joint shaft to the front of the crankshaft (Figure 10.32). The gear box uses a step-up gear ratio, which causes the impeller of the pump to turn faster than the engine. This ratio is set to match the torque curve of the engine to the rotation speed required for the impeller to deliver the pump's rated capacity. It is usually between 1½ and 2½:1. The maximum capacity pump that can be used depends on the limitations of the engine driving it, but typically it can go as high as 1,250 gpm (5 000 L/min) without major modifications by the transmission, pump, or apparatus manufacturer. To use a front-mount pump, a chassis should have a front-mount PTO option that provides a coupling to the front

Figure 10.31 Front-mount pumps are preferred in some jurisdictions.

of the crankshaft. The chassis should also have an opening in the radiator for the driveshaft when required.

One disadvantage of the front-mount pump is that the pump and gauges are more susceptible to freezing in cold climates than are pumps that are contained within the body of the apparatus. This can be overcome through the use of external lines that circulate radiator coolant through the pump body. Gauges and connecting pipes, on the other hand, may suffer damage during extremely cold weather. When possible, it is desirable to enclose the gauges and protect them from the weather.

The front-mount pump is in a vulnerable position in the event of a collision. For this reason, it is important that the chassis and frame be modified to provide a solid base of protection for the pump. Where they have been properly installed, front-mount pumps have been known to withstand accidents that have caused considerable damage to the apparatus.

Like PTO-driven pumps, front-mount pumps can be used for pump and roll operations. Again, if the unit is to be used for pump and roll, a pressure gauge is needed inside the cab so that the driver can use the gauge for a reference instead of using the speedometer.

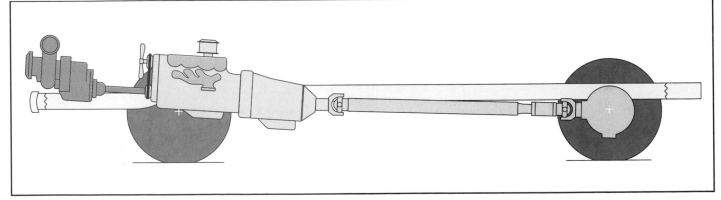

Figure 10.32 The drive arrangement for front-mount pumps.

The front-mount pump is engaged and controlled from the pump location itself. This puts the driver/operator in a vulnerable spot, namely, standing in front of the vehicle while operating the pump. Therefore, it is essential that a lock be provided to prevent the road transmission from being engaged while the pump is operating. This lock is needed with either a manual or an automatic transmission and should always be used when the pump is in service. Because the operating lever that engages the pump is located at the pump, a warning light in the cab should be provided to alert the driver that the pump is engaged. If the vehicle is driven with the pump turning and no water being discharged, damage to the pump results. Damage occurs either to the packing from running the pump dry, or to the pump itself from overheating the water by churning it inside the pump with no water flowing.

Midship Transfer Drive

Most fire department pumpers have the fire pump mounted laterally across the frame behind the engine and transmission (Figure 10.33). Power is supplied to the pump through the use of a split-shaft gear case (transfer case) located in the drive line between the transmission and the rear axle (Figure 10.34). By shifting a gear and collar arrangement inside the gear box, power can be diverted from the rear axle and transmitted to the fire pump. The pump is then actually driven by either a series of gears or a drive chain.

Figure 10.33 Midship pumps are the most common type found in the fire service today.

The gear ratio is set to match the engine torque curve to the speed of rotation required for the impeller to deliver the rated capacity of the pump. This ratio is arranged so that the impeller turns faster than the engine, usually one and a half to two and a half times as fast. The maximum capacity that can be obtained by this system is limited only by the engine horsepower and the size of the pump. Most fire pumps can operate over a range of capacities. These capacities are determined by the piping arrangement and gear ratio used. For example, a particular two-stage centrifugal pump popular in the fire service can be rated anywhere from 500 to 1,500 gpm (2 000 L/min to 6 000 L/min) with no major changes to the pump itself.

The normal arrangement is for the transfer case to be controlled from inside the cab of the apparatus. This can be done by a mechanical linkage or by electrical, hydraulic, or air-operated controls. If the pumper is so equipped, the operator should engage the pump and put the road transmission in the proper gear before leaving the cab. If the road transmission is not placed in the correct gear, the pump does not turn at the needed rpm to operate effectively. With the transmission turning, most apparatus registers the engine speed in road miles (km) per hour on the speedometer when the pump is operating. Check to see that the transmission is in the correct gear by observing the speedometer reading after the pump is engaged. With the engine idling and the pump engaged, most speedometers read between 10 to 15 mph (16 km/h to 24 km/h), depending on the apparatus. Some newer apparatus may be designed so that the speedometer does not go above 0 mph (km/h) when the pump is engaged. In this case, the only way to visually tell if the transmission is in the correct gear is to take a tachometer reading off the rear of the transfer case. Obviously, this is not practical for most fire service observations. The information on proper pumping procedures contained in Chapter 11 of this manual gives practical information on how to make sure that the transmission is in the proper gear in these cases.

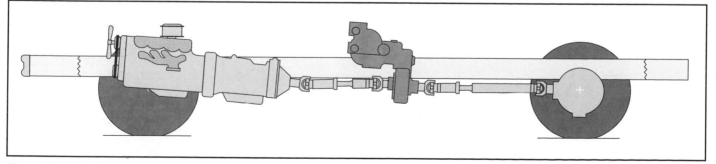

Figure 10.34 The drive arrangement for a midship transfer fire pump.

To prevent damage to the gears, disengage the clutch and place the road transmission in neutral while the power transfer is made. The transmission should be allowed to slow down to idle speed after shifting to neutral to help with a smooth shift in and out of pump. Some power transfer arrangements are provided with a manual override in case of difficulty with the power unit. Operators should frequently practice using this manual override so that they will be prepared for an emergency.

Whereas engaging the pump involves transferring the power from the rear axle coupling shaft to the pump drive gears, it is not possible to transmit power to the rear axle while the pump is engaged. This eliminates the possibility of a pump and roll operation with a conventional gearing arrangement.

To prevent an automatic transmission gear selector from moving during a pumping operation or a manual transmission from slipping out of gear, a lock is provided on the transmission or shift lever to hold it in the proper gear for pumping (Figure 10.35). It is also possible to operate the pump shift control and not have the gears complete their travel. If this happens, the vehicle may begin to move as the engine rpm is increased to build pressure. To prevent movement, some apparatus are equipped with an indicator light on the dash, and the pumping operation should not begin until the green indicator light goes on. Later model automatic transmissions for fire apparatus are equipped with a gear lockup to prevent improper shifting of an automatic transmission while the pump is engaged. Hydraulic circuitry ensures that the transmission is shifted to the correct gear for direct drive of the fire pump.

It should be noted that on some automatic transmissions, it is possible for the transmission to be momentarily engaged in the first and highest (fourth or fifth, depending on the model) forward ranges while internal clutches are being applied and released. The transmission output shaft may begin to rotate at a low torque and speed during the shifting process. This is why it is important that the transmission be in neutral during this entire sequence. If the shift mechanism is not in neutral, the transmission may lock up and the engine may stall.

Rear-Mount Pumps

In recent years, it has become increasingly popular for fire departments to specify fire department pumpers with rear-mount pumps (Figure 10.36). There are a number of advantages to having the pump on the rear of the apparatus. First, it helps provide a more even weight distribution on the apparatus chassis. As well, it typically allows the apparatus to have more compartment space for tools and equipment than a similar sized vehicle with

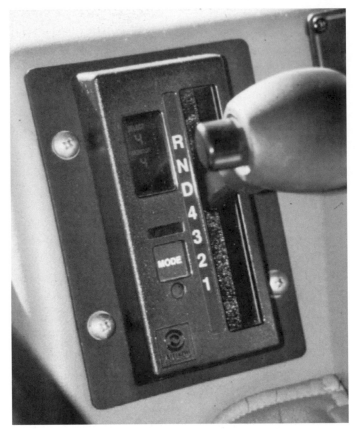

Figure 10.35 The automatic transmission locks in place when the pump is in operation.

Figure 10.36 Rear-mount pumps have gained acceptance in recent years. *Courtesy of Ron Jeffers.*

a midship pump. A disadvantage of the rear-mount pump is that the driver/operator may be more directly exposed to oncoming traffic than in other pump positions. This can be somewhat offset by placing the pump controls on one of the rear sides of the vehicle.

Rear-mount pumps may be powered by either a split-shaft transmission or by a power take-off, depending on the manufacturer or fire department's preference (Figure 10.37). In either case, a driveshaft of appropriate length and size is connected between the transmission

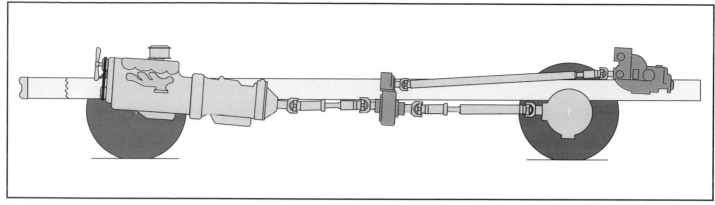

Figure 10.37 The drive arrangement for a rear-mount pump.

and the pump. Other than the location being the rear of the apparatus, the operation of these pumps is the same as previously described for PTO and split-shaft (midship transfer) drive pumps.

Pump Piping and Valves

An integral part of the fire pump system on a fire apparatus is the piping valves that are attached to the pump (Figure 10.38). The primary components of the piping systems are intake piping, discharge piping, pump drains, and valves. According to NFPA 1901, all components of the piping system must be of a corrosion-resistant material. Most piping systems are constructed of cast iron, brass, stainless steel, or galvanized steel. Some systems may include rubber hoses in certain locations. It is not uncommon for an apparatus to have a combination of these materials in the overall piping system. The piping system (as well as the pump itself) must be able to withstand a hydrostatic test of 500 psi (3 450 kPa) before being placed into service. In order to minimize pressure loss within the apparatus, all piping and hoses should be designed so that they run as straight as possible, with a minimum of bends or turns.

Intake Piping

There are two primary ways that water may enter the fire pump. The first is through piping that connects the pump and the onboard water tank. The second is piping that is used to connect the pump to an external water supply.

The majority of all fires are initially fought with the water carried in the fire apparatus tank. Therefore, it is important that the piping from the tank to the pump be large enough to allow for adequate streams to make an effective initial attack. NFPA 1901 states that piping should be sized so that pumpers with a capacity of 500 gpm (1 900 L/min) or less should be capable of flowing 250 gpm (950 L/min) from their booster tank. Pumpers with capacities greater than 500 gpm (1 900 L/min) should be able to flow at least 500 gpm (1 900 L/min).

Figure 10.38 The fire pump supplies a variety of discharge piping and valves. *Courtesy of Hale Fire Pump Co.*

Many pumpers today are equipped with tank-to-pump lines as large as 4 inches (100 mm) in diameter. Mobile water supply units may have multiple tank-to-pump lines. All modern pumps are equipped with check valves in these lines. The check valves prevent damage to the tank if the tank-to-pump valve inadvertently opens when water is being supplied to the pump under pressure, such as during a relay. If this line has a check valve in it, it is impossible to fill the tank through the pump by opening the tank-to-pump valve. The tank-to-pump valve must be maintained in good condition. If it leaks, priming the pump when the tank is empty is impossible because air from the tank is drawn into the pump and no vacuum is established. It is also possible for the water to be drawn out of the tank during a drafting operation. Prime is lost when the tank is emptied.

The pump must be capable of being supplied by water sources that are external to the apparatus. These sources may be pressurized or static sources. The pump must be primed when drafting from a static water supply source. This involves removing all or most of the air from the pump thus lowering the atmospheric pressure within the pump casing. The primer is tapped into the pump at a high point on the suction side or the impeller eye and a valve (priming valve) is used. Any air trapped in the pump during the priming operation can prevent a successful drafting operation. For this reason, all intake lines to a centrifugal pump are normally located below the eye of the impeller, and no part of the piping is above this point. The single exception to this may be the tank-to-pump line where the water is moving under the natural pressure of gravity if an electric primer is used or midline if a mechanical primer is used.

The primary intake into the fire pump is through large-diameter piping and connections. Intake piping is round in shape at the point where the intake hose connects to it. As the piping nears the pump itself, it typically tapers to a square shape. This is to eliminate the vortex that may occur in water that flows through circular piping. This vortex could result in air entering the pump if it was not eliminated. On a front-mount pump, this pipe and connection extend from the lower portion of the pump (Figure 10.39). On midship pumps, there usually is a large intake connection on either side of the apparatus (Figure 10.40).

Additional large-diameter intakes may be piped to the front or the rear of the apparatus (Figure 10.41). However, because of the number of bends and length of pipe, front and rear intakes may not allow the pump to supply its rated capacity. In some cases, these intakes may allow maximum flows that are several hundred gpm (L/min) lower than the rated capacity of the pump. These front and rear intakes should be considered auxiliary intakes in the same manner as the pump-panel-mounted gated intakes. Pumps that have a capacity of 1,500 gpm (6 000 L/min) or greater may require more than one large intake connection at each location (Figure 10.42). Table 10.1 shows the intake sizes for various size pumps as required by NFPA 1901.

Figure 10.40 The main intakes on midship pumps are usually found on either side of the pump.

Figure 10.41 Front or rear intakes on midship pumps are really auxiliary intakes.

Figure 10.39 The main intake on a front-mount pump extends from the lower portion of the pump.

Figure 10.42 Large volume pumpers, such as this 3,000 gpm (12 000 L/min) industrial pumper, will have multiple large intakes on each side of the apparatus. *Courtesy of Celanese Corp., Clear Lake, Texas Plant.*

Table 10.1
Required Number of Intake Hoses and Size For Fire Pumps

Rated Capacity (gpm)	(L/min)	Maximum Suction Hose (in.)	(mm)	Maximum Number of Suction Lines	Maximum Lift (ft)	(m)
250	950	3	76	1	10	3
300	1136	3	76	1	10	3
350	1325	4	100	1	10	3
450	1700	4	100	1	10	3
500	1900	4	100	1	10	3
600	2270	4	100	1	10	3
700	2650	4	100	1	10	3
750	2850	4½	113	1	10	3
1000	3785	5	125	1	10	3
1250	4732	6	150	1	10	3
1500	5678	6	150	2	10	3
1750	6624	6	150	2	8	2.4
2000	7570	6	150	2	6	1.8
2250	8516	8	150	3	6	1.8
2500	9463	8	150	3	6	1.8
2750	10,410	8	150	4	6	1.8
3000	11,356	8	150	4	6	1.8

Reprinted with permission from NFPA 1901, Automotive Fire Apparatus, Copyright © 1996. National Fire Protection Association, Quincy, MA 02269. This reprinted material is not the complete and official position of the National Fire Protection Association, on the referenced subject which is represented only by the standard in its entirety.

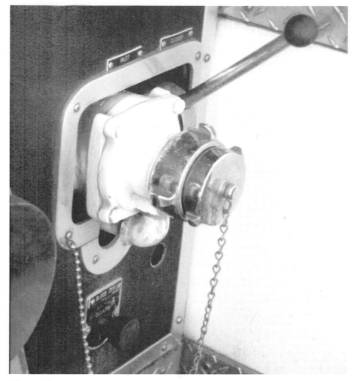

Figure 10.43 Most pumpers are equipped with gated auxiliary intakes for MDH hose.

Additional intake lines, usually gated, are provided for use in relay operations or anytime water is being received through small-diameter supply lines (Figure 10.43). Most of these intake openings have threads to use 2½-inch (65 mm) hose couplings. The amount of flow that can be obtained is determined by the diameter of the pipe from the valve to the pump inlet and the straightness of its routing. If the pipe is no larger than 2½ inches (65 mm) and if it contains 90-degree bends or T-fittings, the friction loss may be enough to limit the flow through these intakes to 250 gpm (950 L/min). If 3-inch (77 mm) pipe is used and care is given to the fittings, it may be possible to flow as much as 450 gpm (1 700 L/min) through one of these intake openings.

Discharge Piping

According to NFPA 1901, enough 2½-inch (65 mm) or larger discharge outlets must be provided in order to flow the rated capacity of the fire pump (Figure 10.44).

Figure 10.44 A typical 2½-inch (65 mm) pump discharge connection.

Table 10.2 lists the allowable discharge rates by outlet size specified in NFPA 1901. (**NOTE:** Keep in mind that in fireground applications, each of these discharges is capable of flowing considerably more water than the amounts listed on this table.) As a minimum, all fire apparatus with a rated pump capacity of 750 gpm (2 850 L/min) or greater must be equipped with at least two 2½-inch (65 mm) discharges. Pumps that are rated less than 750 gpm (2 850 L/min) are only required to have one 2½-inch (65 mm) discharge. Discharges that are larger than 2½ inches (65 mm) may not be located directly on the pump operator's panel.

The apparatus may also be equipped with discharges that are less than 2½ inches (65 mm) in size. Smaller discharges are commonly provided in a variety of positions on the apparatus for use with preconnected attack lines (Figure 10.45). Discharges to which 1½-, 1¾-, and 2-inch (38 mm, 45 mm, and 50 mm) handlines are attached must be supplied by at least 2-inch (50 mm) piping.

Discharge piping is constructed of the same materials as previously described for intake piping. Discharges are usually equipped with a locking ball valve, and they should always be kept locked when they are open to prevent movement. This is especially true when it has been necessary to "gate down" a line to reduce the discharge pressure being supplied from it. All valves should be designed so that they are easily operable at pressures of up to 250 psi (1 724 kPa).

When multiple attack lines are in use with different pressures required, the only way to supply them is to set the engine pressure to the highest pressure needed. Then, each of the other lines should have the valve partially closed until the reduced flow through the valve is sufficient to provide the desired pressure at the hoseline. Individual line pressure gauges or flowmeters are essential to do this properly. Without individual line gauges, providing good fire streams becomes a matter of guesswork or constant feedback from the nozzle.

A tank fill line (sometimes referred to as a pump-to-tank line) should also be provided from the discharge side of the pump (Figure 10.46). This allows the tank to be filled without making any additional connections when the pump is being supplied from an external supply source. It also provides a means of replenishing the water carried in the tank after the initial attack has been made from the water tank on the apparatus. In some multiple-stage fire pumps, this line is taken from the first stage, which provides for reduced pressures to the tank, as a measure of safety. NFPA 1901 requires that apparatus with a water tank that is less than 1,000 gallons (3 785 L) in size to have a tank fill line that is at least 1-inch (25 mm)

Table 10.2
Discharge Rates By Outlet Size

| Outlet Size | | Flow Rates | |
(in.)	(mm)	(gpm)	(L/min)
2½	65	250	950
3	76	375	1420
3½	89	500	1900
4	100	625	2365
4½	113	750	2850
5	125	1000	3785

Reprinted with permission from NFPA 1901, Automotive Fire Apparatus, Copyright © 1996. National Fire Protection Association, Quincy, MA 02269. This reprinted material is not the complete and official position of the National Fire Protection Association, on the referenced subject which is represented only by the standard in its entirety.

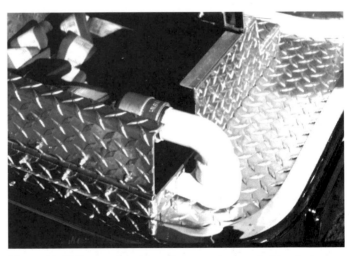

Figure 10.45 Preconnected hoselines may be supplied by discharges that are less than 2½-inches (65 mm) in diameter.

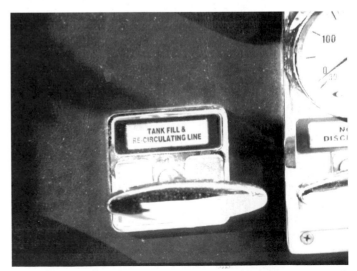

Figure 10.46 The tank fill line is operated by a control on the pump panel.

in diameter. Apparatus with tanks that are 1,000 gallons (3 785 L) or larger must have at least a 2-inch (51 mm) tank fill line.

The tank fill line can be used to circulate water through the pump to prevent overheating when no lines are flowing. In a two-stage pump with the fill line coming off the first stage, overheating can still result in the second stage portion of the pump. A circulator valve is more effective for preventing overheating (Figure 10.47). The circulator valve is connected to the discharge side of the pump and enables water to be dumped into the tank or outside the tank on the ground. One valve that has been used in the past has a tank position and spill position to indicate what happens to the water, but there is no closed position designated on the valve. Either position allows air to be introduced into the pump, which makes drafting difficult if not impossible. Although it is not so designated, this particular valve is closed when the handle is in the center position, and it should be in that position when drafting.

Some pumpers have a booster line cooling valve that serves the same function as the circulator valve by diverting a portion of the discharge water into the tank. While either of these arrangements serves the purpose for normal pumping operations, the piping is done with small copper tubing, and the flow is limited to approximately 10 to 20 gpm (40 L/min to 80 L/min). During prolonged operations with intermittent flows, or when operating at very high pressures, this may not be enough water to keep the pump cool. To do this, it may be necessary to discharge water through some type of waste or dump line.

Valves

Most of the intake and discharge lines from the pump are controlled by valves. When the pump is new, valves must be airtight. Even though the valves are constructed to resist wear (in some cases, they are self-adjusting), they do require repair as they age and are subjected to frequent use.

The most common type of valve is the ball-type valve that permits full flow through the lines with a minimum of friction loss (Figure 10.48). The most common actuators for ball-type valves are either push-pull handles (commonly called T-handles) or quarter-turn handles (Figures 10.49 a and b). The push-pull valve handle uses a sliding gear-tooth rack that engages a sector gear connected to the valve stem. This gear arrangement provides a mechanical advantage that makes the valve easier to operate under pressure. A gear arrangement of this type also allows precise values of pressure to be set when adjusting individual lines. The operating linkage can be designed to allow the valve to be mounted in a location remote from the pump panel. The push-pull lever usu-

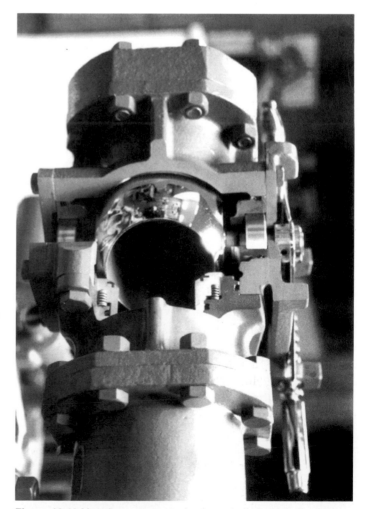

Figure 10.48 Most fire pump pumping is controlled with ball valves.

Figure 10.47 The circulator valve is used to keep the pump from overheating when water is not being discharged.

Figure 10.49a Push-pull valves are used on many fire apparatus.

Figure 10.49b Quarter-turn valve controls may be used on some apparatus.

Figure 10.50 The quarter-turn valve is locked by turning the handle in the specified direction.

ally has a flat handle that can be used to lock the valve in any position by a 90-degree twist of the handle. When operating the push-pull lever, it must be pulled straight out, in a level manner. Otherwise, binding of the shaft occurs, and the valve is inoperable from that point on.

The quarter-turn handle has a simpler mechanical linkage with the handle mounted directly on the valve stem. The valve is opened or closed by a 90-degree movement of the handle. Some of the older quarter-turn valves are locked in position by raising and lowering the handle, but the more modern versions lock by rotating the handle in a clockwise direction (Figure 10.50). Some valves lock automatically when the handle is released, but the majority of valves in use require a positive action to lock the valve in position anytime the line is in use.

Newer apparatus may be equipped with valves that are hydraulically, pneumatically, or electrically controlled. These valves use a ball-type valve that is operated by a toggle switch on the pump operator's panel (Figure 10.51). A visual display shows a readout of how far the valve is opened. Markings on the panel indicate which direction to operate the switch in order to open or close the valve.

Figure 10.51 Some modern pumpers have hydraulic, pneumatic, or electrical valves that are controlled by a toggle switch.

Gate or butterfly valves are also used on fire apparatus (Figures 10.52 a and b). These are most commonly used on large-diameter intakes and discharges. Gate valves are most commonly operated by a handwheel. Butterfly valves are most commonly operated by quarter-turn handles. Both may be equipped with hydraulic, pneumatic, or electric actuators. These are commonly used as remote-operated dump controls on water tenders (tankers).

NFPA 1901 requires all valves on intakes or discharges that are 3 inches (77 mm) or greater to be equipped with slow acting valve controls. This prevents movement from the fully opened to fully closed positions (or vice versa) in less than three seconds. The purpose of this is to minimize the risk of damage caused by water hammer when large volumes of water are being moved.

Pump Drains

Most connections to the pump are equipped with drain valves on the line side of the control valve. On the discharge fittings, these drain valves provide a way for the driver/operator to relieve the pressure from the hoseline after the discharge valve and nozzle have both been closed. This is useful when the hoseline was not bled off by opening the nozzle and the nozzle is a great distance from the apparatus. If there were no other way to get rid of the pressure trapped in the line, the driver/operator would have to leave the apparatus and travel to the nozzle to bleed the line. Otherwise, it would be very difficult to break the hose connection at the discharge. By use of the drain valve, the unused line can be drained and the hose disconnected, even while the pump is still in service.

The bleeder line on the gated intake serves another purpose (Figure 10.53). If a supply line is connected to the gated intake of a pump while the attack lines are being supplied from the tank, it is desirable to make the changeover to the supply line without interrupting the fire streams. Because the hoseline is full of air when it is dry, the water coming through the line would force air into the pump, probably causing it to lose its prime. At the very least, it would cause fluctuations in the nozzle pressure. By opening the bleeder valve on the line side of the intake valve, the air can be forced out of the hose through the bleeder as the line fills with water. When all the air has been evacuated from the line and the bleeder is discharging a steady stream of water, the drain valve can be closed, the intake valve opened, and the tank-to-pump valve closed. These actions occur on a coordinated basis, and the transition can be made with no interruption to the flow.

Another purpose for pump and piping drains is to remove all the water from the system in climates where

Figure 10.52a Intake gate valves may be equipped with a handwheel.

Figure 10.52b Butterfly valves typically have a quarter-turn control handle.

Figure 10.53 All gated intakes are equipped with a bleeder valve that allows air to be removed from the system before it enters the fire pump.

freezing might occur. Water expands in size when it freezes and can damage pump and piping components, in addition to the fact that it creates barriers to water flow if the pump is needed. Drains must be supplied at the lowest point on the pump and at the lowest point on each line connected to it. A number of these drains are connected to a common master drain valve. When the control handle is operated, all the drains are opened simultaneously, and the pump can be drained with one operation. This drain valve should not be opened when the pump is operating either with pressure or vacuum on the intake. An "O" ring is used as a gasket to maintain an airtight connection when the valve is closed. If the valve is operated with pressure or vacuum on it, damage to the "O" ring inside the valve could result. Some additional drain valves may be supplied if it is not convenient to connect all the drains to the master valve. It is important that they all be opened for the pump to drain completely. Close all drains immediately after use. Failure to do so could result in an inability to prime the pump and begin drafting operations if necessary. A discharge valve should be opened to allow air to replace the draining water. If not, the vacuum holds the water in the pump.

It may be necessary to manually unwind and drain booster hoselines during freezing weather. Simply opening the drain valve on the booster line supply piping does not allow all water to drain from the hose. Failure to get out all water may result in the water freezing inside the hose, causing it to burst.

Newer apparatus may be equipped with piping from the vehicle's air brake system to the booster reel. When the valve in this line is opened, compressed air is discharged through the booster hose to remove the water. This negates the need to manually remove the hose to drain it.

Automatic Pressure Control Devices

The volume of water moving through the pump may change suddenly when a nozzle is shut down rapidly, or when the setting is changed on a variable gallonage nozzle. While modern nozzles can tolerate less-than-ideal pressures and still maintain an effective fire stream, the firefighter on the nozzle cannot tolerate any sudden changes in pressure. During the critical stages of the attack, a sudden change in pressure and the accompanying change in fog pattern or stream reach can be disastrous. When a pump is supplying multiple attack lines, any sudden flow change in one line can cause a pressure surge on the other. Even an alert operator finds it impossible to compensate for these sudden changes in time to

protect the nozzlemen on the other lines. Some type of automatic pressure regulation is essential to ensure the safety of personnel operating the hoselines.

NFPA 1901 requires some type of pressure control device to be part of any fire apparatus pumping system. The system must be operated within 3 to 10 seconds after the discharge pressure rises at least 30 psi above the set level. A yellow indicator light located on the pump operator's panel must illuminate when the pressure control system is in control of the pressure in the pump (Figure 10.54). Devices that are designed to discharge excess water to the atmosphere must be designed to do so in a manner that does not expose the driver/operator or other personnel to a pressurized water stream.

Relief Valves

There are two basic concepts for pressure relief valves: those that relieve excess pressure on the discharge side of the pump and those that relieve excess pressure on the intake side of the pump.

A discharge pressure relief valve is an integral part of all fire pumps that are not equipped with a pressure governor (Figure 10.55). The main feature of a relief valve is its sensitivity to pressure change and its ability to relieve excessive pressure within the pump discharge. An adjustable spring-loaded pilot valve actuates the relief valve to bypass water from the discharge to the intake chamber of the pump. Although only a small quantity of water is rerouted, the rerouting permits the pump to continue in operation when pressure rises above the working, or set, pressure.

There are many types of relief valves available. One of the most common types uses a spring-controlled pilot valve. When the pump discharge pressure is lower than allowed by the pilot valve setting, water flows from the

Figure 10.54 A yellow light illuminates when the pressure control system is in operation.

Figure 10.55 The pressure relief control valve is prominently located on the pump panel.

pump discharge to the pressure chamber of the pilot valve (Figure 10.56). A diaphragm separates this pressure chamber from the pilot valve. Tension against the diaphragm is regulated by adjusting the handle of the pilot valve against the spring. As long as the hydraulic force in this chamber is less than the force of the spring, the pilot valve stays closed. The water is then transmitted back to the main valve chamber, and the main valve stays closed.

When the pump discharge pressure rises higher than allowed by the pilot valve setting, the spring in the pilot valve compresses (Figure 10.57). This allows the needle valve to move to the left, causing water to dump back into the pump intake. This action in turn reduces the pressure in the tube and behind the main valve. The hydraulic force on the small end of the main valve is now greater than that behind the main valve. Consequently, the main valve opens and permits part of the water to return to the intake side, thus reducing discharge pressure. When the discharge pressure drops below the set pressure in the pilot valve, the pilot valve reseats, pressure increases behind the large end of the main valve, and it closes.

A second type of relief valve, although equipped with a pilot valve, operates in a slightly different manner (Figure 10.58). When the pressure goes above the set pressure, the pilot valve moves, compressing its spring until the opening in the pilot valve housing is uncovered.

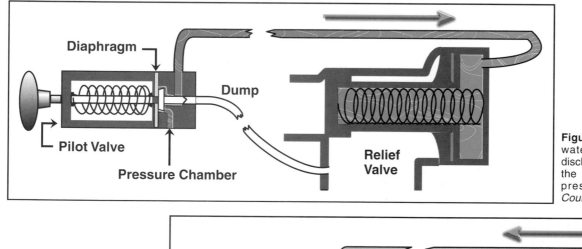

Figure 10.56 A schematic of water flow when the pump discharge pressure is lower than the pilot valve setting of the pressure control system. *Courtesy of Waterous Company.*

Figure 10.57 A schematic of water flow when the pump discharge pressure is higher than the pilot valve setting of the pressure control system. *Courtesy of Waterous Company.*

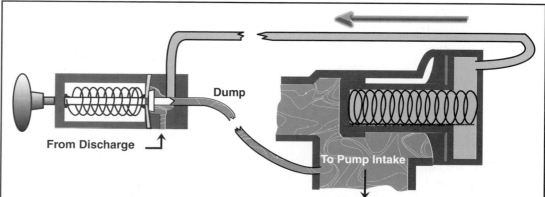

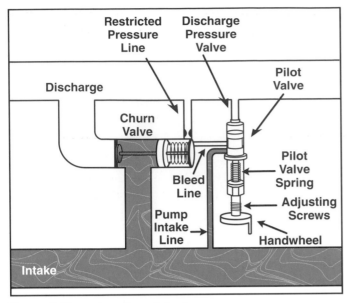

Figure 10.58 An alternative relief valve.

Water flows through this opening and through the bleed line and also into the pump intake. This flow reduces the pressure on the pilot valve side of the churn valve, allowing the higher pressure on the discharge side to force the churn valve open. Water flows from the discharge into the intake, relieving the excessive pressure.

The driver/operator should remember that while discharge pressure relief valves are quick to react to over-pressure conditions, they are somewhat slower to reset back to "all-closed" positions. Therefore, it takes a short time for the pump to return to normal operation.

Intake pressure relief valves are intended to reduce the possibility of damage to the pump and discharge hoselines caused by water hammer when valves/nozzles are closed too quickly. There are two basic kinds of intake pressure relief valves. One type is supplied by the pump manufacturer and is an integral part of the pump intake manifold. The second type is an add-on device that is screwed onto the pump intake connection (Figure 10.59). In either case, these devices are set by the driver/operator to allow a maximum amount of pressure into the fire pump. If the incoming pressure exceeds the set level, the valve activates and dumps the excess pressure/water until the water entering the pump is at the preset level. It is generally recommended that intake relief valves be set to open when the intake pressure rises more than 10 psi (70 kPa) above the desired operating pressure.

Most screw-on intake pressure relief valves are also equipped with a manual shut-off valve that allows the water supply to the pump to be stopped if so desired. Bleeder valves on the intake pressure relief valve allow air to be bled off as the incoming supply hose is charged (Figure 10.60). This is particularly important when using

Figure 10.59 Many jurisdictions that use LDH have external intake relief valves screwed onto their main pump intakes.

Figure 10.60 External intake relief valves have bleeder valves that allow air to be removed from the hose without it entering the fire pump.

these devices in conjunction with large diameter supply hose. A large amount of air pushes through these hoses until a solid column of water reaches the valve.

Pressure Governor

Pressure can also be regulated on centrifugal pumps by a mechanical or electronic governor that is pressure activated to adjust the engine throttle. The main feature of a pressure governor is that it regulates the power output of the engine to match pump discharge requirements. When the pressure in the discharge piping of the pump exceeds the pressure necessary to maintain safe fire streams, the excessive pressure must be reduced. Because the speed of the impellers determines the pressure, and the engine speed determines the speed of the impellers, it is only necessary to reduce the engine speed to reduce the pressure.

Excessive pressures are generally caused by shutting down one or more operating hoselines. When excessive pressure builds up, a tube from the discharge side of the pump transmits the resulting pressure rise to a governing device, which then cuts back the throttle. The device varies with each manufacturer's design; it may be attached to either a regular or an auxiliary throttle. A pressure governor device can be used in connection with a throttle control, engine throttle, and/or pump discharge (Figure 10.61).

On older types of pressure control governors, when the control handle is pulled out, it lifts a split nut from a serrated piston shaft (Figure 10.62). The engine throttle is then advanced until the desired pump pressure is reached. The water control valve from the discharge side of the pump is then opened to permit water from the discharge of the pump to enter the cylinder. This hydraulic pressure causes the piston to compress the control and move the serrated shaft. This movement activates the carburetor linkage and butterfly valve to control engine speed.

The control handle must then be pushed in to lower the split nut against the serrated piston shaft (Figure 10.63). Engine speed is then governed by the water

pressure on the discharge side of the pump. If a hoseline is shut off, the increase in pressure travels through the pipe from the pump discharge into the cylinder, moving the piston from its set position. This movement causes the carburetor control linkage to adjust the butterfly valve and reduce the engine speed. When a discharge line is reopened, the reverse occurs and the piston moves back to increase engine speed and returns to the preset pump pressure position.

Another type of governor is the piston assembly governor. The speed of the engine is controlled by a

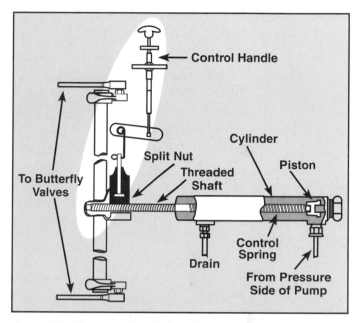

Figure 10.62 The control handle is pulled out to disengage the split nut from the serrated shaft.

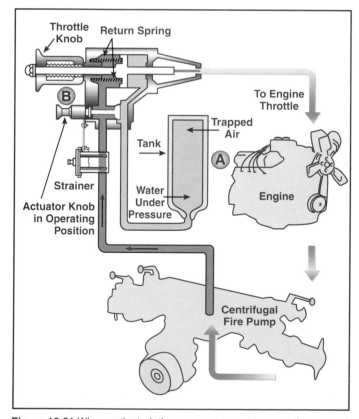

Figure 10.61 When activated, the pressure governor uses the trapped air and water in Area A as a reference pressure for the piston that controls the engine throttle. A change in pump pressure effecting Area B moves the piston, thus adjusting the engine speed. *Courtesy of Hale Fire Pump Company.*

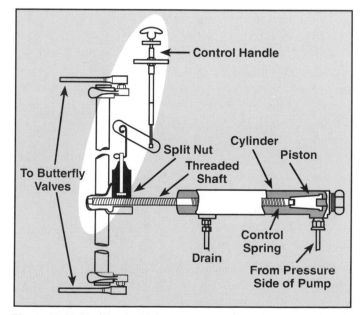

Figure 10.63 After the desired pressure is reached, the control handle is pushed back in, setting the governor for that pressure.

governor assembly (Figure 10.64). This assembly fits onto the carburetor (gasoline engines) or throttle link (diesel engines) and reduces or increases the engine speed under the control of a rod connected to a piston in a water chamber. The governor chamber contains a flexible diaphragm to isolate the piston and spring assembly from the water while exposing the face of the piston to the water pressure. Pressure comes into the water chamber through a strainer from the pressure side of the pump. Water returning from the water chamber goes through the governor valve assembly to a two-position governor return valve. This valve allows the water from the pump to be discharged on the ground or return to the intake side of the pump.

When drafting or pumping from the tank, the draft (or closed) position is used; therefore, the water returns to the intake side of the pump. When pumping from a hydrant (or external pressure source), the hydrant (or open) position is used so that the water is discharged onto the ground. This hydrant drainline to the ground can be run through a check valve to the tank to prevent water from continuously running on the ground.

Newer apparatus may be equipped with electronic governors (Figure 10.65). These use a pressure sensing element that is connected to the discharge manifold. This element controls the action of an electronic amplifier that compares the pump pressure to an electrical reference point. When necessary the element changes the throttle setting by adjusting the amount of fuel supplied to the engine, thus bringing the pump to the desired operating pressure. This governor maintains any pressure that is set on the control knob above 50 psi (345 kPa). It also returns the engine to idle speed anytime the pressure drops below 50 psi (345 kPa). They also have a cavitation protection mode that returns the engine to idle when the intake pressure drops below 30 psi (210

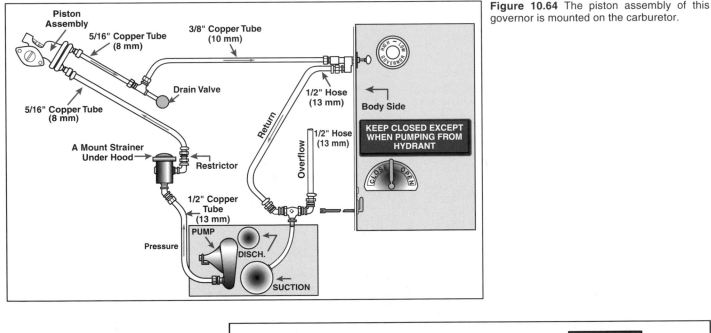

Figure 10.64 The piston assembly of this governor is mounted on the carburetor.

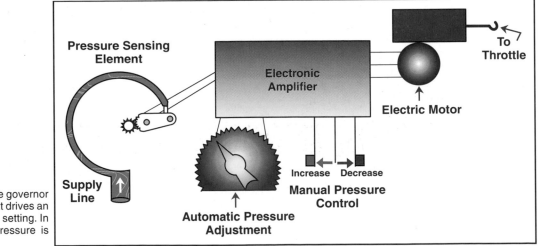

Figure 10.65 An electronic pressure governor has a pressure sensing element that drives an electric motor to change the throttle setting. In this way, the desired operating pressure is achieved.

kPa). Electronic governors are so accurate and quick to respond that they virtually eliminate the need for the pump to be equipped with a relief valve.

A major drawback to mechanical governors is that a closed discharge line can reduce the rpm of the powerplant to lower the pressure. This in turn lowers, for a short time, all discharge line working pressures. This condition does not occur with electronic governors. Electronic governors adjust the engine rpm, but maintain the same discharge pressure on the lines that are still flowing.

Priming Methods and Devices

Successfully drafting water depends upon the ability to create a lower pressure within the pump and the intake hose than exists in the atmosphere. This results in water being forced up into the intake hose and fire pump. As stated earlier in this chapter, the centrifugal fire pumps in use today are not capable of creating this condition by themselves. Some other device is required to create the vacuum that allows drafting to occur. The devices that make drafting possible are called priming devices, or simply *primers*. Primers fall into three categories: positive displacement, exhaust, and vacuum.

Positive Displacement Primers

Most modern fire apparatus use positive displacement primers. Positive displacement pump theory was discussed earlier in this chapter. Small versions of both the rotary vane and rotary gear type positive displacement pumps are used as priming pumps. The rotary vane primer requires a relatively high rpm as compared to a rotary gear primer and can be driven by either mechanical means from the pump transfer gear case or by an electric motor (Figure 10.66). Electric motors are the most common choice by most manufacturers and fire departments. Use of an electric motor as the driver allows for maximum flexibility in mounting and application to different chassis and pumps. As well, electric primers can be operated effectively, regardless of engine speed. Some apparatus have primers that are driven off the transfer case of the transmission; however, these are not as common as electric-driven primers.

The primer pump inlet is connected to a primer control valve that is in turn connected to the fire pump. If there are high spots in the way the pump is constructed or mounted, the line from the priming pump may be connected to the pump in several different locations. If the pump is driven electrically, the priming valve usually incorporates a switch into the valve. In this way, only one operation is necessary to prime the pump.

Most primers use an oil supply or some other type of fluid (Figure 10.67). The oil/fluid serves two purposes. As the pump wears, the clearances between the gears and the case increase, and the pump loses efficiency. A thin film of oil/fluid is drawn into the pump and seals the gaps between the gears and the case. The oil/fluid fills any irregularities in the housing caused by pumping contaminated water, and it improves the efficiency of the primer. The oil/fluid also acts as a preservative and minimizes deterioration of the metal parts by inhibiting corrosion when the pump is not in use. To derive the maximum benefit from the oil/fluid, it is necessary to operate the primer periodically so that a coating of oil/fluid forms on all the metal parts.

There is a vent in the oil/fluid line from the reservoir to the priming pump. Because the tank is frequently mounted higher than the priming pump, a siphon action drains the tank after the primer has been used. This vent

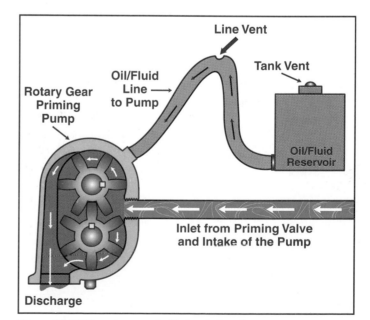

Figure 10.66 A rotary vane primer with an electric motor. *Courtesy of Waterous Company.*

Figure 10.67 The primer control is located on the pump.

breaks the siphon. It must be large enough to serve the purpose, but small enough to allow the priming pump to draft the oil out of the tank when the primer is in use. This vent should be checked frequently to ensure that it is free of dirt.

The desired engine rpm for transfer case, gear-driven primer operation depends on the construction of the primer, the gear ratio of the transfer case, and other features that are unique to a particular installation. The operating manual for the fire pump or apparatus should specify the desirable engine speed (rpm) for priming, but in general, it should operate with an engine rpm around 1,000 to 1,200. It is best that these primers be activated with the engine at idle speed and then the throttle increased to indicated rpms. This minimizes the wear on the mechanical clutch pack.

Exhaust Primers

Exhaust primers are still found on some older pieces of apparatus. The exhaust primer operates on the same principle as a foam eductor. Exhaust gases from the vehicle's engine are prevented from escaping to the atmosphere by the exhaust deflector. The gases are diverted to a chamber where the velocity of the gases passing through a venturi creates a vacuum. This chamber is connected through a line and a priming valve to the intake of the pump (Figures 10.68 a and b). There the air is evacuated into the venturi chamber and then discharged along with the exhaust gases into the atmosphere.

Creating a vacuum by using exhaust requires high engine rpm to generate the maximum velocity of the exhaust gases. At best, this primer is not very efficient. It is used primarily on portable pumps where price and weight are of prime importance.

The exhaust primer requires a great deal of maintenance because the exhaust gases moving through the primer tend to leave carbon deposits. These deposits decrease the efficiency of the device even further. Use of an exhaust primer requires that any air leaks in the pump be kept to an absolute minimum and that the suction hose and gaskets be kept in good condition.

Vacuum Primers

Probably the simplest type of primer, the vacuum primer, makes use of the vacuum already present in the intake manifold of any gasoline engine. These devices were common on older, gasoline-powered fire apparatus. To prime the pump, it is necessary only to connect a line from the intake manifold of the engine to the intake of the fire pump with a valve connected in the line to control it. This connection creates two dangerous conditions. First, the intake manifold contains explosive gases that could

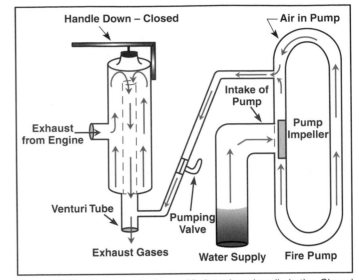

Figure 10.68a An exhaust primer with the primer handle in the *Closed* position. *Courtesy of Bennie Spaulding.*

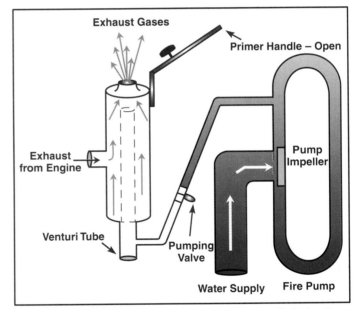

Figure 10.68b An exhaust primer with the primer handle in the *Open* position. *Courtesy of Bennie Spaulding.*

be drawn into the fire pump and cause damage. Second, when the pump is primed, water could be drawn through the pump and into the intake manifold, causing damage to the engine (Figure 10.69). To prevent this possibility, the line from the intake manifold contains a check valve. When this valve is under pressure, such as occurs during a backfire, it closes, cutting off the pump from the engine. Under a vacuum, the valve stays open and allows the pump to prime normally. As the air comes through the line from the intake side of the pump, the pump is primed and fills with water. When the water enters the wet chamber of the primer, the float operates and closes the ball valve. This action opens the vent valve and allows the vacuum line to take air from outside the pump. If the float does not operate, the water continues up into the dry

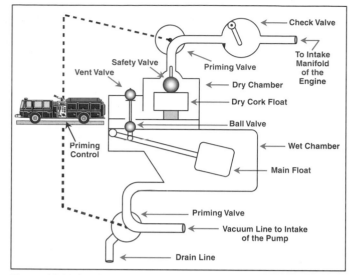

Figure 10.69 A schematic diagram of a vacuum primer system.

chamber, where a cork float rises and closes the safety valve. When the priming valve is closed, the drain opens and allows the water from the primer to drain, and the primer is ready to operate again.

Because engine vacuum is maximum near idle speed, the primer works best at low engine rpm. It is only necessary to speed the engine enough to enable it to keep from stalling when the vacuum line is opened to prime the pump.

Pump Panel Instrumentation

In order to swiftly and efficiently operate the fire pump, the driver/operator must be familiar with all instrumentation located on the pump operator's panel. Some of this instrumentation is specific to the pump operator's panel and some of it is duplications of gauges found in the cab on the dashboard. As a minimum, NFPA 1901 requires the following controls and instruments located on the pump operator's panel:

- Master pump intake pressure indicating device
- Master pump discharge pressure indicating device
- Weatherproof tachometer
- Pumping engine coolant temperature indicator
- Pumping engine oil pressure indicator
- Pump overheat indicator
- Voltmeter
- Pump pressure controls (discharge valves)
- Pumping engine throttle
- Primer control
- Water tank to pump valve (discussed earlier in this

chapter in the "Intake Piping" section)

- Tank fill valve (discussed earlier in this chapter in the "Discharge Piping" section)
- Water tank level indicator

Most of the devices that are listed as "indicators" are found on the pump operator's panel in the form of gauges. Electronic indicators that provide digital readouts are available in some cases, but are not commonly found in use. The exception to this is flowmeters. More information on flowmeters can be found in Chapter 9.

Although not required by NFPA 1901, it is recommended that a pumping engine fuel gauge be contained on the pump operator's panel. The pump operator's panel fuel gauge is important during extended pumping operations when fuel may run low. This allows the driver/operator to keep tabs on the fuel level without having to climb into the apparatus cab to check the gauge. As well, an automatic transmission temperature gauge may be added so that the driver/operator can monitor this to make sure transmission damage does not occur during pumping operations.

Master Intake and Discharge Gauges

The master intake and discharge gauges are the two primary gauges used to determine the water pressure entering and leaving the pump (Figure 10.70). The *master intake gauge* (sometimes referred to as the *vacuum or compound gauge*) must be connected to the intake side of the pump. This gauge must be capable of measuring either positive or negative pressure. The gauge is usually calibrated from 0 to 600 psi (0 kPa to 4 137 kPa) positive pressure and from 0 to 30 inches (0 mm to 762 mm) of vacuum on the negative side. This gauge provides an indication of the vacuum present at the intake of the pump during priming or when the pump is operating

Figure 10.70 The master intake and discharge pressure gauges.

from draft. This provides an approximation of how much pump capacity is not being used. As the flow from the pump increases, the vacuum reading increases because more negative pressure is required to overcome the friction loss in the suction hose. As the vacuum reading approaches 20 inches (508 mm), the pump is near its maximum capacity and is not able to supply any additional lines.

The master intake gauge also provides an indication of the residual pressure when the pump is operating from a hydrant or is receiving water through a supply line from another pump. Because the physical size of the vacuum scale is so small, even a slight error in the zero setting of the gauge can result in a large error in measuring the vacuum on the intake side of the pump. If the gauges used are not designed for use in a vacuum, they can be damaged. Refer to NFPA 1901 for further requirements of the master intake gauge.

A master pump discharge pressure gauge is also required on a pumper. It must be calibrated to measure 600 psi (4 137 kPa) unless the pumper is equipped to supply high-pressure fog streams; in which case, the gauge may be calibrated up to 1,000 psi (6 900 kPa). The pump discharge pressure gauge registers the pressure as it leaves the pump, but before it reaches the gauges for each individual discharge line.

External connections to these gauges must be made to allow installation of calibrated gauges when service tests on the apparatus are performed (Figure 10.71). These connections are also used for standard gauges when the initial acceptance tests of the pump are made.

Tachometer

The tachometer records the engine speed in revolutions per minute (rpm) (Figure 10.72). It can give an experienced driver valuable information about the condition of the pump. The tachometer is useful as a means of trouble analysis when difficulty with the pump is encountered. When the initial acceptance tests of the pump are made, the rpm required to pump its rated capacity is determined, and the information is recorded on an identification plate on the pump panel. A gradual increase in the amount of rpm required to pump the rated capacity indicates wear in the pump and a need for repairs.

Pumping Engine Coolant Temperature Indicator

The pumping engine coolant temperature indicator shows the temperature of the coolant in the engine that powers the fire pump (Figure 10.73). This may be the main vehicle engine, or, in the case of an auxiliary engine-driven pump, the pump engine. The operating temperature of the engine is important: an engine that

Figure 10.71 Test connections for the master intake and discharge pressure gauges should be located somewhere on the pump panel.

Figure 10.72 The tachometer allows the pump operator to monitor engine speed from the pump panel.

operates too cool is not efficient, and an operating temperature that is too high may cause damage to the mechanical parts. Auxiliary cooling devices are provided to compensate for the lack of movement and can be used to maintain the engine temperature within tolerable limits when the pump is in operation. These devices are discussed later in this chapter.

Pumping Engine Oil Pressure Indicator

The pumping engine oil pressure indicator shows that an adequate supply of oil is being delivered to the critical areas of the engine that is powering the fire pump. It is

Figure 10.74 The pump panel may include a visual overheating indicator.

Figure 10.73 The engine coolant temperature gauge alerts the pump operator to engine overheating problems.

Figure 10.75 The voltmeter shows the status of the vehicle's electrical supply system.

not a measure of the oil level in the crankcase, but if the oil level in the crankcase drops too low, it is impossible for the oil pump to maintain the required amount of pressure. While normal operating oil pressures are specified in the maintenance manual, the pressure for a particular engine will vary, so the operator needs to be familiar with the expected reading. Any significant deviation from the normal oil pressure reading is an indication of pending problems.

Pump Overheat Indicator

The pump panel may be equipped with an audible or visual indicator that warns the driver/operator when the pump overheats (Figure 10.74). Overheating may occur when the pump is run for prolonged periods during which no water is being discharged. Methods to prevent pump overheating are discussed in Chapter 11.

Voltmeter

The voltmeter provides a relative indication of battery condition and alternator output by measuring the drop in voltage as some of the more demanding electrical accessories, such as the primer, are used (Figure 10.75). It indicates the top voltage available when the battery is fully charged.

Pump Pressure Indicators (Discharge Gauges)

Pump pressure indicators, more commonly known as discharge gauges, can be connected to the individual discharge fittings of the pump (Figure 10.76). These gauges must be connected to the outlet side of the discharge valve so that the pressure being reported is the pressure actually being applied to the hoselines after the

Figure 10.76 Each pump discharge should be equipped with its own pressure gauge.

valve. This allows pressure in each discharge to be adjusted down from the overall pump discharge pressure if necessary. The gate valve must be readjusted each time the flow at the nozzle being supplied is changed.

Readjustment is necessary because the pressure loss in the valve is determined by the amount of water flowing through it. If the nozzle is shut down on the hoseline being supplied, the individual pressure gauge reads the same as the master pressure gauge because there is no flow through the valve to reduce the pressure. No attempt should be made to readjust the gate valve until water is flowing again.

Individual pressure gauges may also be supplied with master stream devices or the lines that are supplying them. These gauges are very important as it is impossible to maintain effective master streams unless the proper pressure is being supplied to the appliance. With the large flows required, friction loss is high in the supply lines. An individual line gauge is the only way to be certain that the pump has been adjusted properly.

NFPA 1901 allows flowmeter readouts to substitute for individual pressure discharge gauges (Figure 10.77). However, even if a flowmeter system is used, the apparatus is still required to have the master intake and pressure gauges. More information on flowmeters can be found in Chapter 9.

Pumping Engine Throttle

A pumping engine throttle must be contained on the pumper operator's panel. This device is used to increase or decrease the speed of the engine that is powering the fire pump. By increasing or decreasing the speed of the engine, the driver/operator controls the amount of pressure that the fire pump is supplying to the discharge. The most common throttle used on the pump operator's panel is the type of knob that turns (Figure 10.78). The throttle knob is turned one way or another until the desired rpm/pressure is achieved. Automatic throttle controls, such as those that are operated by a toggle switch, are also available and in use on newer apparatus (Figure 10.79).

Primer Control

The primer control is used to operate the priming device when the pump is going to be used to draft from a static water supply (Figure 10.80). This control is generally in the form of a push button, toggle switch, or pull lever. Priming devices and their operation were described earlier in this chapter.

Water Tank Level Indicator

This device is intended to let the driver/operator know how much water is remaining in the onboard water tank. This allows the driver/operator to anticipate how much longer attack hoselines may be supplied before an external water supply source is needed. This is particularly

Figure 10.77 Flowmeters may be used in place of individual discharge gauges.

Figure 10.78 This type of throttle control has been commonly used for many years.

Figure 10.79 Newer apparatus may be equipped with toggle-type throttle controls.

Figure 10.80 The primer control.

Figure 10.81a Multilight water tank level indicators are very common.

Figure 10.81b Another style of a multilight water tank level indicator.

crucial when supplying interior fire attack hoselines with the water tank as the pump's only supply. Interior crews need to be evacuated if the onboard supply is exhausted prior to an external water supply being established.

The most common type of water tank level indicator is one that uses a series of lights on the pump operator's panel (Figures 10.81 a and b). Sensors within the tank send signals that indicate the amount of water in the tank by one-quarter levels (empty, ¼, ½, ¾, full). These lights may be small LCD-type lights on a pump panel plate, or they may be larger lights mounted where they may be viewed by the driver/operator and other personnel on the scene (Figure 10.82). Some apparatus are equipped with site gauges that allow the driver/operator to view the actual level of water in the tank through a clear tube. As the level of water in the tank decreases, so does the level in the tube. Newer apparatus may be equipped with digital readouts that indicate the level of water in the tank.

Auxiliary Cooling Devices

The primary function of auxiliary coolers is to control the temperature of coolant in the apparatus engine during pumping operations. There are two types of auxiliary coolers commonly found on older apparatus: the marine type and the immersion type. The marine type is inserted into one of the hoses used in the engine cooling system so that the engine coolant must travel through it as it circulates through the system (Figure 10.83). The cooler itself contains a number of small tubes similar to the flues in a steam boiler. Surrounding the tubes is a water jacket that is connected to the discharge of the fire pump. When the fire pump is operating, water from the pump can be circulated through the water jacket of the auxiliary cooler. As the coolant from the radiator passes through the tubes in the cooler, the colder water from the fire pump comes

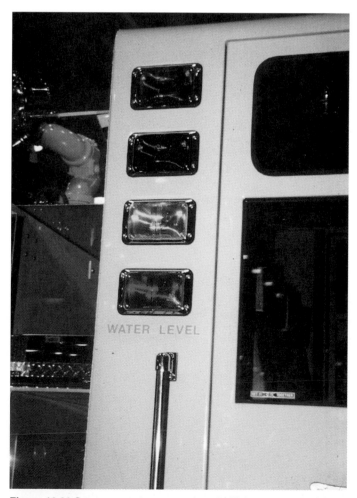

Figure 10.82 Some apparatus are equipped with large water tank level displays that are visible well away from the truck.

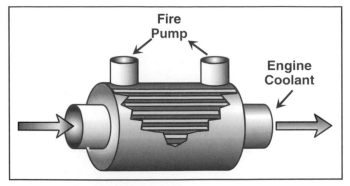

Figure 10.83 A marine-type cooler.

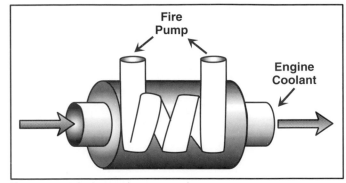

Figure 10.84 An immersion-type cooler.

in contact with the metal tubes. The colder water conducts heat away from the tubes, thus reducing the temperature of the coolant flowing through them. A valve is provided on the pump panel to control the amount of water being supplied to the auxiliary cooler from the pump panel.

The immersion-type auxiliary cooler is mounted in a similar manner to the marine type, with the radiator coolant passing through the body of the cooler. In this case, the water being supplied by the fire pump passes through a coil or some type of tubing mounted inside the cooler so that it is immersed in the coolant. As the cooler water from the fire pump passes through the tubing, some of the heat from the coolant is absorbed by the tubing and dissipated in the water from the fire pump. This type of cooler is also controlled by a valve on the pump operator's panel to regulate the degree of cooling desired (Figure 10.84).

Both types of auxiliary coolers are constructed so that the coolant in the radiator does not come in contact with the cooling water from the fire pump. Thus, the auxiliary cooler can be used without contaminating the engine coolant.

Some manufacturers also supply a radiator fill valve. This valve can be used to refill the radiator if the coolant level drops too low for effective cooling during a pumping operation. The radiator fill valve should only be used in an emergency because whatever type of water is being supplied to the fire pump gets into the radiator and is circulated through the engine and the entire cooling system. If antifreeze is installed in the radiator, it is diluted by the water from the fire pump and may become ineffective.

If it is absolutely necessary to use the radiator fill valve to replenish the coolant in the radiator, the line from the radiator fill valve should be restricted sufficiently to reduce the pressure to the radiator to safe levels. It is still possible, however, that dangerous pressures could build up in the cooling system faster than the overflow line can relieve them. Because excess pressure in the radiator could result in bursting, the radiator fill valve should then be opened a very small amount. The radiator should be watched closely to keep the amount of contaminated water pumped into the system to an absolute minimum. As soon as possible after the pumping operation, the cooling system should be serviced. It should be thoroughly flushed and the system refilled with the correct amount of clean, fresh antifreeze solution.

In situations where a dramatic failure of a radiator hose or other cooling system element occurs while the pumper is supplying hoselines for crews in dangerous positions, a water stream may be directed at the engine to maintain some form of cooling until crews may be withdrawn to safety. This should only be done for a short period of time. Once the crews are safe, the engine should be shut down immediately.

Chapter 11

Operating Fire Pumps

Job Performance Requirements

This chapter provides information that will assist the reader in meeting the following job performance requirements from NFPA 1002, *Standard on Fire Apparatus Driver/Operator Professional Qualifications*, 1998 edition. Particular portions of the job performance requirements (JPRs) that are met in this chapter are noted in bold text.

3-2.1* Produce effective hand or master streams, given the sources specified in the following list, so that the pump is safely engaged, all pressure control and vehicle safety devices are set, the rated flow of the nozzle is achieved and maintained, and the apparatus is continuously monitored for potential problems.

- **Internal tank**
- **Pressurized source**
- **Static source**
- **Transfer from internal tank to external source**

(a) *Requisite Knowledge:* Hydraulic calculations for friction loss and flow using both written formulas and estimation methods, **safe operation of the pump**, problems related to small-diameter or dead-end mains, low-pressure and private water supply systems, hydrant cooling systems, and reliability of static sources.

(b) *Requisite Skills:* **The ability to position a fire department pumper to operate at a fire hydrant and at a static water source, power transfer from vehicle engine to pump, draft, operate pumper pressure control systems, operate the volume/pressure transfer valve (multistage pumps only), operate auxiliary cooling systems, make the transition between internal and external water sources**, and assemble hose lines, nozzles, valves, and appliances.

3-2.2 Pump a supply line of 2½ in. (65 mm) or larger, given a relay pumping evolution the length and size of the line and the desired flow and intake pressure, so that the proper pressure and flow are provided to the next pumper in the relay.

(a) *Requisite Knowledge:* Hydraulic calculations for friction loss and flow using both written formulas and estimation methods, **safe operation of the pump**, problems related to

small-diameter or dead-end mains, low-pressure and private water supply systems, hydrant cooling systems, and reliability of static sources.

(b) *Requisite Skills:* **The ability to position a fire department pumper to operate at a fire hydrant and at a static water source, power transfer from vehicle engine to pump, draft, operate pumper pressure control systems, operate the volume/pressure transfer valve (multistage pumps only), operate auxiliary cooling systems, make the transition between internal and external water sources**, and assemble hose lines, nozzles, valves, and appliances.

3-2.4 Supply water to fire sprinkler and standpipe systems, given specific system information and a fire department pumper, so that water is supplied to the system at the proper volume and pressure.

(a) *Requisite Knowledge*: **Calculation of pump discharge pressure; hose layouts; location of fire department connection; alternative supply procedures if fire department connection is not usable; operating principles of sprinkler systems as defined in NFPA 13**, *Standard for the Installation of Sprinkler Systems*, **NFPA 13D**, *Standard for the Installation of Sprinkler Systems in One- and Two-Family Dwellings and Manufactured Homes*, **and NFPA 13R**, *Standard for the Installation of Sprinkler Systems in Residential Occupancies Up To and Including Four Stories in Height*; **fire department operations in sprinklered occupancies as defined in NFPA 13E**, *Guide for Fire Department Operations in Properties Protected by Sprinkler and Standpipe Systems*; **and operating principles of standpipe systems as defined in NFPA 14**, *Standard for the Installation of Standpipe and Hose Systems*.

6-2.1* Produce effective fire streams, utilizing the sources specified in the following list, so that the pump is safely engaged, all pressure-control and vehicle safety devices are set, the rated flow of the nozzle is achieved, and the apparatus is continuously monitored for potential problems.

- **Water tank**
- **Pressurized source**
- **Static source**

Once the driver/operator understands how the pump works, the next step is to become proficient at operating the pump under a variety of conditions. A knowledge of the operational theory of fire pumps is of no use on the fireground if the driver/operator is unable to put that knowledge to practical use. In this chapter, the driver/operator is instructed in methods for operating the pump under a number of different circumstances.

The first step all driver/operators must take in order to begin any pumping operation is to make the fire pump operational. Once this is accomplished, they may choose one of three sources of water to supply the pump: onboard water tank, pressurized source, or static source. This chapter discusses the procedures for placing the pump into operation using each of these three sources of water. Also included is information on operating the pump at occupancies equipped with automatic sprinkler and standpipe systems.

The information contained in this chapter provides sound, generic advice on the operation of fire department pumpers. While many of the basic principles are the same for all makes and models of pumps and apparatus, each manufacturer's equipment has special nuances with which the driver/operator must be familiar. For that reason, it cannot be overemphasized that the driver/operator must consult the operator's manual for the apparatus and pump he is operating.

Making the Pump Operational

The process of making the fire pump operational, also referred to as "putting the pump into gear," begins after the apparatus has been properly positioned and the parking brake has been set. The proper position for the apparatus is determined by a variety of factors such as tactical use (attack vs. supply) and the water supply source that will be used. Proper apparatus positioning on the fireground is covered in detail in Chapter 5.

Once the apparatus is in position, the parking brake must be set. Most large, modern apparatus are equipped with air brakes. The parking brake control on the apparatus is usually in the form of a push/pull control on the dashboard (Figure 11.1). Follow the chassis

Figure 11.1 Apparatus with air brakes usually have a yellow push/pull control.

manufacturer's directions for activating the brake. Older model, large apparatus may be equipped with a manual (spring-activated) parking brake that is activated by a hand lever (Figure 11.2). Make sure that the lever is locked into position when the brake is set. Apparatus on light-duty truck chassis, such as minipumpers and wildland apparatus, may have parking brakes that are pedal or hand-lever activated (Figure 11.3). Again, follow the chassis manufacturer's directions for activating the brake.

On the majority of apparatus, the remainder of the procedure for making the fire pump operational takes place before the driver/operator exits the cab. However, there are a few exceptions to this rule, and they will be discussed in the appropriate sections. Once the driver/operator does exit the cab, the next step, in all cases except when the apparatus is used for pump-and-roll operations, should be to chock the apparatus wheels (Figure 11.4). *IFSTA recommends that the apparatus wheels be chocked every time the apparatus is parked with the engine running and the driver/operator exits the cab.* The placement of wheel chocks provides an added measure of safety in the event that the apparatus jumps into road gear and overrides the parking brake

system. Tests have indicated that the apparatus may override the parking brake system at engine speeds as low as 1,300 rpm.

The procedure for making the pump operational varies depending on the type of pump drive and the manufacturer of the apparatus. The following sections highlight the basic procedures for engaging each type of pump driver.

Engaging Power Take-Off (PTO) and Front-Mount Pumps

The procedure for engaging power take-off (PTO) and front mount pumps is generally the same. Keep in mind that both of these types of pumps may be operated in either stationary or pump-and-roll modes. The procedure for engaging the pump differs slightly depending on which mode is used. The following sections list the procedures required for placing a PTO or front-mount pump into gear.

Figure 11.3 Smaller vehicles have foot-pedal parking brake controls.

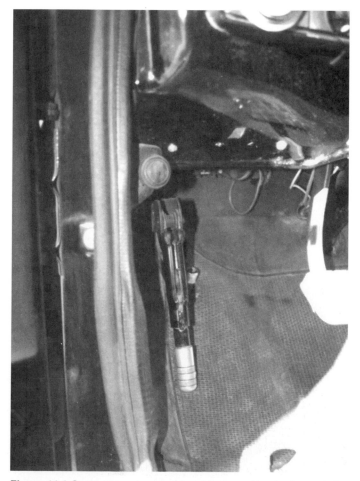

Figure 11.2 Some apparatus have hand-lever parking brake controls.

Figure 11.4 The apparatus wheels should be chocked during pumping operations.

Pump and Roll Operation (Manual Transmission)

Step 1: Bring the apparatus to a full stop and set the parking brake.

Step 2: Disengage the clutch (push in the clutch pedal).

Step 3: Place the transmission into neutral.

Step 4: Operate the PTO control (Figure 11.5). This may be a lever, push/pull control, or other type switch located in the cab. On some front-mount pumps, this may be a lever located in front of the pump. In this case, it is necessary to exit the cab at this time and then reenter the cab to complete the process.

Step 5: Place the transmission into the gear recommended by the manufacturer.

Step 6: Release the parking brake.

Step 7: Engage the clutch (let the clutch pedal out) slowly.

Once the PTO has been properly activated and the pump is engaged, an indicator light on the dashboard should light (Figure 11.6). This signals that the pump is operational. If this light does not illuminate, repeat the procedure.

Pump and Roll Operation (Automatic Transmission)

Step 1: Bring the apparatus to a full stop, and set the parking brake.

Step 2: Place the transmission into neutral or leave in drive, depending on the manufacturer's instructions.

Step 3: Operate the PTO control. This may be a lever, push/pull control, or other type switch located in the cab. On some front-mount pumps, this may be a lever located in front of the pump. In this case, it is necessary to exit the cab at this time and then reenter the cab to complete the process.

Step 4: Place the transmission into the gear recommended by the manufacturer.

Step 5: Release the parking brake (Figure 11.7).

Once the PTO has been properly activated and the pump is engaged, an indicator light on the dashboard should light. This signals that the pump is operational. If this light does not illuminate, repeat the procedure.

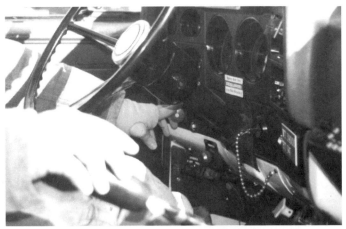

Figure 11.5 The driver/operator operates the PTO control while the transmission is in neutral.

Figure 11.6 The indicator light is lit when the PTO is engaged.

Figure 11.7 Once the transmission is in the proper gear for pump-and-roll operation, release the parking brake.

Stationary Operation (Manual Transmission)

Step 1: Bring the apparatus to a full stop, and set the parking brake.

Step 2: Disengage the clutch (push in the clutch pedal).

Step 3: Place the transmission into neutral (Figure 11.8).

Step 4: Operate the PTO control. This may be a lever, push/pull control, or other type switch located in the cab. On some front-mount pumps, this may be a lever located in front of the pump. In this case, it is necessary to exit the cab at this time and then reenter the cab to complete the process.

Step 5: Engage the clutch (let the clutch pedal out) slowly.

Once the PTO has been properly activated and the pump is engaged, an indicator light on the dashboard should light. As well, the vehicle speedometer should show a speed slightly above 0 mph (km/h), even though the vehicle is not moving. This signals that the pump is operational. If this light does not illuminate, repeat the procedure.

Stationary Operation (Automatic Transmission)

Step 1: Bring the apparatus to a full stop, and set the parking brake.

Step 2: Place the transmission into neutral or leave in drive, depending on the manufacturer's instructions.

Step 3: Operate the PTO control. This may be a lever, push/pull control, or other type switch located in the cab. On some front-mount pumps, this may be a lever located in front of the pump.

Once the PTO has been properly activated and the pump is engaged, an indicator light on the dashboard should light. As well, the vehicle speedometer should show a speed slightly above 0 mph (km/h), even though the vehicle is not moving. This signals that the pump is operational. If this light does not illuminate, repeat the procedure. You may then exit the cab and properly position the wheel chocks.

Disengaging PTO and Front-Mount Pumps

The procedure for disengaging PTO and front-mount pumps is as follows:

Step 1: Use the throttle control to reduce the engine to idle speed.

Step 2: Disengage the clutch (push in the clutch pedal) on manual transmissions, or place the automatic transmission into neutral.

Step 3: Operate the PTO control in the direction opposite of that used to engage the pump.

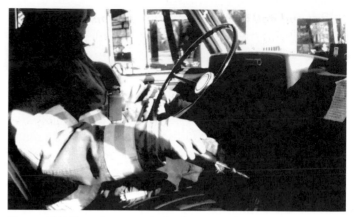

Figure 11.8 Place the transmission into neutral by pushing the clutch with your left foot and moving the gear shifter to the appropriate position.

The PTO should now be deactivated and the pump no longer in gear. The pump indicator light on the dashboard should no longer be illuminated. If it is still lit, repeat the procedures.

Engaging Midship Transfer Driven Pumps

Midship transfer driven pumps require that both the pump and truck transmission be in gear for operation. The engine power is then transferred to the pump instead of to the drive wheels. The drive transmission must be placed into the proper gear to efficiently use the engine power for the pump. The gear used should be the one recommended in the owner's manual. The following sections list the steps for engaging most midship transfer driven pumps:

Manual Transmission

Step 1: Bring the apparatus to a full stop and set the brake. Allow the engine to idle.

Step 2: Shift the drive transmission into neutral. This applies for both standard and automatic transmissions.

Step 3: Operate the pump shift control to transfer power from the drive axle to the pump drive. This control is usually located somewhere on the dashboard (Figure 11.9).

Figure 11.9 The pump shift control for a midship transfer fire pump.

Step 4: Shift the truck transmission into the proper gear for pumping, and lock the shifter into place.

Step 5: Engage the clutch (let out the clutch pedal) slowly.

Automatic Transmissions

Step 1: Bring the apparatus to a full stop. Allow the engine to slow to idle speed. If sufficient time is not allowed for the engine to slow down, a hard shift (also known as a gear slam, grinding gears etc.) occurs.

Step 2: Shift the drive transmission into neutral. Once the transmission is in neutral, set the brake according to manufacturer's instructions.

Step 3: Operate the pump shift control to transfer power from the drive axle to the pump drive. This control is usually located somewhere on the dashboard.

Step 4: Shift the truck transmission into the proper gear for pumping, and lock the shifter into place.

Step 5: Depress the accelerator to ensure the shift has been completed and the apparatus will not "drive away."

Once power has been transferred from the drive transmission to the pump transmission, an indicator light on the dashboard should illuminate to signify that the transfer is complete (Figure 11.10). If the light does not come on, repeat the procedure. Once the transfer has been made, the vehicle speedometer should show a speed somewhat above 0 mph (km/h) even though the vehicle is stationary. Some newer apparatus do not provide this check because the speedometer reading is taken off the rear of the transfer case. In these cases, it will be necessary to rely on the visual indicator to ensure that the pump is engaged.

Disengaging Midship Transfer Driven Pumps

The procedure for disengaging midship transfer driven pumps is as follows:

Step 1: Use the throttle control to reduce the engine to idle speed.

Step 2: Disengage the clutch (push in the clutch pedal) on manual transmissions, or place the automatic transmission into neutral. Wait a few seconds for the pump drive shaft to stop spinning before continuing. Otherwise, the gears could clash and be damaged as the pump shift control is operated.

Step 3: Operate the pump shift control from the PUMP to the ROAD position.

The pump transfer case should now be in the road position. The pump indicator light on the dashboard should no longer be illuminated. If it is still lit, repeat the procedure.

Engaging Auxiliary Engine Driven Fire Pumps

Auxiliary engine driven fire pump arrangements are designed so that the pump is always in gear when the auxiliary engine is running. Thus, the process of engaging these pumps simply involves starting the auxiliary engine. The procedure for starting these engines varies widely depending on the manufacturer of the engine and the design of the apparatus. Some have electric starter switches that are located either in the cab of the apparatus, directly on the engine, or both (Figure 11.11). Others require manual starting using a pull cord located directly on the engine flywheel. Driver/operators of apparatus equipped with auxiliary engine driven pumps must be thoroughly trained on the starting procedure for the apparatus to which they are assigned. The driver/operator must also be aware that if the engine is running and the pump is turning, the water supply to the pump must be opened so that the pump will not run dry for a prolonged period of time.

Figure 11.10 This indicator light illuminates when a proper pump shift is made.

Figure 11.11 The driver/operator simply pushes the engine start button to put the pump into service.

Operating from the Water Tank

Of the three possible types of water supply for the fire pump, most driver/operators operate solely from the onboard water tank at the vast majority of incidents. The need to use an external pressurized or static water supply source occurs much less frequently. In some circumstances, the fire attack begins with water in the tank and then, as the fire progresses, it becomes necessary to make the transition to an external supply source. The pump operator must be able to make the transition from the apparatus tank to an incoming water supply smoothly, with no disruption of the fireground operation. The following sections detail operating from the onboard water tank and making the transition to using an external, pressurized water supply. Making the transition from using the onboard water tank to a drafting operation is extremely rare. In most cases where drafting is necessary, the apparatus will be put into that mode immediately after being positioned in the desired location. Information on draft is covered later in this chapter.

Putting the Pump into Operation

Begin by making the fire pump operational, as described previously in this chapter. Once this is accomplished, the driver/operator exits the apparatus cab, chocks the wheels, and proceeds to the pump operator's panel. Next, the tank-to-pump valve must be opened fully (Figure 11.12). If the valve has a locking mechanism, lock the valve in the OPEN position. This prevents vibration or inadvertent contact from closing it and interrupting the flow of water to the pump. Once the valve is opened, the pump fills with water.

If the pump is a multistage, set the transfer valve to the proper position before pressure builds in the pump. In most cases, the pump should be in a SERIES (PRESSURE)

position if the operation is from the tank because maximum flow is limited by the size of the pump piping. If it appears that the pump will be required to furnish more than 50 percent of its rated capacity, it should be set to the PARALLEL (VOLUME) position from the start. This prevents having to shut down later on to make the changeover from the SERIES position. As stated in Chapter 10, some pump manufacturers may recommend that their pumps remain in the SERIES position until up to 70 percent of capacity is required. Make sure that you know the recommendations for the pump you are operating.

Increase engine rpm by using the hand throttle (Figure 11.13). Observe the master pressure gauge as the throttle is being advanced. If the pump is normally full of water, the master pressure gauge should start to rise as soon as the rpm is increased. If the pump has been drained of water, it will be full of air. The air has to be forced out of the pump by the water as it enters before discharge pressure can build. If this is the case, at least one of the discharge valves or the tank fill line must be open before the air can escape and the pump can fill with water. Operating the primer also speeds the removal of air from the pump. If the master pressure gauge still fails to register a reading, the pump may not be in the proper gear. Immediately decrease the engine speed and return to the apparatus cab and verify that the transmission is in the proper gear or that the pump shift transfer has been made.

The operator has to use judgment based on previous experience and departmental SOPs when determining the amount of pressure to build within the pump at this time. It depends on the number and size of hoselines that are in the process of being deployed. With experience, the operator knows at what pressure the pump needs to be before any lines are actually opened.

Figure 11.12 The tank-to-pump valve is operated as soon as the driver/operator reaches the pump panel.

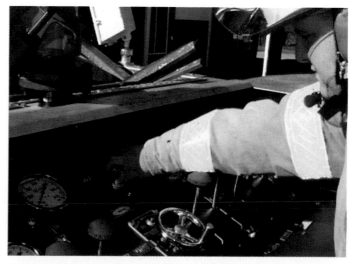

Figure 11.13 Increase the throttle to provide sufficient discharge pressure.

If none of the attack lines are ready to be charged by the time the pump pressure has built up, the pump-to-tank line (tank fill valve) can be partially opened to allow water to circulate. Any air that has been trapped in the pump or piping has been removed, and a stable pressure can be established. If the hoselines are ready for water when the pressure is set, the discharge valve can be opened slowly, locked into position, and the flow initiated (Figure 11.14). Set the automatic pressure regulating device for the operating pressure. Information on setting pressure regulating devices for specific makes of pumps is contained in the pump manufacturer's operations manual.

An alternate method for putting the pump into operation is to open the appropriate discharges before you increase the discharge pressure by throttling up the pump. When following this procedure, open the tank-to-pump valve first. Then, open the discharges to the hoselines that are going to be used. When these valves are open, the hoselines begin to fill with water. As the throttle is increased, the pressure in the hoselines also begin to increase. If the firefighters already have the nozzle(s) open, the air bleeds off the hoseline(s), and the pump discharge pressure may be brought directly to the desired level with a minimum of guesswork. If the nozzles are not open, the driver/operator has to build up a pressure based on previous experience and departmental SOPs. Once the lines are opened, it may be necessary to further adjust the pressure.

While the pump is in operation, carefully observe all gauges, including engine gauges and those gauges associated with operation of the pump. Be ready to take whatever action is required to remedy any abnormal reading. If a flowmeter is being used on the discharge line, the throttle may be adjusted accordingly until the desired flow rate is achieved.

During the initial attack, hoselines operate intermittently. If no water is being used for an extended time but the discharge pressure is maintained at relatively high levels, the pump will overheat. To prevent this, some means of maintaining water movement through the pump must be found. If the pump is equipped with a circulator, or a booster cooling valve, it can be opened and set to the tank position (Figure 11.15). This causes water to circulate through the pump and back into the tank, thus providing some cooling without wasting any water. The tank fill line may also be used to circulate water through the booster tank and maintain a cooling action.

Be conscious of the water level in the tank. As the level drops, warn the officer in charge of the amount of water

Figure 11.14 Once adequate pressure has been developed, the appropriate discharge is opened.

Figure 11.15 The circulator valve should be in the tank position.

remaining in the tank. You should also be able to give an estimate of the amount of time the water will last at the present rate of consumption. If the pump is equipped with a flowmeter, it may give you a tally of how much water has been flowed to that point. Otherwise, you will be limited to observe the water tank level indicator.

Transition to an External Water Supply

If the incident requires more water than is carried in the apparatus onboard water tank, it is necessary to transition to an external supply source before depleting the water in the tank. This eliminates any interruption of water to attack lines. Transition from the water tank to an external water supply source most commonly involves the use of a pressurized source. This will be a hoseline being supplied by another apparatus or from a fire hydrant.

The supply line should be connected into the appropriate intake fitting of the pump. Because in this case, the pump is already in operation, it is necessary to make this connection to an intake that is equipped with a closed gate valve. Air in the empty supply hose will cause problems if it enters the pump while the pump is supplying attack lines under pressure. To purge this air, open the bleeder valve on the gated intake line so that the air can escape as the supply line fills with water (Figure 11.16). When all the air has been forced from the line and a steady stream of water comes from the bleeder valve, close the bleeder valve. The water supply is now at the pump, ready for the transition.

Figure 11.16 Open the bleeder valve on the gated intake to remove air from the system.

To begin the transition, the operator opens the intake valve from the supply line very slowly while observing the pressure on the intake and discharge gauges (Figure 11.17). At the same time the intake valve is opened, the tank-to-pump valve can be closed. Many apparatus are designed so that the pump-to-tank line may be cracked open to refill the water tank without interrupting other pumping operations. This should be done whenever possible. This provides a cushion of water supply in the event that the external water supply is lost unexpectedly.

Continue opening the intake gate valve and adjusting the discharge pressure until the intake is completely open. It may be necessary to adjust the engine throttle to get the desired discharge pressure. When this is complete, it may be necessary to adjust the pressure regulating devices again.

Operating Procedure for Transition to an External Water Supply

The steps for making the transition from the pumper's water tank to an external water supply are as follows:

Step 1: Position the apparatus in a safe position, and immobilize it by setting the parking brake and blocking the wheels.

Step 2: Engage the pump, and select the proper gear in the road transmission. Lock the gear into place.

Step 3: Open the tank-to-pump valve.

Step 4: Set the transfer valve to the SERIES (PRESSURE) position if necessary.

Step 5: Increase the throttle setting to obtain the desired pressure, priming if necessary.

Step 6: Set the relief valve or pressure governor.

Step 7: Open the circulator valve or partially open the tank fill valve.

Step 8: When an external supply becomes available, open the intake valve while closing the tank-to-pump valve. As soon as an adequate supply of

Figure 11.17 Once the air has been bled from the line, the gated intake valve may be opened.

water is available, divert enough of it through the tank fill line to replenish the supply in the tank. On some apparatus, the tank-to-pump line is not equipped with a check valve. This allows the tank to be filled through this line as well.

Step 9: Adjust the engine throttle to achieve the desired discharge pressure if necessary.

See Flowchart 11.1 and Table 11.1 located at the end of this chapter for additional direction on operating the pump from the water tank and for troubleshooting during this operation.

Operating from a Pressurized Water Supply Source

There are two basic pressurized water supply sources that may be used to supply a fire pump: a hydrant or a supply hose from another fire pump. When using either of these two sources, water enters the pump under pressure from the source. As the discharge pressure from the fire pump increases, however, the incoming pressure from the supply source drops due to friction loss in the water system. If the discharge pressure (and the resulting flow volume) is increased too much, the intake pressure from the supply source may be reduced below 0 psi (0 kPa). Operating at a negative pressure from a fire hydrant is dangerous because it increases the possibility of damage to the fire pump and water system due to water hammer if the flow is cut off suddenly. Water heaters or other domestic appliances on a municipal water supply system may also be damaged by a negative pressure.

Operating at a negative pressure when being supplied from another pumper is equally dangerous. If the discharge pressure (and the resulting flow volume) is increased too much, the friction loss in the supply line may cause the residual pressure to drop below 0 psi (0 kPa). If this happens, the supply hose may collapse, resulting in an interruption of the water supply. Operating at or near this point can also cause damage to the pump from cavitation. Cavitation is explained in more detail later in this chapter in the section on Operating from a Static Water Supply Source.

Operating at near-zero residual pressure requires close attention to the master intake pressure (compound) gauge and prompt action before the danger point is reached. As a standard practice, it is not desirable to reduce the incoming pressure from a hydrant or supply pumper below 20 psi (140 kPa). In rare instances, it may be necessary to reduce the intake pressure below 20 psi (140 kPa), but this operation requires close attention on the part of the driver/operator to ensure that the pressure does not drop to zero. Each fire department should have SOPs that provide guidance for driver/operators on this issue.

The remainder of this section on Operating from a Pressurized Water Supply Source focuses primarily on hydrant operations. Chapter 13 contains more detailed information on receiving water from another pumper.

Operating from a Hydrant

Hydrants are a common water supply (Figure 11.18). Whether the pumper directly supplies attack lines or supplies another pumper in a relay, the operation of the pumper on the hydrant is basically the same.

Figure 11.18 In suburban and urban areas, pumpers commonly operate from a hydrant. *Courtesy of Gainesville (FL) Fire-Rescue.*

Choosing a Hydrant

The first consideration in selecting a hydrant is to determine which hydrant is most appropriate in terms of fire fighting and safety needs. The hydrant closest to the fire is not always the best choice. Due to limitations of the water system, the closest hydrant may not be capable of supplying the necessary amount of water. Connecting to such a hydrant would mean that supply lines would have to be laid to another hydrant with a larger capacity. In addition, the closest hydrant may be too close to the fire; connecting to it would put the safety of personnel and equipment in jeopardy.

In order to select the best hydrant, a thorough knowledge of the water system is required. The best hydrants are located on large water mains that are arranged in a grid pattern so that they can receive water from several directions at the same time. Water mains that are interconnected into a grid and that have a relatively high amount of circulation usually have low amounts of sedimentation and deterioration. The worst hydrants typically are those located on "dead-end mains." Single lines used to supply relatively small amounts of water may be partially clogged, which further reduces their capacity.

Fire departments should have access to water department records that indicate the reliability of all hydrants in their jurisdiction. This information should be included in pre-incident planning. It may also be part of the computer dispatch system. This allows dispatchers to provide responding companies with appropriate hydrant location information. Apparatus may also be equipped with map books that the company officer may check for hydrant information and locations while en route to the scene. Fire hydrants may be color coded to indicate the flow that can be expected from them. See NFPA 291, *Recommended Practice for Flow Testing and Marking of Hydrants,* for more information on color coding of hydrants (Table 11.2).

Table 11.2
Hydrant Color Codes

Hydrant Class	Color	Flow
Class AA	Light Blue	1,500 gpm (5 680 L/min) or greater
Class A	Green	1,000 -1,499 gpm (3 785 L/min to 5 675 L/min)
Class B	Orange	500-999 gpm (1 900 L/min to 3 780 L/min)
Class C	Red	Less than 500 gpm (1 900 L/min)

Once a hydrant has been selected and test flowed, the pumper should be hooked to it as quickly as possible. The driver/operator and the pumper crew should be well trained and practiced at making this connection as smoothly as possible. For detailed instructions on making hydrant connections using small and large diameter intake hose(s), see Chapter 5.

Making a Forward Lay

There are two primary ways that a fire department pumper may use a hydrant. The first involves stopping at the hydrant, dropping the end of one or more supply lines at the hydrant, and proceeding to the fire location (Figure 11.19). This process is called making a *forward lay*.

The procedures used once the forward lay has been made vary from department to department. In some cases, the supply line(s) merely lay there until a second pumper comes along to hook up to the hydrant and supply the line(s). Other departments connect those lines directly to the hydrant and begin to work off hydrant pressure (Figure 11.20).

Many departments that prefer forward lays use a four-way hydrant valve to aid this process (Figure 11.21). The four-way hydrant valve allows the original supply line that was forward laid by the first pumper to be immediately charged using hydrant pressure. The valve has a second discharge outlet, usually 4½ or 5 inches (115 mm or 125 mm) in diameter, that is equipped with a shutoff valve. This enables a second pumper to be connected to the hydrant without interrupting the flow to the original

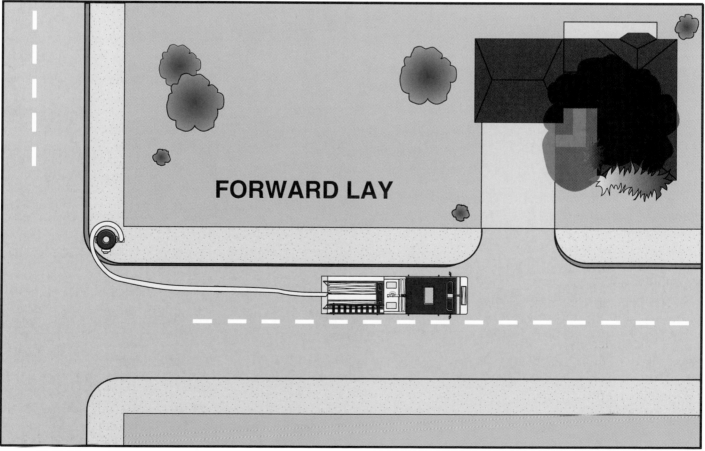

Figure 11.19 A forward lay.

Figure 11.20 Some jurisdictions commonly operate off hydrant pressure. This is not recommended for long lays of MDH.

Figure 11.21 Hydrant valves are commonly used by departments that routinely make forward lays.

Figure 11.22 The apparatus that made the forward lay is being supplied by hydrant pressure. *Courtesy of Gainesville (FL) Fire-Rescue.*

Figure 11.23 The next-arriving pumper connects to the large discharge of the hydrant valve. *Courtesy of Gainesville (FL) Fire-Rescue.*

Figure 11.24 The second pumper can now boost the pressure of the original supply line. *Courtesy of Gainesville (FL) Fire-Rescue.*

supply line. A second intake connection on the valve allows the pumper connecting to the valve to boost the pressure in the original supply line laid by the first pumper.

Figures 11.22 through 11.24 illustrate the operation of one type of four-way valve. In Figure 11.22, the valve is connected to a hydrant. The original supply line laid by the first pumper has been connected to the supply line (forward lay) outlet. The hydrant has been opened, and the clapper valve has operated to allow the water to flow into the supply line. The second pumper then connects to the large diameter pumper intake connection on the four-way valve (Figure 11.23). The pumper intake shutoff is opened, and the water from the hydrant fills the pump without interfering with the flow through the original supply line. The pump operator connects one of the pumper discharge outlets to the second intake of the hydrant valve (Figure 11.24).

As the pressure in the pump builds, it overcomes the pressure from the hydrant that keeps the clapper valve open. When the pump pressure gets high enough, the clapper valve closes. When this happens, the original supply line is being fed from the pump, not directly from the hydrant. Figures 11.25 a and 11.25 b give an overview of the entire operation.

It is recommended that gate valves be attached to the unused hydrant discharges (Figure 11.26). This allows other connections to be made to the hydrant at a later time without having to interrupt the flow in the original line(s).

When using 2½- or 3-inch (65 mm or 77 mm) hoselines to supply the pumper directly off hydrant pressure, it is

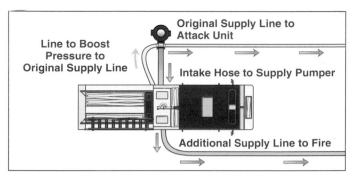

Figure 11.25b A completed hydrant valve connection.

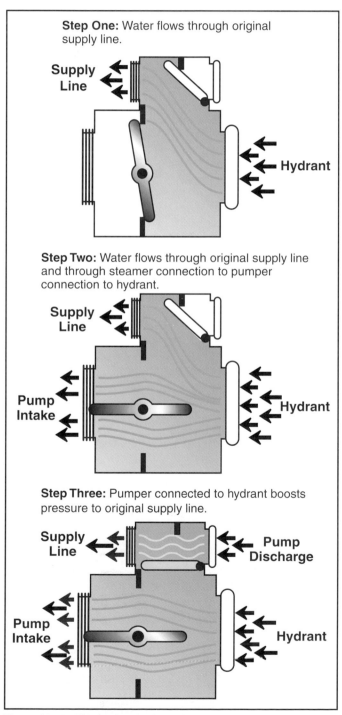

Step One: Water flows through original supply line.

Step Two: Water flows through original supply line and through steamer connection to pumper connection to hydrant.

Step Three: Pumper connected to hydrant boosts pressure to original supply line.

Figure 11.25a This shows the flow of water through a hydrant valve.

Figure 11.26 Some jurisdictions require a gate valve to be attached to the unused hydrant discharge. This allows a second line to be charged at a later time without shutting down the hydrant. *Courtesy of Rich Mahaney.*

recommended that the lines be no longer than 300 feet (90 m). Beyond this point, the amount of friction loss in the lines is too great to provide an adequate supply of water. This distance may be lengthened somewhat if the hydrant is on a special high-pressure main. The 300-foot (90 m) maximum also does not apply when using large diameter supply hose (4-inch [100 mm] or larger). These lines may be operated off hydrant pressure for significantly longer distances than smaller hose. The exact distances vary with the size and manufacturer of the hose, the required flow, and the strength of the water system supplying the hydrant.

Making a Reverse Lay

With the *reverse lay*, hose is laid from the fire to the water source. This method is used when a pumper must first go to the fire location so that a size-up can be made before laying a supply line (Figure 11.27). It is also the most expedient way to lay hose if the apparatus that lays the hose must stay at the water source such as when drafting or boosting hydrant pressure to the supply line. If threaded couplings are used, hose beds set up for reverse lays should be loaded so that the first coupling to come off is male (Figure 11.28).

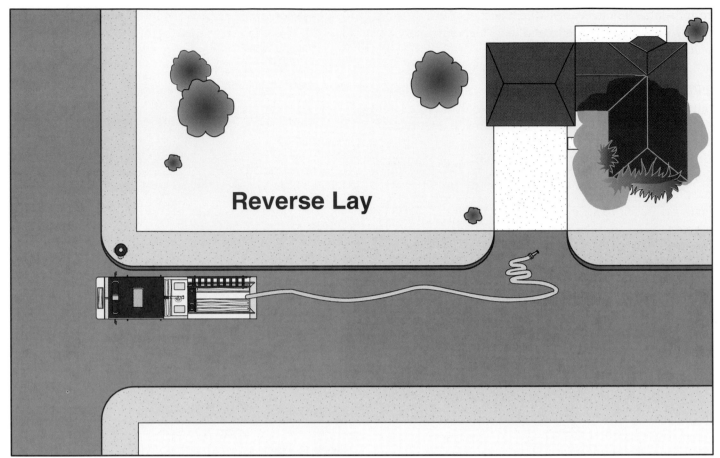

Reverse Lay

Figure 11.27 A reverse lay.

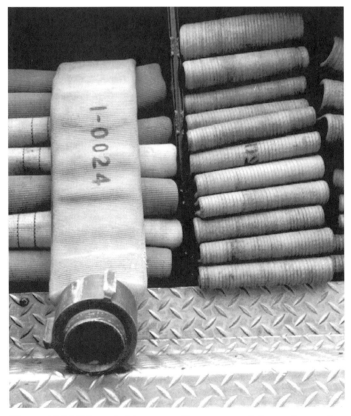

Figure 11.28 If threaded hose couplings are used, the male coupling must come off the hose bed first for a reverse lay.

Laying hose from the incident scene back to the water source has become a standard method for setting up a relay pumping operation when using medium diameter hose as a supply line. With medium diameter hose, it is necessary in most cases to place a pumper at the hydrant to supplement hydrant pressure to the supply hose. It is, of course, always necessary to place a pumper at the water source when drafting. The reverse lay is the most direct way to supplement hydrant pressure and perform drafting operations.

A disadvantage with the reverse lay for single engine company operations, however, is that essential fire fighting equipment, including attack hose, must be removed and placed at the fire location before the pumper can proceed to the water source. This causes some delay in the initial attack. The reverse lay also obligates one person, the pump operator, to stay with the pumper at the water source, thus preventing that person from performing other essential fire location activities.

A common operation involving two pumpers — an attack pumper and a water supply pumper — calls for the first-arriving pumper to go directly to the scene to start an initial attack on the fire using water from its tank, while the second-arriving pumper lays the supply line from the attack pumper back to the water source. This is a relatively simple operation because the second pumper needs only to connect its just-laid hose to a discharge outlet, connect a suction hose, and begin pumping.

When reverse laying a supply hose, it is not necessary to use a four-way hydrant valve. One can be used, however, if it is expected that the pumper will later disconnect from the supply hose and leave the hose connected to the hydrant. This situation may be desirable when the demand for water diminishes to the point that the second pumper can be made available for response to other incidents. As with a forward lay, using the four-way valve in a reverse lay provides the means to switch from pump pressure to hydrant pressure without interrupting the flow.

The reverse lay is also used when the first pumper arrives at a fire and must work alone for an extended period of time. In this case, the hose laid in reverse becomes an attack line. It is often connected to a reducing wye so that two smaller hoses can be used to make a two-directional attack on the fire (Figure 11.29).

Once the pumper that has made the reverse lay reaches the supply source, the driver/operator hooks the hose that was just laid to a pump discharge and then connects the pumper to the water supply source. The procedures for connecting to water supplies was outlined in Chapter

5. Operation of the pump at this point follows the same procedures as described for other operations in this chapter.

Getting Water into the Pump

After the connections to the hydrant are made, certain preliminary checks should be performed before opening the hydrant. The tank-to-pump valve must be closed if the intake is not equipped with a shutoff valve. Many of the newer pumps have check valves in the tank-to-pump line. These check valves prevent water from entering the tank under pressure from the pump intake. Many older pumps do not have this type of protection. If water is allowed to enter the tank under pressure, the venting may not be adequate to allow the pressure to dissipate, and damage to the tank may result. If the intake is equipped with a shutoff valve, it is acceptable to charge the intake line and bleed off the air while still pumping from the water tank (Figure 11.30). Whether or not a particular apparatus has a check valve in the tank-to-pump line or a shutoff valve on the intake, the driver/operator should always make sure that the tank-to-pump valve is closed when switching to an external water supply.

When opening a dry barrel hydrant, it must be opened all the way. If it is not opened fully, the drain valve at the base of the hydrant may be open at the same time water

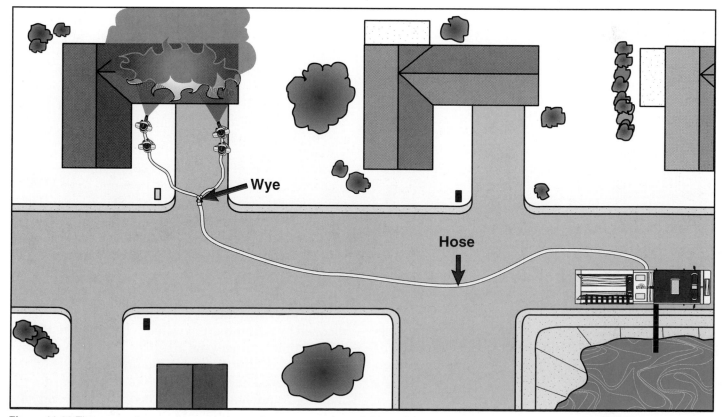

Figure 11.29 The reverse lay may be used when only one pumper is available and the water supply is distant from the fire scene.

Figure 11.30 Air from the supply line may be bled off at the intake valve.

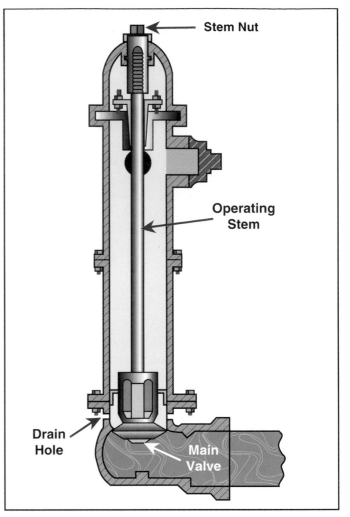

Figure 11.31 A dry barrel hydrant schematic.

is coming in from the main (Figure 11.31). This flow of water (coming from the system and leaving through the drain valve) washes away the gravel that is supporting the body of the hydrant. Loss of support will, of course, eventually damage the hydrant. It is generally recommended that the hydrant valve be opened until the valve stem will spin no further. Do not jam the valve stem past the point where it easily stops turning.

After the pump is full of water and the pressure in the system has stabilized, a reading of the pressure on the master intake gauge indicates the static pressure in the water supply system. This reading is very important for estimating the remaining capacity of the hydrant as the water begins to move. Therefore, it may be desirable for the pump operator to record or remember the static pressure before the operation begins. Without knowing the static pressure, it is impossible to establish a reference for determining whether additional hoselines or pumpers can be supplied.

Putting the Pump in Service

When operating from any pressurized water supply source, it is not advisable to engage the pump drive system before leaving the cab if there will be an extended period of time before water is introduced into the pump. This is because the pump would be running dry with no lubrication or cooling while the connections are being made. As well, the hydrant may be found to be inoperable, and the apparatus will have to be relocated. If it appears that water will be introduced into the pump within a few minutes, the pump may be engaged before exiting the cab.

If you are operating a two-stage pump, set the transfer valve to the proper position before increasing the throttle to build discharge pressure. The proper position to use is dictated by the amount of water to be supplied. Even if only one line is initially put in service, it may be necessary to supply additional lines as the fire progresses. If the pump is placed into the SERIES (PRESSURE) position, it may not be possible to do this. Because the pump on a hydrant or in the middle of a relay is usually called on to supply large amounts of water, it is good operating practice to set the transfer valve to the PARALLEL (VOLUME) position. An exception to this rule would be if very high pressures are required for very long supply lines. Obviously, if operating a single-stage pump, none of these considerations are applicable.

Open discharge valves slowly, particularly when using LDH. The lines are empty, and they must fill before pressure can be established. While observing both the master intake and discharge gauges, increase the engine rpm. If the master intake gauge drops below 20 psi (140 kPa), do not increase the throttle further because the pump is in danger of cavitation. If there is adequate supply, then advance the throttle until the desired pressure and flow have been established. The relief valve or governor can now be set.

The pump operator at the hydrant or in the middle of a relay may be some distance from the actual fireground operation, and it may be difficult to determine how much water is being used or even if any of the lines are flowing. Because of the intermittent use of water in fire fighting operations, it is necessary to keep the pump from overheating by ensuring a continued minimum flow. Some methods for preventing overheating are as follows:

- Pull some of the booster line off the reel and securely tie off the nozzle to a solid object. Open the valve that supplies the booster reel, and discharge water in a direction that will not harm people or damage property (Figure 11.32).

Figure 11.32 A booster line may be tied off to an object and continuously flowed to prevent the pump from overheating.

- Open a discharge drain valve. These should be designed to discharge the water in a manner that will not harm people or damage property. Some are equipped with threaded outlets that allow a hose to be connected and then routed away from the apparatus to a safe discharge point. This is advantageous when parking on nonpaved surfaces. In this case, it is better to discharge water away from the apparatus so that it does not become stuck in the mud that could be created by discharging water directly beneath it.

- Partially open the tank fill valve or pump-to-tank line. Even if the water tank becomes full and overflows through the tank vent, this cooling action is better than allowing the pump to overheat.

- Use a bypass or circulator valve if the apparatus is so equipped.

The flow should be sufficient to cool the pump without seriously diminishing the amount of water available for fire suppression.

Additional Water Available from a Hydrant

When a pumper is connected to a hydrant and is not discharging water, the pressure shown on the intake gauge is the *static pressure*. When the pumper is discharging water, the intake gauge reading is the *residual pressure*. The difference between these two pressures is used to determine how much more water the hydrant can supply. There are three methods that can be used for this determination:

- Percentage method

- First-digit method

- Squaring the lines method

You may note that if you apply each of these methods to the same problem, slightly different answers may result. None of these methods provide precise answers under a full range of conditions. In each case, the answers provided by these methods provide a reliable enough figure upon which fireground operations may be based.

Percentage Method

To use the percentage method, first calculate the drop in pressure as a percentage. The easiest way to do this is by using the following formula:

EQUATION I

$$\text{Percent Drop} = \frac{(\text{Static - Residual})(100)}{\text{Static}}$$

Table 11.3 indicates how much more water is available from a hydrant and how many additional hoselines can be supplied. If the percentage is 10 or less, three additional lines with the same flow as the line being used may be added: 11-15 percent, two lines; 16-25 percent, one line. If the percentage is over 25 percent, more water may be available, but not as much as is flowing through the line being used.

Example 1

A pumper is supplying one line with 250 gpm flowing. The static pressure was 70 psi, and the residual pressure reading 63 psi. Determine how many additional lines can be added.

$$\text{Percent Drop} = \frac{(\text{Static - Residual})(100)}{\text{Static}}$$

$$\text{Percent Drop} = \frac{(70 - 63)(100)}{70}$$

$$\text{Percent Drop} = \frac{(7)(100)}{70} = \frac{(700)}{70}$$

$$\text{Percent Drop} = 10$$

From Table 11.3, we see a 10 percent drop means three times the amount being delivered is available. This means three additional lines of 250 gpm may be flowed.

First-Digit Method

The first-digit method for calculating available water is somewhat quicker and easier than the percentage method. One disadvantage of this method, however, is that it cannot be used by those fire departments using metrics.

Step 1: Find the difference in psi between the static and residual pressures.

Step 2: Multiply the first digit of the static pressure by 1, 2, or 3 to determine how many additional lines of equal flow may be added as explained below.

- If the psi drop is equal to or less than the first digit of the static pressure multiplied by one (1), three additional lines of equal flow may be added.

- If the psi drop is equal to or less than the first digit of the static pressure multiplied by two (2), two additional lines of equal flow may be added.

- If the psi drop is equal to or less than the first digit of the static pressure multiplied by three (3), one additional line of equal flow may be added.

The following example demonstrates how to compute additional water at a hydrant by the first-digit method.

Example 2

A pumper is supplying one line with 250 gpm flowing. The static pressure was 65 psi, and the residual pressure reading 58 psi. Determine how many additional lines can be added.

Difference in psi = Static Pressure - Residual Pressure

Difference in psi = 65 - 58

Difference in psi = 7

First digit of static pressure x 1

6 x 1 = 6

7 is not less than 6 but is less than 12 (2 x 6), so two more lines at 250 gpm each can be added.

Squaring the Lines Method

Another method for determining the additional amount of water that is available at a hydrant is known as the squaring the lines method. In order to successfully use this method, the driver/operator must note the static pressure on the water system before any pump discharges are open. An alternative to this would be knowing the usual static pressure on that water supply system under normal circumstances. This information would be obtained from pre-incident plans or previous experience.

Table 11.3 Additional Water Available at a Hydrant	
Percent Decrease of Pumper Intake Pressure	**Additional Water Available**
0-10	3 times amount being delivered
11-15	2 times amount being delivered
16-25	Same amount as being available
25 +	More water might be available, but not as much as is being delivered.

Table 11.4 Assumed Flow Rates for Squaring the Lines Method	
Hose Size in inches (mm)	**Flow Rate in gpm (L/min)**
1½ (38)	125 (500)
1¾ (45)	175 (700)
2 (50)	200 (800)
2½ (65)	250 (1 000)

In addition to knowing the static pressure on the water supply system, the driver/operator must also have a relatively close idea of the volume of water that is initially being flowed by the pumper. This is simple if the pumper is supplying a fixed gallonage nozzle. If an automatic nozzle is being supplied, the driver/operator will have to assume a reasonable flow based on the size of the hose being used. Table 11.4 shows reasonable flow rates that may be assumed for various sized attack lines.

Once these figures have been established, the additional amount of water can be determined by squaring the number of lines currently flowing and multiplying this by the original pressure drop. The following example shows how this is done.

Example 3

A fire department pumper connects to a fire hydrant that has a static pressure of 60 psi. When a 250 gpm handline is opened, the intake pressure drops to 52 psi. Determine how many more 250 gpm handlines may be operated without lowering the residual pressure in the water system below 20 psi.

Difference in psi = Static Pressure - Residual Pressure

Difference in psi = 60 - 52

Difference in psi = 8

If a second 250 gpm line were to be added, the pressure drop would be as follows:

(Number of lines)^2 = Multiplication factor

$2^2 = 4$

(Multiplication factor) x (original pressure drop) = resultant pressure drop in system at the new flow rate

4 x 8 = 32 psi drop

If the pressure drop is 32 psi, this means that the residual pressure in the system will be 28 psi (60 − 32). It is unlikely that a third hoseline could be added under this circumstance.

You may note that when using this method, the pressure drop in the system is quadrupled each time the flow rate is doubled. The driver/operator may wish to remember this simple fact in order to do quick mental calculations.

Shutting Down the Hydrant Operation

Due to the danger of water hammer and pressure surges on water systems, all changes in flow should be made smoothly, with no sudden actions. This is especially true when shutting down hydrant operations. Hydrants have been broken and pushed out of the ground by water hammer when a large flow has been stopped suddenly. Use the following procedure to shut down the pump:

Step 1: Gradually slow the engine rpm to reduce the discharge pressure.

Step 2: Take the pressure control device out of service if in use.

Step 3: Slowly and smoothly close the discharge valves.

Step 4: Place the drive transmission into neutral, and disengage the pump control device.

Close the hydrant slowly and completely to prevent water hammer (Figure 11.33). The last turn of the operating nut opens the drain holes in the base of a dry barrel hydrant, allowing the water to drain from it. This is particularly important in areas subject to freezing temperatures. In these areas, water left in the hydrant barrel freezes, making the hydrant unusable. Frozen water in the hydrant may also cause permanent damage to the hydrant. After a hydrant is closed and before the caps are put in place, the operator can visually check to be sure the water is draining from the hydrant (Figure 11.34). If it does not drain on its own, it may be necessary to manually drain/pump out the hydrant.

For more information on operating the pump from a pressurized water supply, particularly when being supplied by another pumper, see Chapter 13, Relay Pumping Operations. See Flowchart 11.2 and Table 11.1 located at the end of this chapter for additional direction on operating the pump from a hydrant and for troubleshooting during this operation.

Figure 11.33 Close the hydrant slowly and completely to prevent water hammer.

Figure 11.34 Check the dry barrel hydrant to make sure that all the water has drained from it.

Operating from a Static Water Supply Source

All fire department pumpers should be capable of pumping water from a static water supply. In most cases, the static water supply will be located at a lower level than the fire pump. Because one drop of water will not stick to another, it is not possible to pull water into the pump from a lower level. However, it is possible to evacuate some of the air inside the fire pump. This creates a pressure differential (partial vacuum), which allows atmospheric pressure acting on the surface of the water to force water into the fire pump. In order to accomplish this, an airtight, noncollapsible waterway (hard intake hose) is needed between the fire pump and the body of water to be used.

In Figure 11.35, enough air has been evacuated to reduce the atmospheric pressure inside the fire pump and the intake hose to 12.7 psi (86 kPa). Because atmospheric pressure at sea level is about 14.7 psi (100 kPa), a negative pressure of -2 psi (-14 kPa) is measured on the intake (compound) pressure gauge as 4 inches (100 mm) of vacuum. This vacuum causes water to rise 4.6 feet (1.4 m) into the intake hose from the surface of the water. The weight of the water, combined with the reduced air pressure acting on its surface, creates a balance (Figure 11.36).

As the water begins to move through the pump, additional pressure losses are encountered. Any type of fire hose, strainer, or appliance creates a certain amount of friction loss. The amount of friction loss is proportional to the amount of water moving through it. The inertia of water is an additional factor in pressure loss. As the water begins to move through the pump, a certain amount of energy is consumed in getting the water at rest to begin to move and increase its velocity sufficiently to supply the amount of water needed.

The amount of friction loss in the hard intake hose is dependent upon the diameter of the hose. The smaller the hose, the greater the friction loss, and accordingly, less water will be supplied to the pump. To allow for this, the size of the intake hose is increased for larger capacity pumps (Table 11.5). It is possible to increase the capacity of a certain-sized pump if the size of the intake hose is increased. For example, a pumper rated at 750 gpm (3 000 L/min) is normally supplied with a 4½-inch (115 mm) intake hose, which allows the pumper to reach its capacity. However, by equipping the same pumper with a 5-inch (125 mm) hose, the capacity can be increased to 820 gpm (3 280 L/min) if all other factors remain the same.

The total pressure available to overcome all these pressure losses is limited to the atmospheric pressure at sea level (14.7 psi or 100 kPa). Atmospheric pressure

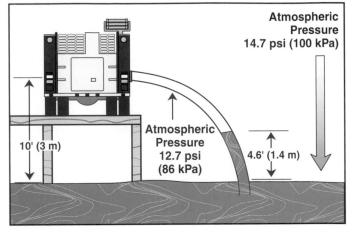

Figure 11.35 In this example, enough air has been evacuated to reduce the atmospheric pressure by 2 psi (14 kPa). The resulting vacuum causes the water to rise 4.6 feet (1.4 m) into the intake hose.

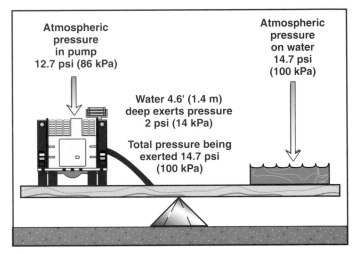

Figure 11.36 As the pressure inside the intake hose changes, the level goes up or down to maintain a balance at all times.

decreases 0.5 psi (3.5 kPa) for each 1,000 feet (305 m) of elevation gain. Thus, the atmospheric pressure in a city that is located 5,000 feet (1 525 m) above sea level is 12.2 psi (72.5 kPa). Because the same 14.7 psi (100 kPa) of atmospheric pressure must overcome the elevation pressure as well as the friction loss, increasing the height of the lift also decreases total pump capacity. If the lift were increased from 10 to 16 feet (3 m to 5 m), the vacuum needed to handle the increased lift would increase from 9 to 14 inches (229 mm to 356 mm) of mercury, which leaves 5 inches (125 mm) less to overcome friction loss. Using the previous example of the 750 gpm (3 000 L/min) pumper, increasing the lift to 16 feet (5 m) with the 4½-inch (115 mm) hard intake hose would reduce the pump capacity to 585 gpm (2 340 L/min).

While the pump is moving water, the vacuum reading on the master intake gauge provides an indication of the remaining pump capacity. The maximum amount of vacuum that most pumps develop is approximately 22

Table 11.5
Number and Size of Intake Hoses Based on Pump Capacity

Rated Capacity (gpm)	Rated Capacity (L/min)	Maximum Suction Hose Size (in.)	Maximum Suction Hose Size (mm)	Maximum Number of Suction Lines
250	1 000	3	77	1
300	1 200	3	77	1
350	1 400	4	100	1
450	1 800	4	100	1
500	2 000	4	100	1
600	2 400	4	100	1
700	2 800	4	100	1
750	3 000	4½	115	1
1000	4 000	5	125	1
1250	5 000	6	150	1
1500	6 000	6	150	2
1750	7 000	6	150	2
2000	8 000	6	150	2
2250	9 000	8	150	3
2500	10 000	8	150	3
2750	11 000	8	150	4
3000	12 000	8	150	4

Reprinted with permission from NFPA 1901, Automotive Fire Apparatus, Copyright © 1996. National Fire Protection Association, Quincy, MA 02269. This reprinted material is not the complete and official position of the National Fire Protection Association, on the referenced subject which is represented only by the standard in its entirety.

inches of mercury (560 mm/hg). A reading approaching this figure is a warning to the pump operator that the pump is getting close to the limit of its ability. Any further water supply requirements necessitate some other means of delivery. If an attempt is made to increase the discharge from the pump beyond the point of maximum vacuum on the intake, cavitation results.

Cavitation can be described as that condition where, in theory, water is being discharged from the pump faster than it is coming in. Cavitation is sometimes expressed as "the pump running away from the water." Cavitation occurs when air cavities are created in the pump or bubbles pass through the pump. They move from the point of highest vacuum into the pressurized section where they collapse or fill with fluid.

The high velocity of the water filling these cavities causes a severe shock to the pump. In extreme cases or over prolonged usage, this shock results in damage to the pump. A more scientific and accurate explanation is as follows: As the pressure drops below atmospheric, the boiling point of water drops to the point that the liquid changes to a vapor and creates a bubble of water vapor or steam. As the vapor passes through the impeller of the pump, the pressure increases, the vapor condenses, and water rushes in to fill the void. It is evident that the temperature of the water, the height of the lift, and the amount of water being discharged affect the point at which cavitation begins.

There are a number of indications that a pump is cavitating. The hose streams will fluctuate, as will the pressure gauge on the pump. A distinctive sound described as a popping or sputtering may be heard as the water leaves the nozzle. In cases of severe cavitation, the pump itself will be noisy, sounding rather like gravel is passing through it. The best indication of cavitation, however, is the lack of reaction on the pressure gauge to changes in the setting of the throttle.

When a pump reaches the point of cavitation, it is discharging all of the water that the atmospheric pressure or other pressure source can force into the intake. Because discharge pressure builds when water is supplied from the pump faster than it can be taken away, increasing pump rpm will not increase discharge pressure when there is no more water available to be supplied (Figures 11.37 a and b).

Cavitation often results when a pump has been equipped with inadequate piping from the water tank. Because a pump operates most often from its own tank, the pump can be severely damaged from habitual attempts to pump more water from the tank than the piping allows to flow into the pump. Such damage can be especially severe when the pump is operating at relatively high pressures supplying long attack lines. Cavitation can occur while operating from a poor hydrant system, but it most often occurs during drafting operations. *Simply put, you can only discharge the amount of water that has been taken in on the intake side of the pump.*

Successfully operating a fire department pumper from draft is one of the most challenging tasks pump operators face. It requires a thorough knowledge of the principles involved in drafting as well as a familiarity with the apparatus. All driver/operators should master drafting even if it is unlikely that they will be required to do so in an actual emergency operation.

Selecting the Drafting Site
The first consideration in establishing a successful drafting operation is selecting the site. If the purpose of the

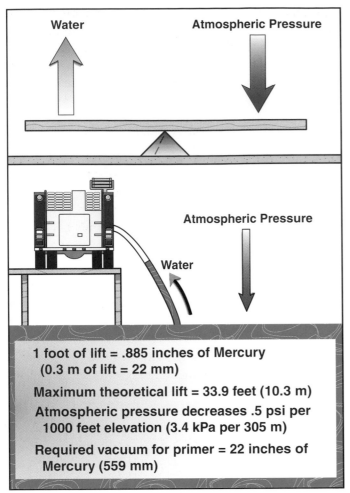

Figure 11.37a The principles of lift.

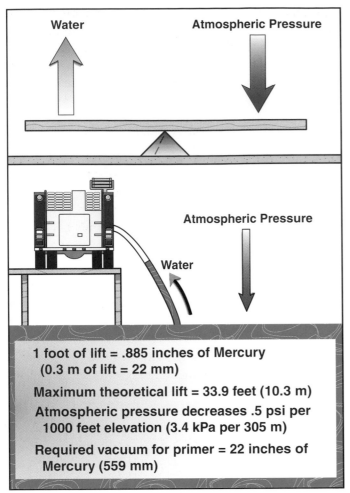

Ratio of Lift, Vacuum, and Air Pressure		
33.9' (10.3 m)	30.00" (762 mm)	14.7 psi (101 kPa)
32.2' (9.8)	29.04" (738)	14 psi (96.5)
29.9' (9.1)	26.96" (685)	13 psi (89.6)
27.6' (8.4)	24.88" (632)	12 psi (82.7)
25.3' (7.7)	22.8" (579)	11 psi (75.8)
23.0' (7.0)	20.72" (526)	10 psi (69.0)
20.7' (6.3)	18.64" (473)	9 psi (62.1)
18.4' (5.6)	16.56" (420)	8 psi (55.2)
16.1' (4.9)	14.48" (367)	7 psi (48.3)
13.8' (4.2)	12.4" (314)	6 psi (41.4)
11.5' (3.5)	10.32" (262)	5 psi (34.5)
9.2' (2.8)	8.24" (209)	4 psi (27.6)
6.9' (2.10)	6.12" (155)	3 psi (20.7)
4.6' (1.4)	4.08" (104)	2 psi (13.8)
2.3' (.7)	2.04" (52)	1 psi (6.90)
Lift Water Feet (meters)	Vacuum Mercury Inches (mm)	Air Pressure psi (kPa)

Figure 11.37b This chart shows the ratios between lift, vacuum in mercury, and air pressure.

operation is to supply water to a fireground directly or through a relay, there may not be much choice as to where to set up. If the draft is being established to supply water tenders (tankers) for a shuttle operation, there may be several choices. The choice is dictated by the following factors:

• Amount of water

• Type of water

• Accessibility of water

Amount of Water Available

The most important factor in the choice of the draft site is the amount of water available. If the draft is from a static body of water, such as a pool or lake, the size of the body becomes significant. A backyard swimming pool holding only 12,000 gallons (48 000 L) of water will not last long if large streams are in use. A small pond may not have a large capacity, but the rate of replenishment may make up for lack of volume. Even a rather small stream can provide a good supply if the water is moving rapidly.

In order for a pumper to approach its rated capacity using a traditional strainer, there should be a minimum

of 24 inches (610 mm) of water over the strainer. It is also desirable to have 24 inches (610 mm) of water all around the strainer (Figure 11.38). This helps to ensure maximum capacity and to avoid drawing foreign objects, such as sand and gravel, into the pump. If there are not 24 inches (610 mm) above the strainer, the rapid movement of the water into the intake strainer causes a whirlpool to form. In extreme cases, this whirlpool can allow air to get into the intake hose and cause the pump to lose its prime. A wooden board, capped plastic bottle, or beach ball may be placed above the strainer to prevent a whirlpool effect (Figure 11.39).

To draft from a swiftly moving shallow stream, a dam can be constructed from available material to raise the level, the bottom can be dug out to form a pool of greater depth, or a combination of the two can be used (Figure 11.40). A better solution is the use of a "floating" strainer. With this type of strainer, the end of the intake hose floats on the surface, and the water is drawn into the intake hose through a series of holes on the bottom of the

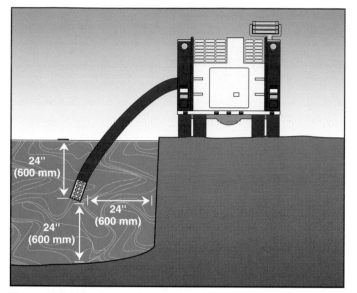

Figure 11.38 There should be a minimum of 24 inches (600 mm) all around the strainer.

Figure 11.39 A beach ball, capped plastic bottle, or similar object may be floated above the strainer to prevent a whirlpool from forming.

Figure 11.40 In some cases, it may be possible to dam a small stream in order to make it more suitable for drafting.

strainer (Figure 11.41). Because the water has to enter the hose from the bottom and come around the float, there is no way that a whirlpool can develop. In order for a floating strainer to work properly, it must float free in the water and not be constrained by the rigidity of the intake hose. A disadvantage of the floating strainer is that it is limited to taking water in on only one side. This may limit the ability of the pumper to reach its rated discharge capacity.

When drafting from a portable water tank or swimming pool, a low-level strainer may be used. Low-level strainers are designed to sit directly on the bottom of the tank or pool (Figure 11.42). They are capable of allowing the water to be drafted down to a depth of about 2 inches (50 mm).

An additional consideration in jurisdictions that use ocean water as a static water supply source is the tidal movement. What may be an acceptable drafting location at one time of the day may not be usable when the tide goes out. This could even occur during an extended pumping operation. Concessions to the water movement must be made in these cases.

Figure 11.41 Floating strainers may be used in shallow water.

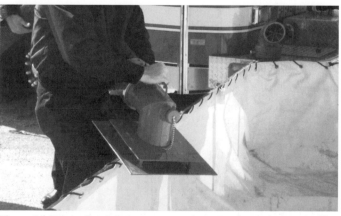

Figure 11.42 Low-level strainers are most commonly used when drafting from portable water tanks.

The temperature of the water may also be a consideration when choosing the type of water to draft. Water that is below 35°F (2°C) or above 90°F (32°C) may adversely impact the pump's ability to reach capacity. Water that is close to freezing is commonly encountered in cold weather climates. In many situations, using this cold water is unavoidable, although water several feet below the surface may be 5°F to 10°F (3°C to 5°c) warmer. Heated water is most commonly found near the discharges of power plant cooling towers. Avoid using this source if possible.

Type of Water

In an emergency, any type of water can be used for fire fighting; however, pumping nonpotable (untreated) water can be harmful to the pump. In some areas, most of the available water is salt water. Salt water can cause corrosion and a gradual deterioration of the pump if it is not thoroughly flushed after each use. Sulfur water is common in the vicinity of coal mines. Other chemicals may be present near industrial plants. Flush the pump with fresh water each time it is used to pump nonpotable water (Figure 11.43). When possible, avoid filling the onboard water tank with nonpotable water.

Drafting operations can be seriously affected by foreign material and debris clogging the strainer. An open body of water may contain a high concentration of algae or other marine vegetation that blocks the strainer and prevents effective operation. Leaves or other debris floating on the water can also block the strainer. Using a floating dock strainer or covering the end of the strainer with a bucket helps prevent this. However, if high flow rates are attempted, the water source capacity may be limited by unavoidable blockages.

The most common type of contamination, and possibly the most damaging, is dirty or sandy water. If dirt particles are too small to be caught in the strainer, they pass into the pump where they can cause serious deterioration. As the water passes into the eye of the impeller, the sand acts as an abrasive in the area between the clearance rings and the hub of the impeller. Such abrasion quickly increases the spacing to an intolerable amount, causing increased slippage from the discharge back into the intake and reducing the capacity of the pump.

As the dirty water passes through the pump, it is forced into the packing by the discharge pressure. When this occurs, the packing becomes contaminated by the dirt and can no longer be adjusted to form a good seal. This causes air leaks at the intake of the pump and reduces its ability to draft.

Accessibility

Accessibility to the water source is another important consideration in selecting a drafting site. Because drafting is accomplished by evacuating the air from the pump and allowing the atmospheric pressure to push water into it, a maximum of 14.7 psi (100 kPa) is available. This 14.7 psi (100 kPa) has to overcome elevation pressure and friction loss in the intake hose. As the amount of lift increases, the following occur:

- Elevation pressure increases
- Less friction loss can be overcome
- Capacity of the pump is decreased

All fire pumps meeting NFPA and Underwriter's Laboratories requirements are rated to pump their capacity at 10 feet (3 m) of lift (Figure 11.44). If the lift is less, the capacity is higher; if the lift is greater, the capacity de-

Figure 11.43 The pump should be thoroughly flushed with clear water after drafting.

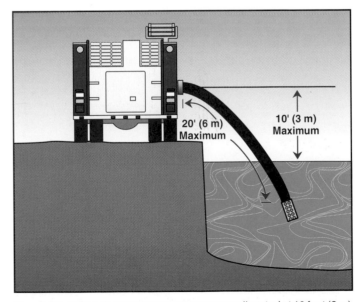

Figure 11.44 Fire department pumpers are usually rated at 10 feet (3 m) of lift.

20' (6 m) Maximum

10' (3 m) Maximum

creases. While a pumper in good condition can lift water approximately 23 or 24 feet (7 m or 7.3 m), all available atmospheric pressure is required to overcome the lift. As a result, the tolerable amount of friction loss is so low that the capacity is too small to be of practical value. For effective operation, the maximum lift considered reasonable for most fire department pumpers is about 20 feet (6 m). At 20 feet (6 m) of lift, the amount of water that can be supplied is only about 60 percent of the rated capacity of the pump. In selecting a drafting site, it is important to keep the lift as low as possible. It would be better to lay out an extra 100 feet (30 m) of supply line to set up the draft in a location where the lift will be lower and more water can be supplied.

Other considerations in selecting a drafting site are the stability of the ground, the time of the year, the convenience of connecting hoselines, and the safety of the operator. A well-trained operator will be familiar with the potential drafting sites in the area as well as the limitations and special considerations needed in using each site. It is a good practice to do some advance preparation at the most strategic drafting sites to improve their accessibility and usability before they are actually needed.

For more detailed information on static water supply sources, see Chapter 12, Static Water Supply Sources.

Connecting to the Pump and Preliminary Actions

Once the desired drafting position has been chosen, the driver/operator should move the apparatus toward that position. In some cases, it is possible to place the apparatus directly at the drafting location and deploy the intake hose from that spot. If this is the case, set the parking brake and place the wheel chocks once the apparatus is parked. If the pumper is close to a road, leave the warning lights operating. Because there will be some delay before getting water into the pump, it should not be engaged until all connections are made and it is ready to put into operation.

In other cases, the area around the actual drafting spot may be limited. This requires the driver/operator to temporarily park the apparatus short of the final drafting spot. The hard intake hose and strainer arrangement are connected to the apparatus at this location. Once connected, the driver/operator slowly eases the apparatus into the final drafting spot while other firefighters carry the hose and strainer and put them into the desired position (Figure 11.45).

Before connecting sections of hard intake hose, check the gaskets to make sure that they are in place and that no dirt or gravel has accumulated inside the coupling (Figure 11.46). The strainer and the necessary sections of hard intake hose must be coupled and made airtight. The key to doing this is getting each section of hose in line with the other before the coupling is turned. If the gaskets are in good condition and the coupling is connected properly, it should be possible to get an airtight connection when the couplings are handtight. If necessary, use a rubber mallet to make the connections tighter (Figure 11.47). It is important to have enough help to connect the hard intake hose without putting it on the ground because dirt may get into the coupling. Fasten a rope to the end of the strainer (Figure 11.48). This aids in handling the hard intake hose after it is connected and in positioning the strainer properly.

Figure 11.45 Once everything is connected, the pumper is eased into its final draft position.

Figure 11.46 Make sure that the gaskets are in place and in good condition.

Figure 11.47 It may be necessary to hit the coupling lugs with a rubber mallet to ensure the connection is airtight.

Figure 11.48 A rope may be attached to the end of the hose to assist with positioning the strainer.

Once the sections of hard intake hose are connected with the strainer and rope attached, it is usually easier to connect the hose to the intake fittings by first putting the strainer in the water and pulling the apparatus into position (Figure 11.49). In other cases, as previously described, the hard intake hose may be connected to the truck and then both moved into position at the same time. If a butterfly-type intake valve is normally connected to the pump intake, it should be removed before connecting the intake hose to the apparatus. This is necessary because the butterfly valve obstructs the waterway which reduces the drafting capacity.

Improper positioning of the hard intake hose can result in the formation of an air pocket that can prevent effective drafting. For example, routing the hard intake hose over a fence or railing that is higher than the level of the pump intake may leave a high point that traps air in the hose (Figure 11.50).

If a barrel strainer is used, the rope that was tied to it can be used to suspend the strainer above the bottom by tying it to the pumper or to a tree or other natural object. If the bottom slopes away steeply from the edge, it may be possible to put a roof ladder in the water and rest the intake hose on it (Figure 11.51). Under extremely adverse conditions, a shovel or some other piece of flat metal can be placed on the bottom to protect the strainer.

Once the apparatus, hard intake hose, and strainer are in place, the driver/operator follows the procedures described earlier in this chapter for making the fire pump operational.

Priming the Pump and Beginning Operation

Once the fire pump has been made operational, the draft operation is started by priming the pump. If operating a two-stage pump, the transfer valve should be in the VOLUME (PARALLEL) position during priming. This is necessary because when the pump is in the PRESSURE (SERIES) position, air may be trapped inside the pump, and it will be more difficult to remove. Before you begin priming, make sure that all drains and valves are closed and that all unused intake and discharge openings are capped to make the pump as airtight as possible.

If the primer is a positive displacement pump that is driven by the transfer case, the engine rpm should be set according to the manufacturer's instructions. Most priming pumps are intended to work best when the engine is set between 1,000 and 1,200 rpm. If the priming pump is driven by an electric motor, the exact rpm is not critical; however, 1,000 to 1,200 rpm should be sufficient to keep the alternator charging and prevent loss of prime once the pump fills with water. If the apparatus uses a vacuum-type primer, the engine rpm should be kept as low as possible without causing the engine to stall.

Figure 11.49 In some cases, the intake hose is connected to the apparatus after the other end of the hose has been placed in water.

Figure 11.50 If the suction spot has a high point in it, drafting may not be possible. *Courtesy of Mike Wilbur.*

Figure 11.51 A ground ladder may be used to support the intake hose during drafting.

After the pump has been made airtight and the engine rpm set, operate the primer control (Figure 11.52). As the air is evacuated from the pump, the master intake gauge registers a vacuum reading. This reading should be 1 inch (25 mm) for each 1 foot (0.3 m) of lift. This vacuum, which is measured from the surface of the water to the eye of the impeller of the pump, is required to overcome the elevation pressure and cause the water to enter the pump.

Figure 11.52 Once the hose is in position, the primer control should be operated.

As the primer operates, the vacuum reading increases and water is forced into the hard intake hose. As the hose fills with water, the weight of the water causes the hard intake hose to drop. When the body of the pump fills with water, the primer discharges water onto the ground under the apparatus. At first, there will not be a steady stream as the water entering the pump mixes with the remaining air that is being removed.

The priming action should not be stopped until all the air has been removed and the primer is discharging a steady stream of water. As the pump fills with water, a pressure indication shows on the master discharge gauge.

The entire priming action typically requires 10 to 15 seconds from start to finish, but it should not take more than 30 seconds (45 seconds in pumps larger than 1,250 gym [5 000 L/min]) to accomplish. If water has not been obtained in 30 seconds, stop priming and check to find out what the problem is. The most common cause of inability to prime is an air leak that prevents the primer from developing enough vacuum to successfully draft water.

Open drains and valves are the most common source of air leaks. The drains should all be closed during priming and drafting operations. If the pump is equipped with a circulator valve or booster cooling valve, air can leak if

the valve is not turned off completely. An intake relief valve is another frequent offender. Older intake relief valves may be equipped with a shutoff valve. Make sure this valve is closed as well. Newer apparatus have a threaded intake relief discharge that can be capped. Cap the outlet before attempting to prime the pump again (Figure 11.53).

The next area to check for an air leak is the couplings. Make sure that the couplings on the hard intake hose are airtight. This can be done by hitting each of them a few times with a rubber mallet.

If the previous measures are taken and priming is still not possible, examine the following possible causes for inability to prime:

- Insufficient fluid in the priming reservoir

- Engine speed (rpm) is too low

- Lift is too high

- A high point in the hard intake hose creating an air pocket

After the pump has been successfully primed, increase the setting of the throttle before attempting to open any of the discharges. Increasing the setting of the throttle is needed to boost the pressure to somewhere between 50 and 100 psi (350 kPa and 700 kPa). Open the desired discharge valve slowly while observing the discharge pressure. If the presssure drops below 50 psi (350 kPa), pause for a moment to allow it to stabilize before opening the valve further. If the valve is opened too quickly, air may enter the pump and cause it to lose its prime.

If the pressure continues to drop, momentarily operating the primer may eliminate the air still trapped in the pump and restore the pressure to the original value. Do not try to set the discharge pressure to the desired value until water is flowing. If the hoselines are not ready for

Figure 11.53 Capping the intake relief discharge may help the priming process.

water, another discharge or a booster line can be opened, allowing water to discharge back into the source.

Constant movement of water through the pump is necessary to prevent overheating, but there is a more important reason to maintain flow at draft. The vacuum to initially prime the pump comes from the primer, but once the pump begins to operate, the primer is no longer used. The vacuum is now maintained by the movement of water through the pump. When no water is being discharged, even the smallest air leak may result in a loss of vacuum. This can be prevented by using the same methods for preventing overheating of the pump discussed earlier in this chapter. Perhaps the most common method used in this circumstance is to discharge a booster line back into the static water supply source during the entire time the pump is operating (Figure 11.54). Once the desired pressure has been established and hoselines are in service, set the relief valve or pressure governor.

Operating the Pump from Draft

Operating from draft is the most demanding type of pump operation, both from the standpoint of the apparatus and the operator. It demands a careful monitoring of the gauges associated with the motor as well as those associated with the pump. The engine temperature gauge must be kept at the normal operating temperature by use of the auxiliary cooler. Only under the most critical conditions should the radiator fill valve be used while operating at draft. The reason for this is because it is impossible to be certain of the quality of the water being pumped, and damage to the engine could result. Other engine gauges are not as critical, but any deviation from normal engine temperature is a signal that another pumper should be used for drafting.

The problems that can occur while operating from draft fall into one of three categories:

- Air leak on the intake side of the pump

- Whirlpool allowing air to enter the pump

- Air leakage due to defective packing in the pump

Air leaks on the intake side of the pump are the most common problems. If the discharge pressure gauge begins to fluctuate with a corresponding loss of vacuum on the intake gauge, air is probably coming into the pump along with the water. The first place to check is the intake hose. Even if the connections were tight enough initially, vibration may have caused them to leak. If the water tank is empty, the tank-to-pump line may be a source of leakage. In addition, check all drains and intake openings for leakage.

Figure 11.54 Discharge a stream back into the water source at all times to avoid a loss of prime.

If there is not enough water over the strainer, a whirlpool may be allowing air to enter the pump. This can be corrected by placing a beach ball or other floating object above the strainer. A floating dock strainer can also be used in shallow water where whirlpooling is a problem.

Defective packing in the pump may cause air leakage. If there is an excessive amount of water leaking from the packing, that is, if it is leaking a steady stream instead of dripping, the packing is probably the cause. If this is the case, nothing can be done about it while the pump is operating. If the problem is severe enough to cause the pump to lose its prime, the pumper will have to be replaced on the fireground and put out of service for repair.

While the pump is operating, a gradual increase in the vacuum reading may be noted with no change in the flow rate. This is an indication that a blockage is developing. Blockage most often occurs after the pump has been operating satisfactorily for sometime at a high rate of discharge, which creates a maximum velocity of the water that is entering the strainer. In extreme cases, the pump may go into cavitation, resulting in a fluctuation and a gradual decrease of the discharge pressure. An immediate cure is to decrease the engine rpm until the pressure drops. A drop in pressure indicates that the flow has decreased below the point of cavitation.

Because the reduced pressure may not be adequate to maintain the desired flow, try to clear the blockage and restore the operation to normal. The most common place for blockage to develop is at the strainer. If leaves or debris are present in the water being pumped, an accumulation against the strainer may be partially blocking the flow. This can be cleared by physically cleaning it if the water is shallow enough to allow access to the strainer (Figure 10.55). An alternate method is to charge the booster line under pressure and attempt to wash away the material blocking the strainer (Figure 10.56).

Figure 11.55 A pike pole may be used to clear debris from the strainer.

Figure 11.56 If the strainer is not too deep, the pressure from a hose stream may cause enough turbulence to clear the strainer.

Blockage may also occur against the strainer that is located in the intake of the pump. This blockage cannot be cleared without interrupting the water flow because it is necessary to remove the intake hose to clear this strainer.

Blockage can occur for other reasons. Because the intake hose is operating under a vacuum, the inner liner can become detached from the hose itself and collapse, thus blocking a portion of the waterway. This may not be immediately apparent as the vacuum is removed when the hose is disconnected and the liner may return to the proper position.

Exceeding the capacity of the pump also leads to cavitation. This condition is also accompanied by a high reading of the vacuum gauge. Cavitation can occur at a flow well below the rated capacity of the pump if there is a high lift involved. The actual capacity of the pump in a given situation is determined by the size and length of the intake hose as well as the lift and the strainer design.

Shutting Down the Operation

When shutting down after a drafting operation, slowly decrease the engine speed to idle, take the pump out of gear, and allow the pump to drain. Stabilize the engine temperature in the same manner as described in Chapter 4 for shutting down the apparatus. After the pump is drained and the connections have been removed, operate the mechanical primer for a few seconds until primer oil or fluid comes out of the discharge from the priming pump. This lubricates the parts of the primer and helps to preserve them in good condition. Unless the pump has been pumping very clean uncontaminated water, it should be thoroughly flushed when a supply of fresh water is available.

See Flowchart 11.3 and Table 11.1 located at the end of this chapter for additional direction on operating the pump from a draft and for troubleshooting during this operation.

Sprinkler and Standpipe Support

Fixed fire protection systems are a building's first line of defense against uncontrolled fire within that occupancy. One of the most important functions with which the driver/operator is charged is providing adequate support to these systems when they are being used to suppress a fire. In order to provide adequate support to fixed fire suppression systems, the driver/operator must understand the information contained in the following sections.

Supporting Automatic Sprinkler Systems

Properly installed and maintained fire sprinklers have a long history of providing reliable, automatic protection to all types of occupancies. The water supply for sprinkler systems is designed to supply only a fraction of the total number of sprinklers actually installed on the system. If a large fire should occur, or if a pipe breaks, the sprinkler system will need an outside source of water and pressure in order to do its job effectively.

Fire department pre-incident planning activities should identify all occupancies in the community that have automatic sprinkler systems. Included in the pre-incident planning information should be the location of the fire department connection (FDC), the nearest hydrant or water supply source to the FDC, and any special requirements for pump pressure required to supply the system.

Although individual fire department standard operating procedures (SOPs) may vary, one procedure must be the same for all departments: Upon arrival at a sprinklered property, preparations should be immediately made to supply the FDC. Fire department connections consist of a siamese with at least two 2½-inch (65 mm) female connections or one large diameter sexless connection that is connected to a clappered inlet (Figure 11.57). As water flows into the system, it first passes through a check valve. This valve prevents water flow from the sprinkler system back into the fire department connection (Figure 11.58). However, it does allow water from the FDC to flow into the sprinkler system.

Depending on local SOPs, the first-arriving or other first-alarm engine company should locate the fire department connection and the nearest suitable hydrant. If there is any indication of an actual fire, such as smoke or the ringing of a sprinkler alarm, a minimum of two 2½-inch (65 mm) hoselines or one 3-inch (77 mm) hoseline should be connected to the fire department connection. Additionally, the engine company should lay supply lines to the hydrant and make all the appropriate connections (Figure 11.59). It is a general rule of thumb that one 1,000 gpm (4 000 L/min) rated pumper should supply the FDC for every 50 sprinklers that are estimated to be flowing.

The interior attack crews should locate the fire and determine whether charging of the sprinkler system is necessary. It may be argued that it is prudent to charge the system immediately upon arrival of the fire companies. Obviously, this action is appropriate if a fire is evident. However, in some situations the sprinklers may have extinguished the fire, or the system activation may be a malfunction. In most cases, it is desirable to confirm the presence of fire before pumping into the system.

If the system needs to be charged, the driver/operator should slowly develop the amount of pressure needed to supply the system. Multistage pumps should be in the VOLUME (PARALLEL) position. Some occupancies have the suggested discharge pressure printed on a plate at the FDC. In other cases, a recommended pressure may be contained in pre-incident planning information. If none of this information is available, the general rule of thumb is to discharge 150 psi (1 050 kPa) into the FDC. From outside the building, it is usually difficult to determine how many sprinklers are operating in the fire area and the amount of friction loss in the system. If it becomes obvious that the fire is spreading, additional lines should be connected to the fire department connection and charged. A rapid size-up may be hindered by low visibility because of the smoke being cooled by the sprinklers. This cooling causes the smoke to lose its normal buoyancy.

When possible, a firefighter or crew familiar with the building should immediately check the control valves, if they are accessible, to ensure that they are open. Closed

Figure 11.57 Fire protection systems receive pumping support through the fire department connection.

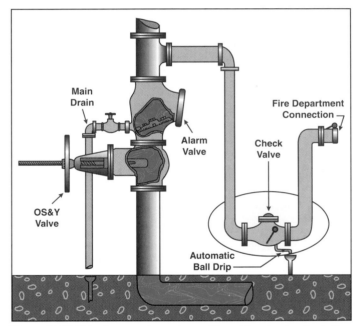

Figure 11.58 This diagram shows the position of the check valve in the fire department connection arrangement.

Figure 11.59 The pumper should be hooked to a water supply source and the fire department connection.

Figure 11.60 A firefighter, or crew of firefighters, should visually ensure proper system operation in the system control/riser room.

valves should be opened except in the case where it is known that the building or area has been undergoing construction or renovation affecting the sprinklers. Opening valves under these circumstances would cause a severe loss of water to the system. The firefighter or crew assigned to this task should carry a flashlight and a portable radio. If the sprinkler system is supplied by a fixed fire pump on the property, the firefighter should also ensure that the pump is running (Figure 11.60). Control valves are frequently located in the pump room so that both functions can be performed by the same firefighter or crew.

Supporting Standpipe Systems

Standpipes are used to speed fire attack in multistory or single-story buildings with large floor areas. Fire attack teams attach lines to the 2½- or 1½-inch (65 mm or 38 mm) connections provided on each floor (Figure 11.61). House or standpipe lines should not be used by fire personnel unless a hose testing program similar to the one used to test all fire department hose is implemented (Figure 11.62). These lines are often unlined single-jacket

cotton hose that have not been tested or removed since they were installed. Fire attack crews should bring attack lines with them when they enter a property protected with standpipes to ensure an attack without the fear of house lines bursting (Figure 11.63).

Standpipes may be wet or dry, depending upon owner preference or local code requirements. Wet-pipe systems contain water under pressure and are ready to be used as soon as lines can be attached to the outlet. Dry-pipe systems must be supplied with water from a pumper that attaches to a standpipe FDC outside the building. Standpipe FDCs should be clearly identified to prevent confusion between sprinkler and standpipe connections (Figure 11.64). Wet standpipes also have an FDC that should be supplied with water under pressure to supplement the system's water supply.

Figure 11.63 Firefighters should carry all their own equipment, including hose, into the building.

Figure 11.61 Firefighters should connect their hose bundles to the discharges that are most commonly found in the stairwell.

Figure 11.62 Fire department personnel should not use house or standpipe lines.

Figure 11.64 Fire department connections should be clearly marked as to their intended use.

The hose lays previously described for sprinkler system FDCs apply here as well. Pump discharge pressure depends on the following:

- Friction loss (25 psi [150 kPa]) in the standpipe
- Friction loss in the hose lay from the pumper to the fire department connection
- Friction loss in the hose on the fire floor
- Nozzle pressure for the type of nozzle employed
- Elevation pressure due to the height of the building

Generally, the friction loss in the standpipe is small unless the flows are large, such as when two 2½-inch (65 mm) lines are being supplied from the same riser. Allowance must also be made for pressure losses in the fire department connection and pipe bends. The driver/operator should also be familiar with the hose and nozzles that are used in high-rise packs that are carried to the fire floor by firefighters. This allows for more accurate hydraulic calculations. Refer to Chapter 8 for details on how to make these hydraulic calculations.

Add approximately 5 psi (35 kPa) to the desired nozzle pressure for each floor above the standpipe connection that will have operating fire streams. Also include friction loss for the attack line(s), standpipe piping, and layout from the pumper to the standpipe connection. These are lengthy calculations that cannot be made each time a standpipe line is needed. The fire department should have a planned pump discharge pressure or develop a rule of thumb for each building in the area equipped with a standpipe. Pump discharge pressures in excess of 200 psi (1 400 kPa) are not encouraged unless the standpipe system has been designed to withstand higher pressures.

When a standpipe system is known to be equipped with pressure-reducing valves, the elevation pressure used must be based on the *total height* of the standpipe or the zone being used (Figure 11.65). This is because the pressure-reducing valve generally acts to reduce whatever pressure is presented to it. If the pressure at a particular pressure-reducing valve is less than that for which the valve was adjusted, the result will be inadequate pressure for the hoselines.

Frequently, vandals or curious individuals may open the hose valves in dry standpipes and leave them in an open position. When the standpipe is charged, water will discharge on levels below the fire floor, and it will be necessary for a firefighter to go down the stairwell to close the valves.

Often when dry standpipes are charged, there will be a time delay before the delivery of water to the hose valve because of the amount of air that must be expelled from the system. Air can also be trapped and become compressed so that when a valve is opened, the air under high pressure is released.

A number of simple operations can be used by the fire department to overcome various standpipe impairments. For example, if a fire department connection has a frozen swivel, a double male can be used with a double female (Figure 11.66). If a fire department connection is totally unusable because of vandalism, the standpipe riser can be charged at the first-floor level by attaching a double female to a hose valve at the first-floor level. If an individual hose valve on an upper floor is found inoperative, a valve on the next floor down can be used. To overcome a problem in a single-riser building where the standpipe is totally unserviceable, it may be necessary to hoist a line up the outside of the building. Often, this can be accomplished by unrolling hose down the side of the building from every second or third floor rather than attempting to hoist the line several floors. Standpipes in adjacent buildings can also be used to protect exposures. As a last resort, supply hose can be laid up the interior stairwell to take the place of the standpipe.

For more information on operations at occupancies containing automatic sprinkler and standpipe systems, consult the IFSTA **Private Fire Protection and Detection** manual.

Figure 11.65 Pressure-reducing valves are common in tall high-rises.

Figure 11.66 A double male and double female may be used to make connections to a frozen FDC swivel go more easily.

Table 11.1
TROUBLESHOOTING DURING PUMPING OPERATIONS

Problems Common to all Types of Operation			
Problem	**Symptoms**	**Probable Cause**	**Possible Solutions**
Unable to get a reading on the pressure gauge when the pump is put in service.	Green light indicating that the pump shift transfer is complete is not illuminated.	Pump drive system is not fully engaged.	Check the position of the shift transfer control. If it is in the proper position, release the transfer control, allow the gears to turn, then operate it again. Automatic transmission: repeat pump shifting procedure to ensure power transfer to pump operation.
	Green light is on, no mph (km/h) reading registers on the speedometer.	Vehicle clutch is not engaged.	Check the remote clutch control on the pump operator's panel.
		Road transmission is not in the proper gear. Automatic transmission selector is in wrong position.	Check the shift lever: lock it in the proper position for pump operation.
	Speedometer reading is normal for pump operation. All indications are correct and rpm reading is as specified.	There is no water in the pump.	Check the water supply. Ensure that all applicable valves are open. Primer pump may need to be operated to eliminate air in the main pump.
		Gauge is defective. Pressure may be there but not reading on the master pressure gauge.	Check the shutoff (snubber) dampening valve associated with the gauge. Open the gate valve to one of the capped discharge outlets and look for pressure reading on the individual line gauge. Open an uncapped discharge outlet from the pump to see if water is discharged under pressure.
Pump will not develop sufficient pressure.	The rpm reading on the tachometer is normal when compared with the UL plate.	Two-stage pump: • Pump transfer valve is in the wrong position.	The transfer valve should be in the SERIES or PRESSURE position anytime more than 200 psi (1 400 kPa) is needed.
		• Swing check valve may be leaking if the pump is in the SERIES position.	Set the discharge pressure to 50 psi (350 kPa), change from PARALLEL to SERIES, and listen for the metallic sound of the valve operating. If it is blocked, remove the strainer from the large intake and attempt to clear the valve seat of any debris.
		• The transfer valve has not completed its travel and is only partially operated.	Check all mechanical and electrical indications as well as observing pressure gauge readings as the valve is operated. Use the manual override controls to complete the operation of the valve. It may be possible to assist the action of the valve mechanically if the power transfer mechanism is faulty.
		Automatic transmission: transmission not staying in pumping gear lockup and is downshifting as load increases.	Remove pump from service and repair or adjust so apparatus will operate in correct pumping gear.

Problems Common to all Types of Operation			
Problem	**Symptoms**	**Probable Cause**	**Possible Solutions**
Pump will not develop sufficient pressure.	The rpm reading on the tachometer is normal when compared with the UL plate.	Wear on the clearance rings inside the pump may cause excessive slippage.	Take the pumper out of service until it can be repaired.
	Relief valve is operating and the indicator light is on.	Relief valve pressure adjustment is set too low.	Increase the operating pressure of the relief valve by turning the adjustment control clockwise until it closes.
	The indicator light shows that the relief valve is closed.	The relief valve may be stuck open or may not be properly seated, allowing water to bypass back to the intake.	Turn the relief valve operating control to the OFF position.
			Increase the setting of the adjustment control to maximum clockwise position.
			Exercise the valve by rapidly turning the control valve on and off or alternately increasing and decreasing the pressure adjustment control to cause the valve to operate.
	Engine rpm cannot be raised to the value required as determined by the UL plate, even at full throttle. Tachometer reading is low, pressure gauge reading is too low.	Flow requirements may be exceeding the capacity of the pump.	The capacity of the pump in the SERIES or PRESSURE position is limited to 50 percent of its rated capacity at 250 psi (1 700 kPa) and 70 percent at 200 psi (1 400 kPa) net pump pressure.
		Throttle linkage may have slipped or be stuck.	Check the action of the linkage. It may be possible to override the action of the throttle by using the accelerator or hand throttle in the cab or by manually operating the linkage at the carburetor.
		Severe engine overheating can reduce the power available to drive the pump.	Check the engine temperature gauge. Adjust the auxiliary cooling valve to maintain the proper operating temperature.
			Check the level of the coolant in the radiator. If it is too low and the water that is being pumped is clean and pure, open the radiator fill valve to bring the coolant up to the proper level.
		Reduced engine power.	If additional pressure is essential, use another pumper. Take the unit out of service until it can be repaired.
The pump is unable to supply its rated capacity.	The rpm reading on the tachometer is normal when compared to the UL plate.	Two-stage pump: • Transfer valve is in the wrong position.	The transfer valve should be in the PARALLEL or VOLUME position anytime 50 percent or more of the capacity of the pump is needed.
		• The swing check valve is not opening completely.	Remove the strainer from the large intake opening on each side of the pump. Make sure that the valve swings freely by inserting a rod and pushing on the face of the valve.

Problems Common to all Types of Operation			
Problem	**Symptoms**	**Probable Cause**	**Possible Solutions**
The pump is unable to supply its rated capacity.	The rpm reading on the tachometer is normal when compared to the UL plate.	Two-stage pump: • The swing check valve is not opening completely.	Remove the strainer from the large intake opening on each side of the pump. Make sure that the valve swings freely by inserting a rod and pushing on the face of valve.
	The intake gauge registers 0 or has a positive pressure indicated.	Blockage in the waterways of the pump. An object lodged inside the impeller can reduce the capacity.	Thoroughly back-flush the pump by connecting a supply line to the highest discharge outlet and opening the large intake fittings.
		Wear in the pump, usually the clearance rings or relief valve, allowing slippage from the discharge back to the intake.	Remove the pumper from service until it can be repaired.
	The intake compound gauge is registering a high vacuum, and the discharge pressure gauge is fluctuating (cavitation).	Blockage of the strainer at the intake fitting of the pump.	Disconnect the intake line and clean any accumulated debris from the strainer.
		Inadequate water supply or supply lines.	Connect an additional supply line to the intake of the pump.
			Reduce the amount of discharge lines being supplied.
			Lower the discharge pressure to reduce the amount of water flowing through the lines.
	Unable to develop enough engine rpm at full throttle to supply the rated capacity.	See items listed under "Pump will not develop sufficient pressure with low engine rpm."	
Pump overheating while in operation.	Pump overheating warning light or physical observation.	Inadequate flow through the pump while operating under pressure.	Open the booster cooling valve or set the circulator valve to TANK or SPILL position as appropriate.
			Open the tank fill valve if it is connected to the discharge side of the pump.
			Use the booster line to maintain a minimum flow of water while pumping.
		Excessive throttle on relief valve-equipped pumps.	Reduce engine rpm.
Relief valve is inoperative or slow acting.	Pressure surges are excessive when individual hoselines are shut down.	Strainer in the pressure line to the pilot valve is dirty.	Open the flush line while pumping clean water to back-flush the in-line strainer.
			Remove the strainer element and wash in clear water.
		Relief valve is corroded or dirty.	If the relief valve is equipped with a shut-off, set the discharge pressure on the pump to 150 psi (1 050 kPa), set the relief valve adjustment control to minimum and alternately turn the valve off and on for 60 seconds. If no control valve is provided, set the pump pressure to 150 psi (1 050 psi) and cause the relief valve to operate by cranking the adjustment control in and out rapidly.

Problems Common to all Types of Operation			
Problem	**Symptoms**	**Probable Cause**	**Possible Solutions**
Relief valve is inoperative or slow acting.	Pressure surges are excessive when individual hoselines are shut down.	Defective relief valve	Dismantle the valve and clean all working parts, replacing all defective items as needed.
Operating from the Tank			
Unable to establish an adequate operating pressure or a loss of pressure occurs when the first discharge valve is opened.	Pressure increases with the engine rpm up to a point, then holds steady or fluctuates.	Some air is trapped in the pump as it fills with water.	Operate the primer until all of the air is removed from the pump. Transfer the pump from SERIES to PARALLEL and back several times while flowing water.
			If only small lines are being supplied, opening the pump to tank valve and increasing the velocity of the water through the pump for a short time may remove the rest of the air.
		Automatic transmission: • Intermittent transmission slippage.	Adjust transmission.
		• Low transmission fluid level.	Increase fluid level.
		• Transmission not remaining in correct pumping gear.	Adjust transmission.
Fluctuation of the pressure gauge and a reduction of discharge pressure when additional lines are put in service.	High vacuum reading on the intake compound gauge.	The tank-to-pump valve may not have been fully opened or has vibrated to a partially CLOSED position.	Check the tank-to-pump valve and lock it in the OPEN position if the valve has a locking mechanism.
		Tank-to-pump piping may be too small to supply the amount of water required by the hoselines in service.	Shut down one of the hoselines if fire-ground conditions permit.
			Reduce the discharge pressure until fluctuations stop and pressure gauge begins to drop.
While pumping, the discharge pressure drops to a very low value and water supply is interrupted.	Compound gauge on the intake reads 0 or fluctuates. Engine speed increases.	An air leak in the pump.	Check all caps and valves on the intake side of the pump.
	Water gauge reads empty.	The water supply from the tank is exhausted or nearly so.	Reduce the pressure until the gauge becomes steady and flow resumes.
			Arrange for another water supply as soon as possible.
Operating from Draft			
The suction line collapses when the discharge valve to a hoseline is opened.	The intake pressure drops to less than 0. Discharge pressure also drops.	Kinks in suction lines can cause excessive friction loss when the water begins to move.	Rearrange the suction line to eliminate any kinks or restrictions.

Operating from a Hydrant

Problem	Symptoms	Probable Cause	Possible Solutions
The suction line collapses when the discharge valve to a hoseline is opened.	The intake pressure drops to less than 0. Discharge pressure also drops.	Suction line may be too small for the amount of flow needed.	3-inch (77 mm) suction hose can supply approximately 500 gpm (2 000 L/min), but gated 2½-inch (65 mm) intake fittings may limit the flow to less that 300 gpm (1 200 L/min). If 2½- or 3-inch (65 mm) or 77 mm) suction line is used, it should be brought into the large intake, either through a suction siamese or bell reducer.
	Water coming out of the ground around the barrel of the hydrant.	Hydrant not fully opened.	Turn the hydrant wrench in a counter-clockwise direction until it reaches the limit of its travel.
While supplying water, the suction line collapses and the pump begins to cavitate.	Intake pressure drops to less than 0. Discharge pressure fluctuates and decreases.	Additional water being demanded by the hoselines that are being supplied.	Reduce the number of hoselines in service or the flow settings on the nozzles.
			Reduce the discharge pressure on the pump by reducing the throttle setting until cavitation stops.
		Additional demands on the water system may have reduced the residual pressure in the system.	Decrease the amount of water being used by changing the nozzle settings or taking attack lines out of service.
			Obtain a supplementary supply from another hydrant, a relay, or from a water shuttle.

Operating from Draft

Pump will not prime.	Unable to get water into the pump through the hard suction hose. No vacuum reading is registered on the intake compound gauge.	Drain valve left open.	Check the master drain valve to make sure it is fully closed. Check the individual drain valves for governor, auxiliary cooler, and so on.
		Intake valves left open or caps not airtight.	Tighten caps on large suction fittings that are not being used.
			Make sure the tank-to-pump valve is closed if the tank is empty.
		Booster line cooler valve or circulator valve left in OPEN position (Barton American pump).	Shut the booster line cooling valve.
			Put the circulator valve in the Vertical position, between the TANK and SPILL settings.
		Intake relief valve may be leaking.	If the relief valve is equipped with a shut-off valve, close it. If it has fire hose threads on the discharge opening, put a cap on it.
		Suction hose connections are not airtight.	Listen for air leaks at each connection. Tighten with a rubber mallet if any connection appears to be leaking.

Operating from Draft

Problem	Symptoms	Probable Cause	Possible Solutions
Pump will not prime.	Unable to get water into the pump through the hard suction hose. No vacuum reading is registered on the intake compound gauge.	Suction hose connection to the floating strainer not airtight.	Tighten the coupling with a rubber mallet or a spanner wrench.
		Pump packing is too loose and leaking air.	Take pumper out of service until the packing can be adjusted or the pump repacked.
		Not operating the primer long enough to get rid of the air.	A typical fire pump requires 15 seconds to prime through 20 feet (6 m) of suction hose but may take as long as 30 seconds to prime completely.
		Tank-to-pump valve not sealing with an empty tank.	Partially fill tank (temporary). Repair valve (permanent).
			Engage the pump, build up 100 psi (700 kPa), and discharge water from the booster line. While keeping the end of the suction hose submerged, take the cap off the end and install the strainer. Slowly close the tank-to-pump valve while continuing to flow water from the booster line. The pump is now primed.
		Engine rpm is too low.	The engine rpm should be as specified in the manufacturer's instructions but is usually 1,000 to 1,200 rpm.
		No oil in the reservoir for the priming pump.	A supply of oil should be carried on any pumper that uses oil for priming. If this is the case, oil can be put in the reservoir to replenish the supply.
			If the regular priming oil is not available, putting water in the reservoir may enable the primer to operate, but it should be thoroughly cleaned out and a supply of oil put in at the first opportunity.
	The electric motor will not operate to drive the primer.	A bad battery or a poor connection between the cable and battery terminals.	Inspect the battery terminals. If they are corroded, it may be possible to loosen the fastening and clean the terminal sufficiently to run.
	Very little air is being discharged from the primer.	Defective primer.	If the primer will not operate for mechanical reasons, the pump can be primed by connecting the hard suction hose to the pump and installing the cap on the end of the hose, then submerging the hose under water. Open the tank-to-pump valve, allowing the suction hose to fill with water as well as the pump.
The pump loses its prime when the first discharge valve is opened and water begins to flow.	The discharge pressure gauge drops sharply.	Valve may have been opened too rapidly.	Carefully observe pressure gauge while slowly opening the discharge valve. If the pressure begins to drop suddenly, pause and allow it to stabilize before continuing.

Operating from Draft			
Problem	**Symptoms**	**Probable Cause**	**Possible Solutions**
The pump loses its prime when the first discharge valve is opened and water begins to flow.	The discharge pressure gauge drops sharply.	The pump was not completely primed and still had some air trapped in it when the primer was released.	Allow the primer to continue to operate until a steady stream of water is discharging from it before closing the priming valve.
			If pressure drops suddenly while opening a discharge valve, operate the primer momentarily to remove any remaining air from the pump.
		The pump may not be turning fast enough to sustain the prime when water begins to flow.	Adjust the discharge pressure to 75 psi (525 kPa) or more before opening the discharge valve.
		The priming valve can stick causing air to leak into the pump.	Check the priming valve control to see that it is in the CLOSED position. Exercise the valve to clear any debris from the valve seat.
		Rock or debris in impeller.	Take pumper out of service until it can be repaired.
	The reading on the pressure gauge drops sharply and the intake gauge returns to the 0 reading.	A high spot in the suction line has trapped a quantity of air when the pump was primed. This air has been drawn into the pump when the water began to move through the suction hose.	Attempt to eliminate the high spot by moving the suction hose and prime the pump again.
			If the high spot cannot be eliminated, it may be possible to scavenge the air from the suction hose by operating the primer again each time the pressure begins to drop.
The pump loses its prime during the course of a pumping operation.	The pump loses it prime when all nozzles are closed and no water is flowing.	An air leak on the intake side of the pump.	Check all connections to see that they are airtight.
		The packing may be mis-adjusted, allowing air to leak into the pump around the shaft.	Open a discharge outlet and allow water to flow at all times to maintain a higher vacuum reading in the intake side of the pump. If the problem becomes acute, take the pumper out of service until it can be repaired.
	The pump loses its prime when it is operating near its maximum capacity. The vacuum reading on the intake gauge is near 0 and is fluctuating.	A whirlpool over the strainer is allowing air to get into the pump through the suction hose.	Put a board or other object over the whirlpool to break up the whirling motion and stop the air from getting into the suction hose.
		Air leak on the intake side of the pump.	Check all connections.
The pump goes into cavitation when the flow increases.	The intake gauge registers more than 22 inches (559 mm) of vacuum, the pressure gauge fluctuates and decreases reading.	The flow exceeds the capacity of the pump at the lift that is required.	The capacity of the pump decreases as the lift increases. At a 20-foot (6 m) lift, the capacity of the pump is only 60 percent of what it would be with a 10-foot (3 m) lift.

Operating from Draft			
Problem	**Symptoms**	**Probable Cause**	**Possible Solutions**
The pump goes into cavitation when the flow increases.	The intake gauge registers more than 22 inches (559 mm) of vacuum, the pressure gauge fluctuates and decreases reading.	The suction line may be partially blocked.	Debris blocking the strainer on the suction hose. Clean the strainer manually.
			Debris is trapped in the strainer at the intake of the pump. Shut down the pump, remove the suction hose and clear the strainer.
		Inner rubber liner of suction hose has become separated from hose. The result is restriction caused by the "bubble" of inner liner.	The suction hose can collapse. It will have to be replaced if the reduced capacity is unacceptable.
			Replace suction hose.
Operating in Relay			
Intake supply line collapses when the throttle setting is increased to establish the initial discharge pressure as required.	The intake pressure gauge reading is negative, that is, reading vacuum instead of pressure.	The dump line or uncapped discharge used to waste water while establishing the relay may still be open.	Close the valve to any uncapped discharge or dump line.
			If the pumper is operating as the terminal unit in a relay, adjust the valve on the dump line to bring the residual pressure at the intake of the pump to 50 psi (350 kPa).
		The terminal pumper may be attempting to take more water from the relay than it can supply.	Notify the water supply officer that the relay is unable to supply the amount of water being called for so that additional supply lines can be put in service.
			Reduce the discharge pressure until the intake gauge registers a positive reading.
While the relay is operating, the intake pressure increases above 50 psi (350 kPa).	The intake pressure gauge is reading above 50 psi (350 kPa) and the discharge pressure also increases accordingly.	Changes in flow of the attack line cause the friction loss to decrease and the residual pressure to increase.	No action is necessary unless the pressure increase becomes dangerous. Minor variations are to be expected in a relay and frequent adjustments are undesirable.
While the relay is operating, the intake pressure increases dangerously.	Intake pressure gauge is reading above 150 psi (1 050 kPa), the discharge pressure is above 200 psi (1 050 kPa).	Hoselines have been shut down with no corresponding dumping of excess water.	Open the uncapped discharge or dump line until the intake residual pressure returns to 50 psi (350 kPa).

Courtesy of Bill Eckman.

FLOWCHART 11.1
OPERATING A PUMPER FROM THE TANK ON THE APPARATUS
Step 1: Positioning the Apparatus

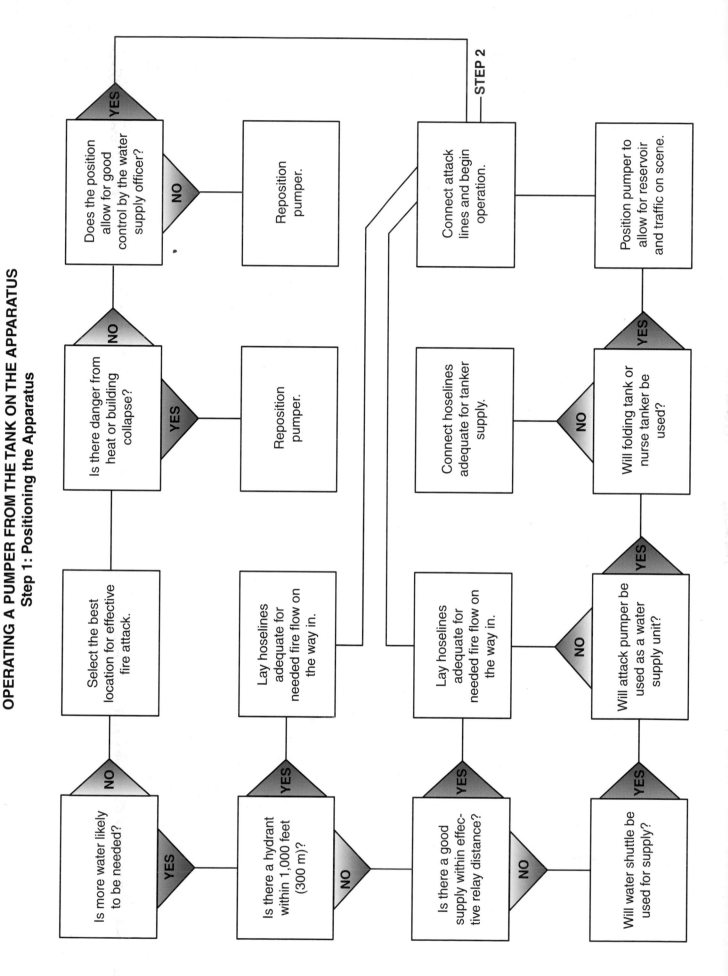

OPERATING A PUMPER FROM THE TANK ON THE APPARATUS
Step 2: Put Pump in Service and Establish Operating Pressure

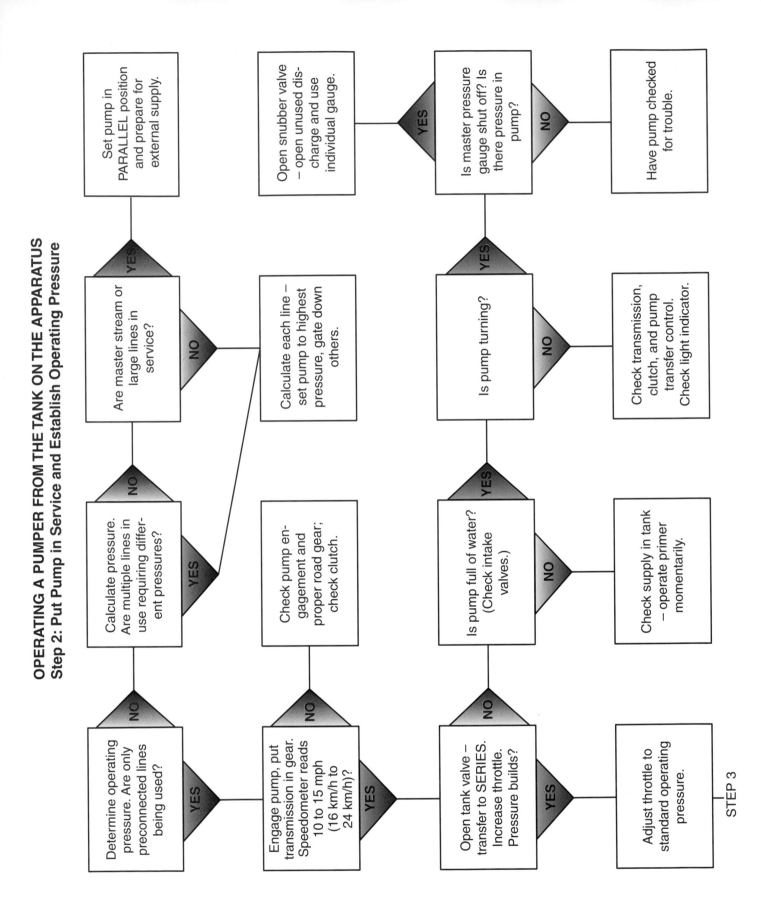

STEP 3

OPERATING A PUMPER FROM THE TANK ON THE APPARATUS
Step 3: Maintaining Good Operating Pressure

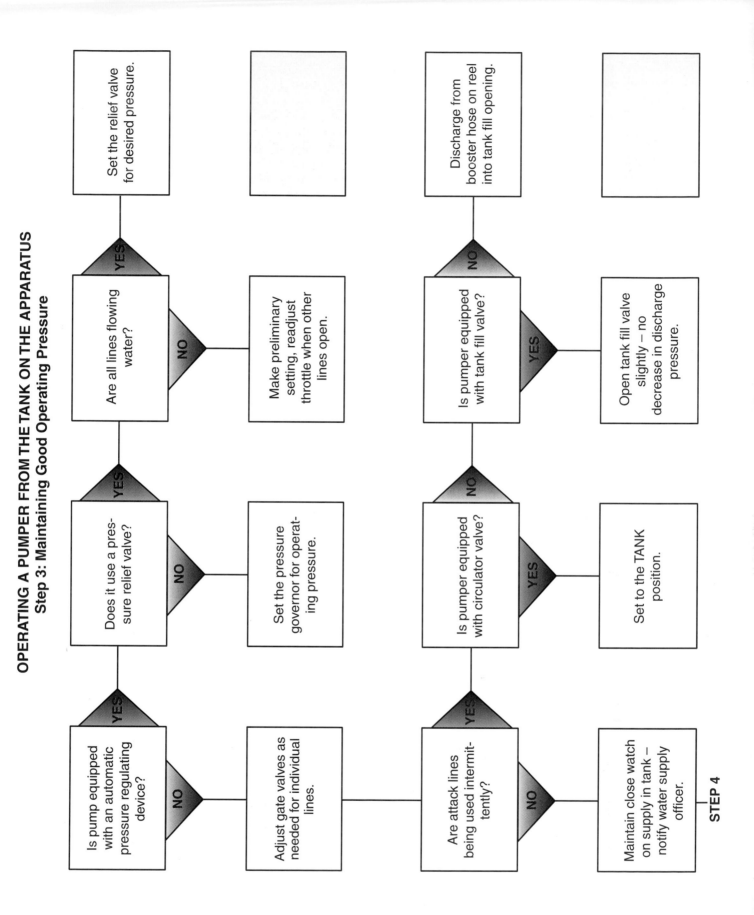

Is pump equipped with an automatic pressure regulating device?

YES → Does it use a pressure relief valve?

NO → Adjust gate valves as needed for individual lines.

Does it use a pressure relief valve?

YES → Are all lines flowing water?

NO → Set the pressure governor for operating pressure.

Are all lines flowing water?

YES → Set the relief valve for desired pressure.

NO → Make preliminary setting, readjust throttle when other lines open.

Are attack lines being used intermittently?

YES → Is pumper equipped with circulator valve?

NO → Maintain close watch on supply in tank – notify water supply officer.

Is pumper equipped with circulator valve?

YES → Set to the TANK position.

NO → Is pumper equipped with tank fill valve?

Is pumper equipped with tank fill valve?

YES → Open tank fill valve slightly – no decrease in discharge pressure.

NO → Discharge from booster hose on reel into tank fill opening.

STEP 4

OPERATING A PUMPER FROM THE TANK ON THE APPARATUS
Step 4: Making the Transition to an External Supply

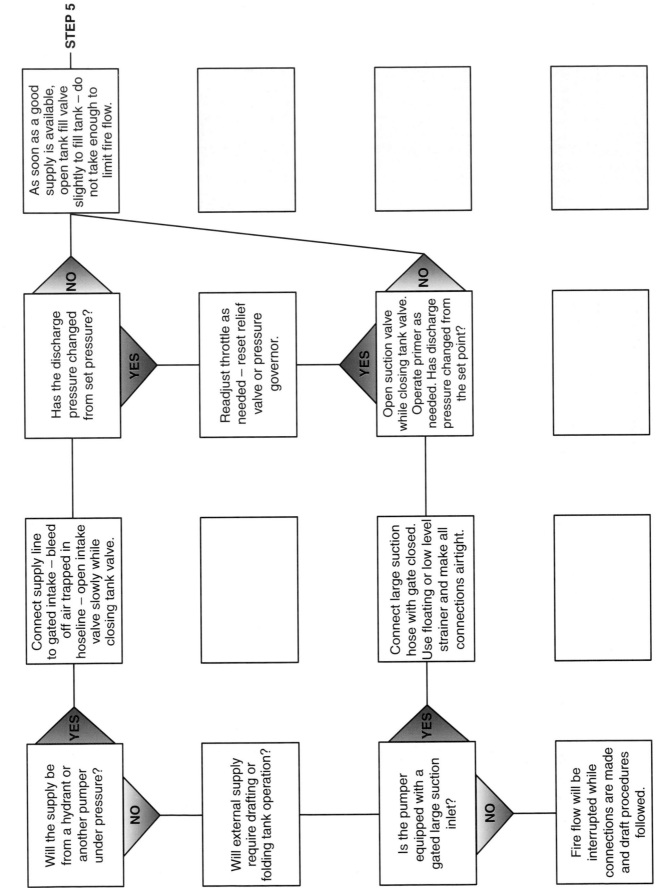

STEP 5

As soon as a good supply is available, open tank fill valve slightly to fill tank – do not take enough to limit fire flow.

NO — Has the discharge pressure changed from set pressure? — **YES** → Readjust throttle as needed – reset relief valve or pressure governor. — **YES** — Open suction valve while closing tank valve. Operate primer as needed. Has discharge pressure changed from the set point? **NO**

Connect supply line to gated intake – bleed off air trapped in hoseline – open intake valve slowly while closing tank valve.

Connect large suction hose with gate closed. Use floating or low level strainer and make all connections airtight.

YES — Will the supply be from a hydrant or another pumper under pressure? **NO** → Will external supply require drafting or folding tank operation? → **YES** — Is the pumper equipped with a gated large suction inlet? **NO** → Fire flow will be interrupted while connections are made and draft procedures followed.

OPERATING A PUMPER FROM THE TANK ON THE APPARATUS
Step 5: Shutting Down

Slowly close valve on unneeded line. Are other lines in service?

YES → Open individual line drain valve to bleed off pressure – disconnect hoseline.

NO → Reduce throttle to idle, disengage pump, take transmission out of gear. → Remove all hose connections and let pump drain through gated intake openings. → Open all drains.

FLOWCHART 11.2

OPERATING FROM A HYDRANT
Step 1: Positioning the Pumper and Connecting to the Hydrant

```
Is there a hydrant          ──YES──►  Will the closest          ──YES──►  Is the selected          ──YES──►  Position and immo-
within 200 feet                       hydrant supply the                  hydrant a safe                     bilize the pumper –
(60 m) of the fire                    needed fire flow?                   distance from the                  will suction hose
location?                                                                 fire?                              reach to steamer
                                                                                                             connection on hydrant?
    │                                     │                                   │                                   │
    NO                                    NO                                  NO                                 YES   NO
    │                                     │                                   │                                   │     │
    ▼                                     ▼                                   ▼                                   │     ▼
Select the most             Arrange for a              Select another                        Connect short
accessible hydrant          supplemental water         hydrant – set up a                    section of 3-inch
for the fire location.      supply.                     relay if needed.                      (77 mm) hose from
                                                                                              hydrant to suction
                                                                                              siamese or gated
                                                                                              inlet.
```

```
Is the selected             ──YES──►  Lay supply line on                   Will the attack          ──NO──►  Lay sufficient
hydrant within 1,000                  arrival. Use 4-way                    pumper also be used               hoselines from the
feet (300 m) of the                   hydrant valve and                    as the water supply               water supply pumper
fire location?                        gate valve to provide                pumper?                           to the attack pumper
                                      supported supply                                                       for needed fire flow.
    │                                 line.                                     │
    NO                                                                        YES
    │                                                                          │
    ▼                                                                          ▼
Set up a relay to get                                         Connect soft sleeve
the water from the                                            or supply hose from
hydrant to fire                                               steamer connection
location.                                                     on hydrant to pumper.
```

```
Open hydrant, note          ───── STEP 2
static pressure and
begin operation.
```

OPERATING FROM A HYDRANT
Step 2: Putting the Pumper in Service

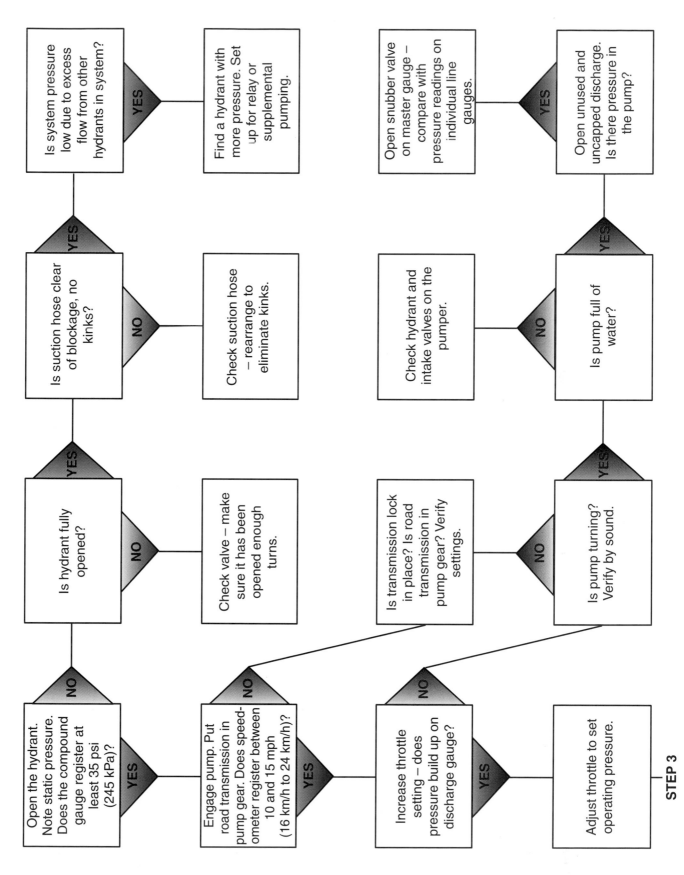

Is system pressure low due to excess flow from other hydrants in system?

YES → Find a hydrant with more pressure. Set up for relay or supplemental pumping.

Is suction hose clear of blockage, no kinks?

NO → Check suction hose – rearrange to eliminate kinks.

Is hydrant fully opened?

NO → Check valve – make sure it has been opened enough turns.

Open the hydrant. Note static pressure. Does the compound gauge register at least 35 psi (245 kPa)?

Engage pump. Put road transmission in pump gear. Does speedometer register between 10 and 15 mph (16 km/h to 24 km/h)?

Is transmission lock in place? Is road transmission in pump gear? Verify settings.

Is pump turning? Verify by sound.

NO → Increase throttle setting – does pressure build up on discharge gauge?

YES → Adjust throttle to set operating pressure.

Open snubber valve on master gauge – compare with pressure readings on individual line gauges.

YES → Open unused and uncapped discharge. Is there pressure in the pump?

Check hydrant and intake valves on the pumper.

NO → Is pump full of water?

STEP 3

OPERATING FROM A HYDRANT
Step 3: Maintain Operating Pressure

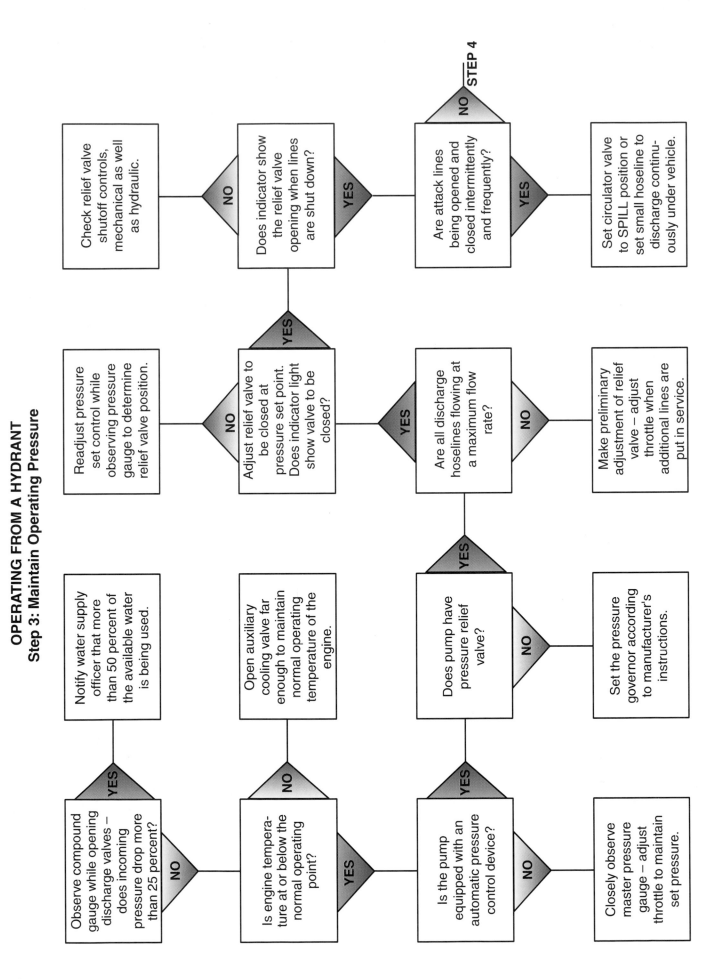

Observe compound gauge while opening discharge valves – does incoming pressure drop more than 25 percent?

YES → Notify water supply officer that more than 50 percent of the available water is being used.

NO →

Is engine temperature at or below the normal operating point?

NO → Open auxiliary cooling valve far enough to maintain normal operating temperature of the engine.

YES →

Is the pump equipped with an automatic pressure control device?

NO → Closely observe master pressure gauge – adjust throttle to maintain set pressure.

YES →

Does pump have pressure relief valve?

NO → Set the pressure governor according to manufacturer's instructions.

YES →

Are all discharge hoselines flowing at a maximum flow rate?

NO → Make preliminary adjustment of relief valve – adjust throttle when additional lines are put in service.

YES →

Adjust relief valve to be closed at pressure set point. Does indicator light show valve to be closed?

NO → Readjust pressure set control while observing pressure gauge to determine relief valve position.

YES →

Does indicator show the relief valve opening when lines are shut down?

NO → Check relief valve shutoff controls, mechanical as well as hydraulic.

YES →

Are attack lines being opened and closed intermittently and frequently?

YES → Set circulator valve to SPILL position or set small hoseline to discharge continuously under vehicle.

NO → **STEP 4**

OPERATING FROM A HYDRANT
Step 4: Operating Practices

Does residual pressure drop below 0 psi (0 kPa) (vacuum reading)?

YES → Reduce throttle to reduce pressure until residual is more than 0 psi (0 kPa) – notify water supply officer.

NO →

Does residual pressure drop below 20 psi (140 kPa)?

YES → Notify water supply officer that critical point on hydrant has been reached — supply may fail.

NO →

Observe residual pressure while operating. Does residual pressure drop 25 percent or more from static?

YES → Notify water supply officer that water from this hydrant will be limited — 50 percent of capacity is being used.

NO →

Continue to closely observe gauges while operating.

STEP 5

OPERATING FROM A HYDRANT
Step 5: Shutting Down

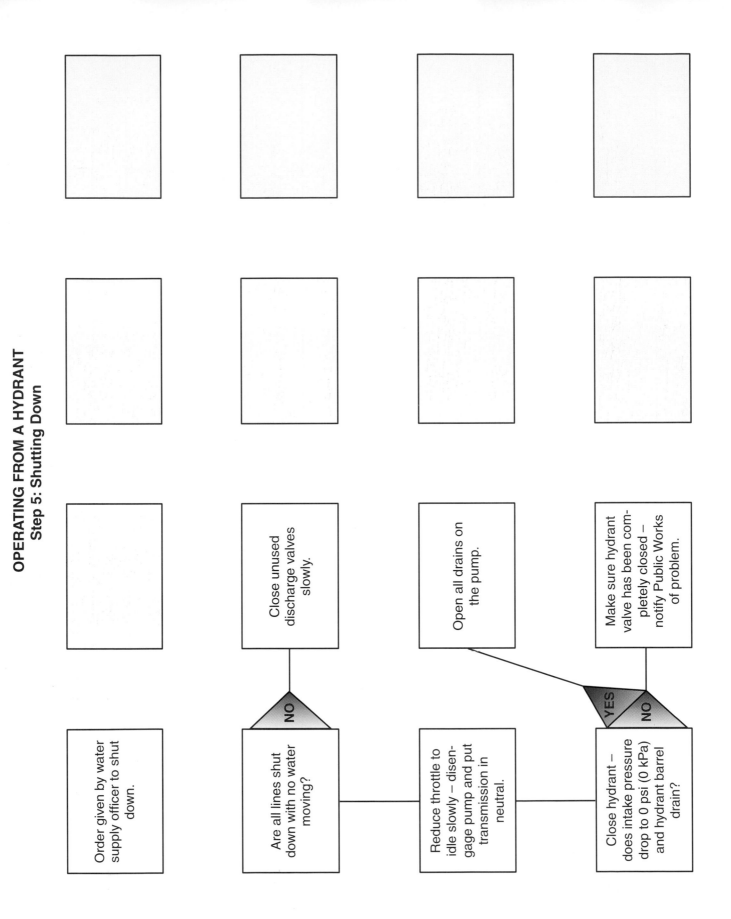

Order given by water supply officer to shut down.

Are all lines shut down with no water moving? — **NO** → Close unused discharge valves slowly.

Reduce throttle to idle slowly – disengage pump and put transmission in neutral.

Open all drains on the pump.

Close hydrant – does intake pressure drop to 0 psi (0 kPa) and hydrant barrel drain? — **YES** → Make sure hydrant valve has been completely closed – notify Public Works of problem. / **NO**

FLOWCHART 11.3

OPERATING FROM DRAFT

Step 1: Positioning, Connecting, and Priming the Pumper

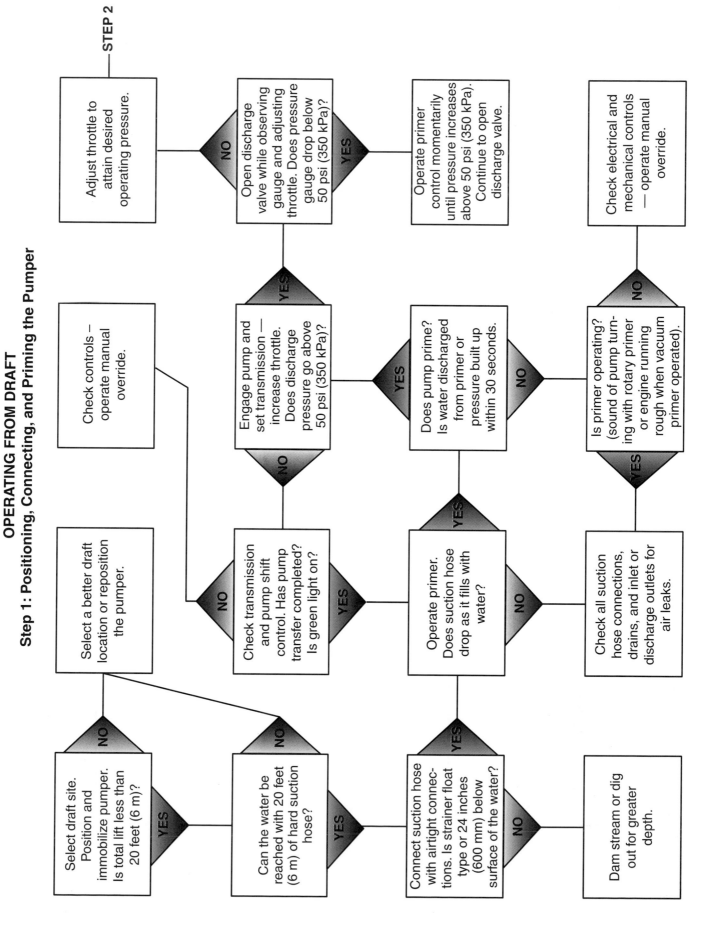

Select draft site. Position and immobilize pumper. Is total lift less than 20 feet (6 m)?

NO — Select a better draft location or reposition the pumper.

YES — Can the water be reached with 20 feet (6 m) of hard suction hose?

NO — Select a better draft location or reposition the pumper.

YES — Connect suction hose with airtight connections. Is strainer float type or 24 inches (600 mm) below surface of the water?

NO — Dam stream or dig out for greater depth.

YES — Operate primer. Does suction hose drop as it fills with water?

YES — Does pump prime? Is water discharged from primer or pressure built up within 30 seconds.

NO — Check all suction hose connections, drains, and inlet or discharge outlets for air leaks.

Is primer operating? (sound of pump turning with rotary primer or engine running rough when vacuum primer operated).

YES — Check all suction hose connections, drains, and inlet or discharge outlets for air leaks.

NO — Check electrical and mechanical controls — operate manual override.

YES (pump prime) — Engage pump and set transmission — increase throttle. Does discharge pressure go above 50 psi (350 kPa)?

NO — Check transmission and pump shift control. Has pump transfer completed? Is green light on?

YES (green light) — Operate primer control momentarily until pressure increases above 50 psi (350 kPa). Continue to open discharge valve.

NO (green light) — Check controls — operate manual override.

YES (discharge pressure) — Open discharge valve while observing gauge and adjusting throttle. Does pressure gauge drop below 50 psi (350 kPa)?

YES — Operate primer control momentarily until pressure increases above 50 psi (350 kPa). Continue to open discharge valve.

NO — Adjust throttle to attain desired operating pressure.

— STEP 2

OPERATING FROM DRAFT
Step 2: Putting the Pumper in Service

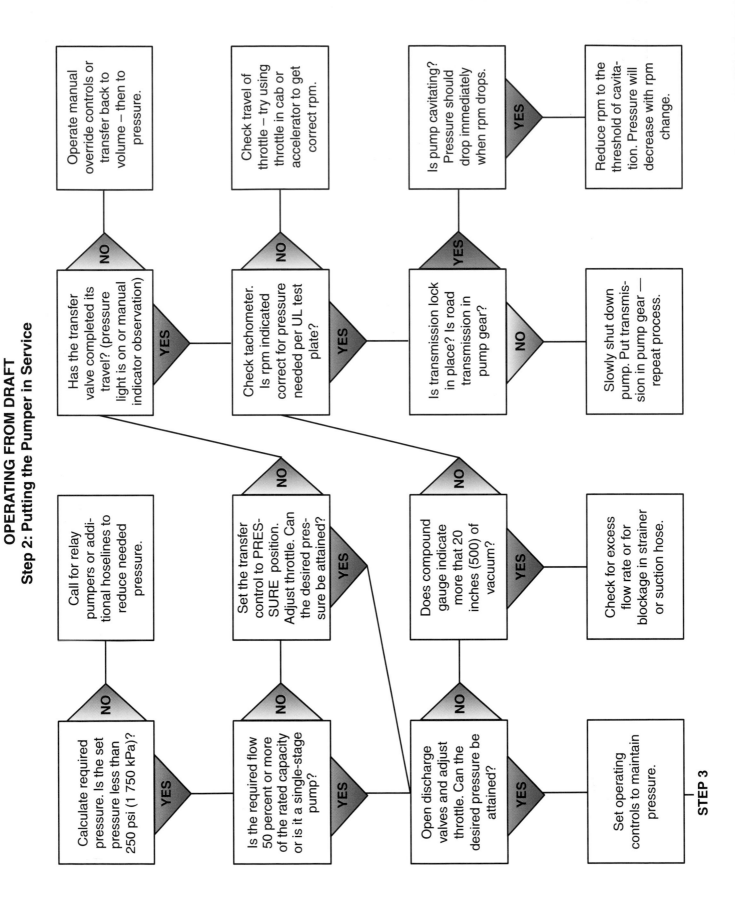

OPERATING FROM DRAFT
Step 3: Maintain Operating Pressure

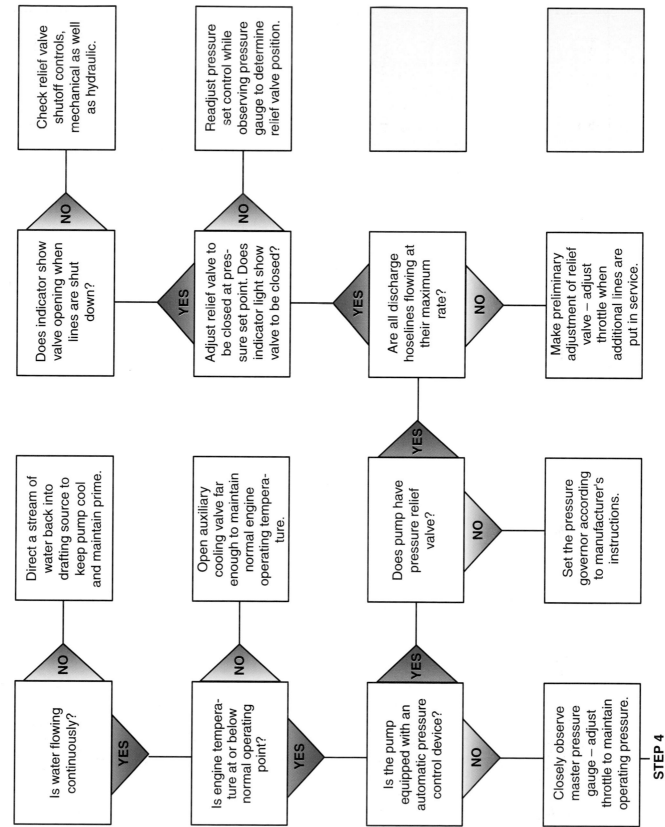

Check relief valve shutoff controls, mechanical as well as hydraulic.

Does indicator show valve opening when lines are shut down? — **NO**

Readjust pressure set control while observing pressure gauge to determine relief valve position.

Adjust relief valve to be closed at pressure set point. Does indicator light show valve to be closed? — **NO** / **YES**

Are all discharge hoselines flowing at their maximum rate? — **YES**

Make preliminary adjustment of relief valve – adjust throttle when additional lines are put in service. — **NO**

Direct a stream of water back into drafting source to keep pump cool and maintain prime.

Is water flowing continuously? — **NO** / **YES**

Open auxiliary cooling valve far enough to maintain normal engine operating temperature.

Is engine temperature at or below normal operating point? — **NO** / **YES**

Does pump have pressure relief valve? — **YES**

Set the pressure governor according to manufacturer's instructions. — **NO**

Is the pump equipped with an automatic pressure control device? — **YES**

Closely observe master pressure gauge – adjust throttle to maintain operating pressure. — **NO**

STEP 4

Reduce throttle slowly to idle. Remove all hose connections.	Take pump out of gear. Open all drains.	Operate primer for 10 seconds with pump dry.	Cap all openings. Close all drains.

Job Performance Requirements

This chapter provides information that will assist the reader in meeting the following job performance requirements from NFPA 1002, *Standard on Fire Apparatus Driver/Operator Professional Qualifications*, 1998 edition. Particular portions of the job performance requirements (JPRs) that are met in this chapter are noted in bold text.

3-2.1* Produce effective hand or master streams, given the sources specified in the following list, so that the pump is safely engaged, all pressure control and vehicle safety devices are set, the rated flow of the nozzle is achieved and maintained, and the apparatus is continuously monitored for potential problems.

- Internal tank
- Pressurized source
- **Static source**
- Transfer from internal tank to external source

(a) *Requisite Knowledge:* Hydraulic calculations for friction loss and flow using both written formulas and estimation methods, safe operation of the pump, problems related to small-diameter or dead-end mains, low-pressure and private water supply systems, hydrant cooling systems, and **reliability of static sources.**

(b) *Requisite Skills:* **The ability to position a fire department pumper to operate at a fire hydrant and at a static water source,** power transfer from vehicle engine to pump, draft, operate pumper pressure control systems, operate the volume/pressure transfer valve (multistage pumps only), operate auxiliary cooling systems, make the transition between internal and external water sources, and assemble hose lines, nozzles, valves, and appliances.

3-2.2 Pump a supply line of 2½ in. (65 mm) or larger, given a relay pumping evolution the length and size of the line and the desired flow and intake pressure, so that the proper pressure and flow are provided to the next pumper in the relay.

(a) *Requisite Knowledge:* Hydraulic calculations for friction loss and flow using both written formulas and estimation methods, safe operation of the pump, problems related to small-diameter or dead-end mains, low-pressure and private water supply systems, hydrant cooling systems, and **reliability of static sources.**

(b) *Requisite Skills:* **The ability to position a fire department pumper to operate at a fire hydrant and at a static water source,** power transfer from vehicle engine to pump, draft, operate pumper pressure control systems, operate the volume/pressure transfer valve (multistage pumps only), operate auxiliary cooling systems, make the transition between internal and external water sources, and assemble hose lines, nozzles, valves, and appliances.

6-2.1* Produce effective fire streams, utilizing the sources specified in the following list, so that the pump is safely engaged, all pressure-control and vehicle safety devices are set, the rated flow of the nozzle is achieved, and the apparatus is continuously monitored for potential problems.

- Water tank
- Pressurized source
- **Static source**

(a) *Requisite Knowledge:* Hydraulic calculations for friction loss and flow using both written formulas and estimation methods, safe operation of the pump, proper apparatus placement, personal safety considerations, problems related to small-diameter or dead-end mains and low-pressure and private water supply systems, hydrant cooling systems, and **reliability of static sources.**

(b) *Requisite Skills:* **The ability to position a wildland fire apparatus to operate at a fire hydrant and at a static water source,** properly place apparatus for fire attack, transfer power from vehicle engine to pump, draft, operate pumper pressure control systems, operate the volume/pressure transfer valve (multistage pumps only), operate auxiliary cooling systems, make the transition between internal and external water sources, and assemble hose lines, nozzles, valves, and appliances.

6-2.2 Pump a supply line, given a relay pumping evolution the length and size of the line and pumping flow and desired intake pressure, so that adequate intake pressures and flow are provided to the next pumper in the relay.

(a) *Requisite Knowledge:* Hydraulic calculations for friction loss and flow using both written formulas and estimation

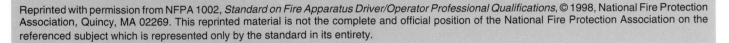
In an ideal world, all areas that fire departments are responsible for protecting would be serviced by a reliable municipal water supply system. This would allow pumpers to connect to a hydrant and receive an adequate supply of water at a usable pressure. Rural areas, as well as some urban and suburban areas, often do not have this convenience and must depend on other sources to provide sufficient water for fire fighting. In some cases, urban and suburban jurisdictions do have a water supply system, but it is not sufficient for major fire fighting operations. This requires that other sources be used. As well, natural disasters, such as tornadoes or earthquakes, may render the municipal water supply system inoperable, again forcing fire departments to use alternative supply sources.

The alternative to pressurized water supply systems is static water supply sources such as lakes, rivers, oceans, cisterns, streams, portable water tanks, and swimming pools. All fire departments must be familiar with the static water supply sources available within their jurisdiction. In rural jurisdictions, static water supply sources may be the primary source of water for fire protection. In other jurisdictions, they may be considered auxiliary water supplies. In either case, driver/operators must understand how to evaluate the usability and reliability of the static water supply source from which they may be expected to pump.

This chapter examines the critical issues that determine the usability and reliability of static water supply sources. Also covered is the procedure for determining the net pump discharge pressure required when pumping from a static water supply source.

Principles of Lift

As was discussed in the previous chapter, the process of raising water from a static source to supply a pumper is known as *drafting* (Figure 12.1). In order to evaluate the usability of static water supply sources and to operate effectively from them, the driver/operator must understand some basic principles related to lift.

During drafting operations, there is generally an elevation difference between the static water source and the center of the pump (the pump is usually higher than the water supply source). This elevation difference is known as *lift* (Figure 12.2). During the process of drafting, the primer pumps the air out of the intake hose and the fire pump. This creates a pressure differential between the inside of the pump and intake hose and the atmosphere. The pressure in the pump and intake hose will be less than the atmospheric pressure. This results in water being forced up into the hose and pump. This process is sometimes referred to as creating a *vacuum*. In reality, it is just a partial vacuum, as total vacuums are impossible to create in fire service equipment.

Because the pressure is now greater outside the intake hose, water is forced to rise inside the intake hose (Figure 12.3). This rise continues until either the pump is full of water or the head within the intake hose

Figure 12.1 *Drafting* is the process of a pumper taking water from a static water supply source.

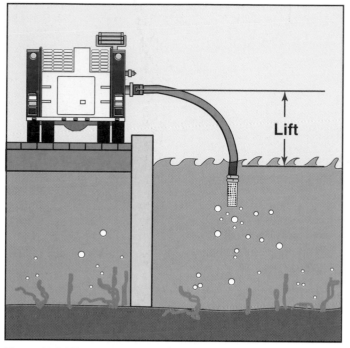

Figure 12.2 *Lift* is the distance between the fire pump and source of the water.

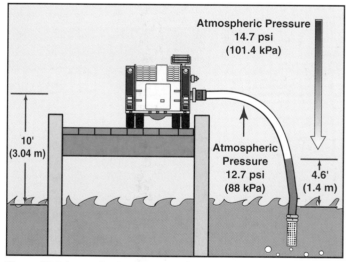

Atmospheric Pressure
14.7 psi
(101.4 kPa)

Atmospheric
Pressure
12.7 psi
(88 kPa)

10'
(3.04 m)

4.6'
(1.4 m)

Figure 12.3 Atmospheric pressure forces water into the intake hose and eventually into the fire pump.

matches the negative pressure. The height of possible lift is not affected by the angle of the intake hose, but it is affected by the amount of vacuum that the pump can produce and by the atmospheric pressure. If the water does not rise to the level of the pump, drafting is not possible.

Theoretical Lift

Although not possible with a fire department pumper, a total vacuum would allow water to be raised by atmospheric pressure to a height in accordance with this pressure. In the U.S system of measurement, at sea level a pump could theoretically lift water 33.8 feet (14.7 psi x 2.3 ft/psi). This theoretical lift decreases as the altitude of

drafting operations increases. For every 1,000 feet of altitude, atmospheric pressure decreases by about 0.5 psi.

In the metric system of measurement, at sea level a pump can theoretically lift water 10 m (100 kPa x 0.1 m/kPa). This theoretical lift decreases as the altitude of drafting operations increases. For every 100 m of altitude, atmospheric pressure decreases by about 1 kPa.

Keep in mind that because a total vacuum is not possible in field conditions (in fact it is virtually impossible under laboratory conditions), fire department pumpers cannot be expected to draft water that is located 33.8 feet (10 m) below the level of the pump. With that in mind, there are a few other lift concepts that the driver/operator must understand.

Maximum Lift

In fire service terms, *maximum* lift is defined as the maximum height to which any amount of water may be raised through a hard intake hose to the pump. The maximum lift varies depending on the atmospheric pressure and the condition of the fire pump and primer. In most circumstances, the maximum lift is no more than 25 feet (7.5 m) (Figure 12.4). Keep in mind that as you approach maximum lift all the available atmospheric pressure is being used to overcome the lift. As a result, the tolerable amount of friction loss in the intake hose is so low that the volume of water available for the fire pump is too low to be of practical value.

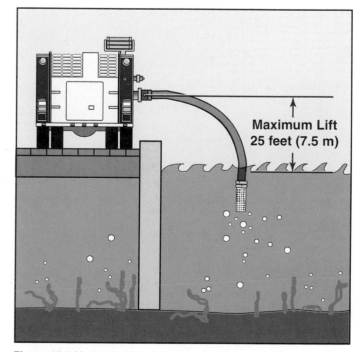

Maximum Lift
25 feet (7.5 m)

Figure 12.4 Maximum lift is determined by the available atmospheric pressure.

To determine the maximum lift that a pumper can achieve in the U.S. system of measurement, the vacuum reading in inches of mercury (mm of mercury) at the intake gauge is read and multiplied using the following formula:

EQUATION J
L = 1.13 Hg

Where: L = Height of lift in feet

1.13 = A constant

Hg = Inches of mercury

Example 1
Determine the maximum height to which water will be lifted when the intake gauge reads 14 inches of mercury.

L = 1.13 Hg

L = 1.13 (14)

L = 15.82 feet maximum lift

To determine the maximum lift that a pumper can achieve in the metric system of measurement, the vacuum reading in millimeters of mercury (mm of mercury) at the intake gauge is read and multiplied using the following formula:

EQUATION K
L =(0.013 56)(Hg)

Where: L = Height of lift in meters

0.013 56 = A constant

Hg = mm of mercury

Example 2
Determine the maximum height to which water will be lifted when the intake gauge reads 250 mm of mercury.

L = (0.013 56)(Hg)

L = (0.013 56)(250)

L = 3.4 meters maximum lift

Dependable Lift
While it is important to understand the concepts of theoretical and maximum lift, the driver/operator must be more concerned with the dependable lift. *Dependable lift* is the height a column of water may be lifted in sufficient quantity to provide a reliable fire flow. Taking

Table 12.1a
Discharge at Various Lifts

Rated Capacity, Pump			500 gpm		750 gpm		1000 gpm		1250 gpm	1500 gpm		
Intake Hose Size (inches)			4	4½	4½	5	5	6	6	6	Dual 5	Dual 6
Lift in Feet	20' Intake Hose (Two Sections)	4	590	660	870	945	1160	1345	1435	1735	1990	2250
		6	560	630	830	905	1110	1290	1375	1660	1990	2150
		8	530	595	790	860	1055	1230	1310	1575	1810	2040
		10	500	560	750	820	1000	1170	1250	1500	1720	1935
		12	465	520	700	770	935	1105	1175	1410	1615	1820
		14	430	480	650	720	870	1045	1100	1325	1520	1710
		16	390	430	585	655	790	960	1020	1225	1405	1585
	30' Intake Hose (Three Sections)	18	325	370	495	560	670	835	900	1085	1240	1420
		20	270	310	425	480	590	725	790	955	1110	1270
		22	195	225	340	375	485	590	660	800	950	1085
		24	65	70	205	235	340	400	495	590	730	835

NOTES:

1-Net pump pressure is 150 psi. Operation at a lower pressure will result in an increased discharge; operation at a higher pressure, a decreased discharge.

2-Data based on a pumper with ability to discharge rated capacity when drafting at not more than a 10-foot lift. Many pumpers will exceed this performance and therefore will discharge greater quantities than shown at all lifts.

CONDITIONS: Operating at Net Pump Pressure of 150 psi; Altitude of 1000 feet; Water Temperature of 60°F; Barometric Pressure of 28.94" Hg (poor weather conditions).

Table 12.1b
Discharge at Various Lifts

Rated Capacity, Pump		2 000 L/min		3 000 L/min		4 000 L/min		5 000 L/min	6 000 L/min		
Intake Hose Size (mm)		100	115	115	125	125	150	150	150	Dual 125	Dual 150
Lift in Meters	1.2 (6 m Intake Hose Two Sections)	2 233	2 498	3 293	3 577	4 391	5 091	5 432	6 568	7 532	8 517
	1.8	2 119	2 385	3 142	3 426	4 201	4 883	5 205	6 283	7 532	8 139
	2.4	2 006	2 252	2 990	3 255	3 994	4 656	4 959	5 962	6 851	7 722
	3.0	1 893	2 120	2 839	3 104	3 785	4 428	4 732	5 678	6 511	7 325
	3.7	1 760	1 968	2 650	2 915	3 539	4 183	4 449	5 337	6 113	6 889
	4.3	1 628	1 817	2 451	2 725	3 293	3 956	4 163	5 016	5 754	6 473
	4.9	1 476	1 628	2 214	2 479	2 990	3 634	3 861	4 637	5 318	6 000
	5.5 (9 m Intake Hose Three Sections)	1 230	1 400	1 874	2 120	2 536	3 161	3 407	4 107	4 694	5 375
	6.1	1 022	1 173	1 608	1 817	2 233	2 744	2 990	3 615	4 202	4 807
	6.7	738	852	1 287	1 419	1 836	2 233	2 498	3 028	3 596	4 107
	7.3	246	265	776	890	1 287	1 514	1 874	2 233	2 763	3 161

NOTES:

1-Net pump pressure is 1 000 kPa. Operation at a lower pressure will result in an increased discharge; operation at a higher pressure, a decreased discharge.

2-Data based on a pumper with ability to discharge rated capacity when drafting at not more than a 3 meter lift. Many pumpers will exceed this performance and therefore will discharge greater quantities than shown at all lifts.

CONDITIONS: Operating at Net Pump Pressure of 1 000 kPa; Altitude of 300 meters; Water Temperature of 15.5°C; Barometric Pressure of 735 mm Hg (poor weather conditions).

into consideration the surrounding atmospheric pressure and friction loss in the intake hose, every fire pump in good repair should have a dependable lift of at least 14.7 feet (4.5 m). Tables 12.1 a and b list minimum discharges that can be expected from a pumper operating at draft at various lifts.

It should be noted that all fire department pumping apparatus have their fire pumps rated when drafting from a lift of 10 feet (3 m) through 20 feet (6 m) of hard intake hose (Figure 12.5). As the lift or length of intake hose is increased, the capacity of the pump decreases accordingly. The pump is only able to deliver 70 percent of its capacity at a 15 foot (4.5 m) lift and 60 percent at a 20 foot (6 m) lift.

Determining Net Pump Discharge Pressure

Net pump discharge pressure takes into account all factors that contribute to the amount of work that the pump must do to produce a fire stream. When at draft, the net pump discharge pressure is more than the pressure shown on the discharge gauge. This pressure is the sum of the

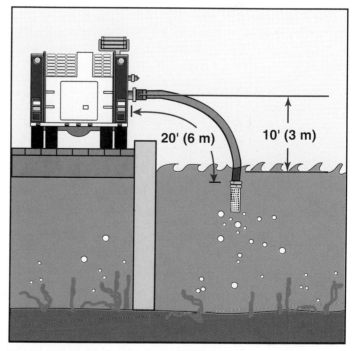

Figure 12.5 All fire department pumpers are rated at a lift of 10 feet (3 m) through 20 feet (6 m) of intake hose.

pump discharge pressure and the intake pressure correction. The intake pressure correction must take into account friction loss in the intake hose and the height of the lift. Tables 12.2 a and b give these allowances. They are then used in the following formulas to calculate the pressure correction.

EQUATION L

Pressure Correction (U.S.) =

$$\frac{\text{lift} + \text{total intake hose friction loss}}{2.3}$$

EQUATION M

Pressure Correction (Metric) =

$$\frac{\text{lift} + \text{total intake hose friction loss}}{0.1}$$

After completing this correction, the net pump discharge pressure at draft can be calculated from the following equation:

EQUATION N

$NPDP_{Draft}$ = PDP + Intake pressure correction

Where: $NPDP_{Draft}$ = Net pump discharge pressure at draft

PDP = Pump discharge pressure in psi or kPa

Example 3

A 1,000 gpm pumper operating from draft is discharging water at 120 psi. The required lift is 9 feet through 20 feet of 5-inch intake hose. Calculate the net pump discharge pressure.

$$\text{Pressure correction} = \frac{\text{lift} + \text{total intake hose friction loss}}{2.3}$$

$$\text{Pressure correction} = \frac{9 + (4.5 + 1.5)}{2.3}$$

$$\text{Pressure correction} = \frac{15}{2.3}$$

Intake pressure correction = 6.5 psi

$NPDP_{Draft}$ = PDP + Intake pressure correction

= 120 psi + 6.5 psi

= **126.5 psi net pump discharge pressure at draft**

Example 4

A 4 000 L/min pumper operating from draft is discharging water at 830 kPa. The required lift is 3 m through 6 m of 125 mm intake hose. Calculate the net pump discharge pressure.

$$\text{Pressure correction} = \frac{\text{lift} + \text{total intake hose friction loss}}{0.1}$$

$$\text{Pressure correction} = \frac{3 + (2.4 + 0.45)}{0.1}$$

$$\text{Pressure correction} = \frac{5.85}{0.1}$$

Intake pressure correction = 58.5 kPa

$NPDP_{Draft}$ = PDP + Intake pressure correction

= 830 kPa + 58.5 kPa

= **888.5 kPa net pump discharge pressure at draft**

Table 12.2a
Allowances for Friction Loss in Intake Hose

Rated Capacity of Pumper gpm	Diameter of Intake Hose in Inches	For 10 ft. of Intake Hose	Allowance (feet) for Each Additional 10 ft. of Intake Hose
500	4	6	plus 1
	4½	3½	plus ½
750	4	7	plus 1½
	5	4½	plus 1
1000	4½	12	plus 2½
	5	4½	plus 1½
	6	4	plus ½
1250	5	12½	plus 2
	6	6½	plus ½
1500	6	9	plus 1
1500	4½ (dual)	7	plus 1½
1500	5 "	4½	plus 1
1500	6 "	2	plus ½
1750	6	12½	plus 1½
1750	4½ (dual)	9½	plus 2
1750	5 "	6½	plus 1
1750	6 "	3	plus ½
2000	4 ½ "	12	plus 1½
2000	5 "	8	plus 1½
2000	6 "	4	plus ½

NOTE: The allowance computed above for the capacity test should be reduced by 1 psi for the allowance on the 200-psi test and by 2 psi for the allowance on the 250-psi test.

Table 12.2b
Allowances for Friction Loss in Intake Hose

Rated Capacity of Pumper L/min	Diameter of Intake Hose in mm	For 3 m of Intake Hose	Allowance (feet) for Each Additional 3 m of Intake Hose
2 000	100	1.8	plus 0.3
	115	1.1	plus 0.15
3 000	115	2.1	plus 0.45
	125	1.4	plus 0.3
4 000	115	3.7	plus 0.75
	125	2.4	plus 0.45
	150	1.2	plus 0.15
5 000	125	3.8	plus 0.61
	150	2	plus 0.15
6 000	150	2.7	plus 0.3
	115 (dual)	2.1	plus 0.45
	125	1.4	plus 0.3
	150	0.6	plus 0.15
7 000	6	3.8	plus 0.45
	115 (dual)	2.9	plus 0.61
	125	2	plus 0.3
	150	0.9	plus 0.15
8 000	115	3.7	plus 0.75
	125	2.4	plus 0.45
	150	1.2	plus 0.15

NOTE: The allowance computed above for the capacity test should be reduced by 7 kPa for the allowance on the 1 350 kPa test and by 14 kPa for the allowance on the 1 700 kPa test.

Natural Static Water Supply Sources

There are two primary types of static water supply sources that firefighters may draw from if necessary: natural and man-made. The following sections examine natural static water supply sources.

Natural static water supply sources include lakes, ponds, streams, rivers, and oceans (Figures 12.6 a and b). Driver/operators must be familiar with all the potential drafting sources within their jurisdiction. These sources should be identified during pre-incident planning and appropriate approach routes and drafting spots should be noted. This eliminates the need to "guess" about where a reliable water supply source can be obtained under emergency conditions. When evaluating a natural static water supply source, there are two primary factors that must be considered: adequacy and accessibility.

Figure 12.6a A large lake is a reliable water supply.

Figure 12.6b Most large rivers, if they are accessible, are good water sources.

Adequacy of the Natural Static Water Supply Source

The adequacy of large natural static water supply sources, such as major rivers, lakes, and oceans, is generally not a major issue. Although periods of draught may result in lake and river levels dropping and tides receding in ocean waters (creating more of an accessibility issue), they still generally provide an endless supply of water from a fire protection standpoint.

However, fire department personnel must evaluate small streams and ponds with a little more caution when determining their adequacy for fire protection (Figures 12.7 a and b). Serious tactical implications may result from committing fire apparatus to a static water supply source that ends up being insufficient for the necessary fire flow. The driver/operator must also keep in mind that these smaller static sources may be more susceptible to fluctuations in adequacy during periods of draught than would larger bodies of water. In other words, while a major river may still provide an adequate supply during a draught, a small stream may drop below the usability level. The driver/operator must always be cognizant of current conditions before committing to a particular location.

Figure 12.7a Small ponds may supply adequate water for some fire fighting operations.

Figure 12.7b Small streams can provide water for drafting if adequate water depth is available and the water is moving fast enough.

The adequacy of a small stream may be evaluated using the following formula:

EQUATION O

Q (U.S.) = A x V x 7.5

Where Q = Flow in gpm

A = Area in ft² (width x depth)

V = Velocity in ft/min

7.5 = A constant (the number of gallons per ft³)

Example 5

Evaluate the adequacy of a small stream that is 10 feet wide with an average depth of 2 feet flowing approximately 15 feet per minute.

Q = A x V x 7.5

Q = (10 x 2)(15)(7.5)

Q = (20)(15)(7.5)

Q = 2,250 gpm

EQUATION P

Q (Metric) =A x V x 1000

Where Q = Flow in L/min

A = Area in m² (width x depth)

V = Velocity in m/min

1000 = A constant (the number of liters per m³)

Example 6

Evaluate the adequacy of a small stream that is 3 meters wide with an average depth of 0.6 meters flowing approximately 4.5 meters per minute.

Q = A x V x 1000

Q = (3 x 0.6)(4.5)(1000)

Q = (1.8)(4.5)(1000)

Q = 8 100 L/min

In order to effectively evaluate the adequacy of a pond or small lake, the driver/operator must have an approximate idea of the surface area and depth of the water. The rule of thumb for evaluating pond and small lake capacity is that every 1 foot (0.3 m) of depth for an area of 1 acre (0.4 ha) (approximately the size of a football field) provides 1,000 gpm (4 000 L/min) for 5 hours (Figure 12.8).

Accessibility of the Natural Static Water Supply Source

The second major factor in the usability of a static water supply source is the ability of the fire apparatus to get close enough to it to set up an effective drafting operation. Driver/operators must be conscious of the various types of problems related to water source accessibility

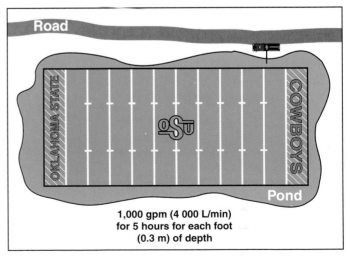

Figure 12.8 An area about the size of a football field and 1 foot (0.3 m) deep supplies 1,000 gpm (4 000 L/min) for 5 hours.

and the solutions for each. Common problems include the following:

- Inability to reach the water with a pumper
- Wet or soft ground approaches
- Inadequate depth for drafting
- Silt and debris
- Freezing weather

Inability to Reach the Water with a Pumper

Areas are often encountered where pumpers are not able to reach the water due to the type of terrain encountered. This includes problems such as the following:

- Bridges that are too high above the water's surface (Figure 12.9)
- Bridges that will not support the weight of fire apparatus
- Extremely high banks (Figure 12.10)
- Terrain that will not allow the apparatus close enough to reach the water with intake hoses

Figure 12.9 Obviously, this bridge is too far above the water's surface to be useful for drafting. Note the flood depth gauge on the bridge support.

Fire departments can avoid these problems by using public boat launching facilities, by constructing gravel drives, by installing dry hydrants, or by clearing brush to drafting points of high usage (Figure 12.11). Appropriate drafting sites should be noted in pre-incident plans. Bridges that will support apparatus and that are reachable by intake hose should especially be noted. In circumstances where the inaccessible water supply source is *the only* possible source, it may be necessary to position portable pumps at the water supply source. The portable pumps may then supply a nearby pumper that in turn is supplying a relay operation or water shuttle filling operation (Figure 12.12).

Figure 12.10 The drafting depth in the foreground would be unsuitable, However, if the pumper were positioned a few hundred feet (meters) farther down the bridge, drafting would be possible.

Figure 12.11 Boat ramps make excellent drafting sites.

Figure 12.12 In some cases, it may be necessary to use portable pumps to pump from an otherwise inaccessible static water supply source.

Wet or Soft Ground Approaches

Weather conditions can also limit access to static water sources. Wet ground approaches are troublesome for fire department apparatus. Wet approaches can trap apparatus, effectively blocking access to the water source (Figure 12.13). Grass and vegetation can hide soft ground spots. Even on ground that is marginally traversable, after a vehicle stops for a period of time, settling may occur. Personnel and towing apparatus will then have to be called to free the stuck vehicle.

Driver/operators who operate in cold weather climates must remember that frozen ground that allows the apparatus to be safely driven across at first may later thaw out and cause the apparatus to sink in place. Two common causes for ground thawing are heat from the apparatus exhaust and warm water in the apparatus water tank spilling onto the ground. If the apparatus is in an effective position and pumping water when the sinking starts to occur, operations should continue until the need for water has ended. Arrangements can then be made to free the apparatus.

Even in favorable weather conditions, the terrain around the water supply source may be too soft to support the weight of the apparatus. This includes land that is very marshy or that has a high sand content. Driver/operators must be familiar with the normal soil conditions within their jurisdiction.

Inadequate Depth for Drafting

The water depth needed for drafting is an important operational consideration. A depth of 2 feet (0.5 m) of water above and below a barrel-type strainer is a good rule of thumb for minimum drafting depth, although lesser depths have been used successfully. If the water is not deep enough for drafting with a barrel-type strainer, special low-level or floating strainers may be used. Low-level strainers are most commonly used to draft from portable water tanks and can draw water down to a 1- to 2-inch (25 mm to 50 mm) depth. Floating strainers also allow safe drafting from water as shallow as 1-foot (0.3 m) deep.

In small streams where the water is running fast but the minimum draft depth is not adequate, a ladder and a salvage cover can be used to dam the stream and raise the water level to permit drafting. Use the following procedure to construct the dam:

Step 1: Spread a salvage cover on the ground. Place the ladder on one of the long sides (Figure 12.14).

Step 2: Roll up the ladder in the cover. Leave about 4 feet (1.2 m) of cover free to form a flap (Figure 12.15).

Step 3: Place the ladder and cover assembly across the stream, preferably at a point where the stream bottom is level (Figure 12.16).

Figure 12.13 The approach to the draft site must be solid. This site is fine during dry weather, but it could become a problem if it is wet and muddy.

Figure 12.14 Place the ladder on the tarp.

Figure 12.15 Roll up the ladder in the tarp.

Figure 12.16 Place the ladder and tarp in the stream.

Step 4: Stretch the flap upstream and anchor with rocks or with straight bars stuck through the grommets (Figure 12.17).

It is sometimes necessary to do some tucking at the ends to prevent serious leakage. It may be necessary to support the ladder near its midpoint to prevent the weight of the water buildup from bowing the ladder (Figure 12.18). This may not create a deep enough pocket of water for drafting with a fire department pumper, but it will be adequate for drafting with a portable pump. In this case, the small amount of water is better than having no water at all.

Silt and Debris

A number of operational problems are presented by static water supply sources that contain excess silt and debris, including the following:

- Clogging the strainer, resulting in reduced water intake

- Seizing-up or damage to fire pumps

- Clogging fog stream nozzles

All hard intake lines should have strainers attached when drafting from a natural source. The intake hose should be located and supported so that the strainer does not rest on or near the bottom. Either a single ladder or a roof ladder can be used for this purpose. The intake hose with strainer attached is brought through the two bottom rungs or through the second and third rungs from the bottom depending on the steepness of the bank (Figure 12.19). This procedure tilts the strainer toward horizontal and keeps it off the bottom.

A preferable method for avoiding silt and debris is by the installation of a dry hydrant (Figure 12.20). A dry

Figure 12.18 If the water is causing the ladder to bow, place some type of support behind the middle of the ladder.

Figure 12.19 The strainer may be kept off the bottom by weaving the hose and strainer between two ladder rungs.

Figure 12.17 It may be necessary to hold down the tarp flap with rocks.

Figure 12.20 Dry hydrants speed drafting operations.

hydrant allows easy access to natural sources without the set-up time required for a regular drafting operation. It also avoids the wet ground approach problem and can circumvent the problem of silt and debris when installed properly. The proper installation of dry hydrants was discussed in Chapter 5.

Regardless of intake method, any operation that causes salt water or dirty water to be drawn into the pump requires that the pump be flushed with clean fresh water after the water supply operation is completed. This helps prevent corrosion to the pump and its components. It also ensures that all debris that could clog piping, strainers, relief valves, or nozzles is removed from the system.

Freezing Weather

A major problem in many jurisdictions is that winter freezing can effectively block the majority of static supplies. Surface freezing can make access difficult at best, while shallow ponds and lakes may freeze to near bottom, rendering them useless. Over the years, a variety of methods have been used to attempt to aid access to frozen ponds and lakes. These include the following:

- Barrels filled with an antifreeze solution are floated on the water's surface before the water freezes. If the need to draft arises, the top and bottom of the barrel may be knocked out to provide an access hole for the intake hose and strainer.

- Wooden plugs or plastic garbage cans are stabilized at a location so that they may be driven through the ice if the need to draft arises.

Both of these operations require the fire department to have these devices in place before the seasonal freeze. They are also dependent on local or state environmental regulations that may prohibit them from being performed.

In cases where no provisions for quick access through the ice have been made, it may be necessary to cut a hole through the ice for the intake hose and strainer. The traditional tools for doing this include the axe, chain saw, and power auger (Figure 12.21). Whichever is used, firefighters should be extremely cautious when operating on the ice. Make sure that the strength of the ice structure is not compromised to the point that it may fail and cast firefighters into the water. Provisions for making an ice rescue should be on the scene before firefighters begin work on the ice. If the area being breached is more than chest deep, it may be advisable for the firefighters on the ice to wear personal flotation devices (PFD) (Figure 12.22).

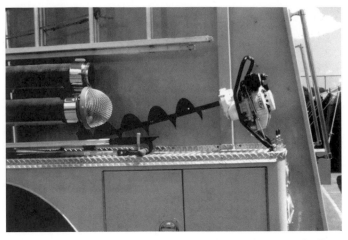

Figure 12.21 Power augers may be carried on the apparatus to facilitate breaching ice for drafting. *Courtesy of Rich Mahaney.*

Figure 12.22 Firefighters who are on the ice should wear PFDs if they are not certain that the water beneath them is less than waist deep.

Man-Made Static Water Supply Sources

Man-made static water supply sources provide a reliable source of water for fire protection. In some cases, these sources were located and constructed with fire protection concerns specifically in mind. Common types of man-made static water supply sources include the following:

- Cisterns
- Private water storage tanks
- Ground reservoirs
- Swimming pools
- Agricultural irrigation systems

Cisterns

Cisterns are underground water storage receptacles that are usually found in areas that are not serviced by a hydrant system (Figure 12.23). Cisterns typically receive their water from wells or rainwater runoff. Their primary purposes are usually for domestic or agricultural use; however, some are placed specifically for fire department use. Some urban and suburban jurisdictions have cisterns placed throughout their communities as a backup in the event of a failure of the main water supply system. Fire departments that are located in regions subject to harsh winter weather often have cisterns beneath the apparatus bay floors of their fire stations. Lacking any other accessible water supply source in their jurisdiction, this allows the apparatus to be pulled inside the fire station and quickly filled during inclement weather.

Cisterns vary in capacity; however, sizes from 10,000 to 100,000 gallons (40 000 L to 400 000 L) are common. Water in the cistern may be accessed by firefighters in one of two manners. Some are simply equipped with a manhole-type cover that is removed to provide access for the intake hose and strainer. Others may have a dry hydrant arrangement attached to them that allows for quick connection by a fire department pumper.

Freezing can be a problem with cisterns if they are not below the frost line. Placing the intake below the frost level in a manner similar to the dry hydrant intake placement allows year-round access to the source.

Private Water Storage Tanks

Private water storage tanks are commonly found on residential, industrial, and agricultural properties. These tanks range in size from several hundred to many thousands of gallons (liters). The tanks may be at ground level or elevated. Technically, elevated water tanks are not a static source as they have a head pressure associated with them in proportion to their height.

Fire department personnel should not be overreliant on private water storage tanks, particularly smaller ones (Figure 12.24). Each of these tanks should be identified in pre-incident planning and its capacity noted. Keep in mind that they may not be totally full at all times, further reducing their reliability. Fire departments should encourage the owners of these tanks to make sure that they are equipped with the appropriate connections that allow fire apparatus to hook into them in an expedient manner.

Ground Reservoirs

Ground reservoirs are man-made impoundments that have the same characteristics of a pond or small lake.

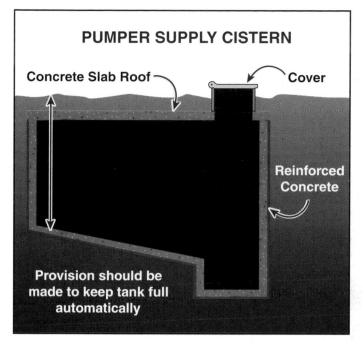

Figure 12.23 A typical cistern layout.

Figure 12.24 Some private properties have small elevated water tanks. This tank would not provide enough water for an extended fire fighting operation.

They are most commonly found on commercial or industrial properties and at municipal water treatment facilities (Figure 12.25). They typically contain many millions of gallons (liters) of water. Ground reservoirs tend to be more accessible than regular ponds or lakes. They commonly have improved roads servicing one or more sides of the reservoir. Dry hydrants may be provided to speed the use of the reservoir by fire department personnel. Otherwise, standard drafting operations using intake hose and strainers may be used to draw water from these sources.

Swimming Pools

Swimming pools provide another source of clean water for fire fighting. In some communities, property owners who seek permits to construct swimming pools must sign an agreement that allows the fire department access to and use of the water in the pool any time it is deemed necessary.

Pools have a unique access problem in that they are often fenced for security purposes (Figure 12.26). Pre-incident plans should contain information on the best way to access pools that may be used for fire protection purposes. Driver/operators must take extra care to avoid excess property damage when trying to reach the access points of swimming pools. Once the apparatus is in place, drafting may be performed in the same manner as it would in any other static water supply source. If the pool is inaccessible to the pumping apparatus, floating or stationary portable pumps may be used to pump the water to the apparatus.

When including swimming pools in pre-incident planning, the capacity of the pool should be included as a fire protection consideration. Most pools contain enough water to extinguish a fire in a single-family dwelling. However, they probably would not contain enough water to support an extended operation during a larger incident. The capacity of a pool can be calculated by one of the following formulas:

Square/Rectangular Pools

EQUATION Q

Capacity in gallons = L x W x D x 7.5

> Where: L = Length in feet
>
> W = Width in feet
>
> D = Average depth in feet
>
> 7.5 = Number of gallons per cubic foot

Figure 12.25 Ground reservoirs are commonly found at water treatment facilities.

Figure 12.26 Fences may make pools difficult to access.

Example 7

Calculate the volume capacity of a swimming pool that is 40 feet wide, 100 feet long, and has an average depth of 6 feet.

> Capacity in gallons = L x W x D x 7.5
>
> Capacity in gallons = (40)(100)(6)(7.5)
>
> **Capacity in gallons = 180,000 gallons**

EQUATION R

Capacity in liters = L x W x D x 1000

> Where: L = Length in meters
>
> W = Width in meters
>
> D = Average depth in meters
>
> 1 000 = Number of liters per cubic meter

Example 8
Calculate the volume capacity of a swimming pool that is 12 meters wide, 30 meters long, and has an average depth of 2 meters.

Capacity in liters = L x W x D x 1000

Capacity in liters = (12)(30)(2)(1000)

Capacity in liters = 720 000 liters

Round Pools

EQUATION S
Capacity in gallons = π x r² x D x 7.5

Where: π (Pi) = 3.14

r = radius or Z\x the diameter in feet

D = Average depth in feet

7.5 = Number of gallons per cubic foot

Example 9
Calculate the volume capacity of a swimming pool that is 24 feet in diameter with an average depth of 4 feet.

Capacity in gallons = π x r² x D x 7.5

Capacity in gallons = (3.14)(12)²(4)(7.5)

Capacity in gallons = 13,565 gallons

EQUATION T
Capacity in liters = π x r² x D x 1000

Where: π (Pi) = 3.14

r = radius or ½ diameter in meters

D = Average depth in meters

1 000 = Number of liters per cubic meter

Example 10
Calculate the volume capacity of a swimming pool that is 8 meters in diameter with an average depth of 1.3 meters.

Capacity in liters = π x r² x D x 1000

Capacity in liters = (3.14)(4)²(1.3)(1000)

Capacity in liters = 65 312 liters

Keep in mind that swimming pools may be partially or totally drained during winter months or during repairs. Thus, they cannot typically be relied upon as a year-round water supply source.

Some larger indoor and outdoor swimming pools may be equipped with dry hydrant-type connections that allow fire apparatus to be connected to them as a water supply source. For indoor pools, this connection may have an appearance similar to that of a fire department connection for a sprinkler or standpipe system. This connection should be clearly marked. Directions for opening any necessary valves should also be included.

The pump should be flushed with clear water after drafting from a swimming pool. This is necessary to remove any excess chlorine from the pump and piping to avoid damage.

Agricultural Irrigation Systems
Agricultural irrigation systems are another potential fire protection water source. Systems in some locations may flow in excess of 1,000 gpm (4 000 L/min). Irrigation water is generally transported by two methods: open canals and piped systems. Open canals may run through a property and may have several accessible points from which pumpers can draft. Piped systems are the more common type of irrigation water system. As with canals, several points may be accessible to fire department pumpers, but the piping must have connections that are usable at these locations (Figure 12.27). Special threaded adapters and specialized tools may be needed to use an irrigation system.

Figure 12.27 Fire departments in agricultural areas may carry adapters that allow them to connect to irrigation systems for water supply.

Chapter 13

Relay Pumping Operations

Job Performance Requirements

This chapter provides information that will assist the reader in meeting the following job performance requirements from NFPA 1002, *Standard on Fire Apparatus Driver/Operator Professional Qualifications*, 1998 edition. Particular portions of the job performance requirements (JPRs) that are met in this chapter are noted in bold text.

3-2.2 Pump a supply line of 2½ in. (65 mm) or larger, given a relay pumping evolution the length and size of the line and the desired flow and intake pressure, so that the proper pressure and flow are provided to the next pumper in the relay.

(a) *Requisite Knowledge:* **Hydraulic calculations for friction loss and flow using both written formulas and estimation methods, safe operation of the pump, problems related to small-diameter or dead-end mains, low-pressure and private water supply systems, hydrant cooling systems, and reliability of static sources.**

(b) *Requisite Skills:* **The ability to position a fire department pumper to operate at a fire hydrant and at a static water source, power transfer from vehicle engine to pump, draft, operate pumper pressure control systems, operate the volume/pressure transfer valve (multistage pumps only), operate auxiliary cooling sys-**

tems, make the transition between internal and external water sources, and assemble hose lines, nozzles, valves, and appliances.

6-2.2 Pump a supply line, given a relay pumping evolution the length and size of the line and pumping flow and desired intake pressure, so that adequate intake pressures and flow are provided to the next pumper in the relay.

(a) *Requisite Knowledge:* **Hydraulic calculations for friction loss and flow using both written formulas and estimation methods, safe operation of the pump, problems related to small-diameter or dead-end mains and to low-pressure and private water supply systems, hydrant cooling systems, and reliability of static sources.**

(b) *Requisite Skills:* **The ability to position a wildland apparatus to operate at a fire hydrant and at a static water source, transfer power from vehicle engine to pump, draft, operate pumper pressure control systems, operate the volume/pressure transfer valve (multistage pumps only), operate auxiliary cooling systems, make the transition between internal and external water sources, and assemble hose lines, nozzles, valves, and appliances.**

Reprinted with permission from NFPA 1002, *Standard on Fire Apparatus Driver/Operator Professional Qualifications,* © 1998, National Fire Protection Association, Quincy, MA 02269. This reprinted material is not the complete and official position of the National Fire Protection Association on the referenced subject which is represented only by the standard in its entirety.

In many fire situations, the water source is remote to the fire scene. In order to supply water to the fire, relay pumping operations must be employed. A *relay operation* uses a pumper at the water supply source to pump water under pressure through one or more hoselines to the next pumper in line. This pumper, in turn, boosts the pressure to supply the next pumper, and so on, until water reaches the fireground apparatus. This chapter provides information on the various compo-

nents of a relay operation, factors that influence the capability of the relay operation, and various methods for establishing a relay operation.

Relay Apparatus and Equipment

A variety of different types of apparatus, hose, and equipment may be used to establish a relay pumping operation. Most commonly, standard fire department pumpers are used to lay hose and perform the pumping duties in a relay operation. However, any type of apparatus equipped with a sufficient fire pump, including pumper-tankers (pumper-tenders) and aerial apparatus, may be placed in the middle of a relay to assist with pumping chores.

For ease of explanation, we use the following terms to describe the various functions of the apparatus used in a relay pumping operation (Figure 13.1):

- The *source pumper* (also called the *supply* pumper) is that pumper connected to the water supply at the beginning of the relay operation. The water supply may be a fire hydrant or a static water supply source. The source pumper then pumps water to the next apparatus in line.

- The *relay pumper*, sometimes also referred to as the *in-line pumper*, is a pumper or pumpers connected within the relay that receives water from the source pumper or another relay pumper, boosts the pressure, and then supplies water to the next relay pumper or the attack pumper.

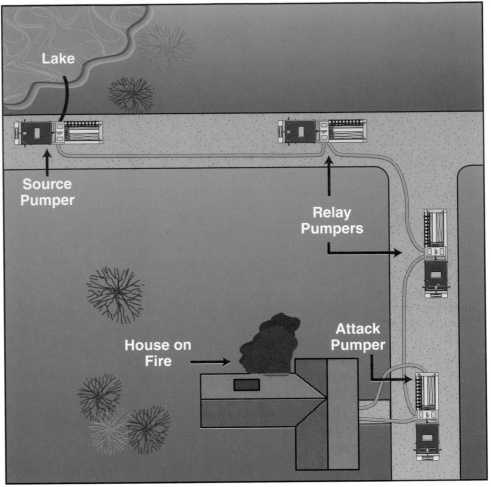

Figure 13.1 The various names for pumpers in a relay.

- The *attack pumper* is a pumper located at the fire scene that will be receiving water from the relay and supplying attack lines and appliances as needed for fire suppression.

Some fire departments use hose tenders to assist in the long hose lays associated with relay pumping operations (Figure 13.2). Hose tenders may or may not also be equipped with a fire pump that allows them to participate in the pumping operation once the hose is laid (Figure 13.3). Hose tenders usually carry a mile (1.6 km) or more of large diameter (4-inch [100 mm] or larger) hose (Figure 13.4). This hose may be carried in a traditional style hose bed or on a large, mechanically operated reel. Hose tenders also generally carry a wide assortment of relay valves, discharge manifolds, and other special water supply equipment used for relay pumping operations.

Both medium and large diameter hose may be used for relay pumping operations. Medium diameter hose (MDH) includes 2½- and 3-inch (65 mm and 77 mm) hoselines.

Figure 13.2 A hose tender.

When MDH is used for relay operations, generally two or three of these hoselines are laid. LDH ranges in size from 3½ to 12 inches (90 mm to 300 mm), with 4- and 5-inch (100 mm and 125 mm) being the most common sizes.

There are a variety of hose and pump appliances that may be used to assist with relay pumping operations. *Intake pressure relief valves*, also called *relay relief valves*, are intended to reduce the possibility of damage to the pump and discharge hoselines caused by water hammer when valves are closed too quickly or intake pressures otherwise rise dramatically. There are two basic kinds of intake pressure relief valves. One type is supplied by the

Figure 13.3 Hose tenders that are equipped with a fire pump may participate in relay pumping.

Figure 13.4 Hose tenders often carry several thousand feet (meters) of large diameter hose. This unit carries 1 mile (1.6 km) of 5-inch (125 mm) hose.

Figure 13.5 Relay pumpers are often equipped with external intake relief valves.

Figure 13.6 The valve operates when excess pressures are received at the intake.

Figure 13.7 Use the bleeder to discharge air from the hose as it is charged with water.

pump manufacturer and is an integral part of the pump intake manifold. The second type is an add-on device that is screwed onto the pump intake connection (Figure 13.5). In either case, these devices are preset to allow a set amount of pressure into the fire pump. If the incoming pressure exceeds the preset level, the valve activates and dumps the excess pressure/water until the water entering the pump is at the preset level (Figure 13.6). These valves should be set within 10 psi (70 kPa) of the static pressure of the water system supplying the pumper or 10 psi (70 kPa) above the discharge pressure of the previous pumper in the relay.

Most screw-on intake pressure relief valves are also equipped with a manual shut-off valve that allows the water supply to the pump to be stopped if so desired. Bleeder valves on the intake pressure relief valve allow air to be bled off as the incoming supply hose is charged (Figure 13.7). This is particularly important when using these devices in conjunction with large diameter supply hose. A large amount of air is pushed through the hose until a solid column of water reaches the valve. Bleeder valves may be located directly in the intake piping to the pump itself.

Relays dependent upon later-arriving mutual aid companies can set up an initial relay of lesser volume and greater spacing with in-line relay valves placed in the relay line for the incoming pumpers (Figures 13.8 a and b). These valves allow the late-arriving pumpers to hook up after the relay is operating and boost the pressure (and corresponding volume) without interrupting operations (Figure 13.9).

If an LDH relay pumping operation is intended to support more than one attack pumper at the fire scene, a discharge manifold may be used to break down the LDH into two or more hoselines that may then be connected to the attack pumpers (Figure 13.10). There are a number of different designs for these discharge manifolds. They typically have one LDH inlet and any combination of MDH and LDH discharges (Figure 13.11). If multiple MDH lines are being used in the relay, each line may be used to support a different attack pumper at the scene (Figure 13.12).

Figure 13.8a A relay valve. *Courtesy of Jaffrey Fire Protection.*

Figure 13.8b Some jurisdictions preconnect relay valves to the hose in the apparatus bed.

Figure 13.9 This diagram shows how a relay valve works.

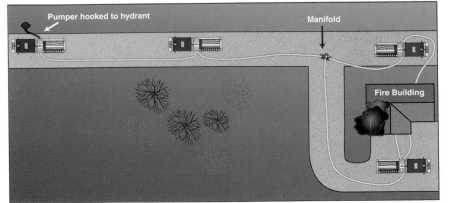

Figure 13.10 Some relays supply a manifold that, in turn, sends smaller supply hoses to multiple attack pumpers.

Figure 13.11 An LDH manifold.

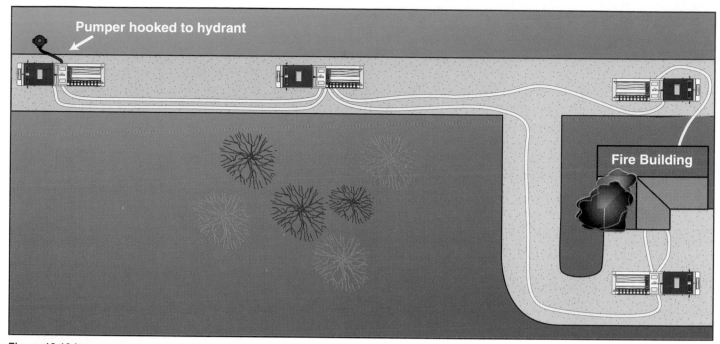

Figure 13.12 In some cases, each hoseline may supply a different attack pumper.

Relay Pumping Operational Considerations

Before entering into a discussion on the different types of relay pumping operations that may be used, it is necessary to cover some important operational considerations that apply to all types of relays. In the most basic sense, a relay operation is based on two things:

- The amount of water required at the emergency scene
- The distance from the emergency scene to the water source

In some cases, the relay must supply the total water necessary to complete the fire fighting operation. In other cases, the relay is used to supplement an inadequate municipal water supply system. In either case, the amount of water you need to flow has a major impact on the design of the relay.

The distance the water must be relayed is also very important. The longer the relay distance is, the more hose that will be necessary. More hose equates to more friction loss. As discussed in Chapters 7 and 8, friction loss is directly affected by the amount of water being flowed, the size of the hose being used, and the distance between pumpers. In simple terms, if it is desirable to increase the amount of flow through a relay, one of three things will be necessary:

- Increase the size of the hose or number of hoselines used in the relay.
- Increase the pump discharge pressure of the pumpers operating in the relay.
- Increase the number of pumpers in the relay.

There are certain limitations to accomplishing each of these three options. With regard to hose size, it would generally be impractical to shut down an existing relay to replace the hose being used with a larger hose (the largest hose available should have been laid at the beginning of the operation). It may be possible to have engine companies or hose tenders not already committed to pumping in the relay to lay an additional hoseline between the relay pumpers. Each pumper may have this additional hoseline attached to its pump and may begin flowing the hose when all pumpers are ready.

While it may be possible to have pumpers increase their pump discharge pressure, this does not necessarily mean that the volume of water through the relay will be increased. All fire department pumpers are rated to pump their maximum volume capacity at a net pump discharge pressure of 150 psi (1 000 kPa). If the pump operates at a pressure higher than 150 psi (1 000 kPa), the volume capability of the pumper is reduced proportionally. Depending on the length of the hose lay and the volume of water being flowed, you will eventually get to a point where increasing the pressure will not increase the volume.

When considering increasing the pressure, keep in mind that you are also limited by the pressure to which the fire hose you are using is annually tested. Tables 13.1 and 13.2 show the annual service test and recommended maximum operating pressures for various types of fire hose. These figures were obtained from NFPA 1961, *Standard on Fire Hose* and NFPA 1962, *Standard for the Care, Use, and Service Testing of Fire Hose Including Couplings and Nozzles.*

Table 13.1
Annual Service Test and Maximum Operating Pressures
for Fire Hose Manufactured Prior to 1987

Type of Hose	Annual Service Test Pressure in psi (kPa)	Maximum Operating Pressure in psi (kPa)
Attack Hose	250 (1 720)	225 (1 550)
Relay Hose – 3½- to 5-inch (90 mm to 125 mm)	200 (1 380)	185 (1 275)
Relay Hose – 6-inch (150 mm) and larger	150 (1 030)	135 (930)

Table 13.2
Annual Service Test and Maximum Operating Pressures
for Fire Hose Manufactured after 1987

Type of Hose	Annual Service Test Pressure in psi (kPa)	Maximum Operating Pressure in psi (kPa)
Attack Hose	300 (2 070)	275 (1 895)
Relay Hose – 3½- to 5-inch (90 mm to 125 mm)	200 (1 380)	185 (1 275)
Relay Hose – 6-inch (150 mm) and larger	200 (1 380)	185 (1 275)

At no time should a relay pumping operation result in discharge pressures that exceed the maximum operating pressure for the hose that is being used. Fire departments may specify and purchase hose that is designed for higher pressures than these NFPA minimums. If that is the case, the hose should be pumped at pressures that do not exceed 90 percent of the annual service test pressure.

Elevation pressure is also a factor in relay pumping operations. If the relay operation is pumping uphill, the pressure loss on the system is greater than that caused simply by friction loss. The reverse is true if the operation is going downhill. Elevation pressure is not affected by the amount of water to be moved, only by the topography.

Increasing the flow in a relay may be accomplished by placing additional pumpers in the relay. By shortening the length of hose each pumper has to supply, maximum pressures (and accordingly, maximum flows) may be maintained within the hose assembly. The down side to this possibility is that if in-line relay valves were not placed in the hose lay from the outset, it will be necessary to shut down the relay when the additional pumper(s) is (are) added.

On the other hand, in situations where low flow rates are required and LDH is available, the required spacing between pumpers may be so great that it exceeds the amount of hose carried on each pumper. In these cases, it may be necessary to call pumpers solely to lay their hose but not actually to participate in the pumping process.

Types of Relay Pumping Operations

Each fire department, or group of fire departments in a particular region, should have a standard operating procedure for the type of relay pumping operation they will use should the need arise. Agencies that have the potential to work together on relay pumping operations should perform relays in a training environment on a regular basis.

Some metropolitan, county, and other regional fire organizations have organized procedures for providing relay pumping capabilities at an emergency scene. In these situations, when the incident commander determines that it is necessary to use a relay to provide an adequate amount of water to the scene, he notifies the dispatch center to respond a relay task force or strike team. At this point, three to five pumpers, each with large capacity fire pumps and usually large diameter hose, are dispatched to the scene. These companies come in and establish the water supply independent of the companies already operating on the scene. Hopefully, the com-

panies that operate as part of the relay task force or strike team train together on a regular basis. There are two basic designs for relay pumping operations: the maximum distance relay method and the constant pressure relay method.

Maximum Distance Relay Method

The *maximum distance relay method* involves flowing a predetermined volume of water for the maximum distance that it can be pumped through a particular hose lay. By using Tables 13.3 or 13.4, the driver/operator can determine the distance that a certain flow may be pumped through the type of hose carried on the apparatus. Built into the figures in these tables is the consideration that a 20 psi (140 kPa) residual pressure is available at the next pumper in the relay. As well, the figures in these charts were based on a discharge pressure of 200 psi (1 400 kPa)

Table 13.3
Maximum Distance Relay Lengths in Feet

Flow in gpm	Hose size in inches						
	One 2½	One 3	One 4	One 5	Two 2½	One 2½ & One 3	Two 3s
250	1,440	3,600	13,200	33,000	5,760	9,600	14,400
500	360	900	3,300	8,250	1,440	2,400	3,600
750	160	400	1,450	3,670	640	1,050	1,600
1000	90	225	825	2,050	360	600	900
1250	50	140	525	1,320	200	375	500

Table 13.4
Maximum Distance Relay Lengths in Meters

Flow in L/min	Hose size in mm						
	One 65	One 77	One 100	One 125	Two 65s	One 65 & One 77	Two 77s
1 000	440	1 100	4 260	9 420	1 770	2 980	4 670
2 000	110	275	1 070	2 360	443	740	1 160
3 000	49	122	470	1 050	200	330	520
4 000	28	69	270	590	110	180	290
5 000	18	44	170	380	70	120	190

for 2½- and 3-inch (65 mm and 77 mm) hose and 185 psi (1 300 kPa) for 4- and 5-inch (100 mm and 125 mm) hose.

When considering the distances in Tables 13.3 and 13.4, the driver/operator must keep one other thing in mind. As we will learn more in depth in Chapter 16, all fire department pumpers are rated to flow their maximum volume at 150 psi (1 000 kPa), 70 percent of their maximum volume at 200 psi (1 350 kPa), and 50 percent of their maximum volume at 250 psi (1 700 kPa). Because Tables 13.3 and 13.4 are based on discharge pressures of 185 and 200 psi (1 300 kPa and 1 400 kPa), the following are the minimum pump capacities that must be used to achieve the flows/distances on the tables:

- 250 and 500 gpm (1 000 L/min and 2 000 L/min) flows: a 750 gpm (3 000 L/min) rated pumper

- 750 gpm (3 000 L/min) flow: a 1,250 gpm (5 000 L/min) rated pumper

- 1,000 gpm (4 000 L/min) flow: a 1,500 gpm (6 000 L/min) rated pumper

- 1,250 gpm (5 000 L/min) flow: a 1,750 gpm (7 000 L/min) rated pumper

If the local procedure is to use distances less than those on the chart, it will be possible to use smaller capacity pumps than those listed or remove one or more pumpers from the relay.

Using the figures in Tables 13.3 and 13.4, the number of pumpers needed to relay a given amount of water can be determined by using the following formula:

EQUATION U

$$\frac{\text{Relay Distance}}{\text{Distance From Table 13.3 or 13.4}} + 1 = \text{Total Number of Pumpers Needed}$$

Note that when using this formula, you always need to **round up** to the nearest whole number. For example, if the answer is 3.2, you actually need four pumpers to achieve the flow.

Example 1

If a single line of 3-inch hose is used, how many pumpers will be needed to supply 1,000 gpm to a fire scene that is 1,000 feet from the water source (Figure 13.13)?

From Table 13.3, it can be seen that the maximum distance water will flow at 1,000 gpm through 3-inch hose is 225 feet. Divide this figure into the distance to the fire; then add 1 for the attack pumper:

$$\frac{1,000}{225} = 4.4 + 1 = \textbf{5.4 or 6 pumpers needed}$$

Example 2

If a single line of 77 mm hose is used, how many pumpers will be needed to supply 4 000 L/min to a fire scene that is 300 meters from the water supply source (Figure 13.13)?

From Table 13.4, it can be seen that the maximum distance water will flow at 4 000 L/min through 77 mm hose is 69 meters. Divide this figure into the distance to the fire; then add 1 for the attack pumper:

$$\frac{300}{69} = 4.3 + 1 = \textbf{5.3 or 6 pumpers needed}$$

Example 3

If two lines of 3-inch hose are used, how many pumpers will be needed to supply 750 gpm to a fire scene that is 2,000 feet from the water source (Figure 13.14)?

Find the distance in Table 13.3 for two 3-inch hoses at 750 gpm. Then divide this distance into the total distance:

$$\frac{2,000}{1,600} = 1.25 + 1 = \textbf{2.25 or 3 pumpers needed}$$

Example 4

If two lines of 77 mm hose are used, how many pumpers will be needed to supply 3 000 L/min to a fire scene that is 600 meters away from the water source (Figure 13.14)?

Find the distance in Table 13.4 for two 77 mm hoses at 3 000 L/min. Then divide this distance into the total distance:

$$\frac{600}{520} = 1.15 + 1 = \textbf{2.15 or 3 pumpers needed}$$

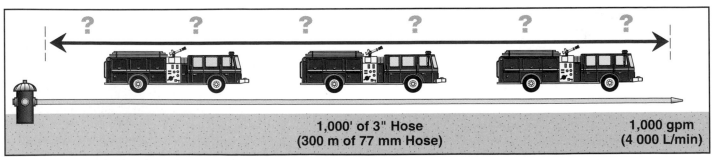

Figure 13.13 Examples 1 and 2.

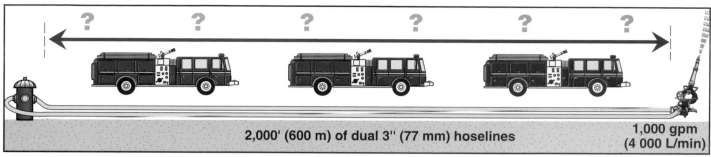

Figure 13.14 Examples 3 and 4.

When using this method to establish a relay, common sense must be used. For example, by checking the tables, we can see that 750 gpm may be flowed for 1,600 feet (520 meters) through dual 3-inch (77 mm) hoses. If the distance to the fire scene is 2,000 feet (600 m) away, it would not make sense to place 1,600 feet (520 m) between the source pumper and the relay pumper, with only 400 feet (80 m) remaining between the relay pumper and the attack pumper. In this case, it would make more sense to place the relay pumper more toward the middle of the supply hose.

As well, if there are engine companies with different sizes of supply hose on board, the distances between the pumpers may be varied accordingly. For example, if several engine companies are setting up a 1,000 gpm (4 000 L/min) relay using 5-inch (125 mm) hose, each pumper can be 2,050 feet (590 m) apart. If an engine shows up with a dual lay of 3-inch (77 mm) hose, it can still participate as long as the dual 3-inch (77 mm) lay is limited to 900 feet (290 m).

Constant Pressure Relay Method

The second type of relay pumping operation that fire departments may choose to use is the constant pressure relay. This relay method establishes the maximum flow available from a particular relay setup by using a constant pressure in the system. The constant pressure relay depends on a consistent flow being provided on the fireground. The attack pumper can maintain this flow by using an open discharge or secured waste line to handle the excess beyond the flow being used in the attack lines.

A relay using constant pressure has several advantages if all driver/operators are properly trained in its use. Some of these advantages are as follows:

- It speeds relay activation. Each driver/operator knows exactly how much hose to lay out and how to pump it without awaiting orders.
- It requires no complicated calculations to be made on the emergency scene.
- Radio traffic and confusion between pump operators are reduced.

- The attack pumper driver/operator is able to govern fire lines with greater ease.
- Operators in the relay only have to guide and adjust pressure to one constant figure.

Forming the Constant Pressure Relay

Step 1: Position the attack pumper at the fire. The incident commander makes the size-up and determines the quantity of water needed. An initial attack may be made with the water carried on the pumper.

Step 2: Position the largest capacity pumper available at the water source if possible. The pumper's crew begins making the necessary connections to the water supply.

Step 3: Lay out the hose load from the relay pumpers according to the procedures used in your jurisdiction. Figure 13.15 shows the flows that may be achieved through 750 feet (225 m) of various hose lays. This length may be increased or decreased according to local policy. Always leave at least two sections of hose in reserve in the hose bed in the event that a hose failure occurs during the operation.

Step 4: Connect all supply lines to the pumpers in the relay.

Step 5: *The driver/operator for each pumper, except the source pumper:* Open an unused discharge gate if the pump does not have a relay relief valve. This allows the air from the hoselines to escape as the water advances up the hoseline.

Step 6: Pump 175 psi (1 200 kPa) from the pumper at the water source.

Step 7: *The driver/operator at the first relay pumper:* Close the unused discharge gate once a steady stream of water flows from it, then advance the throttle until 175 psi (1 207 kPa) is developed. Each successive driver/operator follows the same procedure.

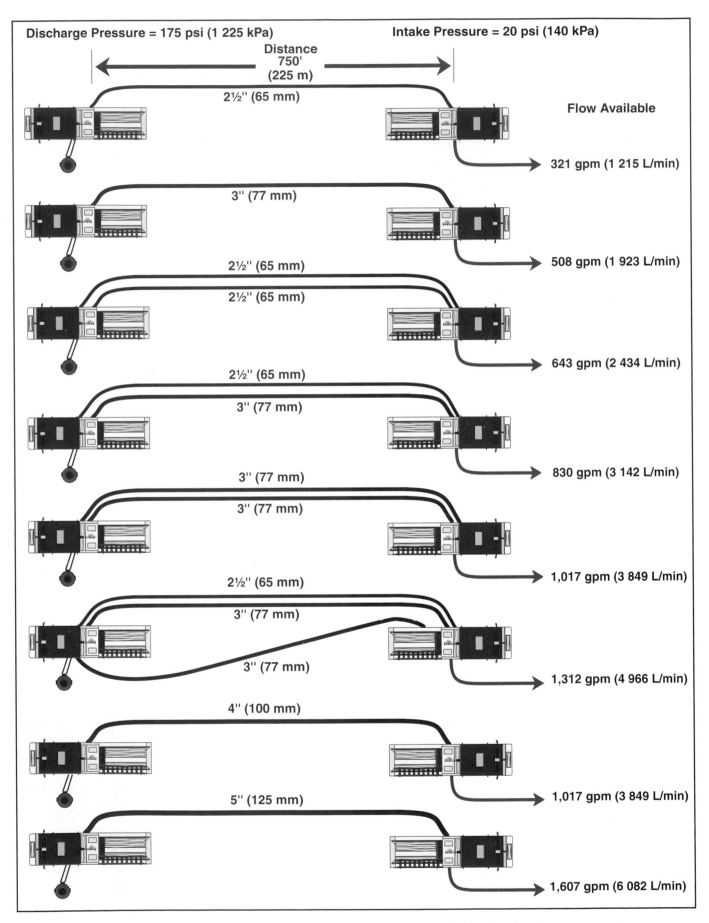

Discharge Pressure = 175 psi (1 225 kPa)   **Intake Pressure = 20 psi (140 kPa)**

Distance
750'
(225 m)

2½" (65 mm)

Flow Available

321 gpm (1 215 L/min)

3" (77 mm)

508 gpm (1 923 L/min)

2½" (65 mm)
2½" (65 mm)

643 gpm (2 434 L/min)

2½" (65 mm)
3" (77 mm)

830 gpm (3 142 L/min)

3" (77 mm)
3" (77 mm)

1,017 gpm (3 849 L/min)

2½" (65 mm)
3" (77 mm)
3" (77 mm)

1,312 gpm (4 966 L/min)

4" (100 mm)

1,017 gpm (3 849 L/min)

5" (125 mm)

1,607 gpm (6 082 L/min)

Figure 13.15 This chart shows some reasonable estimates of the flows that are available for the given hose lays.

Step 8: *Each driver/operator:* Set the pressure regulating device.

Step 9: *The attack pumper driver/operator:* Adjust the discharge pressure(s) to supply the attack line(s).

Step 10: Maintain the flow from the attack pumper during temporary shutdowns by using one or more discharge gates as waste or dump lines. Do not shut down attack lines unless absolutely necessary. If a hoseline bursts, open a discharge gate on the relay pumper before the rupture to dump water until the length is replaced.

Step 11: Lay additional hoselines between the apparatus in the relay if additional water supplies are needed on the fireground.

Driver/operators in a constant pressure relay should keep correcting their pump discharge pressure to 175 psi (1 200 kPa) until:

• Intake pressure from pressurized sources drops to 20 psi (150 kPa). If the intake pressure drops below 20 psi (150 kPa), there is a danger that the pump will go into cavitation. Cavitation can be recognized by the fact that increasing the engine rpm does not result in an increase in discharge pressure. This is a signal that the relay's maximum capacity has been reached. The results of cavitation can be pump damage and/or disruption of the flow to the fireground.

• Operating the hand throttle does not result in an increase in rpm. (Engine has reached governed speed.)

The constant pressure figure of 175 psi (1 207 kPa) can be modified as needed for:

• Variations in relay pumper spacing (increase pressure for greater spacing, decrease for lesser spacing)

• Severe elevation differences between source and fire (decrease discharge pressure when pumping downhill, increase when pumping uphill)

• Increases in needed fire flow

• Large diameter hose (requires lower discharge pressure to supply the same volume of water)

When increasing the relay pressure, the source pumper is adjusted until the desired pressure is reached. Each successive pumper is similarly adjusted. When a decrease in the flow is required, the attack pumper throttles down. One way to do this is by opening the dump line to relieve excessive water. The source pumper should discharge its dump line back into the water supply source. Successively, the relay pumpers toward the water source throttle down to the desired pressure. The water supply officer or incident commander must realize the flow and

pressure limitations of a given relay setup and should not attempt to exceed the capabilities of apparatus and hose.

There are some basic guidelines that should be followed for all relay pumping operations, regardless of which type is chosen for use by a particular jurisdiction. All driver/operators must be very familiar with these guidelines in order to be successful in participating in a relay pumping operation.

Putting a Relay into Operation

A relay pumping operation always begins with the source pumper. As a rule, the largest capacity pumper should be used at the source. If the relay is being supplied from draft, the source pumper will have to develop a higher net pump discharge pressure than the other pumpers in the relay. This higher net pump discharge pressure is needed because the relay pumpers will have a residual pressure at the intake to reduce the amount of pressure needed from the pump. It is important to remember that the maximum capacity of the relay is determined by the capacity of the smallest pump and the smallest hoseline used within the relay.

Once the water supply has been established, the source pumper opens an uncapped discharge or allows water to waste through a dump line until the first relay pumper is ready for water (Figure 13.16). This is particularly important if operating from a static water supply source. Failure to keep water through the pump could result in a loss of prime, thus delaying the operation. The discharge pressure is built up to the desired value by increasing the throttle. While doing this, the dump line discharge valve should be slowly closed to keep from wasting all the water from the water supply.

The relay pumper should be waiting for water with the dump line or discharge open and the pump out of gear.

Figure 13.16 The source pumper should keep a waste line flowing to ensure that prime is not lost.

When both the source pumper and the relay pumper are ready, the discharge supplying the hoseline on the source pumper is opened while the valve on the dump line is closed in a coordinated action. The discharge to the supply line must be opened slowly to prevent a sudden discharge into the empty hoseline. The water then begins to move from the source pumper to the relay pumper. As the water fills the line, the air will be forced through the pump and out the open dump line of the relay pumper. When water comes out of the dump line, the pump on the relay pumper can be engaged.

NOTE: If the waiting period for receiving water from the source pumper or another relay pumper is only going to be a few minutes, the pump on the waiting pumper may be engaged before receiving the water. This saves time in the overall setup of the relay pumping operation. Be alert for delays that might necessitate taking the pump out of gear to avoid damage caused by running it dry for too long. See Chapter 11 for more information.

Another option would be to start the relay, or at least fill the hoselines, with water from the apparatus water tank. This works best on short relays using MDH. On long lays of LDH, it may not even be possible to completely fill the hose between the two pumpers. Follow local policies on using tank water to start a relay operation.

It is most desirable to maintain an intake pressure of 20 to 30 psi (150 kPa to 200 kPa). If the relay pumper is receiving an intake pressure greater than 50 psi (350 kPa), the valve to the dump line on the relay pumper must be adjusted to limit the residual to the 50 psi (350 kPa) maximum. The pump discharge pressure increases as the throttle setting on the relay pumper is increased; therefore, the valve to the dump line will have to be gated down to maintain the 50 psi (350 kPa) residual pressure. If the dump line is allowed to flow unrestricted, the friction loss would increase in the hoseline from the source pumper to the point that the pump would go into cavitation.

Once the pump discharge pressure on the relay pumper has reached the desired pressure with the water being discharged, this portion of the relay has been established and no further adjustments should be necessary. When the next relay pumper is ready for water, the same procedure will be followed. The first relay pumper opens the discharge valve supplying the next pumper while closing the dump line on a coordinated basis. The operator does this while carefully observing the intake gauge to maintain the 50 psi (350 kPa) residual pressure. The next relay pumper allows the water to discharge through the dump line and follows the same procedure used by the first relay pumper in receiving water from the source pumper.

This action can go on until the relay is complete, building even the most complex relay section by section with a minimum of delay.

When the water reaches the attack pumper, the operator should bleed out the air from the supply line by opening the bleeder valve on the intake being used. The intake valve on the attack pumper can then be opened and a water supply established through the relay. There is a need for a dump line on the attack pumper as well. When one of the attack lines is shut down, an alert operator can open the dump line to allow water to flow, thus preventing a dangerous pressure buildup in the relay.

Operating the Relay

Once the relay is in operation and water is moving, all pump operators set their automatic pressure control devices to an appropriate level. The use of automatic pressure control devices is essential when operating in a relay due to the cumulative nature of pressure increases when changes in flow occur. The auxiliary cooler can be adjusted as necessary to maintain the proper engine operating temperature over the extended periods of time that are often necessary during a relay.

If the pumper is equipped with an intake relief valve, it should be put in service by taking off any caps on the outlet or by opening any valves associated with it. If the valve is readily adjustable, it should be set to discharge at 10 psi (70 kPa) above the static pressure of the water system it is attached to or 10 psi (70 kPa) above discharge pressure of the previous pumper in the relay. At no time should the relief valve be set for a higher amount than the safe working pressure of the hose. At this setting, the valve will not open and cause excessive fluctuations when minor changes in flow occur.

If the attack pumper is equipped with a readily adjustable intake relief valve, set it between 50 and 75 psi (350 kPa and 525 kPa) to establish a stable operating condition for the attack pumper. If an attack line is shut down or the amount of discharge changes, the friction loss in the supply line decreases and the residual pressure increases. The intake relief valve opens, allowing water to dump out of the intake. When this happens, the flow through the supply line and the entire relay increases and pressures return to their original settings. Additional flow requirements by the attack pumper (another attack line is opened) reduces the residual pressure and causes the relief valve to close. The dumping action then stops, allowing the pressure to again return to the original setting.

There is a tendency to overreact when operating in relay. Small variations in pressure are not significant and no attempt should be made to maintain exact pressures. As long as the intake pressure does not drop below 10 psi (70 kPa) or increase above 100 psi (700 kPa), no action should be required. Changing the pressure at any of the pumpers in a relay operation has some effect on the others. Excessive changes can result in constantly varying pressure throughout the relay. In some cases, it takes a long period of time for a pressure change to actually occur in a long relay. This time delay is often responsible for overcorrection errors that can have negative effects for the entire relay operation.

Effective relay operations, like all other fireground operations, require good communications. Each unit in the relay must be aware of the actions of other units so that operations can be coordinated properly. Radios are an obvious means of communication but must be used cautiously (Figure 13.17). With all the activity on the fire scene, too much radio usage in establishing a relay can hamper not only fire fighting activities but also efforts to establish the water supply. When pumpers are within sight of each other, hand signals can be used. In extreme cases, messengers on foot can still be effective.

Where additional radio frequencies are available, one channel should be dedicated to coordination of the water supply operation. Once the water is moving, however, a minimum of communication should be required. When units involved in the relay are equipped with radios that are incompatible, portable radios may be useful. Ambulances or utility units that are radio equipped and not otherwise occupied can be used to establish communications throughout the relay.

Shutting Down the Relay

Relay operations should be shut down from the fire scene first. If the source pumper is shut down while the rest of the relay is still operating, the pumpers will run out of water and cavitation can result. Starting with the attack pumper, each operator should slowly decrease the

Figure 13.17 Headsets make it easier for the driver/operator to monitor radio traffic.

throttle, open the dump line, and take the pump out of gear. Once all of the pumpers are shut down, the hose may be drained and readied for reloading.

For more information on relay pumping and troubleshooting these operations, refer to Table 11.1 in Chapter 11 and Chart 13.1 at the end of this chapter.

OPERATING IN RELAY

Step 1: Positioning Pumpers; Connecting Lines

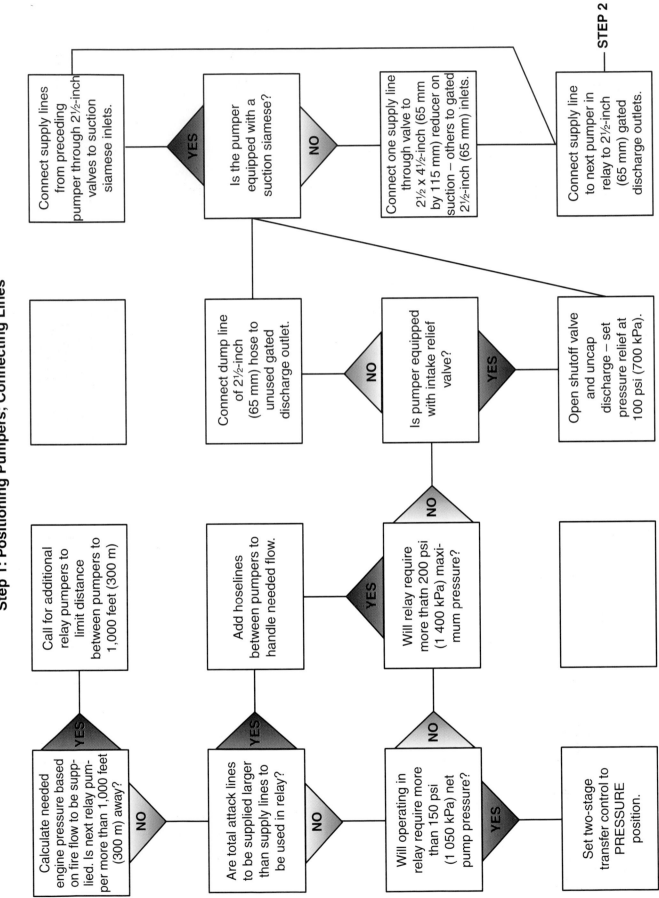

OPERATING IN RELAY
Step 2: Putting the Relay in Operation

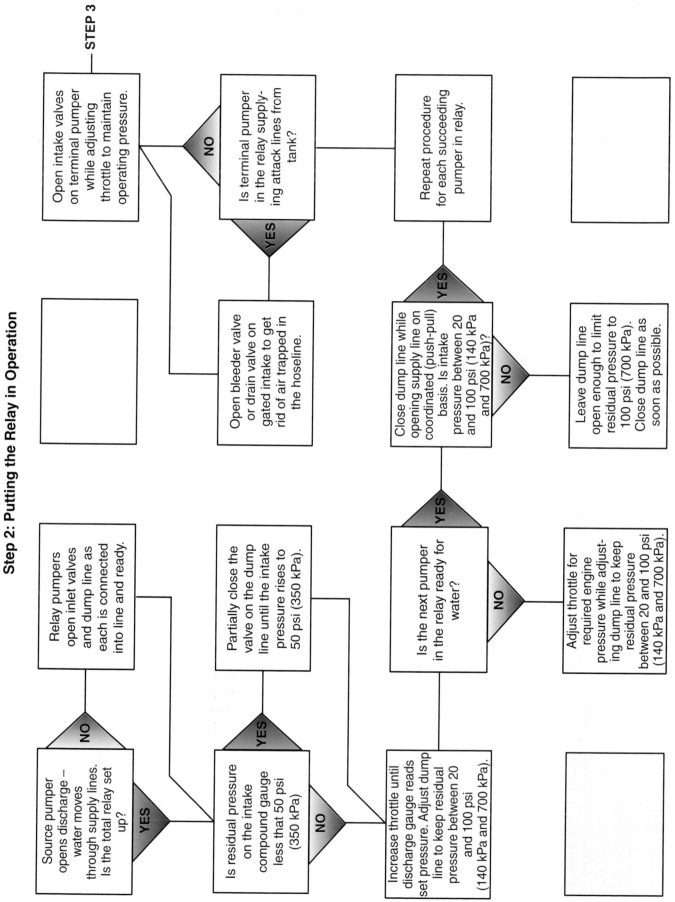

STEP 3

Open intake valves on terminal pumper while adjusting throttle to maintain operating pressure.

Is terminal pumper in the relay supplying attack lines from tank?

NO / **YES**

Open bleeder valve or drain valve on gated intake to get rid of air trapped in the hoseline.

Repeat procedure for each succeeding pumper in relay.

Close dump line while opening supply line on coordinated (push-pull) basis. Is intake pressure between 20 and 100 psi (140 kPa and 700 kPa)?

YES / **NO**

Leave dump line open enough to limit residual pressure to 100 psi (700 kPa). Close dump line as soon as possible.

Relay pumpers open inlet valves and dump line as each is connected into line and ready.

Source pumper opens discharge — water moves through supply lines. Is the total relay set up?

NO / **YES**

Partially close the valve on the dump line until the intake pressure rises to 50 psi (350 kPa).

Is residual pressure on the intake compound gauge less that 50 psi (350 kPa)

YES / **NO**

Is the next pumper in the relay ready for water?

YES / **NO**

Increase throttle until discharge gauge reads set pressure. Adjust dump line to keep residual pressure between 20 and 100 psi (140 kPa and 700 kPa).

Adjust throttle for required engine pressure while adjusting dump line to keep residual pressure between 20 and 100 psi (140 kPa and 700 kPa).

OPERATING IN RELAY
Step 3: Maintain Operating Pressure

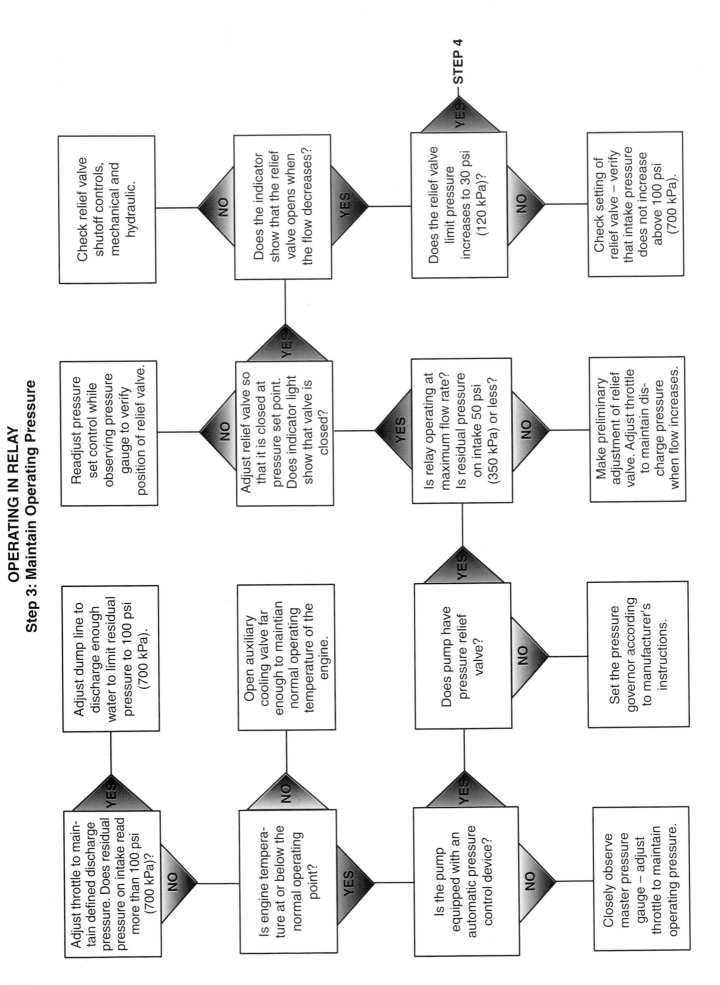

Adjust throttle to maintain defined discharge pressure. Does residual pressure on intake read more than 100 psi (700 kPa)?

— YES → Adjust dump line to discharge enough water to limit residual pressure to 100 psi (700 kPa).

— NO →

Is engine temperature at or below the normal operating point?

— NO → Open auxiliary cooling valve far enough to maintian normal operating temperature of the engine.

— YES →

Is the pump equipped with an automatic pressure control device?

— YES → Does pump have pressure relief valve?

— NO → Closely observe master pressure gauge – adjust throttle to maintain operating pressure.

Does pump have pressure relief valve?

— YES → Is relay operating at maximum flow rate? Is residual pressure on intake 50 psi (350 kPa) or less?

— NO → Set the pressure governor according to manufacturer's instructions.

Is relay operating at maximum flow rate? Is residual pressure on intake 50 psi (350 kPa) or less?

— YES → Adjust relief valve so that it is closed at pressure set point. Does indicator light show that valve is closed?

— NO → Make preliminary adjustment of relief valve. Adjust throttle to maintain discharge pressure when flow increases.

Adjust relief valve so that it is closed at pressure set point. Does indicator light show that valve is closed?

— YES → Does the indicator show that the relief valve opens when the flow decreases?

— NO → Readjust pressure set control while observing pressure gauge to verify position of relief valve.

Does the indicator show that the relief valve opens when the flow decreases?

— YES → Does the relief valve limit pressure increases to 30 psi (120 kPa)?

— NO → Check relief valve shutoff controls, mechanical and hydraulic.

Does the relief valve limit pressure increases to 30 psi (120 kPa)?

— YES → STEP 4

— NO → Check setting of relief valve – verify that intake pressure does not increase above 100 psi (700 kPa).

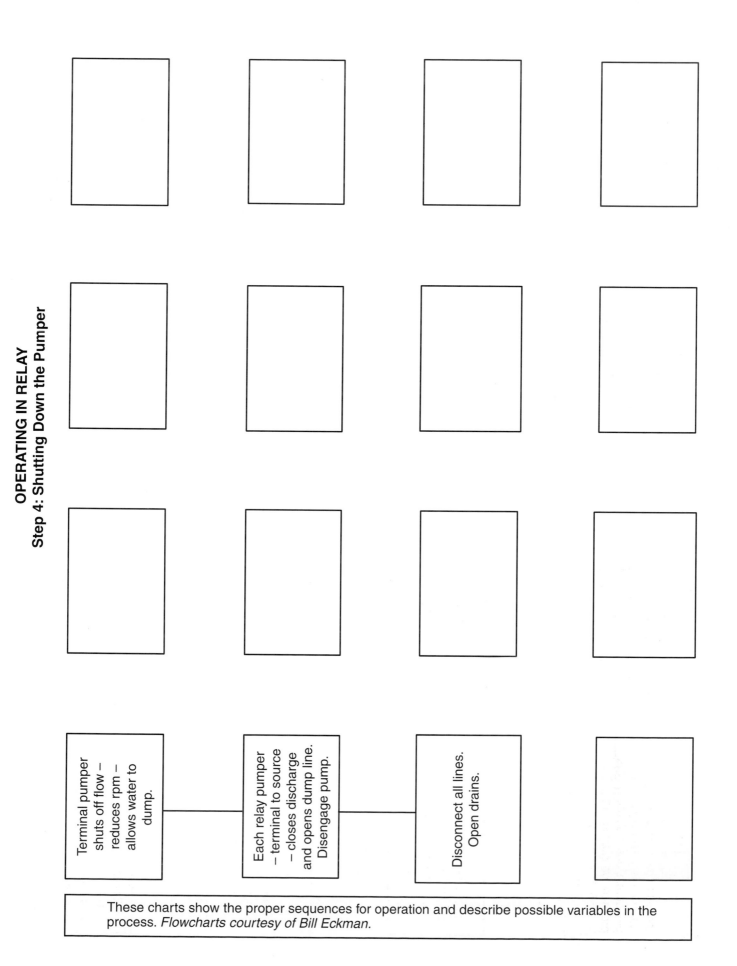

OPERATING IN RELAY
Step 4: Shutting Down the Pumper

Terminal pumper shuts off flow – reduces rpm – allows water to dump.

Each relay pumper – terminal to source – closes discharge and opens dump line. Disengage pump.

Disconnect all lines. Open drains.

These charts show the proper sequences for operation and describe possible variables in the process. *Flowcharts courtesy of Bill Eckman.*

Chapter 14

Water Shuttle Operations

Job Performance Requirements

This chapter provides information that will assist the reader in meeting the following job performance requirements from NFPA 1002, *Standard on Fire Apparatus Driver/Operator Professional Qualifications*, 1998 edition. Particular portions of the job performance requirements (JPRs) that are met in this chapter are noted in bold text.

8-2.1* Maneuver and position a mobile water supply apparatus at a water shuttle fill site, given a fill site location and one or more supply hoses, so that the apparatus is properly positioned, supply hoses are attached to the intake connections without having to stretch additional hose, and no objects are struck at the fill site.

(a) *Requisite Knowledge*: Local procedures for establishing a water shuttle fill site; **method for marking the stopping position of the apparatus; the location of the water tank intakes on the apparatus.**

(b) *Requisite Skills*: The ability to determine the appropriate position for the apparatus; maneuver apparatus into proper position; and avoid obstacles to operations.

8-2.2* Maneuver and position a mobile water supply apparatus at a water shuttle dump site, given a dump site and a portable water tank, so that all of the water being dis- charged from the apparatus enters the portable tank and no objects are struck at the dump site.

(a) *Requisite Knowledge*: Local procedures for operating a water shuttle dump site and **location of the water tank discharges on the apparatus.**

(b) *Requisite Skills*: The ability to determine the appropriate position for the apparatus, maneuver apparatus into proper position, avoid obstacles to operations, and operate the fire pump or rapid water dump system.

8-2.3* Establish a water shuttle dump site, given two or more portable water tanks, low-level strainers, water transfer equipment, fire hose, and a fire apparatus equipped with a fire pump, so that the tank being drafted from is kept full at all times, the tank being dumped into is emptied first, and the water is transferred efficiently from one tank to the next.

(a) *Requisite Knowledge*: Local procedures for establishing a water shuttle dump site and the **principles of water transfer between multiple portable water tanks**.

(b) *Requisite Skills*: **The ability to deploy portable water tanks, connect and operate water transfer equipment,** and connect a strainer and suction hose to the fire pump.

Reprinted with permission from NFPA 1002, *Standard on Fire Apparatus Driver/Operator Professional Qualifications*, © 1998, National Fire Protection Association, Quincy, MA 02269. This reprinted material is not the complete and official position of the National Fire Protection Association on the referenced subject which is represented only by the standard in its entirety.

Water shuttle operations, sometimes referred to as tender shuttles or tanker shuttles, are used to supply water to emergency scenes that are so remote from the water supply source that relay pumping is not practical. *Water shuttles* involve a process in which tenders (tankers) deliver their load of water to the emergency scene, travel to a filling site, reload with water, and then return to the emergency scene to dump again (Figure 14.1). During the course of a major emergency or large fire, each tender may make as many as a dozen or more round trips.

Relay pumping operations (covered in the previous chapter) and water shuttle operations are the two primary methods of providing water to remote emergency scenes. Each has its advantages and disadvantages. From a tactical standpoint, when given a choice between the two, relay pumping operations generally provide a more reliable continuous water supply with fewer apparatus. For example, four pumpers with 5-inch hose are needed to relay 1,000 gpm (4 000 L/min) to a fire scene that is 1 mile (1.6 km) from the water supply source. To provide continuous 1,000 gpm (4 000 L/min) for 1 mile (1.6 km)

Figure 14.1 Water shuttle operations are used when relay pumping is not feasible.

using a water shuttle, more apparatus — including at least two pumpers and several tenders — are required.

There is also a safety factor to consider. Once a relay pumping operation is established, there are no "moving parts." Other than the very remote possibility of a dramatic hose failure, there is little potential for danger associated with relay pumping. On the other hand, a water shuttle operation relies on the constant movement of apparatus between the emergency scene and the water supply source. The more apparatus movement that is required, the greater chance there is for a collision.

There is one factor in favor of the water shuttle: Once the incident is over, it is certainly less labor intensive to pack up a few portable water tanks and their associated hardware than it is to reload a mile (1.6 km) of large diameter hose. If multiple medium diameter (MDH) supply lines were laid, the amount of work to reload the hose is increased proportionally.

The choice between relay pumping operations and water shuttle operations is typically based on tradition and the type of equipment that a jurisdiction and its mutual aid agencies have to work. If they have traditionally relied on water shuttles as their supply method, this is what they will most likely choose for their next emergency. On extended operations, a water shuttle may be used early in the incident while a long relay pumping operation is being set up. Once the relay is in operation, the water shuttle may be discontinued. This situation requires coordinated planning so that the hose laying operation does not interfere with the shuttle route, causing an interruption in the water supply.

In this chapter, we cover the types of apparatus that are used in a water shuttle and the methods used to evaluate their effectiveness. Also examined are the three major components of a water shuttle operation: the fill site, the shuttle route, and the dump site.

There are many methods for performing a water shuttle operation. The methods and procedures described in this chapter have been found to provide the optimum volume of water supply. Departments that choose to deviate from these procedures may not be able to achieve the same flow rates that the methods in this chapter provide. Individual jurisdictions will have to adjust their procedures based on the available equipment and level of training they possess.

Water Shuttle Apparatus

There are two primary types of apparatus required to operate an effective water shuttle operation: pumpers and tenders (tankers). Each plays an important role in specific portions of the overall water shuttle operation.

Pumpers

An effective water shuttle operation requires at least two pumpers to be successful. One pumper positions at the water supply source and is used to fill empty tenders. This location is formally known as the *fill site*. The pumper that is located here is generally referred to as the *fill site pumper* (Figure 14.2). Depending on the number of tenders in the shuttle and the capacity of the water supply source, more than one fill site or fill site pumper may be used. This allows for two or more tenders to be filled simultaneously.

The second pumper is located at or near the emergency scene and is used to draft water from the portable water tanks. This location is commonly referred to as the *dump site*. The pumper that is located here is typically called the *dump site pumper* (Figure 14.3). Depending on the dump site's proximity to the emergency scene, the dump site pumper can also be the attack pumper or simply be relaying water to another pumper that is supplying the attack lines. This is discussed in greater detail later in this chapter.

Figure 14.2 The fill site pumper loads tenders that have arrived empty from the dump site.

Figure 14.3 The dump site pumper can supply attack lines directly or relay water to an attack pumper at the fire scene.

Figure 14.4 Some jurisdictions have custom-designed small fill site pumping apparatus. *Courtesy of Rich Mahaney.*

The pumpers used in a water shuttle operation can be the standard type of fire department pumper that was described in Chapter 3. Fill site pumpers — operating from static water supply sources — and dump site pumpers need to be equipped with hard intake hose and strainers so that they will be able to draft. Fill site pumpers should have a minimum pump capacity of 1,000 gpm (4 000 L/min) because NFPA 1901, *Standard for Automotive Fire Apparatus*, requires that tenders be designed to be filled at a rate of at least 1,000 gpm (3 785 L/min).

Jurisdictions that frequently perform water shuttle operations may have special pumping apparatus that is specifically designed for filling tenders. These special apparatus are usually light-duty trucks equipped with large-volume, auxiliary-powered, irrigation or trash pumps that discharge through large diameter hose. These fill site pumpers have capacities to 1,600 gpm (6 400 L/min) at a maximum of 80 psi (550 kPa) (Figure 14.4).

Other jurisdictions design fill site pumpers that are on a larger, all-wheel drive chassis. These units often have front-mounted fire pumps or front-intake connections that allow the apparatus to be nosed into static water supply sources (Figure 14.5).

Tenders/Tankers

Tenders (tankers) are the backbone of the water shuttle operation (Figures 14.6 a and b). As discussed in Chapter 2 and according to NFPA 1901, *tenders* are apparatus that carry at least 1,000 gallons (3 785 L) of water. The majority of water tenders in use today carry from 1,500 to 3,000 gallons (6 000 L to 12 000 L) of water, although some smaller and larger units are in service. Each department must choose the size of the tank for its apparatus based on local water requirements, road conditions, and bridge weight restrictions. Vehicle weight restrictions generally limit single, rear-axle trucks to a maximum tank capacity

Figure 14.5 Full-size pumpers, with all-wheel drive capabilities, make excellent fill site pumpers. *Courtesy of Warren Gleitsmann.*

Figure 14.6a Elliptical tenders are designed solely for shuttling water. *Courtesy of Joel Woods.*

Figure 14.6b Square body tenders resemble fire department pumpers and, in fact, may be used as such.

of 1,500 gallons (6 000 L). Where tanks of 1,500 gallon (6 000 L) capacity or larger are used, tandem rear axles, tri-axles, or semitrailer construction is required.

When designing or purchasing a tender, members of the department must consider exactly how that truck will be used in order to determine the appropriate tank size and other features needed on the apparatus. Tenders that have water tanks with less than 2,500 gallons (10 000 L) and quick unloading times are the most efficient tenders for use in water shuttle operations. Larger apparatus have longer filling and unloading times. They also are slower when traveling down the road and are more difficult to maneuver around the fill and dump sites. Tenders that have water tanks that exceed 2,500 gallons (10 000 L) are better suited for stationary water supply operations. Examples of these include a tender that is used as a nurse tanker to directly supply a pumper at a fire scene or a tender that refills brush fire apparatus or water-drop aircraft at a large wildland fire.

Tenders that are strictly used for water shuttle operations do not require a fire pump if they are equipped with a suitable gravity dump system. However, most departments choose to put some type of fire pump on the tender. These range from small PTO-driven pumps capable of flowing only 250 gpm (1 000 L/min) up to standard midship transfer-driven pumps that can flow 2,000 gpm (8 000 L/min). Tenders with 750 gpm (3 000 L/min) or larger fire pumps are often called pumper-tenders or pumper-tankers (Figure 14.7).

NOTE: In some regions, the term *pumper-tender* (*pumper-tanker*) is used to specifically describe an apparatus that has the primary function of a pumper with a large water tank. In contrast, the term *tender-pumper* (*tanker-pumper*) is applied to an apparatus that has the primary function to haul water but that is also equipped with a large fire pump. There is no standards-driven basis for the differentiation in these terms, and these are not universally accepted definitions.

As discussed in Chapter 4, some fire departments that cannot afford to purchase custom-manufactured tenders can convert other types of tank trucks to fire service use. Trucks that may be converted into tenders include petroleum tankers, milk trucks, vacuum trucks, and surplus military vehicles (Figure 14.8). This is a viable option to the purchase of manufactured pieces, as long as the weight limit and other capabilities of the vehicle chassis are not exceeded. These chassis frequently are not designed for the weight of water that is carried on them. Keep in mind that a gallon (liter) of water (8.33 pounds [1 kg]) weighs more than a gallon (liter) of gasoline or fuel oil (5.6 pounds [0.67 kg] and 7.12 pounds [0.85 kg] respectively). Multiplied by a thousand gallons (4 000 L) or more, this weight difference becomes significant. Another problem is that in many cases the tanks on these apparatus are improperly baffled for fire service use. Thus, when the vehicle is driven with a partially filled tank, liquid surges within the tank can result in the vehicle becoming out of control.

If an existing tank is being retrofitted to add a large diameter, direct tank discharge valve, follow manufacturer's instructions to ensure that the tank is not structurally damaged as a result of the installation. Departments that construct their own tenders should attempt to meet the requirements contained in NFPA 1901 as closely as possible.

Some of the more important tender requirements that are contained in NFPA 1901 are as follows:

- Tenders equipped with a fire pump must have a tank-to-pump line that is capable of supplying the pump with 500 gpm (1 900 L/min) until at least 80 percent of the water in the tank is emptied. The pump-to-tank fill line should be at least 2 inches (50 mm) in diameter.

Figure 14.7 Pumper-tenders may operate either as a shuttle or attack apparatus. *Courtesy of Joel Woods.*

Figure 14.8 Many departments with limited financial resources convert fuel delivery or military surplus vehicles into tenders.

- Tenders must be equipped with at least one external fill connection that is connected directly into the tank (Figure 14.9). This external fill connection should allow the tank to be filled at a minimum rate of 1,000 gpm (3 785 L/min). The external fill connection should be equipped with a valve, strainer, and a 30-degree elbow. Valves that are 3 inches (77 mm) or larger must be of the slow-closing design.

- Tenders must have at least one large tank discharge that is capable of discharging 90 percent of the tank volume at an average rate of 1,000 gpm (3 785 L/min) (Figure 14.10).

As we will learn later in this chapter, the two primary ways we can increase the efficiency of water shuttle operations are to decrease the amount of time required to fill and the amount of time to unload the tenders. All the other parts of the water shuttle, such as response time and travel time between the dump and fill sites, are fairly fixed constants. Attempting to make up time during road travel is dangerous and can have serious consequences. When designing a tender for water shuttle operations,

much attention must be paid to filling and dumping capabilities. Tenders that use MDH supply lines for filling should have at least two external fill connections piped directly to the tank (Figure 14.11). If LDH is used, one fill connection to the tank is adequate (Figure 14.12).

Depending on the apparatus manufacturer's design or purchaser's preference, the direct tank fill inlet(s) may be designed so that the tank is filled either from the

Figure 14.11 Two MDH lines are used to fill the tender.

Figure 14.9 A direct tank fill does not go through the pump.

Figure 14.10 Direct tank discharges provide for the fastest method of unloading the water tank.

Figure 14.12 Some jurisdictions use one LDH line to fill tenders.

bottom or top (Figures 14.13 a and b). Hydraulically, there is little advantage to one design or the other. Because most tender water tanks are less than 6 feet (2 m) tall, the back pressure created by a nearly full tank on a bottom fill inlet is less than 3 psi (0.5 kPa). This has no measurable slowing affect of the fill process. Tank fill inlets that are designed to fill the tank from the top should be designed so that the hose may be connected at a reasonable level to the ground. Fixed piping should then carry the water to the top of the tank. Hose connections that are near the top of the tank are difficult to connect to and may pose a fall hazard for firefighters assigned to make the hose connection (Figure 14.14).

The increased use of LDH in recent years has led many jurisdictions to using a single LDH line to fill tenders. In reality, this may have little advantage in fill site operations. After the LDH is used to fill the first tender, it is partially full of water when each subsequent tender arrives at the fill site. LDH is extremely heavy and difficult to move when partially full of water. Any advantages that LDH has in increased flow rates may be negated by longer handling times.

Although NFPA 1901 only requires one large tank discharge, more commonly referred to as a *dump valve*, be installed on a tender, it is highly recommended that each tender be equipped with at least three. One should discharge water from the rear of the apparatus, and there should be one on each side (Figure 14.15). Some apparatus may be equipped with front-dump valves, but these are difficult to design into most apparatus and typically have problems meeting the 1,000 gpm (3 785 L/min) flow requirement (Figure 14.16).

There are two primary types of large tank discharges used on tenders: gravity dumps and jet-assisted dumps. *Gravity dumps* simply rely on nature's gravitational pull to empty water from the tank. Gravity dumps usually employ 8-inch (200 mm) or larger round or square piping with a valve that extends to the exterior of the apparatus (Figure 14.17). On the outside of the apparatus there may be an extension of the dump piping that assists in directing the water into a portable tank. The actual valve on the gravity dump may be designed to be opened manually from the discharge location or remotely from the cab of the apparatus. Remotely operated valves may be pneu-

Figure 14.13a Some direct tank inlets are at the bottom of the tank.

Figure 14.14 LDH can be difficult to handle for fill site operations.

Figure 14.13b Some direct tank inlets are piped to the top of the tank. There is little hydraulic advantage of one method over the other.

Figure 14.15 This dump valve can be adjusted to dump from any one of three sides.

Figure 14.16 Some specialized tenders have been designed with front dumps.

Figure 14.17 Gravity dumps do not use any mechanized pressure source.

matically, electrically, or hydraulically actuated. Remote-controlled dump valves increase the safety of the operation by eliminating the need for someone to climb to the top of the apparatus tank. It also reduces the chance of someone being pinched between the apparatus and the portable tank.

Jet-assisted dumps, sometimes simply called *jet dumps*, employ the use of a small diameter in-line discharge that is inserted into the piping of the large tank discharge. The discharge is supplied by the fire pump on the apparatus. The in-line discharge creates a venturi effect that increases the water flow through the large tank discharge (Figure 14.18). There are four primary disadvantages associated with jet-assisted dumps:

- The apparatus must be equipped with a fire pump.
- The fire pump must be engaged before dumping the water from the tank. This adds time to the dumping operation. In most cases, the time saved by the increased flow from the tank is negated by the additional time needed to put the dump in operation.

- Water can still be discharged if the pump is not operating, but it will be a considerably lower rate than if the dump valve were designed for gravity dumping.
- They increase the cost of purchasing the apparatus.

In recent years, most departments purchasing or designing new tenders have chosen larger gravity dumps as opposed to jet-assisted dumps. Their operation is easier, and they are less expensive to install.

Regardless of what methods are used to fill and dump tenders, it is important that the water tank also be equipped with adequate tank vents (Figure 14.19). Failure to have adequate venting during quick-filling operations could result in a dramatic pressure failure of the tank. Failure to have adequate venting during dumping operations could result in a suction effect that collapses

Figure 14.18 A jet-assisted dump.

Figure 14.19 A typical tank vent. *Courtesy of Bob Esposito.*

the tank (Figure 14.20). The driver/operator must always make sure that the vents are completely open when filling and dumping is taking place. Some tenders are equipped with remote-controlled vent hatches. Others require a firefighter to climb to the top of the tank and manually open them. From a safety standpoint, the remote controlled vent hatches are preferable as they eliminate the fall hazard associated with climbing on top of the apparatus.

Figure 14.20 Although this is a milk tanker, this photo shows the dramatic damage that can occur if a tank is improperly vented during dumping. *Courtesy of David Grupp.*

Setting Up a Water Shuttle

The success or failure of a water shuttle quite often hinges on several crucial decisions that must be made at the very beginning of the incident. These decisions include:

- Location of the dump site

- Location of the fill site

- Route of travel for the tenders between dump and fill sites

It is preferable that many of these decisions be made in pre-incident planning for target hazards and geographical areas within each jurisdiction. The pre-incident plan can contain the best fill site, alternative fill sites, the dump site, and desired route of travel for all shuttle apparatus (Figure 14.21). By determining these in pre-incident planning, the Incident Commander or water supply group/sector supervisor does not have to make these decisions under emergency conditions. As well, the Insurance Services Office (ISO) gives extra rating credit to jurisdictions that have automatic aid agreements for water shuttle operations.

Selecting the Dump Site Location

The location of the dump site obviously should be in close proximity to the incident scene. However, the front and center of the incident may not always be the best location for the dump site. An example of this situation is when the fire scene is located down a narrow lane, driveway, or dead-end street. In these cases, it would be advantageous to locate the dump site at the intersection where

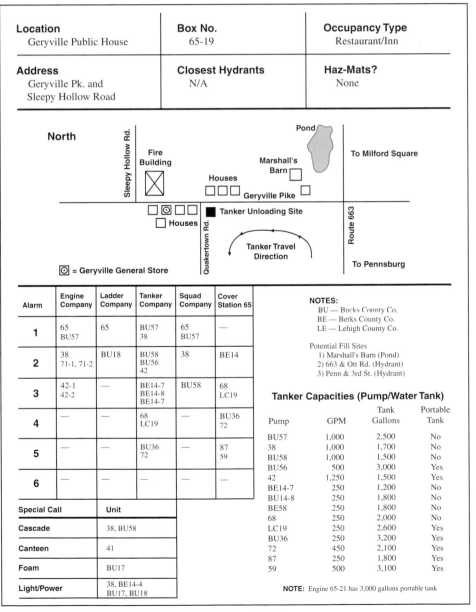

Location	Box No.	Occupancy Type
Geryville Public House	65-19	Restaurant/Inn
Address	**Closest Hydrants**	**Haz-Mats?**
Geryville Pk. and Sleepy Hollow Road	N/A	None

NOTES:
BU — Bucks County Co.
BE — Berks County Co.
LE — Lehigh County Co.

Potential Fill Sites
1) Marshall's Barn (Pond)
2) 663 & Ott Rd. (Hydrant)
3) Penn & 3rd St. (Hydrant)

Alarm	Engine Company	Ladder Company	Tanker Company	Squad Company	Cover Station 65
1	65 BU57	65	BU57 38	65 BU57	—
2	38 71-1, 71-2	BU18	BU58 BU56 42	38	BE14
3	42-1 42-2	—	BE14-7 BE14-8 BE14-7	BU58	68 LC19
4	—	—	68 LC19	—	BU36 72
5	—	—	BU36 72	—	87 59
6	—	—	—	—	—

Special Call	Unit
Cascade	38, BU58
Canteen	41
Foam	BU17
Light/Power	38, BE14-4 BU17, BU18

Tanker Capacities (Pump/Water Tank)

Pump	GPM	Tank Gallons	Portable Tank
BU57	1,000	2,500	No
38	1,000	1,700	No
BU58	1,000	1,500	No
BU56	500	3,000	Yes
42	1,250	1,500	Yes
BE14-7	250	1,200	No
BU14-8	250	1,800	No
BE58	250	1,800	No
68	250	2,000	No
LC19	250	2,600	Yes
BU36	250	3,200	Yes
72	450	2,100	Yes
87	250	1,800	Yes
59	500	3,100	Yes

NOTE: Engine 65-21 has 3,000 gallons portable tank

Figure 14.21 A sample water shuttle preincident plan.

the lane, driveway, or dead-end street meets a thorough-fare (Figure 14.22). The dump site pumper then relays water to the attack pumper located at the fire scene.

Even when the fire scene is located on a through street, the front of the fire scene may be blocked by early arriving apparatus. Because they have committed hoselines or aerial devices, it is not practical to reposition the trucks to provide through access for a water shuttle. In this case, the dump site may be established at an intersection close to the scene, again with the dump site pumper supplying water to attack apparatus at the scene (Figure 14.23). Large parking lots near the fire scene also make excellent dump sites (Figure 14.24).

Selecting the Fill Site Location

Each fire department should have knowledge of appropriate fill sites in its jurisdiction before an incident occurs. Driver/operators and fire officers should have a good knowledge of all the water system hydrants, dry hydrants, and suitable drafting locations within their response district. When the need to establish a water shuttle occurs, the Incident Commander or water supply group/sector supervisor should select the closest suitable water supply source to the scene. Depending on the situation, the closest *suitable* water supply source may not necessarily be the actual closest water supply source. For reasons of travel safety or water flow requirements, it sometimes may be better to establish a fill site at a location that is somewhat farther from the dump site than the closest source.

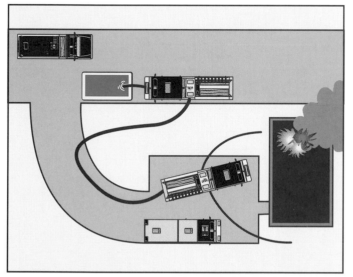

Figure 14.22 It may be more tactically sound to place the dump site at the end of the driveway and pump water to the attack pumper.

Figure 14.24 Parking lots are ideal locations for dump sites.

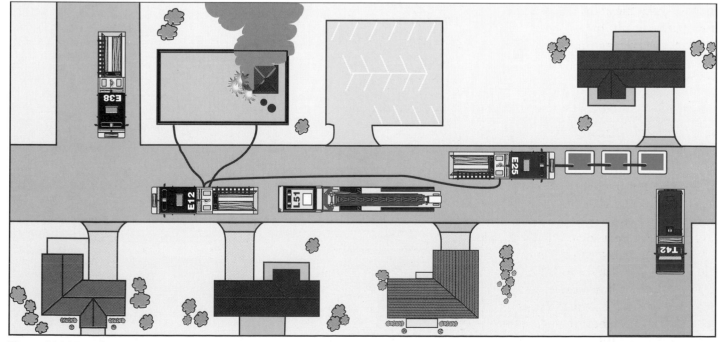

Figure 14.23 Locating the dump site at an intersection may minimize apparatus jockeying.

For example, suppose a fire incident calls for a water shuttle to provide a flow of 750 gpm (3 000 L/min) for an extended period of time. The two closest water supply sources to the dump site are a 250 gpm (1 000 L/min) rural water system hydrant that is 1 mile (1.6 km) from the scene and a well-maintained dry hydrant that is supplied from a large lake that is 2 miles (3.2 km) away. In this case, it probably is better to establish the fill site at the dry hydrant located 2 miles (3.2 km) from the scene. This is because the rural water system hydrant is not able to fill tenders fast enough to sustain the needed water supply. The extra mile driven in each direction is easily made up by the time saved in filling the tanks. When possible, select a fill site that is capable of supplying at least 1,000 gpm (4 000 L/min); this is the rate in which NFPA 1901 specifies that the tender should be filled.

When selecting a fill site or a dump site, try to pick a location that requires a minimum of maneuvering or backing of tenders when they arrive at that location. This speeds the operation and lessens the chance of a collision occurring at either site. The best fill and dump sites are those in which the tenders drive straight in from one direction, fill or dump, and then proceed straight out the other end (Figure 14.25). If some maneuvering of the apparatus is unavoidable, remember that it is always easier to maneuver apparatus before the tank is filled than after.

On large-scale water shuttle operations, it may be advantageous to use multiple fill and dump sites. This is particularly true when the incident scene is very large and water is needed on more than one end of the scene. In actuality, this may require two completely separate shuttle operations to supply each end of the incident.

Selecting the Route of Travel

A key decision in setting up the water shuttle operation is establishing the route of travel for tenders going between the fill and dump sites. Driving the shuttle route is one of

the most hazardous tasks for tender driver/operators. The route of travel for the shuttle operation should take both safety and operational efficiency into consideration.

A circular route of travel is considered to be the optimum method for conducting a water shuttle operation (Figure 14.26). When a circular pattern is employed, the full tankers leaving the fill site follow one route of travel toward the dump site. The empty tenders leave the dump site and proceed to the fill site using a different route of travel. This method eliminates the possibility of large trucks needing to pass each other on narrow, rural roads. This method can also be employed on incidents that occur on limited access or divided highways.

When using a circular pattern, the direction of travel for each leg is not particularly important, unless there is

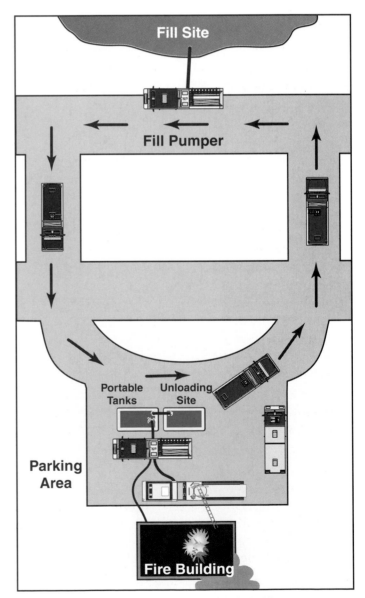

Figure 14.26 A circular shuttle route is most desirable.

Figure 14.25 Drive-through fill sites make for very efficient shuttle operations.

a substantial hill or grade on one or both of the legs. In this case, it is most desirable to have the full tenders travel downhill and the empty ones travel uphill. This speeds the travel time between the fill and dump sites.

If possible, roadways used during shuttle operations should be closed to all traffic other than emergency vehicles. This is particularly important when it is not possible to use a circular shuttle pattern. In these cases, fire apparatus traveling back and forth on the same road will cause a lot of confusion for members of the public driving on the same road, and driver/operators in the shuttle must exercise additional caution.

A number of other safety issues must be considered when selecting a particular route of travel. Some of these include the following:

- *Narrow roads.* The problems posed by these roads include difficulty passing other vehicles and the possibility of getting the apparatus tires off the road surface and causing a collision.

- *Long driveways.* These driveways often require tight maneuvering of the apparatus by the driver/operator. Improper coordination can also result in apparatus approaching each other from opposite directions. At worst, this could result in a serious accident. At best, one of the rigs would be forced to back out all the way because of its inability to pass the other rigs. The dump site section later in this chapter provides information on how to better deal with long driveways.

- *Blind curves and intersections.* Vehicles may cross the centerline on blind curves and enter the path of another vehicle. Blind intersections pose an extreme danger when the driver/operator cannot see oncoming cross traffic. The reverse may also be true when drivers of civilian vehicles cannot see oncoming apparatus. When possible, use police officers to control the flow of traffic at dangerous intersections along the shuttle route.

- *Winding roads.* Winding roads require a lot of concentration on the part of the driver/operator. One slight slip in attention level can result in a collision.

- *Steep grades.* Steep grades, both uphill and downhill, can cause problems for driver/operators. Uphill grades slow the shuttle operation and cause excessive wear on the vehicle. Driving on downhill grades can also be dangerous. Brake fade (discussed in Chapter 4) can result in the driver/operator being unable to slow or stop the vehicle at the bottom of the hill.

- *Inclement weather conditions.* Roads that have not been cleared of ice, snow, standing water, mud, or storm debris should be avoided.

Water Shuttles in the Incident Management System

When setting up a water shuttle operation, it is important to understand how the shuttle fits into the overall command structure of the incident operations. Many fire departments find that it is most effective to view the water shuttle operation and the fireground operation as two independent operations, both supervised by one Incident Commander. On fire incidents, the Incident Commander personally directs fire fighting operations and creates a water supply group or sector to handle the water shuttle operation (Figure 14.27). Whether you call it a water supply group or water supply sector depends on which Incident Management System (IMS) terminology your jurisdiction chooses to use. Both mean the same thing.

When the water supply group/sector is established, the Incident Commander must select a person to be in charge of that group/sector. This person is known as the water supply group/sector supervisor (Figure 14.28). Of all the companies and apparatus in the water shuttle, only the supervisor communicates directly with the Incident Commander. If the Incident Commander chooses to activate the operations section of the IMS, then the water supply group/sector supervisor reports to the operations section chief.

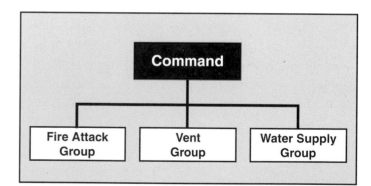

Figure 14.27 The water shuttle's place in the incident management structure.

Figure 14.28 The water supply officer should be easily identifiable.

Jurisdictions that have multiple radio frequencies available to them find it helpful to switch the water shuttle operation to a different channel than that of the fireground operation. This reduces radio clutter and confusion. When the supervisor needs to communicate with the Incident Commander or operations chief, he may simply switch over to the fireground frequency. The radio designation for the water supply group/sector supervisor typically is simply the term "water supply." The person who is selected to be the water supply group/sector supervisor should have considerable experience in pumper and tender operations, shuttle operations, and incident management.

Once a water supply group/sector supervisor has been appointed and a plan for the water shuttle operation has been formulated, the supervisor should appoint individuals to be in charge of the fill site and the dump site. These may be the driver/operator or company officer of the pumpers that are stationed at those locations, although any experienced member of the fire department can be assigned this function. The persons assigned these positions would have the radio designations "fill site" and "dump site." The individuals in charge of the fill and dump sites should be in constant communication with each other and the supervisor. Each must relay to the other when one apparatus is finished at his site and heading for the other site. For example, when one tender has completed dumping its load, "dump site" should contact "fill site" and advise that the tender is en route to be refilled.

The supervisor should monitor the water demand at the dump site very closely and try to anticipate problems before they occur. If the demand is exceeding the supply, additional tenders should be requested. The Incident Commander should be in close contact with the water supply group/sector supervisor in the event that conditions or tactics demand a significant change in the amount of water that is used. The water supply group/sector supervisor can then adjust the resources to meet the demand.

When a significant amount of water is required for an emergency operation, it may be necessary to establish two or more independent water shuttle operations. In this case, there are separate fill and dump sites for each water shuttle. Tenders should be assigned to a specific leg of the operation and remain in that pattern. When two or more shuttles are required, it is necessary to establish a water supply branch. The person in charge of this area is called the water supply branch director. Each individual shuttle operation has a supervisor. Local policies for naming each shuttle varies. For example, if one shuttle is basically on the east side of an incident and the other is on the west, they may be called the "east shuttle" and "west shuttle." The individuals in charge of each would be the east shuttle supervisor and the west shuttle supervisor. Each supervisor appoints someone to head the fill and dump site operations under his command (Figure 14.29). For more information on incident management of emergency operations, see the FPP **IMS Model Procedures Guide for Structural Fire Fighting**.

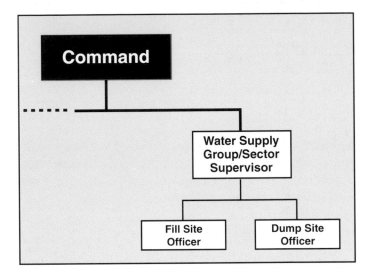

Figure 14.29 One manner of organizing the fill and dump sites within IMS.

Fill Site Operations

The purpose of the fill site operation is to reload empty tenders as expediently as possible. A number of methods may be used to fill tenders during water shuttle operations. Some methods are more efficient than others. The water supply group/sector supervisor should always select the best method for the particular incident being handled.

Positioning the Fill Site Pumper

As discussed earlier in this chapter, water supplies for a shuttle operation come from either a fire hydrant or a static water supply source. *IFSTA recommends that a pumper be used to fill tenders at all fill site operations, regardless of whether a hydrant or static source is used.* Using a separate pumper to provide the water supply maximizes the flow from a hydrant and eliminates the need for the tender to deploy drafting equipment at a static source in order to fill its own tank. The use of the pumper adds a high degree of safety by controlling the movement of water.

In general, the fill site pumper is positioned at the water supply source in the same manner as was described in Chapter 5. The driver/operator must determine the best position for drafting or hydrant connection that also allows maximum access for the tenders to be filled (Figure 14.30). Ideally, the pump panel should be located so that the driver/operator can view both the source and fill operations. This is no problem for top-mount pump panels. However, apparatus with side-mount pump panels generally have the pump panel away from the water with the intake discharge hoses on the opposite side, allowing the driver/operator to view the tender filling operations. Front- and rear-mounted pumps may also require the apparatus to be positioned in a manner that obstructs the driver/operator's view of the tender filling operation. Ideally, each tender should be filled at a minimum rate of 1,000 gpm (4 000 L/min).

When positioning at a hydrant, the driver/operator should connect, at a minimum, a large diameter intake hose between the large pump intake and the steamer connection on the hydrant. If the hydrant is on a particularly strong main, MDH lines may also be connected between the 2½-inch (65 mm) hydrant outlet(s) and the auxiliary intake(s) on the pump (Figure 14.31). This maximizes the amount of water that the pumper is able to flow to the tender being filled.

When positioning at draft, seek a spot that requires a minimum amount of lift or hard intake hose. This maximizes the amount of water that can be supplied to the tender. Remember that the number of discharge lines to the tender fill spot can always be increased or lengthened if necessary. The fill site pumper does not necessarily have to be located directly next to where the tenders are filled.

There may be static water supply sources that are completely inaccessible to fire department pumpers. In these cases, two or more high-volume portable pumps may be used to relay water to the fill site pumper, if it is less than 100 feet (30 m) between the source and the pumper. The fill site pumper may receive water from the portable pump(s) in two primary ways. The first is to hook the discharge hose(s) from the portable pump(s) directly into the pumper's intake connection(s). The second method involves setting up a portable water tank next to the fill site pumper. The discharge line(s) from the portable pump(s) is used to fill the portable tank (Figure 14.32). The fill site pumper then drafts water from the tank to fill incoming water tenders.

Regardless of whether a hydrant or static source is used, a booster line or some other type of small discharge line that is within view of the driver/operator should be continuously flowed from the fill site pumper. This ensures that loss of prime does not occur during drafting operations. When operating at a hydrant, it ensures that pump overheating does not occur when the discharge

Figure 14.30 A fill site pumper.

Figure 14.31 If the fill site is on a particularly strong hydrant, multiple intake hoses may be connected between the hydrant and the pumper.

Figure 14.32 Portable pumps may be used to fill a portable tank, which in turn may be drafted from by the fill site pumper.

lines are not being flowed to fill tenders. When operating from a static supply, the discharge line should return water to the source.

Fill Site Layout

Once the fill site pumper has been positioned at the water supply source, lay out the remainder of the hose and appliances needed to operate at the fill site. The actual makeup of the discharge hose(s) used to fill tenders varies depending on local equipment design and policy. In this manual, we assume that each tender has two 2½-inch (65 mm) direct tank fill connections or one LDH direct tank fill connection on the rear of the vehicle. These are the most common and efficient types of fill connections found on most tenders.

The first thing that must be determined is exactly where the tender driver/operator parks the tender when it arrives at the fill site. It should be positioned so that a minimum of hose is required from the fill site pumper. However, this must be balanced with a number of other considerations. The ideal fill position is one that allows the driver/operator to enter from one direction and exit from another, without the need to turn or back up. The position should also allow room for empty tenders to line up and wait their turn to be filled without creating a traffic hazard (Figure 14.33).

Once the exact fill spot has been established, a traffic cone or similar marker may be used to denote the stopping point for the tender driver/operator. The cone should be positioned on the side of the road or at a safe spot in a parking lot. The tender driver/operator should be instructed to pull the vehicle adjacent to the cone at the point where the driver's side door window is opposite the cone (Figure 14.34). This allows the fill hose to be located at the rear of the vehicle. This reduces the need to move hoselines back and forth when each vehicle arrives.

If the tenders being filled have two 2½-inch (65 mm) direct tank fill connections, two MDH hoselines should be laid to the approximate location of the rear of the tender when it is parked for filling. If available, 3-inch (77 mm) hose should be used as it allows for quicker filling of the tank. Each hoseline should have a gate valve installed between the last section and second to the last section of hose. These valves are used to open the lines once they are connected to the tender for filling. If gate valves are not available, hose clamps may be used (Figure 14.35).

Many jurisdictions prefer to equip the inlets on direct tank fill connections with Storz (sexless) couplings (Figure 14.36). These couplings speed the connection of hose to the apparatus. If the apparatus is equipped with Storz connections, it may be necessary to install Storz adapters on the end of the hoses that are used to fill the

Figure 14.34 Pull the apparatus forward until the driver's door is even with the traffic cone.

Figure 14.35 Hose clamps may be used on fill lines.

Figure 14.33 Stage empty tenders close to the fill site so that they may be pulled into position when a fill spot comes open.

Figure 14.36 Storz fittings speed fill line connections.

tenders. Other types of connections, such as cam-locks, may also be used. The main point here is that all apparatus operating in the shuttle should use the same type of connection so that adjustments do not need to be made each time a different apparatus is filled.

If the tenders being filled have LDH direct tank fill intakes, it generally is only necessary to lay one LDH from the fill site pumper to the fill location. If a large in-line gate valve is not available for the LDH, it may be possible to place an LDH manifold between the last two sections of hose to act as a gate valve. Generally, it is not practical to control the flow of water for filling operations with LDH using a hose clamp. LDH clamps are usually of the screw-down design. These take a considerable amount of time to open and close. If no gate valve or manifold is available, it is best to open and close LDH fill lines at the pump panel of the fill site pumper.

If some of the tenders to be filled have 2½-inch (65 mm) intakes and others have LDH intakes, it is necessary to lay out hose that is easy to connect to either. The best method for accomplishing this is to lay one LDH to the area near the fill site. Connect a manifold to this LDH line. One LDH and two MDH should be connected from the manifold. The gate valves on the manifold can then be used to operate whichever lines are being used at a given time. The manifold should be supervised by a radio-equipped firefighter familiar with water movement and preferably by a pump operator.

Top Fill Methods

The previous method for laying out a fill site is considered to be very efficient under most circumstances. However, depending on local traditions and equipment design, a number of other methods may be used to fill tenders. In some jurisdictions, it is more common to fill the tender through the fill or vent opening on the top of the apparatus. This may be accomplished using either fixed or portable filling equipment. Obviously, if fixed equipment is used, the fill site is limited to that location.

One method of top filling uses permanent or portable overhead pipes that are used at a static water supply source. Portable fill pipes are generally made of PVC or other lightweight material. Fixed overhead pipes may be constructed of any suitable material. These devices are operated by placing one end of the fill pipe in a static source. A pumper discharges into an in-line jet that is within the fill pipe. This creates a high flow rate through the fill pipe. These in-line jets should be pumped at 150 psi (1 050 kPa) and provide 700 to 800 gpm (2 800 L/min to 3 200 L/min) through 4-inch (100 mm) pipe. Larger diameter piping provide additional flow. These devices are only advantageous when the tenders have no other way of being filled than through the top opening. Otherwise, it is a more efficient use of the pumper supplying the in-line jet to simply draft from the site and pump directly into the tenders.

Another method of overhead pipes uses permanent or portable manifolds. The permanent types are located adjacent to a water source and are fed by the fill site pumper (Figure 14.37). Water sources may be hydrants or a static supply source. With these devices, the fill site pumper connects between the water supply and the fill pipe to provide overhead water when a tender is positioned beneath the fill opening. If the water supply source is a reliable one, high fill rates may be accomplished through these devices. The downside of these devices is that it may take some jockeying (and a resulting loss of time) to get the tank opening positioned directly beneath the fill spout.

Filling a tanker through the top with a portable fill device, or an open hose butt, is not recommended due to reaction of the hoseline. Firefighter safety is a primary concern; firefighters can be thrown off or slip from a tanker during top filling by this method.

Figure 14.37 Some jurisdictions top-fill tenders through permanently mounted fill towers.

Operating the Fill Site

Once the fill site pumper has connected to the water supply source and all of the appropriate hose and equipment have been laid out, fill site operations may begin. As previously stated, it is recommended that the fill site pumper remain in gear with the tender fill lines charged at all times. A booster line or other waste line should continuously be flowed to prevent loss of prime (while drafting) or pump overheating (when on a hydrant).

One firefighter should be assigned to handle each tender fill line that is laid out. These firefighters are referred to as the *make and break* personnel. Their primary responsibilities are to make the fill connection when the tender arrives and to disconnect the hose(s) when the tank is full (Figure 14.38). These personnel remain with these lines until other firefighters replace them.

When the tender reaches the fill site, the driver/operator should cautiously pull into the filling position until the driver's door is parallel to the traffic cone that has been laid out as a stop marker. The make and break personnel should be safely off to the side of the fill site as the apparatus approaches. When the apparatus has come to a complete stop, the make and break personnel connect the fill hose(s) to the direct tank fill intake(s). Once the hose is connected, the intake valve(s) may be opened. At this point, the make and break personnel return to the gate valve(s) or manifold on the fill hose(s) and slowly open the valve(s) to start water flowing. They should remain at the gate valve or manifold until the tender is full.

While filling is taking place, the tender driver/operator should remain in the cab of the apparatus. This is to ensure that the apparatus can immediately be driven away once the tank is full and the hose connections are broken. Any tenders that arrive in the area of the fill site while one tender is being filled should stage a safe distance back, and maintain an orderly line. As the tender driver/operators see the first tender leaving the fill site area, the next one in line may then cautiously proceed into the fill spot.

If the fill site has sufficient room, a second set of fill lines may be laid from the fill site pumper for a second tender. While one tender is being filled, the second tender may be pulled into position and connected to the fill lines (Figure 14.39). When the first tender is full and the lines are shut down, the lines to the second tender may be opened to begin immediate filling. When the first tender drives away from the fill site, the next tender in line may be pulled into that spot and readied to fill. Unless the fill site pumper is connected to a high flow hydrant, it is generally not recommended that both tenders be filled at the same time.

Figure 14.38 Make and break personnel should always remain with their hoselines.

Figure 14.39 If space allows, a second tender may be connected to fill hoses while the first is being filled.

Once the water tank is completely full, the make and break personnel should first slowly close the valve(s) on the gate valve or manifold. They should then proceed to the connection at the direct tank fill inlet and operate the bleeder valve to relieve the pressure on the line(s). Once the pressure has been relieved, the hoses should be disconnected and pulled off to the side of the fill site. The driver/operator should then be signaled to proceed back to the dump site. One of the make and break personnel should always be designated as the signal person. The next tender may then proceed into the fill site, and the process is repeated. Tenders should only move in and out of position when they are signaled to do so by the designated person.

The make and break personnel or fill site officer should monitor the ground conditions around the fill site as operations proceed. It is inevitable that a significant amount of water will be spilled in this area. During freezing conditions, ice may begin to form. If the road is not paved, it may begin to become soft. In either case, it may become necessary to adjust the location of the fill site for both operational and safety reasons.

Shutting Down the Fill Site

The previous procedure should be followed until the Incident Commander determines that the water shuttle operation is no longer necessary. Once that decision has been made, the fill site should remain in operation until all tenders participating in the shuttle have been refilled. Most jurisdictions prefer to fill their tenders before they begin their return to quarters. This ensures that they have adequate water to handle any incident they may encounter on their return. They are also available to immediately respond to another call if needed. The exception to this rule is if the fill site was being supplied by a static water source. In this case, it may be more desirable to wait until the tender can get to a hydrant for filling before returning to quarters. It is generally recommended that the apparatus farthest from the scene be released first.

Once all the tenders have been refilled, the fill site pumper and its equipment may be shut down and prepared to return to service. Remember that if the water supply source was a static source, all pumps and equipment should be flushed to clear any stones or other debris from the system.

Dump Site Operations

In a water shuttle operation, the opposite end from the fill site is the dump site. The dump site is located close to the incident, and its goal is to provide a continuous source of water supply to the apparatus attacking the incident. There are a variety of methods that can be used to run a dump site, as well as a variety of methods that may be employed to discharge water from the tender. As with fill site operations, we mention a number of these methods but focus on those that are generally recognized as the most efficient for long-term, high-volume water shuttle operations.

Dump Site Operational Methods

The following are three primary methods that can be used to operate a dump site:

- Direct pumping operations
- Nurse tender operations
- Portable water tank operations

Direct Pumping Operations

In the direct pumping method, a tender pumps the water from its tank directly into the fire pump of the attack pumper. This method is typically accomplished by having the attack pumper lay out a supply line that ends in a location that is easily accessible to the tenders approaching the scene. When the tender arrives at the "dump site," the supply hose is connected to a discharge on the fire pump of the tender (Figure 14.40). The contents of the tender's water tank are then pumped to the attack pumper. In some cases, a siamese is placed at the dump site that allows two tenders to pump into the supply line.

The only real advantage of the direct pumping method is that it reduces the need for the tender to park directly next to the attack pumper. The disadvantages of this method are many. It is very difficult, if not impossible, to provide a constant supply of water to the attack pumper. There will always be an interruption of flow when tenders are disconnecting and connecting. It requires all tenders to have their own onboard fire pump. Tenders that have small capacity pumps or no pumps at all are basically useless if this method is employed. Tenders with small capacity pumps may not be able to supply water to the attack pumper at the same rate the attack pumper is discharging water on the fire.

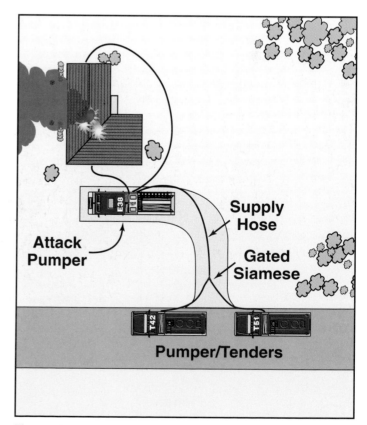

Figure 14.40 Direct pumping operations may be used if portable tanks are not available.

Nurse Tender Operations

The second method is the nurse tender operation. This method generally involves a very large tender that parks immediately adjacent to the attack pumper. The attack pumper is either supplied by a discharge line from the pump of the nurse tender or drafts directly from the tank of the nurse tender (Figure 14.41). Departments that use tractor-trailer tenders typically use them in this manner because they are highly ineffective as shuttle apparatus in most cases.

The primary advantage of this method is that in many cases the nurse tender is so large that the fire is controlled before there is a need to refill its tank. There are several disadvantages to this style of operation. Once again, all shuttle tenders are required to have a sizable fire pump in order to pump their loads into the nurse tender. Even with a sizable fire pump, the dumping time for each tender is significantly higher than if they were able to discharge through their large diameter direct tank discharge (dump) valve.

Portable Water Tank Operations

The most efficient way of operating a water shuttle is to use portable water tanks as the dump site water supply source. When using this method, one or more large portable water tanks are positioned in a strategically sound location near the emergency scene (Figure 14.42). (**NOTE**: Dump site location selection was discussed earlier in this chapter.) Once the tank(s) is(are) positioned, a pumper — called the dump site pumper — deploys hard intake hose with a low-level strainer into one of the tanks and prepares to draft water from the tank (Figure 14.43). When the first tender arrives on

Figure 14.42 Portable tank dump sites are most suitable for continuous water supply operations.

Figure 14.43 The intake hose and low-level strainer are placed into the portable tank.

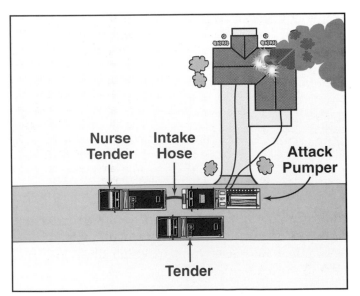

Figure 14.41 Nurse tender operations involve the attack apparatus taking water directly from a tender, as opposed to portable tanks.

the scene, it discharges its water into the portable tank. The dump site pumper may then begin drafting from the tank and supplying water to the attack pumper. In some situations, the dump site may be located directly adjacent to the scene, and the attack pumper may draft from the portable water tank.

This method does not require shuttle tenders to be equipped with a fire pump as long as they have an adequate-sized direct-tank discharge valve (herein referred to as a *dump valve*) and adequate venting. This method is also the easiest of the three methods to ensure a constant supply of water to the attack pumper, although no method is foolproof if not operated properly. The primary disadvantage of this method is that if a multiple tank operation is to be employed, a substantial amount of working space is required for the dump site.

Water Tender Discharge Methods

Water may be unloaded from a tender at the dump site in three basic ways:

- Using a pump on the tender
- Using a dump valve
- Pumping and dumping simultaneously

Tenders may pump their water into a portable tank (or nurse tanker) through one or more discharges or hoselines. This should only be done when the tender is not equipped with a dump valve. In general, it is not possible to pump water out of the apparatus as fast as a dump valve empties the tank. The limiting factors are flow capacity of the tank to pump lines, the pump size, venting capability, and the ability to control hoselines at the delivery point. In addition to the longer amount of time required to empty the tank, this method also requires that the tender driver/operator place the fire pump in gear before discharging water, and take it out of gear before heading back to the fill site. Both of these factors increase the overall time spent at the dump site.

If this method is used, some type of mechanical device must be used to hold the hoses that are discharging into the tanks. Commercial or homemade clamps or fill spouts are available (Figures 14.44 a and b). Some jurisdictions choose to tie a section of hard intake hose to the portable tank and connect the fill lines to it. Any method is suitable as long as firefighters are not required to manually hold the lines.

The most efficient method of unloading tankers is to dump the water into portable tanks through a gravity or jet-assisted dump valve (Figure 14.45). Dumping, with sufficient portable tank capacity, allows tankers to deliver their water rapidly and get back on the road. As mentioned earlier in this chapter, gravity dumps employ large diameter valves and pressure created by the height of the water column in the tank to move the water. Discharge pressure is based on column height:

- 1-foot (0.3 m) column = .434 psi (3 kPa)
- 2-foot (0.6 m) column = .868 psi (6 kPa)
- 3-foot (0.9 m) column = 1.302 psi (9 kPa)
- 4-foot (1.2 m) column = 1.736 psi (12 kPa)

The actual amount of flow through the dump valve is dependent on the design and size of the dump valve, the baffling in the tank, and the venting capability of the tank. Baffles must have sufficiently sized openings to allow free water movement at the bottom of the tank and air movement at the top during rapid filling or unloading. NFPA 1903 requires all tenders to be able to

Figure 14.44a Hose clamps are useful when the tank will be filled by direct pumping.

Figure 14.44b Some agencies have developed special discharge devices for direct pumping portable tank fill operations.

Figure 14.45 The fastest way to fill a portable tank is by dumping the tender's load through a direct tank discharge valve.

dump at a minimum average flow rate of 1,000 gpm (3 785 L/min) for the first 90 percent of the tank. It is most desirable to be able to dump off either side of the truck and off the rear of the truck.

In theory, an apparatus that is equipped with both a fire pump and dump valve could simultaneously pump and dump its load at the same time. In reality, this is not a very timesaving operation. The time spent engaging the fire pump and making any connections necessary for the pumping portion of the evolution generally negates any time advantage that might be gained. It is better to simply use the dump valve on apparatus that also have a fire pump.

Operating the Dump Site

As stated earlier in this chapter, the most efficient method for operating a dump site is to have tenders dump their loads into one or more portable water tanks. A dump site pumper is responsible for drafting from the portable tank and supplying water to the attack pumper. Most jurisdictions find this to be the optimum way to provide a constant high flow of water using a shuttle operation.

Single Portable Tank Operations

The simplest form of a dump site operation is one in which a single portable water tank is used. In this case, the tenders dump their water directly into the tank from which the dump site pumper is drafting (Figure 14.46). A single portable tank works on fires that require relatively low overall flow rates (less than 300 gpm [1 200 L/min]). The portable tank and the dump site pumper must be positioned so that easy in-and-out access is allowed for the tenders that are dumping into the tank.

The most common style of portable water tank is the folding type. This tank is removed from its storage position on the apparatus in much the same manner as a ground ladder would be removed. It is then carried to its deployment position and simply unfolded (Figure 14.47). Portable tank drains should be tucked inside the tank to prevent leakage or dislodging. Some departments choose to place a salvage cover on the ground beneath where the tank is deployed. This saves a little wear on the portable tank liner.

Several other types of portable water tanks may be used in some jurisdictions. One type comes in several sections that must be assembled at the scene. Once the metal framework is assembled, the liner is attached to it and flaked out around the tank. When totally assembled, this type resembles the folding type previously described. Yet another type is the self-supporting or frameless portable tank. This tank is a large bladder that has a floating collar around the opening (Figure 14.48). As water enters the tank, the collar continues to rise until the tank is filled. This style of tank has an intake hose connection in the bottom of the tank from which drafting may be achieved.

Regardless of the style of portable water tank that is used, it is recommended that the portable water tank

Figure 14.46 Single portable tank operations are suitable for small water supply operations.

Figure 14.47 Several firefighters are required to deploy the portable tank.

Figure 14.48 Self-supporting portable tanks are used in some jurisdictions.

have a capacity that is at least 500 gallons (2 000 L) larger than the capacity of the water tank on the apparatus carrying it. This allows the apparatus to dump its entire load into the tank, even when on a slight incline or a road with a high crown.

The dump site pumper should have a low-level strainer attached to the hard intake hose. This allows continuous drafting ability down to a point where only about 2 inches

(50 mm) of water is left in the tank (Figure 14.49). Low-level strainers that are designed for use in portable tanks are commercially available or may be homemade.

Once the tank has been deployed, the first tender may dump its load of water into the tank. A dump site spotter or the dump site officer should wave and guide the tender into position. Adequate guidance must be given to assure the tender driver/operator that the dump valve is properly aligned with the tank. When the tender is in position, the dump valve is opened, and water flows into the portable tank (Figure 14.50). Once the level of water in the portable tank is sufficient for drafting, the dump site pumper should prime its pump and start water flowing. It is recommended that the dump site pumper flow a booster line back into the portable tank or use some other type of waste line arrangement to ensure that prime is not lost when other discharge lines are shut down.

Once the first tender has emptied the contents of its water tank into the portable tank, it should immediately proceed toward the fill site to reload. If there is still space available in the portable tank, the next tender should be brought into position and its water should also be dumped into the portable tank until it is filled. If the second tender is not able to empty its entire load before the portable tank is full, it should remain in position and empty its tank when room in the portable tank becomes available. Additional tenders should be staged and ready to proceed to the portable tank as soon as their water is needed to refill the tank.

Multiple Portable Tank Operations

Incidents that require flow rates in excess of 300 gpm (1 200 L/min) are best served by a multiple portable tank dump site operation. The number of portable tanks that may be used at a dump site is limited only by the number of tanks and the amount of water transfer equipment available at the scene. In reality, the most common multiple portable tank operations used in most jurisdictions range from two to five portable tanks.

When multiple portable tanks are used, each of the tanks is positioned so that water may be transferred from one tank to the next. The ultimate destination of the water in all the tanks is to be routed into the last tank, from which the dump site pumper is drafting water that is being supplied to the attack pumper (Figure 14.51). The basic goal of the multiple tank operations is to keep the final tank in the chain, from which the dump site pumper is drafting, full at all times. Water is constantly being transferred from the preceding tanks in the operation to achieve this goal. When this is done, the upstream tanks empty first and are available for dumping without delay.

There are a number of methods for transferring water between tanks. The simplest method is to connect two

Figure 14.49 The low-level strainer will draft water down to depths of about 2 inches (50 mm).

Figure 14.50 The driver/operator may open the dump valve with a control directly on the valve, or a remote control in the apparatus cab, depending on the design of the apparatus.

Figure 14.51 Multiple portable tank dump sites are required for high-volume water shuttle operations.

tanks by their drain openings (Figure 14.52). However, because most commercially constructed tanks only have one drain, this limits you to a two-tank operation. Using this method maintains the same level of water in both tanks at all times. In general, this method is discouraged by most authorities on water shuttles.

The most efficient method uses jet siphons to move water from one tank to another. The jet siphon is a device that is attached to either a section of hard intake hose or a piece of PVC or aluminum pipe (Figure 14.53). The jet siphon has an inlet for connection of a 1½-inch (38 mm) or larger hose. When the jet siphon, hard intake hose, and 1½-inch (38 mm) hose are assembled, the end with the jet siphon is placed into the tank from which water is transferred (Figure 14.54). Once that tank has a sufficient amount of water in it, the 1½-inch (38 mm) hose is charged. As this flow of water begins through the hard intake hose, water from the portable tank is also drawn in. All of this water flows through the hard intake hose and into the next tank in line (Figure 14.55). If the next tank becomes full, the 1½-inch (38 mm) line is shut down to stop the flow. Care must be taken to ensure that water remains in the apparatus water tank at all times. If the apparatus water tank runs dry, the pumper is not able to supply the jet siphon if the drafting tank becomes empty and pump prime is lost.

If only two or three tanks are going to be used in conjunction with a jet siphon operation, the dump site pumper should be able to supply the hoses to the jet siphons. This requires a little more work on the part of the dump site pumper driver/operator, but it is not unreasonable. If four or more tanks are to be connected using jet siphons, it is advantageous to have a second pumper draft from the tanks solely to supply some or all of the jet siphon lines.

Figure 14.54 The end of the hose with the siphon attached is placed into the tank from which water is transferred.

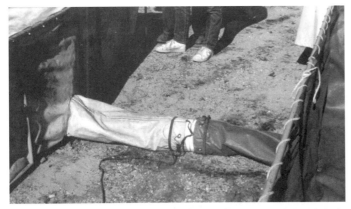

Figure 14.52 Special connectors may be used to connect the tank drain openings.

Figure 14.53 The jet siphon device is supplied by a small hoseline.

Figure 14.55 When the siphon supply line is charged, water is transferred from one tank to the next.

The procedure establishing and running a multiple portable tank shuttle operation is as follows:

Step 1: The first portable tank is deployed in the same manner as described previously in the single-tank operation section. Make sure that the tank drain is on the downhill side of the tank.

Step 2: The dump site pumper driver/operator and crew deploy their hard intake hose and strainer into the first portable tank.

Step 3: The first tender on the scene dumps its water into the first tank. The dump site pumper may then begin the drafting operation.

Step 4: A second portable tank is set up next to the first one (Figure 14.56). A tip-to-tip diamond-shaped arrangement is generally the preferred method for arranging the tanks. If immediately available, use different colored portable tanks to help the dump site officer direct apparatus to the appropriate tank for dumping. However, this is not so crucial that you should wait for a different color tank to arrive on the scene. Whichever tanks are available should be deployed as quickly as possible.

Step 5: The jet siphon equipment is assembled. The end with the siphon is placed in the second portable tank, and the discharge end of the assembly is positioned over the edge of the first portable tank and secured (Figures 14.57 a and b).

Step 6: The next tender dumps its load into the second portable tank. Once the water level is sufficient, the jet siphon supply hose is charged to begin the process of transferring water from the second tank to the first (Figure 14.58).

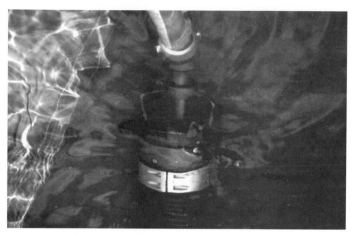
Figure 14.57a The siphon is placed in the tank.

Figure 14.57b The discharge end of the siphon hose is suspended above the top of the receiving tank.

Figure 14.58 Water is effectively moved between the tanks.

Figure 14.56 A diamond-shaped arrangement is usually preferred for multiple portable tank operations.

Step 7: If additional portable tanks are desired, each should be set up and operated as was described for the second tank in Steps 4 through 6. Tenders should always dump into the end-most tank that has room for water (Figure 14.59). As the apparatus is being positioned, the dump site officer should advise the tender driver/operator of which tank to dump in.

Step 8: The dump site pumper driver/operator or jet siphon pumper driver/operator should monitor the level of water in each tank and adjust the siphon lines accordingly.

The dump site officer should monitor the ground conditions around the dump site as operations proceed. It is inevitable that a significant amount of water will be spilled in this area. During freezing conditions, ice may begin to form. It may be necessary to have sand or road salt brought in to improve the road conditions. If the road is not paved, it may begin to become soft. In some instances, it may be necessary to relocate the dump site if this occurs. However, this has a significant negative impact on the fireground operation if this becomes necessary.

If all the portable tanks become empty at some point during the operation, the dump site pumper may continue to support fireground operations using water in its onboard water tank. When this becomes necessary, firefighters in hazardous positions should be withdrawn as loss of water to attack lines may be imminent. Once the portable tanks are refilled, normal operations may resume. The driver/operator of the dump site pumper should refill the apparatus water tank as soon as possible.

Shutting Down the Dump Site

Once the need for a continuous water supply is no longer present, the dump site operation may be disbanded. Before disassembling the operation, it is generally a good idea to make sure that the attack apparatus and dump site pumper have topped off their onboard water tanks. This is because often a little more water is required for overhaul operations. Once the apparatus tanks are full, all the drafting and water transfer equipment can be disassembled, cleaned, and stowed. The portable tank drains may be opened to allow the remaining water to drain out. If there is a lot of sediment or debris left in the tank(s), it should be rinsed out before stowing on the apparatus.

Any tenders that were staged in preparation for dumping may be returned to service or used in any other manner the Incident Commander deems appropriate. In some jurisdictions, one full tender is left at the scene with

Figure 14.59 As long as adequate equipment is available, water may be transferred between several tanks at the same time.

the attack apparatus to provide sufficient water for extended overhaul operations or to provide attack capability in the event of a major flare-up.

Evaluating Tender Performance

For almost as long as the concept of using apparatus with large tanks (tenders or tankers) to supply water for fire fighting has been around, so has the debate over what size tender is most efficient for this application. Some jurisdictions believe that the more water a truck can carry, the better. However, other jurisdictions favor smaller apparatus that can load, unload, and travel back and forth in a more expedient manner than can the larger apparatus.

In reality, there is no general rule of thumb or correct answer to the question of what size tender is the best. Tender performance is based on a number of factors, including loading and unloading time, vehicle condition, drive-train capabilities, and tank size. Thus, it is possible that a large, underpowered tender may not be able to supply water as well as a smaller, well-designed tender.

It is possible to assign a gpm (L/min) figure that each tender is capable of supplying over a variety of distances. This is done by analyzing the filling time, dumping time, and travel time between the dump and fill sites for each tender. This information can be used in pre-incident planning to determine how many tenders are required to provide a desired flow rate for the target hazard. It can also be used by an Incident Commander or water supply supervisor at a fire for which no pre-incident plan was available but for which a desired flow rate has been determined at the scene.

There are two basic methods for developing a flow rating for individual tenders. The first relies solely on actual testing of each tender under realistic water shuttle conditions. Begin with the tender parked in the proper position to dump its load into the portable tank. Start the clock when the dump valve is opened. Keep the clocking running as the tender leaves the dump site, drives to the fill site, and then returns to the dump site. The clock is stopped when the tender's dump valve is back in the proper position to dump its next load. By dividing the amount of water the tender dumped by the time the round trip took, a gpm (L/min) flow rating can be assigned to that tender for a shuttle of that particular distance. For example, suppose a dump site and fill site are located one mile (1.6 km) apart. A 3,000 gallon (12 000 L) tender is able to make the round trip in 12 minutes. Keep in mind that most tenders only actually dump 90 percent of their load before heading off to be refilled. The flow rating for that tender, over a 1 mile (1.6 km) shuttle distance, would be calculated as follows:

$$\text{GPM} = \frac{\text{Tank Size} - 10\%}{\text{Trip Time}} \qquad \text{GPM} = \frac{3,000 - 300}{12 \text{ minutes}}$$

$$\text{GPM} = \frac{2700}{12}$$

GPM = 225

$$\text{L/min} = \frac{\text{Tank Size} - 10\%}{\text{Trip Time}} \qquad \text{L/min} = \frac{12\ 000 - 1\ 200}{12 \text{ minutes}}$$

$$\text{L/min} = \frac{10\ 800}{12}$$

L/min = 900

What this means is that this particular tender is able to supply 225 gpm (900 L/min) over a 1-mile (1.6 km) shuttle route. If the shuttle route is longer, the gpm (L/min) rating drops. This is due to the fact that the travel time between the two sites is extended.

The second method of evaluating tender performance is by using a series of formulas that were originally developed by the ISO. These formulas are used to rate the water supply performance of fire departments that protect rural areas. These formulas divide the shuttle operation into two time elements: travel time and handling time.

EQUATION V (U.S.)
Travel Time in minutes = 0.65 + (1.7)(Distance in Miles)

EQUATION W (Metric)
Travel Time in minutes = $\frac{0.65 + (1.06)}{(\text{Distance in km})}$

These formulas include a built-in factor for acceleration and deceleration times as the tenders leave and approach fill and dump sites. They assume an average travel speed between the fill and dump site of 35 mph (55 km/h). If road conditions allow for faster speeds, the formula has to be adjusted accordingly.

The handling time is computed as follows:

EQUATION X
Handling Time = Fill Site Time + Dump Site Time

The fill site time includes the time spent maneuvering the apparatus into position, the make and break times, and the actual filling time. For the sake of calculation, the flow rate that the fill site pumper is supplying must be established. Each jurisdiction has to determine the fill rate they are most likely to achieve.

The dump site time includes the time spent maneuvering the apparatus into position and the actual dumping time. Again, each tender dumps its load at a different rate. The dumping time can be determined by actual testing or it may be supplied by the manufacturer. Because NFPA 1901 specifies that tenders should be able to be dumped at a minimum of 1,000 gpm (3 785 L/min), we assume that rate in this chapter.

ISO also only allows 90 percent of the tenders total tank capacity to be used for the calculation. The 10 percent loss accounts for water that is spilled or that remains in the tank after the dump valve is closed. Through testing, many departments have found that their tender is actually only dumping 80 percent of its load. In this case, modifications must be done to improve the performance.

One of the most accurate ways to measure the amount of water that is left in the tank after dumping is to weigh the apparatus on a truck scale when it is completely full and then again after the tank has been dumped. The difference in those two weights accounts for the amount of water that was actually dumped. For example, suppose we wish to know how much water is actually dumped from a 2,000 gallon (8 000 L) tender. After weighing the apparatus both times, it is found to be 14,500 pounds (6 577 kg) lighter after the water has been dumped. Because we know that water weighs 8.34 pounds per gallon (1 kg per liter), we can determine that 1,739 gallons (14,500 ÷ 8.34), or 6 577 liters, were dumped. This would account for 87 percent of the capacity of the tender.

Once the travel time and handling time have been determined, the flow rate for that tender may be calculated using the following formula:

EQUATION Y

$$\text{Tender Flow Rate} = \frac{\text{Tender Water Tank Size (in Gallons or Liters)} - 10\%}{\text{Travel Time} + \text{Handling Time}}$$

Example 1

Determine the flow rate for a 2,500 gallon tender that will be shuttling water 3 miles in each direction. Assume fill and dump rates of 1,000 gpm. Also assume that the maneuvering and make and break times at each site total 2 minutes.

Travel Time in minutes = 0.65 + (1.7)(Distance in miles)

Travel Time in minutes = 0.65 + (1.7)(3 miles)

Travel Time in minutes = 0.65 + 5.1

Travel Time in minutes = 5.75

Handling Time = Fill Site Time + Dump Site Time

Handling Time = (2 min. + [tank size ÷ fill rate]) + (2 min. + [tank size ÷ dump rate])

Handling Time = (2 min. + [2,500 ÷ 1,000]) + (2 min. + [2,500 ÷ 1,000])

Handling Time = (2 min. + 2.5 min.) + (2 min. + 2.5 min.)

Handling Time = 4.5 min. + 4.5 min.

Handling Time = 9 minutes

$$\text{Tender Flow Rate} = \frac{\text{Tender Water Tank Size (in Gallons)} - 10\%}{\text{Travel Time} + \text{Handling Time}}$$

$$\text{Tender Flow Rate} = \frac{2,250 \text{ gallons}}{5.75 + 9}$$

$$\text{Tender Flow Rate} = \frac{2,250 \text{ gallons}}{14.75}$$

Tender Flow Rate = 153 gpm on a 3 mile relay

Example 2

Determine the flow rate for a 10 000 liter tender that is shuttling water 5 km in each direction. Assume fill and dump rates of 4 000 L/min. Also assume that the maneuvering and make and break times at each site total 2 minutes.

Travel Time in minutes = 0.65 + (1.06)(Distance in km)

Travel Time in minutes = 0.65 + (1.06)(5 km)

Travel Time in minutes = 0.65 + 5.3

Travel Time in minutes = 5.95

Handling Time = Fill Site Time + Dump Site Time

Handling Time = (2 min. + [tank size ÷ fill rate]) + (2 min. + [tank size ÷ dump rate])

Handling Time = (2 min. + [10 000 ÷ 4 000]) + (2 min. + [10 000 ÷ 4 000])

Handling Time = (2 min. + 2.5 min.) + (2 min. + 2.5 min.)

Handling Time = 4.5 min. + 4.5 min.

Handling Time = 9 minutes

$$\text{Tender Flow Rate} = \frac{\text{Tender Water Tank Size (in Liters)} - 10\%}{\text{Travel Time} + \text{Handling Time}}$$

$$\text{Tender Flow Rate} = \frac{9\ 000 \text{ liters}}{5.95 + 9}$$

$$\text{Tender Flow Rate} = \frac{9\ 000 \text{ liters}}{14.95}$$

Tender Flow Rate = 610 L/min on a 5 km relay

Chapter 15

Foam Equipment and Systems

It used to be that foam fire fighting equipment and systems were primarily limited to industrial and airport fire protection. Municipal fire departments had little, if any, foam application capabilities. If the need arose to use foam to combat an incident, a special call was made for the industrial or airport fire fighting personnel to respond to the scene and handle the situation.

In recent years, foam fire fighting has increased dramatically in the municipal and wildland fire services. There are many reasons why this has occurred, but the major ones are the following:

- Magnitude and frequency of hazardous materials incidents requiring foam for their control

- New advances in foam concentrate technology that have provided products which are more easily used by municipal and wildland firefighters

- Technological improvements in foam proportioning equipment and systems that make their inclusion in the construction of new fire apparatus, or the retrofitting of existing apparatus, feasible for all fire departments

In most cases, the proper assembly and operation of foam fire fighting equipment will be the responsibility of the driver/operator. Although the technological advances in foam equipment design have made the use of these systems somewhat simpler than in the past, they are not foolproof. The driver/operator must still understand the basic principles of foam proportioning and application if the operation is to be successful.

This chapter examines some of the basic concepts in which the driver/operator should be familiar in respect to foam concentrates, portable foam proportioning equipment, apparatus-mounted foam proportioning systems, and foam application equipment. Because there are many manufacturers of this type of equipment, it is impossible to provide specific operational guidelines for each type of system. However, the information in this chapter provides the driver/operator with a sound understanding of the principles of each type of system. For more detailed information on foam fire fighting and equipment, see IFSTA's **Principles of Foam Fire Fighting** manual.

Principles of Foam

There are two basic types of foam: chemical and mechanical. *Chemical foams* are those produced as a result of a reaction between two chemicals. Chemical foams are considered obsolete and are rarely, if ever, found in use today.

Foams in use today are of the mechanical type. *Mechanical foams* must be proportioned (mixed with water) and aerated (mixed with air) before they can be used. To produce quality fire fighting foam, foam concentrate, water, air, and mechanical aeration are needed (Figure 15.1). These elements must be present and blended in the correct ratios. Removing any element results either in no foam production or a poor quality foam. Before discussing types of foams and the foam-making process, it is important to understand the following terms:

- *Foam concentrate* – The raw foam liquid as it rests in its storage container before the introduction of water and air.

- *Foam proportioner* – The device that introduces foam concentrate into the water stream to make the foam solution.

- *Foam solution* – The mixture of foam concentrate and water before the introduction of air.

- *Foam* – The completed product after air is introduced into the foam solution (also known as finished foam).

Aeration should produce an adequate amount of bubbles to form an effective foam blanket. Proper aeration should produce uniform-sized bubbles to provide a longer lasting blanket. A good foam blanket is required to maintain an effective cover over either Class A or Class B fuels for the period of time desired.

To be effective, foam concentrates must also match the fuel to which they are applied. Class A foams are not designed to extinguish Class B fires. Class B fuels are divided into two catagories: hydrocarbons and polar solvents.

Hydrocarbon fuels, such as crude oil, fuel oil, gasoline, benzene, naphtha, jet fuel, and kerosene, are petroleum

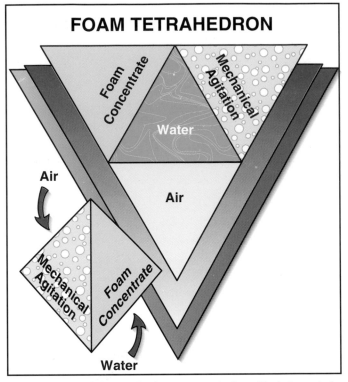

FOAM TETRAHEDRON

Figure 15.1 The foam tetrahedron represents the critical elements for foam production.

based and float on water. Standard fire fighting foam is effective as an extinguishing agent and vapor suppressant because it can float on the surface of hydrocarbon fuels.

Polar solvent fuels, such as alcohol, acetone, lacquer thinner, ketones, and esters, are flammable liquids that are miscible in water — much like a positive magnetic pole attracts a negative pole. Fire fighting foam can be effective on these fuels but only in special alcohol-resistant (polymeric) formulations. It should be noted that many modern fuel blends, which include gasoline with 10 percent or more solvent additives, should be considered polar solvents and handled as such during emergency operations.

Class B foams designed solely for hydrocarbon fires will not extinguish polar solvent fires regardless of the concentration at which they are used. Many foams that are intended for polar solvents may be used on hydrocarbon fires, but this should not be attempted unless the manufacturer of the particular concentrate being used specifically says this can be done. This is why it is extremely important to identify the type of fuel involved before beginning to apply foam.

CAUTION: Failure to match the proper foam concentrate with the fuel results in an unsuccessful extinguishing attempt and could endanger firefighters.

More information on the specific types of foam concentrates is discussed later in this chapter.

How Foam Works

Foam extinguishes and/or prevents fire by the following methods (Figure 15.2):

- *Separating*–Creates a barrier between the fuel and the fire

- *Cooling* – Lowers the temperature of the fuel and adjacent surfaces

- *Suppressing* (sometimes referred to as smothering) – Prevents the release of flammable vapors and therefore reduces the possibility of ignition or reignition.

In general, foam works by forming a blanket on the burning fuel. The foam blanket excludes oxygen and stops the burning process. The water in the foam is slowly released as the foam breaks down. This provides a cooling effect on the fuel.

Foam Proportioning

The term *proportioning* is used to describe the mixing of water with foam concentrate to form a foam solution. Most foam concentrates are intended to be mixed with either fresh or salt water. For maximum effectiveness, foam concentrates must be proportioned at the specific percentage for which they are designed. This percentage rate for the intended fuel is clearly marked on the outside of every foam container. Failure to follow this procedure, such as trying to use 6% foam at a 3% concentration, will result in poor quality foam that may not perform as desired.

Most fire fighting foam concentrates are intended to be mixed with 94 to 99.9 percent water. For example, when using 3% foam concentrate, 97 parts water mixed with 3 parts foam concentrate equals 100 parts foam solution (Figure 15.3). For 6% foam concentrate, 94 parts water mixed with 6 parts foam concentrate equals 100 percent foam solution. Early alcohol-resistant concentrates were sometimes proportioned at a rate as high as 10%, but they are no longer in common usage.

Class A foams are an exception to this rule. The proportioning percentage for Class A foams can be adjusted (within limits recommended by the manufacturer) to achieve specific objectives. To produce a dry (thick) foam suitable for exposure protection and fire breaks, the foam concentrate can be adjusted to a higher percentage. To produce a wet (thin) foam that rapidly sinks into a fuel's surface, the foam concentrate can be adjusted to a lower percentage.

The selection of a proportioner depends on the foam solution flow requirements, available water pressure, cost, intended use (apparatus-mounted or portable), and the agent to be used. Proportioners and delivery devices

Figure 15.2 Foam extinguishes fire in a number of ways.

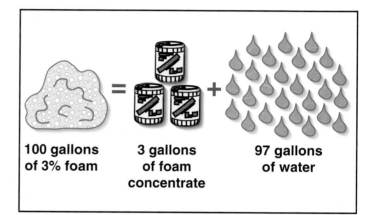

Figure 15.3 Foam that is produced from a 3% concentrate is comprised of 3 parts foam concentrate to 97 parts water.

(foam nozzle, foam maker, etc.) are engineered to work together. Using a foam proportioner that is not compatible with the delivery device (even if the two are made by the same manufacturer) can result in unsatisfactory foam or no foam at all.

There are four basic methods by which foam may be proportioned:

- Induction

- Injection

- Batch mixing

- Premixing

A variety of equipment is used to proportion foam. Some types are designed for mobile apparatus and others are designed for fixed fire protection systems. Each of the common types of foam proportioners is covered in detail later in this chapter.

Induction

The induction (eduction) method of proportioning foam uses the pressure energy in the stream of water to induct (draft) foam concentrate into the fire stream. This is achieved by passing the stream of water through a device called an eductor that has a restricted diameter (Figure 15.4). Within the restricted area is a separate orifice that is attached via a hose to the foam concentrate container. The pressure differential created by the water going through the restricted area and over the orifice creates a suction that draws the foam concentrate into the fire stream. In-line eductors and foam nozzle eductors are examples of foam proportioners that work by this method.

Injection

The injection method of proportioning foam uses an external pump or head pressure to force foam concentrate into the fire stream at the correct ratio in comparison to the flow. These systems are commonly employed in apparatus-mounted or fixed fire protection system applications.

Batch Mixing

Batch mixing is the most simple method of mixing foam concentrate and water. Batch mixing occurs when an appropriate amount of foam concentrate is poured directly into a tank of water (Figure 15.5). It is commonly used to mix foam within a fire apparatus water tank or a portable water tank. It also allows for accurate proportioning of foam. Batch mixing is commonly practiced with Class A foams but should only be used as a last resort with Class B foams. Batch mixing may not be effective on large incidents because when the tank becomes empty, the foam attack lines must be shut down until the tank is

completely filled with water and more foam concentrate is added. Another drawback of batch mixing is that Class B concentrates and tank water must be circulated for a while to ensure thorough mixing before being discharged. The time required for mixing depends on the viscosity and solubility of the foam concentrate.

Premixing

Premixing is one of the more commonly used methods of proportioning. With this method, premeasured portions of water and foam concentrate are mixed in a container. Typically, the premix method is used with portable extinguishers, wheeled extinguishers, skid-mounted twin-agent units, and vehicle-mounted tank systems (Figure 15.6).

Figure 15.5 Batch mixing is the simplest method of proportioning foam concentrate in water.

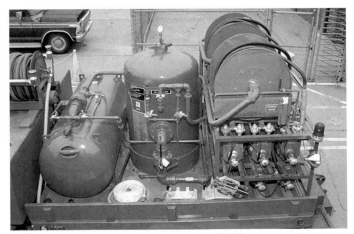

Figure 15.6 Foam is commonly premixed in skid-mounted twin agent units. *Courtesy of Conoco Oil Co.*

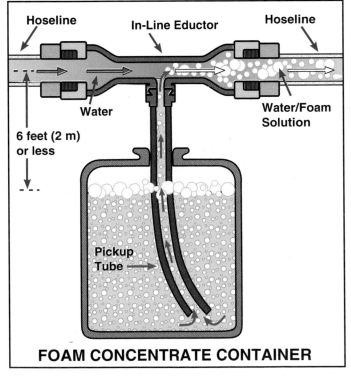

Figure 15.4 Foam is inducted into a fire stream using the pressure differential created in a venturi metering device.

In most cases, premixed solutions are discharged from a pressure-rated tank using either a compressed inert gas or air. An alternative method of discharge uses a pump and a non-pressure-rated atmospheric storage tank. The pump discharges the foam solution through piping or hose to the discharge devices. Premix systems are limited to a one-time application. When used, they must be completely emptied and then refilled before they can be used again.

How Foam is Stored

Foam concentrate is stored in a variety of containers. The type of container used in any particular situation depends on the procedural manner in which the foam is generated and delivered. There are three common foam concentrate storage methods used in the municipal and wildland fire services: pails, barrels (drums), and apparatus tanks.

Pails

Five-gallon (20 L) plastic pails are perhaps the most common containers used by the municipal fire service to ship and store foam concentrate (Figure 15.7). These containers are durable and are not affected by the corrosive nature of foam concentrates. Pails may be carried on the apparatus in compartments, on the side of the apparatus, or in topside storage areas. The containers must be airtight to prevent a skin from forming on the surface of the concentrate. This is particularly true of alcohol-resistant foams. Foam concentrate may be educted directly from the pail when using an in-line or foam-nozzle eductor.

Barrels

If small pails are not convenient, foam concentrate may be shipped and stored in 55-gallon plastic or plastic-lined (220 L) barrels (drums) (Figure 15.8). Some fire departments use these for bulk storage, but they are more common in industrial applications. Foam concentrate can then be transferred to pails or apparatus tanks for actual deployment. Other departments have apparatus that are designed to carry these barrels directly to the emergency scene for deployment. Foam concentrate is educted directly from barrels in the same manner that it is educted from pails.

Apparatus Tanks

Fire apparatus equipped with integral, onboard foam-proportioning systems usually have foam concentrate tanks piped directly to the foam delivery system. This eliminates the need to use separate pails or barrels. Foam concentrate tanks are found on municipal and industrial pumpers, foam tenders, and aircraft rescue and fire fighting (ARFF) apparatus.

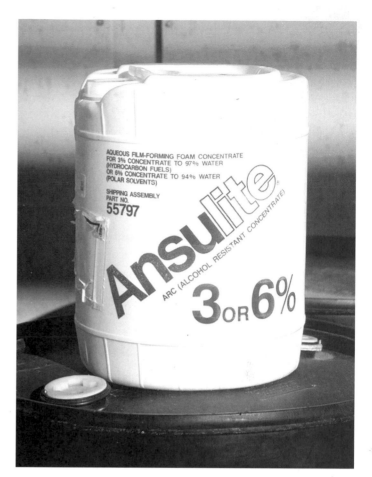

Figure 15.7 Foam concentrate may be purchased in 5-gallon (20 L) pails. *Courtesy of Conoco Oil Co.*

Figure 15.8 Agencies that use large amounts of foam may purchase foam concentrate in 55-gallon (220 L) barrels. *Courtesy of Conoco Oil Co.*

Depending on the design of the apparatus, the type, location, and design of the foam concentrate tank varies. Smaller foam concentrate tanks are located directly above the fire pump area. Newer designs incorporate the foam concentrate tanks as an integral cell within the apparatus water tank. Large foam concentrate tanks may be directly adjacent to the apparatus water tank. Foam tenders and some industrial foam pumpers have a large tank that contains foam concentrate and no water tank (Figure 15.9). Many departments that do not use foam on a regular basis, but have the potential to need a large quantity of concentrate, have special foam concentrate trailers that can be pulled behind other apparatus when needed (Figure 15.10). Regardless of the type of tank, as with foam concentrate pails, it is important that the tank be airtight. Standard vented atmospheric storage tanks are not acceptable for foam concentrate. The tank should also be equipped with a pressure vacuum vent. The NFPA standards provide requirements for foam concentrate storage tank design. Consult the foam concentrate manufacturer for specific requirements on storage tanks.

Foam concentrate tanks on municipal pumpers range from 20 to 200 gallons (80 L to 800 L). Foam pumpers and tenders may carry up to 8,000 gallons (32 000 L) or more of concentrate.

Foam Concentrates

Mechanical foam concentrates can be divided into two general categories: those intended for use on Class A fuels (ordinary combustibles) and those intended for use on Class B fuels (flammable and combustible liquids). The following sections contain information on the concentrates in both of these categories.

Class A Foam

Class A foam has been used since the 1940s; however, only recently has the technology come of age. This agent has proven to be effective for fires in structures, wildland settings, coal mines, tire storage, and other incidents involving similar deep-seated fuels.

> # WARNING
>
> Use Class A foam only on Class A fuels. It is not specifically formulated for fighting Class B fires.

Class A foam is the formulation of specialty hydrocarbon surfactants. The surfactants in this agent reduce the surface tension of the water in the foam solution. Reducing the surface tension provides better penetration of the water, thereby increasing its effectiveness. When used in conjunction with compressed-air foam systems (CAFS), Class A foam has outstanding insulating qualities. These systems entrain large amounts of compressed air and small amounts of water into the foam solution to make foam.

Figure 15.9 Some agencies have large apparatus that are used mainly to transport large quantities of foam concentrate. *Courtesy of Ron Jeffers.*

Figure 15.10 Bulk foam concentrate may be hauled in special foam trailers. *Courtesy of Fire Wagons, Inc.*

Class A foam may be used with fog nozzles; aerating foam nozzles; medium- and high-expansion devices; and compressed-air foam systems using almost any nozzle, including solid stream nozzles (Figure 15.11). The shelf life of Class A foam concentrate can be as much as 20 years. Because this type of foam is used in such small percentages in solution, harm to the environment should not be a concern under ordinary fire suppression conditions. However, there is some evidence that the concentrate may slightly affect aquatic life, so direct application into bodies of water is not recommended. Class A foams approved by the USDA Forest Service have been tested for biodegradability and environmental impact. Class A foam concentrate does have corrosive and supercleaning characteristics, but this should not affect the fire apparatus components exposed to it because of the low percentages used in solution. Proper procedures for flushing application equipment after its use should minimize any adverse effects. Foam concentrate and fire apparatus manufacturers should provide information on flushing requirements for their products.

Figure 15.11 Class A foam may be discharged through a standard fog nozzle. *Courtesy of Mount Shasta (CA) Fire Protection District.*

Proportioning

Class A foam concentrates are mixed in proportions of 0.1% to 3.0%; however, most commonly used concentrates are mixed in proportions of 0.2 to 1.0%. As the expansion ratios are increased, the expansion and drainage characteristics of the foam changes. Drain time increases in proportion with increases in the percentage of the solution. The effect visible to the firefighter is characterized by the foam appearing thicker. Most foam nozzles produce more stable foams at 1.0% concentration than they will at 0.4 to 0.5% concentrations. However, using percentages greater than 0.5% with standard fog nozzles does not appear to increase fire fighting performance. No performance enhancements are gained by proportioning Class A foams higher than their recommended percentages; this only drives up the cost of the operation due to excessive foam concentration use. When performing the following applications, use the listed common guidelines for Class A foam proportioning:

- Fire attack and overhaul with standard fog nozzles — 0.2% to 0.5% foam concentrate

- Exposure protection with standard fog nozzles — 0.5% to 1.0% foam concentrate

- Any application with air aspirating foam nozzles — 0.3% to 0.7% foam concentrate

- Any application with compressed air foam systems (CAFS) — 0.2% to 0.5% foam concentrate

Rates of Application

Application rate refers to the minimum amount of foam solution that must be applied to a fire, per minute, per square foot (square meter) of fire. *The application rate for Class A foam is the same as the minimum critical flow rate for water.* Knowing the correct application rates of

foam solution is of primary importance during emergency operations. These rates should be determined during pre-incident planning. Flow rates should not be reduced when using Class A foam.

Application of Class A Foam

Class A foam can be tailored to meet certain needs, depending on the situation:

- **Areas requiring maximum penetration**. *Wet foam* is very fluid and is desirable for areas requiring maximum penetration. Wet foam has a high water content and a very fast drainage rate (Figure 15.12).

- **Vertical surfaces**. *Dry foam* is a rigid coat that adheres well (Figure 15.13). Its slow drainage rate allows it to cling to these surfaces for extended periods. Dry foam has a very low water content and a high air content. It resembles shaving cream.

Figure 15.12 Class A foam is useful for attacking fires involving densely packed materials, such as hay in a barn fire. *Courtesy of Bob Esposito.*

Figure 15.13 Thick Class A foam can be applied to structures to provide exposure protection during large wildland fires. *Courtesy of Mount Shasta (CA) Fire Protection District.*

- **Areas requiring a balance of penetration and clinging ability.** *Medium foam* has the ability to blanket and wet the fuel equally well. With medium foam, there is a strong need to be able to adjust the air and concentrate ratio.

Drainage, also referred to as drain time, plays a role in foam effectiveness. The bubbles formed in a foam start to break down once they are applied. This breakdown process is what allows the foam to release water and wet the fuel. One of the measurements for stability of the foam is the rate at which bubble breakdown occurs. A short drain time means rapid wetting. A long drain time means that the foam holds water and provides an insulating foam layer for an extended period before the water is released. Several elements affect the breakdown process. Breakdown of the blanket is greatly affected by the heat of the fire (fuel temperature), size of the flame front (wildfires), and, to a lesser extent, the ambient air temperature and wind.

Class B Foam

Class B foam is used to extinguish fires involving flammable and combustible liquids (Figure 15.14). It is also used to suppress vapors from unignited spills of these liquids. There are several types of Class B foam concentrates; each type has its advantages and disadvantages. This section examines the different types of Class B foam concentrates and the characteristics of each.

Class B foam may be proportioned into the fire stream via apparatus-mounted or portable foam proportioning equipment. The foam may be applied either with standard fog nozzles (AFFF and FFFP) or with air-aspirating foam nozzles (all types).

Class B foam concentrates are manufactured from either a synthetic or protein base. Protein-based foams

Figure 15.14 Foam may be needed to control a large flammable liquid fire. *Courtesy of Rich Mahaney.*

are derived from animal protein. Synthetic foam is made from a mixture of fluorosurfactants. Some foam is made from a combination of synthetic and protein bases. The various types of protein-based, synthetic-based, and combination foam concentrates are covered later in the chapter.

Foams that are stored in cool areas will have a greater longevity than foams that are stored in warm areas. Protein-based foams have a normal shelf life of about 10 years. After that time, they may begin to degrade chemically and become less effective. Synthetic-based foams will have a longer shelf life (generally 20 to 25 years) than will protein-based foams.

In general, different manufacturer's foam concentrates should not be mixed together in apparatus tanks, as they may be chemically incompatible. The exception to this would be foam manufactured to U.S. military specifications (Mil-Spec concentrates). The Mil-Specs are written so that mixing can be done with no adverse affects. On the emergency scene, concentrates of a similar type (all AFFFs, all fluoroproteins, etc.), but from different manufacturers, may be mixed together immediately before application.

The chemical properties of Class B foams and their environmental impact vary depending on the type of concentrate and the manufacturer. Generally, protein-based foams are safer for the environment. Consult the data sheets provided by the manufacturer for information on a specific concentrate.

Proportioning

Today's Class B foams are mixed in proportions from 1% to 6%. The proper proportion for any particular concentrate is listed on the outside of the foam container (Figure 15.15). Some multipurpose foams designed for use on both hydrocarbon and polar solvent fuels can be used at different concentrations, depending on which of the two fuels they are used on. These concentrates are normally used at a 1% or 3% rate on hydrocarbons and 3% or 6% on polar solvents, depending on the manufacturer's recommendations.

Older polar solvent mechanical foam concentrates were designed to be used at concentrations of 6% or greater. Depending on the particular fuel being attacked, concentrations as high as 10% were required. These concentrates are not commonly found in use today.

Medium-expansion foams are typically used at either 1½%, 2%, or 3% concentrations. With any Class B foam concentrate, follow the manufacturer's recommendations for proportioning.

Figure 15.15 The required proportioning rate should be clearly marked on the outside of the foam concentrate container. *Courtesy of Conoco Oil Co.*

Foam expansion

Foam expansion refers to the increase in volume of a foam solution when it is aerated. This is a key characteristic to consider when choosing a foam concentrate for a specific application. The method of aerating a foam solution results in varying degrees of expansion, which depends on the following factors:

- Type of foam concentrate used

- Accurate proportioning of the foam concentrate in the solution

- Quality of the foam concentrate

- Method of aspiration

Depending on its purpose, foam can be described by three types: low-expansion, medium-expansion, and high-expansion. NFPA 11, *Standard for Low-Expansion Foam*, states that low-expansion foam has an air/solution ratio up to 20 parts finished foam for every part of foam solution (20:1 ratio). Medium-expansion foam is most commonly used at the rate of 20:1 to 200:1 through hydraulically operated nozzle-style delivery devices. In the high-expansion foams, the expansion rate is 200:1 to 1000:1.

Rates of Application

The rate of application for fire fighting foam varies depending on any one of several variables:

- Type of foam concentrate used

- Whether or not the fuel is on fire (Figure 15.16)

- Type of fuel (hydrocarbon/polar solvent) involved

- Whether the fuel is spilled or in a tank; if the fuel is in a tank, the type of tank will have a bearing on the application rate

The minimum foam solution application rates for ignited fuels are established in NFPA 11. Because there is such a myriad of variables, NFPA 11 contains a series of tables that list the application rate requirements. The following list contains application rates for some of the more common fire scenarios faced by firefighters:

- *Hydrocarbon fuel spill fires (nondiked) using portable extinguishing equipment* – Protein and fluoroprotein foams: 0.16 gpm/ft² (6.5 [L/min]/m²) for 15 minutes; AFFF and FFFP foams: 0.10 gpm/ft² (4.1 [L/min]/m²) for 15 minutes.

Figure 15.16 A typical Class B fire.

- *Polar solvent fuel spill fires (nondiked) using portable extinguishing equipment* – In general, the range is between 0.10 and 0.20 gpm/ft² (4.1 [L/min]/m² and 8.2 [L/min]/m²), depending on the manufacturer's UL listing. Consult the foam concentrate manufacturer's recommendations for specific figures.

- *Hydrocarbon fuel fires in fixed roof storage tanks using portable extinguishing equipment* – The rate is 0.16 gpm/ft² (6.5 [L/min]/m²) for all foams. Combustible liquids require a 50-minute application time; flammable liquids and crude petroleum require a 65-minute application time.

To determine the application rate available from a nozzle, divide the nozzle flow rate by the area of the fire. For example, a 250 gpm [1 000 L/min] nozzle on a 1,000-square-foot (100 m²) fire equals a rate of 0.25 gpm per square foot (10 [L/min]/m²) (Figure 15.17).

Unignited spills do not require the same application rates as ignited spills because radiant heat, open flame, and thermal drafts do not attack the finished foam as they would under fire conditions. No specific rate is given by NFPA 11 for unignited spills. In case the spill does ignite, however, firefighters should be prepared to flow at least the minimum application rate for the specified amount of time based on fire conditions.

All foam concentrate supplies should be on the fireground, at the point of proportioning, before application is started. Once application has started, it should continue uninterrupted until extinguishment is complete. Stopping and restarting may allow the fire and fuel to consume whatever foam blanket has been established. Because polar solvent fuels have differing affinities for water, it is important to know application rates for each type of solvent. These rates also vary with the type and manufacturer of the foam concentrate selected. Foam concentrate manufacturers provide information on the proper application rates as listed by UL. For more complete information on application rates, consult NFPA 11 and the foam manufacturer's recommendations.

Specific Foam Concentrates

Numerous types of foams are selected for specific applications according to their properties and performance. Some foams are thick and viscous and form tough, heat-resistant blankets over burning liquid surfaces; while other foams are thinner and spread more rapidly. Some foams produce a vapor-sealing film of surface-active water solution on a liquid surface. Others, such as medium- and high-expansion foams, are used in large volumes to flood surfaces and fill cavities. The following sections highlight each of the common types of foam concentrates.

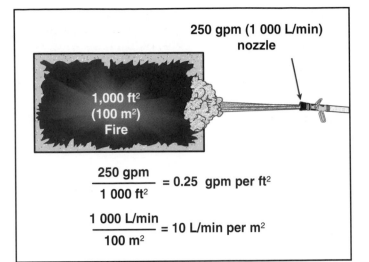

Figure 15.17 Determining the correct application rate is critical for ensuring fire extinguishment.

Regular Protein Foams

Regular protein foam is derived from naturally occurring sources of protein such as hoof, horn, or feather meal. The protein meal is hydrolyzed in the presence of lime and converted to a protein hydrolysate that is neutralized. Other components, such as foam stabilizers, corrosion inhibitors, antimicrobial agents, and freezing point depressants, are then added. Regular protein foam generally has very good heat stability and resists burnback, but it is not as mobile or fluid on the fuel surface as are other types of low-expansion foams. Regular protein foam degrades faster in storage than does synthetic foam (about a 10-year shelf life for regular protein foam as opposed to 25 years for others). Regular protein foam is rarely used in the fire service today.

Fluoroprotein Foam

Fluoroprotein foam, a combination protein-based and synthetic-based foam, is derived from protein foam concentrates to which fluorochemical surfactants are added. The fluorochemical surfactants are similar to those developed for AFFF agents (described later) but are used in much lower concentrations. The addition of these chemicals produces a foam that flows easier than regular protein foam. Fluoroprotein foam provides a strong "security blanket" for long-term vapor suppression. Vapor suppression is especially critical with unignited spills. Fluoroprotein concentrates are still the foam of choice by some fire departments.

Fluoroprotein foam can be formulated to be alcohol resistant by adding ammonia salts that are suspended in organic solvents. Alcohol-resistant fluoroprotein foam maintains its alcohol-resistive properties for about 15 minutes. Alcohol-resistant fluoroprotein foams, like regular fluoroprotein foams, also have a very high degree of heat resistance and water retention.

Film Forming Fluoroprotein Foam (FFFP)

Film forming fluoroprotein foam (FFFP) concentrate is based on fluoroprotein foam technology with aqueous film forming foam (AFFF) capabilities. This film forming fluoroprotein foam incorporates the benefits of AFFF for fast fire knockdown and the benefits of fluoroprotein foam for long-lasting heat resistance. FFFP is available in an alcohol-resistant formulation.

Aqueous Film Forming Foam (AFFF)

AFFF (commonly pronounced A-triple-F) is the most commonly used foam today. Aqueous film forming foam (AFFF) is completely synthetic. It consists of fluorochemical and hydrocarbon surfactants combined with high boiling point solvents and water. Fluorochemical surfactants reduce the surface tension of the water to a degree less than the surface tension of the hydrocarbon so that a thin aqueous film can spread across the fuel.

When AFFF (as well as the previously mentioned FFFP) is applied to a hydrocarbon fire, three things occur (Figure 15.18):

1. An air/vapor-excluding film is released ahead of the foam blanket.

2. The fast-moving foam blanket then moves across the surface and around objects, adding further insulation.

3. As the aerated (7:1 to 20:1) foam blanket continues to drain its water, more film is released. This gives AFFF the ability to "heal" over areas where the foam blanket is disturbed.

Alcohol-resistant AFFFs are available from most foam manufacturers. On most polar solvents, alcohol-resistant AFFFs are used at 3% or 6% concentrations, depending on the particular brand used. Alcohol-resistant AFFFs can also be used on hydrocarbon fires at a 1% or 3% proportion, depending on the manufacturer. Concentrates designed to be used at 3% on hydrocarbon fuels and 6% on polar solvent fuels are commonly referred to as *3 by 6* concentrates. Concentrates used at 3% on both types of fuels are called *3 by 3* concentrates. Concentrates designed to be used at 1% on hydrocarbon fuels and 3% on polar solvent fuels are commonly referred to as *1 by 3* concentrates.

When alcohol-type AFFFs are applied to polar solvent fuels, they

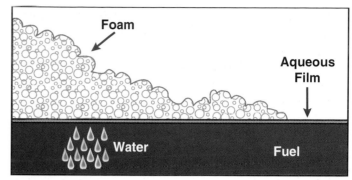

Figure 15.18 This illustration highlights the principles by which AFFF works to control Class B hydrocarbon liquid fires and spills.

create a membrane rather than a film over the fuel. This membrane separates the water in the foam blanket from the attack of the solvent. Then, the blanket acts in much the same way as a regular AFFF. Alcohol-resistant AFFF should be applied gently to the fuel so that the membrane can form first. Alcohol-resistant AFFF should not be plunged into the fuel, but rather it should be sprayed over the top of the fuel.

High-Expansion Foams

High-expansion foams are special-purpose foams and have a detergent base. Because they have a low water content, they minimize water damage. Their low water content is also useful when runoff is undesirable. High-expansion foams have three basic applications:

- In concealed spaces such as basements, in coal mines, and in other subterranean spaces (Figure 15.19)

- In fixed-extinguishing systems for specific industrial uses such as rolled or bulk paper storage

- In Class A fire applications

Figure 15.19 High expansion foam is often used to control basement fires.

High-expansion foam concentrates have expansion ratios of 200:1 to 1,000:1 for high-expansion uses, and 20:1 to 200:1 for medium-expansion uses. (**NOTE**: Whether the foam is used in either a medium- or high-expansion capacity is determined by the type of application device used.)

Low Energy Foam Proportioning Systems

The process of foam proportioning sounds simple: add the proper amount of foam concentrate into the water stream and an effective foam solution is produced. Unfortunately, this process is not as easy as it sounds. The correct proportioning of foam concentrate into the fire stream requires equipment that must operate within strict design specifications. Failure to operate even the best foam proportioning equipment as designed can result in poor quality foam or no foam at all. In general, foam proportioning devices operate by one of two basic principles:

• The pressure of the water stream flowing through an orifice creates a venturi action that inducts (drafts) foam concentrate into the water stream.

• Pressurized proportioning devices inject foam concentrate into the water stream at a desired ratio and at a higher pressure than that of the water.

This section details the various types of low energy foam proportioning devices commonly found in portable and apparatus-mounted applications. A low-energy foam system imparts pressure on the foam solution solely by the use of a fire pump. These systems introduce air into the solution when it either reaches the nozzle or is discharged from the nozzle. High-energy foam systems introduce compressed air into the foam solution before it is discharged into the hoseline. High-energy foam systems are described later in this manual.

Portable Foam Proportioners

Portable foam proportioners are the simplest and most common foam proportioning devices in use today. The three common types of portable foam proportioners are: in-line foam eductors, foam nozzle eductors, and self-educting master stream nozzles.

In-Line Foam Eductors

The in-line eductor is the most common type of foam proportioner used in the fire service. This eductor is designed to be either directly attached to the pump panel discharge or connected at some point in the hose lay. When using an in-line eductor, it is very important to follow the manufacturer's instructions about inlet pressure and the maximum hose lay between the eductor and the appropriate nozzle.

In-line eductors use the Venturi Principle to draft foam concentrate into the water stream (refer to Figure 15.4). As water at high pressure passes over a reduced opening, it creates a low-pressure area near the outlet side of the eductor. This low-pressure area creates a suction effect, called the *Venturi Principle*. The eductor pickup tube is connected to the eductor at this low-pressure point. A pickup tube submerged in the foam concentrate draws concentrate into the water stream, creating a foam water solution (Figure 15.20).

Several very important operating rules must be observed when using eductors. Failure to follow these rules lessens the performance of the eductor.

Rule 1: The eductor must control the flow through the system. In other words, the flow through the eductor should not exceed the rated capacity of the eductor. Exceeding this capacity results in either poor quality foam or no foam at all.

Rule 2: The pressure at the outlet of the eductor (also called *back pressure*) must not exceed 65 to 70 percent of the eductor inlet pressure. Eductor back pressure is determined by the sum of the nozzle pressure, friction loss in the hose between the eductor and the nozzle, and the elevation pressure. If back pressure is excessive, no foam concentrate is inducted into the water.

Figure 15.20 The pick-up tube is placed directly in the foam pail.

Rule 3: Foam solution concentration is only correct at the rated inlet pressure of the eductor, usually 150 to 200 psi (1 050 kPa to 1 400 kPa). Using eductor inlet pressures lower than the rated pressure for the eductor results in rich foam concentrations. Conversely, using inlet pressures greater than the rated inlet pressure produces lean foam concentrations. A too rich or too lean concentration might not work properly.

Rule 4: Eductors must be properly maintained and flushed after each use. Flush the eductor by submerging the foam pickup tube in a pail of clear water and inducting water through it for at least one minute or until clear water is being discharged from the nozzle (Figure 15.21). At the station, thoroughly clean and check the eductor, including the strainer, after each use.

Rule 5: Metering valves must be set to match the foam concentrate percentage and the burning fuel. Failure to do so results in poor quality foam.

Rule 6: The foam concentrate inlet to the eductor should not be more than 6 feet (2 m) above the liquid surface of the foam concentrate (Figure 15.22). If the inlet is too high, the foam concentration will be very lean, or foam may not be inducted at all.

In order for the nozzle and eductor to operate properly, both must have the same rating in gpm (L/min).

Figure 15.21 Thoroughly flush the eductor by drawing clear water through it after every use.

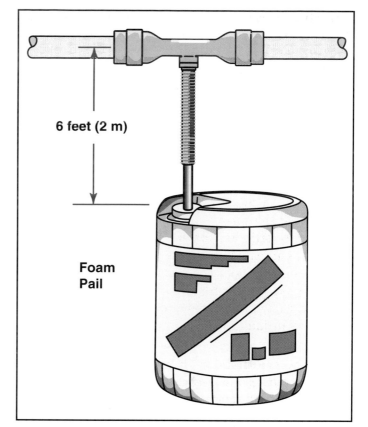

Figure 15.22 The eductor must not be more than 6 feet (2 m) above the surface of the foam concentrate.

Remember that the eductor, not the nozzle, must control the flow. If the nozzle has a lower flow rating than the eductor, the eductor will not flow enough water to pick up concentrate. An example of this situation is a 60 gpm (240 L/min) nozzle with a 95 gpm (380 L/min) eductor.

Using a nozzle with a higher rating than the eductor also gives poor results. A 125 gpm (500 L/min) nozzle used with a 95 gpm (380 L/min) eductor will not result in proper eduction of the foam concentrate. Low nozzle inlet pressure, however, results in poor quality foam. Follow the foam eductor manufacturer's recommendations when using automatic nozzles.

Foam Nozzle Eductors

The foam nozzle eductor operates on the same basic principle as the in-line eductor. However, this eductor is built into the nozzle rather than into the hoseline (Figure 15.23). As a result, its use requires the foam concentrate to be available where the nozzle is operated. If the foam nozzle is moved, the foam concentrate also needs to be moved. The logistical problems of relocation are magnified by the gallons of concentrate required. Use of a foam nozzle eductor compromises firefighter safety: firefighters cannot move quickly, and they must leave their concentrate behind if they are required to back out for any reason.

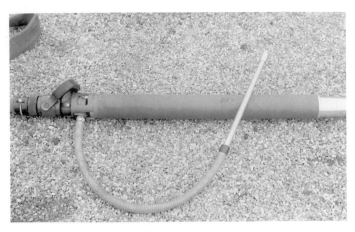

Figure 15.23 A typical handline foam nozzle eductor.

Self-Educting Master Stream Foam Nozzles

The self-educting master stream foam nozzle is used where flows in excess of 350 gpm (1 400 L/min) are required. These nozzles are available with flow capabilities of up to 14,000 gpm (56 000 L/min) (Figure 15.24). The self-educting master stream nozzle uses a modified venturi design to draw foam concentrate into the water stream. The venturi pickup tube is located in the center bore of the nozzle. This results in a "rich" (overproportioned) solution that is diluted at the deflector plates in the nozzle as the solution is discharged. The advantage of this style of foam nozzle eductor is a much lower pressure drop (10 percent or less) than typically associated with standard foam nozzle eductors. This allows for increased stream reach capabilities.

A jet ratio controller (JRC) may be used to supply foam concentrate to a self-educting master stream nozzle. The jet ratio controller is a type of in-line eductor that allows the foam concentrate supply to be as far away as 3,000 feet (900 m) from the self-educting master stream foam nozzle. This allows firefighters, who are involved in operating fire pumps and maintaining the foam concentrate supply, to be a safe distance from the fire. The JRC also allows an elevation change of up to 50 feet (15 m).

The JRC is supplied by a hoseline from the same fire pump that is supplying other hoses to the nozzle (Figure 15.25). The flow of water to the JRC represents about 2½ percent of the total flow in the system. As with a standard in-line eductor, as water flows through the JRC, a venturi is created that draws concentrate through the pickup tube and into the hoseline. The difference is that the JRC proportions the concentrate at a 66½% solution. This rich solution is then pumped to a self-educting master stream foam nozzle where it is further proportioned with the water supplied by the fire pump down to a discharge proportion of 3%. To achieve a proper proportion, it is important that the JRC and nozzle match.

Figure 15.24 Self-educting master stream devices are capable of large flows.

Figure 15.25 The jet-ratio controller somewhat resembles an inline foam eductor.

Apparatus-Mounted Foam Proportioning Systems

Foam proportioning systems are commonly mounted on structural, industrial, wildland, and aircraft rescue and fire fighting apparatus, as well as fire boats. The majority of the foam proportioning systems described in this section can be used for both Class A and Class B foam concentrates. The types of foam proportioning systems covered in this section include the following:

• Installed in-line eductors

- Around-the-pump proportioners
- Bypass-type balanced pressure proportioners
- Variable-flow variable-rate direct injection systems
- Variable-flow demand-type balanced pressure proportioners
- Batch-mixing

Installed In-Line Eductor Systems

Installed in-line eductors use the same principles of operation as do portable in-line eductors. The only difference is that these eductors are permanently attached to the apparatus pumping system (Figure 15.26). The same precautions regarding hose lengths, matching nozzle and eductor flows, and inlet pressures listed for portable in-line eductors also apply to installed in-line eductors. Foam concentrate may be supplied to these devices from either pickup tubes (using 5 gallon [20 L] pails) or from foam concentrate tanks installed on the apparatus.

In many cases, a special version of this device, called a bypass proportioner, is used to reduce the friction loss across the eductor. In the bypass mode, a valve directs the water through a second chamber of the eductor that has no orifice or restrictions (Figure 15.27). This mode is most commonly used when foam is not desired, and the discharge is functioning as a normal plain water attack line.

When foam is desired, move the diverter valve to the foam position; this directs the water flow through the eductor/orifice chamber (Figure 15.28). An adjacent metering valve is usually present to accomodate various types of foam concentrates.

Installed in-line eductors are most commonly used to proportion Class B foams. Take care to adhere to the operating rules for using these devices. Because Class A foam concentrates are normally used at very low concentrations (0.1% to 1.0%), installed in-line eductors are not effective for proportioning Class A foams.

Around-the-Pump Proportioners

This type of proportioner is one of the most common types of built-in proportioners installed in mobile fire apparatus today. The around-the-pump proportioning system consists of a small return (bypass) water line connected from the discharge side of the pump back to

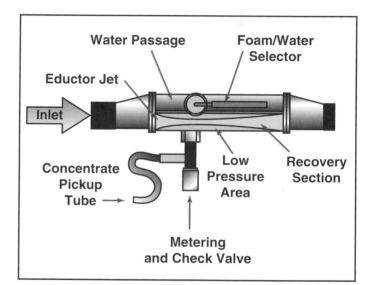

Figure 15.27 In the bypass mode, the pump discharges plain water.

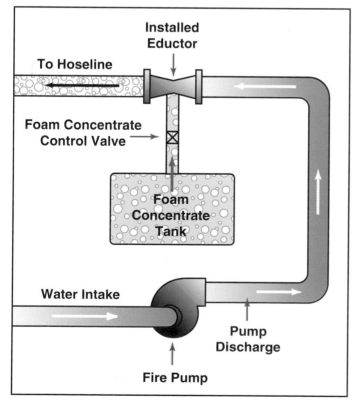

Figure 15.26 The typical arrangement for an installed inline eductor system.

Figure 15.28 The foam system is activated by operating the foam system diverter valve.

the intake side of the pump (Figure 15.29). An in-line eductor is positioned on this bypass line. A valve positioned on the bypass line, just off the pump discharge piping, controls the flow of water through the bypass line. When the valve is open, a small amount of water (10 to 40 gpm [40 L/min to 160 L/min]) discharged from the pump is directed through the bypass piping. As this water passes through the eductor, the resulting venturi effect draws foam concentrate out of the foam concentrate tank and into the bypass piping. The resulting foam solution is then supplied back to the intake side of the pump, where it is then pumped to the discharge and into the hoseline.

Around-the-pump proportioning systems are rated for a specific flow and should be used at this rate, although they do have some flexibility. For example, a unit designed to flow 500 gpm (2 000 L/min) at a 6% concentration proportions 1,000 gpm (4 000 L/min) at a 3% rate.

A major disadvantage of older around-the-pump proportioners is that the pump cannot take advantage of incoming pressure. If the inlet water supply is any greater than 10 psi (70 kPa), the foam concentrate will not enter into the pump intake. The inability of concentrate to enter the pump intake generally means that the proportioner may be used only when operating off the apparatus water tank. Newer units capable of handling intake pressures of up to 40 psi (280 kPa) are now available. Another disadvantage is that the pump must be dedicated solely to foam operation. An around-the-pump proportioner does not allow plain water and foam to be discharged from the pump at the same time.

Bypass-Type Balanced Pressure Proportioners

The bypass-type balanced pressure proportioner is one of the most accurate methods of foam proportioning. It is most commonly used in large-scale mobile apparatus applications such as airport crash vehicles and refinery fire fighting apparatus. The primary advantages of the bypass-type balanced pressure proportioner are its ability to monitor the demand for foam concentrate and to adjust the amount of concentrate supplied. Another major advantage of a bypass-type balanced pressure proportioner is its ability to discharge foam from some outlets and plain water from others at the same time. Thus, a single apparatus can supply both foam attack lines and protective cooling water lines simultaneously.

Apparatus equipped with a bypass-type balanced pressure proportioner have a foam concentrate line connected to each fire pump discharge outlet (Figure 15.30). This line is supplied by a foam concentrate pump separate from the main fire pump. The foam concentrate pump draws the concentrate from a fixed tank. This

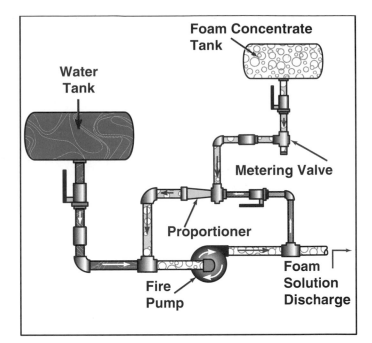

Figure 15.29 Around-the-pump foam proportioners are common on structural fire apparatus.

pump is designed to supply foam concentrate to the outlet at the same pressure at which the fire pump is supplying water to that discharge. The pump discharge and the foam concentrate pressure from the foam concentrate pump are jointly monitored by a hydraulic pressure control valve that ensures the concentrate pressure and water pressure are balanced.

The orifice of the foam concentrate line is adjustable at the point where it connects to the discharge line. If 3% foam is used, the foam concentrate discharge orifice should be set to 3 percent of the total size of the water discharge outlet. If 6% foam is used, the foam concentrate discharge orifice is set to 6 percent of the total size of the water discharge outlet and so on. Because the foam and water are supplied at the same pressure and the sizes of the discharges are proportional, the foam is proportioned correctly.

A limitation of the bypass-type balanced pressure proportioner is its need for a foam pump with PTO or other power source. Also, bypass of concentrate in this system can cause heating, turbulence, and foam concentrate aeration (bubble production in the storage tank).

Variable-Flow Variable-Rate Direct Injection Systems

This type of proportioner operates off power supplied from the apparatus electrical system. Large volume systems may use a combination of electric and hydraulic power. The foam concentrate injection is controlled by monitoring the water flow and controlling the speed of a positive displacement foam concentrate

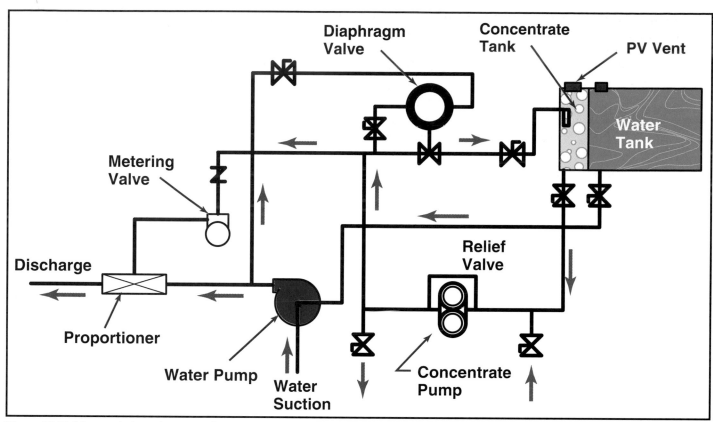

Figure 15.30 A bypass balanced pressure foam proportioning system.

pump, thus injecting concentrate at the desired ratio. Because the water flow governs the foam concentrate injection, water pressure is not a factor. Full flow through the fire pump discharges is possible because there are no flow-restricting passages in the proportioning system.

Variable-flow variable-rate direct injection systems provide foam concentrate rates from 0.1% to 3%. The control unit has a digital display that shows the current water or foam solution flow rate, the total amount of water or solution flowed to this time, the current foam concentrate flow rate, and the amount of foam concentrate used to this time.

These systems can be used with all Class A foam concentrates and many Class B concentrates. These systems may not be used with alcohol-resistant foam concentrates due to the high viscosity of the concentrate. These systems are supplied from atmospheric pressure foam tanks on the apparatus.

There are several advantages to variable-flow variable-rate direct injection systems. One advantage is their ability to proportion at any flow rate or pressure within the design limits of the system. Another advantage is that the system automatically adjusts to changes in water flow when nozzles are either opened or closed. Also, nozzles may be either above or below the pump, without

affecting the foam proportioning. And finally, this system may be used with high-energy foam systems (discussed later in this chapter).

The disadvantage of these systems is that the foam injection point must be within the piping before any manifolds or distribution to multiple fire pump discharges.

Variable-Flow Demand-Type Balanced Pressure Proportioners

The variable-flow demand-type balanced pressure proportioning system, also called a *pumped/demand system*, is a versatile system. In this system, a variable-speed mechanism, which is either hydraulically or electrically controlled, drives a foam concentrate pump. The foam concentrate pump supplies foam concentrate to a venturi-type proportioning device built into the water line (Figure 15.31). When activated, the foam concentrate pump output is automatically monitored so that the flow of foam concentrate is commensurate with the flow of water to produce an effective foam solution.

There are several advantages of the variable-flow demand-type balanced pressure proportioning systems. First, the foam concentrate flow and pressure match system demand. Second, there is no recirculation back to the foam concentrate tank. The system is maintained in

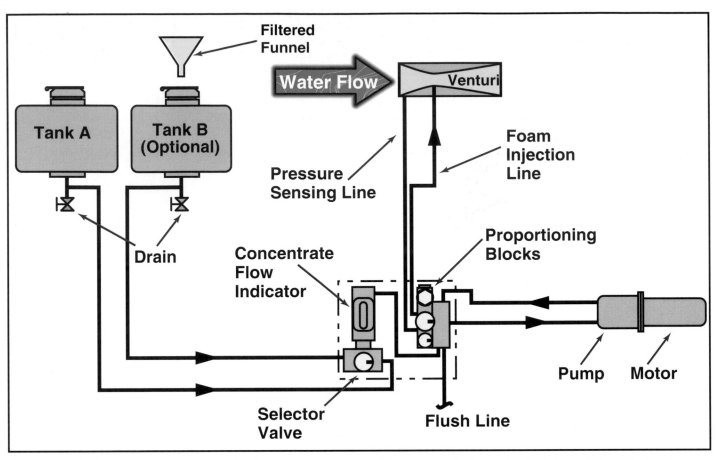

Figure 15.31 The foam concentrate pump supplies foam concentrate to a venturi-type proportioning device that is built into the water line. *Courtesy of KK Products.*

a ready-to-pump condition and requires no flushing after use. Last, water and/or foam solution can be discharged simultaneously from any combination of outlets up to rated capacity.

A limitation of these systems is that the fire pump discharges have ratio controllers (which reduce the discharge area), thus pressure drops across the discharge are higher than those on standard pumpers.

Batch-Mixing

By far, the simplest means of proportioning foam is to simply pour an appropriate amount of foam concentrate into a tank of water. This method is called *batch-mixing* or the *dump-in* method. To do this, the driver/operator pours a predetermined amount of foam concentrate into the tank via the top fill opening at the time foam is needed. The truck is then pumped normally, and foam is discharged through any hoseline that is opened. The amount of foam concentrate needed depends on the size of the water tank and the proportion percentage for which the foam is designed.

The size of the water tank and the proportioning percentage of the foam concentrate dictate how much concentrate must be poured into the water tank. For example,

when using 3% foam concentrate to produce 100 gallons (400 L) of foam solution, 3 gallons (12 L) of foam concentrate must be added to 97 gallons of water. Thus, to form a proper foam solution, a 500-gallon (2 000 L) booster tank would require 15 gallons (60 L) of 3% foam concentrate added to it. Table 15.1 shows the amounts of concentrate required at various percentages for different sizes of water tanks.

In general, batch-mixing is only used with regular AFFF (not alcohol-resistant AFFF concentrates) and Class A concentrates. The AFFF concentrate mixes readily with water, and it will stay suspended in the solution for an extended period of time. When batch-mixing AFFF, the water in the tank has to be circulated for a few minutes before discharge to ensure complete mixing. Batch-mixing can be done at any time, anywhere, and with any equipment. It eliminates the need for more costly foam proportioning equipment.

Class A foam solutions do not retain their foaming properties if mixed in the water for more than 24 hours. There may be further degradation, depending on the product used. Another problem with this method is that lather may form when the water tank is refilled. This may result in either fire pump cavitation or priming difficulty.

Table 15.1					
Amount of Concentrate Needed for Various Sizes of Water Tanks					

Foam Concentrate Proportioning %	Water Tank Size in Gallons (Liters)				
	500 (2 000)	750 (3 000)	1,000 (4 000)	1,500 (6 000)	2,000 (8 000)
1%	5 (20)	7.5m (30)	10 (40)	15 (60)	20 (80)
3%	15 (60)	22.5 (90)	30 (120)	45 (180)	60 (240)
6%	30 (120)	45 (180)	60 (240)	90 (360)	120 (480)

Concentrate to be added in gallons (meters)

Foam solutions are excellent cleansing agents, and they may have a tendency to remove lubricants from ball valve seals. When using batch-mixing, pay special attention to these seals. This method should only be used when no other proportioning system is available.

The disadvantage of this method is that all the water onboard the apparatus is converted to foam solution. This method does not allow for continuous foam discharge on large incidents, as the stream has to be shut down while the apparatus is replenished. It is difficult to maintain the correct concentrate ratio unless the water tank is completely emptied each time.

High-Energy Foam Generating Systems

High-energy foam systems differ from those previously discussed in that they introduce compressed air into the foam solution *prior* to discharge into the hoseline. The turbulence of the foam solution and compressed air going through the piping and/or hoseline creates a finished foam. In addition to simply forming the foam, the addition of the compressed air also allows the foam stream to be discharged considerably greater distances than a regular foam or water fire stream (Figure 15.32).

These systems were used for the first Class A foam applications in the American fire service in the early 1970s. The Texas Forest Service experimented with small compressed-air fire fighting systems that used, as the foaming agent, a pine soap derivative of the paper making process. These systems were formally called Water Expansion Systems (WES) or Water Expansion Pumping Systems (WEPS). The early systems had several inherent limitations, the most important of which was that their compressed air was supplied from air cylinders. This limited the duration at which these systems could operate. Another limitation is that the flow rate of these systems is limited to about 30 gpm (120 L/min). This prevents them from being safely used on structural attacks.

Figure 15.32 A CAFS stream has considerably more reach than does a low energy stream. *Courtesy of Mount Shasta (CA) Fire Protection District.*

In the mid-1980s, the U.S. Bureau of Land Management (BLM) conducted research that led to the development of the type of high-energy Class A foam system that is now becoming common on structural and wildland apparatus. Instead of using air cylinders, the BLM added a rotary air compressor to a standard fire department pumper. This system uses a standard centrifugal fire pump to supply the water. A direct-injection foam-proportioning system is attached to the discharge side of the fire pump. Once the foam concentrate and water are mixed to form a foam solution, compressed air is added to the mixture before it is discharged from the apparatus and into the hoseline. These systems are commonly called a Compressed-Air Foam System (CAFS).

There are several tactical advantages provided by using compressed-air foam:

- The reach of the fire stream is considerably longer than streams from low-energy systems.

- A CAFS produces uniformly sized, small air bubbles that are very durable.

- CAFS-produced foam adheres to the fuel surface and resists heat longer than low-energy foam.
- Hoselines containing high-energy foam solution are lighter than hoselines containing low-energy foam solution or plain water.
- A CAFS provides a safer fire suppression action that allows effective attack on the fire from a greater distance.

A CAFS does have some inherent limitations:

- A CAFS adds expense to a vehicle and adds to the maintenance functions that must be performed on the vehicle.
- Hose reaction can be erratic with a CAFS if foam solution is not supplied to the hoseline in sufficient quantities.
- The compressed air accentuates the hose reaction in the event the hose ruptures.
- Additional training is required for personnel who are expected to make a fire attack using a CAFS or who will operate CAFS equipment.

Most apparatus equipped with a CAFS are also designed to flow plain water should the choice be made to do that. In fact, most CAFS-equipped apparatus only flow foam through preselected discharges. Other discharges may be capable of flowing foam solution or plain water.

The fire pump used on a CAFS-equipped fire vehicle is a standard centrifugal fire pump. The foam proportioning system is some type of automatic, discharge-side proportioning system (Figure 15.33). Foam eductors are typically not used because they are not designed to work at either the 0.1% to 1% eduction rates or the variable flow rates required for Class A foams. Variable-rate automatic-flow sensing proportioners arc neccesary to ensure that foam concentrate is supplied to the fire stream at a proper rate.

In general, 2 cubic feet per minute (cfm) (0.06 m³/min) of airflow per gallon per minute of foam solution flow produces a very dry foam at flows of up to 100 gpm (400 L/min) of foam solution. This produces a large amount of foam at a 10:1 expansion ratio. Most structural and wildland fire attacks using CAFS are done with an air flow rate of 0.5 to 1.0 cfm (0.015 m³/min to 0.03 m³/min) per gallon of foam solution. This rate allows for adequate drainage of solution from the blanket to wet the fuel and prevent reignition. It also prevents smoldering from occurring beneath the foam blanket.

Portable Foam Application Devices

Once the foam concentrate and water have been mixed together to form a foam solution, the foam solution must

Figure 15.33 In addition to foam proportioning controls, compressed air controls are located on the pump panel. *Courtesy of Mount Shasta (CA) Fire Protection District.*

then be mixed with air (aerated) and delivered to the surface of the fuel. With low-energy foam systems, the aeration and discharge of the foam are accomplished by the foam nozzle, sometimes referred to as a foam maker. Low-expansion foams may be discharged through cither handline nozzles or master stream devices. While standard fire fighting nozzles can be used for applying some types of low-expansion foams, it is best to use nozzles that produce the desired result (such as fast-draining or slow-draining foam). The following sections highlight portable foam application devices. (**NOTE**: Foam nozzle eductors and self-educting master stream foam nozzles are considered portable foam nozzles, but they have been omitted from this section because they were covered earlier in the chapter.)

Handline Nozzles

IFSTA defines a handline nozzle as "any nozzle that one to three firefighters can safely handle and that flows less than 350 gpm (1 400 L/min)." Most handline foam nozzles flow considerably less than that figure. The following sections detail the handline nozzles commonly used for foam application.

Solid Bore Nozzles

With respect to foam application, the use of solid bore nozzles is limited to Class A, compressed-air foam system (CAFS) applications. In these applications, the solid bore nozzle provides an effective fire stream that has maximum reach capabilities (Figures 15.34 a and b). Tests indicate that the reach of the CAFS fire stream can be greater than twice the reach of a low-energy fire stream. When using a solid bore nozzle in conjunction with a CAFS, disregard the standard rule of thumb that the discharge orifice of the nozzle be no greater than one-half the diameter of the hose. Tests show that a 1½-inch (38 mm) hoseline may be equipped with a nozzle tip up to 1¼-inch (29 mm) in diameter and still provide an effective fire stream.

Fog Nozzles

Either fixed-flow or automatic fog nozzles can be used with foam solution to produce a low-expansion, short-lasting foam (Figure 15.35). This nozzle breaks the foam solution into tiny droplets and uses the agitation of water droplets moving through air to achieve its foaming action. Its best application is when it is used with regular AFFF and Class A foams. These nozzles cannot be used with protein and fluoroprotein foams. These nozzles may be used with alcohol-resistant AFFF foams on hydrocarbon fires but should not be used on polar solvent fires. This is because insufficient aspiration occurs to handle the polar solvent fires. Some nozzle manufacturers have foam aeration attachments that can be added to the end of the nozzle to increase aspiration of the foam solution (Figure 15.36).

Air-Aspirating Foam Nozzle

The air-aspirating foam nozzle inducts air into the foam solution by a venturi action (Figure 15.37). These nozzles must be used with protein and fluoroprotein concentrates. They may also be used with Class A foams in

Figure 15.35 Film forming foams may be effectively discharged through regular fog nozzles. *Courtesy of Conoco Oil Co.*

Figure 15.34a A typical solid stream nozzle that may be used with a CAFS stream. *Courtesy of Mount Shasta (CA) Fire Protection District.*

Figure 15.36 This attachment may be added to the fog nozzle to increase foam aeration. *Courtesy of KK Products.*

Figure 15.34b The CAFS stream through a solid stream nozzle is a powerful tool. *Courtesy of Mount Shasta (CA) Fire Protection District.*

Figure 15.37 A typical aspirating foam nozzle.

wildland applications. These nozzles provide maximum expansion of the agent. The reach of the stream is less than that of a standard fog nozzle.

Master Stream Foam Nozzles

Large-scale flammable and combustible liquid emergencies are beyond the scope of those that can be handled using handlines. Master stream nozzles are required to deliver adequate amounts of foam in these emergencies (Figure 15.38). As is the case with handline nozzles, standard fixed-flow or automatic fog stream nozzles may be used to deliver foam, when necessary. Their performance would be much the same as described in the section on handline nozzles. Large-scale industrial foam apparatus and ARFF vehicles may be equipped with special aerating foam master stream nozzles (Figure 15.39).

Medium- and High-Expansion Foam Generating Devices

Medium- and high-expansion foam generators produce a high-air-content, semistable foam. For medium-ex-

pansion foam, the air content ranges from 20 parts air to 1 part foam solution (20:1) to 200 parts air to 1 part foam solution (200:1). For high-expansion foam, the ratio is 200:1 to 1,000:1. There are two basic types of medium- and high-expansion foam generators: the water-aspirating type and the mechanical blower.

The water-aspirating type is very similar to the other foam-producing nozzles except that it is much larger and longer (Figure 15.40). The back of the nozzle is open to allow airflow. The foam solution is pumped through the nozzle in a fine spray that mixes with air to form a moderate-expansion foam. The end of the nozzle has a screen, or series of screens, that further breaks up the foam and mixes it with air. These nozzles typically produce a lower-air-volume foam than do mechanical blower generators.

A mechanical blower generator is similar in appearance to a smoke ejector. It operates on the same principle as the water-aspirating nozzle except that the air is forced through the foam spray by a powered fan instead of being pulled through by water movement. This device produces a foam with a high air content and is typically associated with total-flooding applications (Figure 15.41). Its use is limited to high-expansion foam.

Figure 15.38 A foam master stream device. The truck-mounted nozzle is capable of flowing up to 8,000 gpm (32 000 L/min) and is supplied by up to six 5-inch (125 mm) hoses.

Figure 15.40 A high-expansion foam tube.

Figure 15.39 Some apparatus are equipped with foam turrets.

Figure 15.41 Mechanical blowers generate massive amounts of high-expansion foam. *Courtesy of Walter Kidde, Inc.*

Assembling a Foam Fire Stream

To provide a foam fire stream, the driver/operator must be able to correctly assemble the components of the system. The following procedure describes the steps for placing a foam line in service using an in-line proportioner. As mentioned earlier, this is one of the most common methods of foam production used in the municipal fire service.

Step 1: Select the proper foam concentrate for the burning fuel involved.

Step 2: Check the eductor and nozzle to make sure that they are hydraulically compatible (rated for same flow) (Figure 15.42).

Step 3: Check to see that the foam concentrate percentage listed on the foam container matches the eductor rating/setting. If the eductor is adjustable, set it to the proper concentration setting.

Step 4: Attach the eductor to a hose capable of efficiently flowing the rated capacity of the eductor and the nozzle (Figure 15.43).

- Avoid kinks in the hose.

- If the eductor is attached directly to a pump discharge outlet, make sure that the ball valve gates are completely open. In addition, avoid connections to discharge elbows. This is important because any condition that causes water turbulence adversely affects the operation of the eductor.

Step 5: Attach the attack hoseline and desired nozzle to the discharge end of the eductor. The length of the hose from the eductor to the nozzle should not exceed the manufacturer's recommendations.

Step 6: Open enough buckets of foam concentrate to handle the task. Place them at the eductor so that the operation can be carried out without any interruption in the flow of concentrate.

Step 7: Place the eductor suction hose into the concentrate (Figure 15.44). Make sure that the bottom of the concentrate is no more than 6 feet (2 m) below the eductor.

Step 8: Increase the water supply pressure to that required for the eductor. Be sure to consult the manufacturer's recommendations for your specific eductor. Foam should now be flowing.

Troubleshooting Foam Operations

There are a number of reasons for failure to generate foam or for generating poor quality foam. The most common reasons for failure are as follows:

- Failure to match eductor and nozzle flow, resulting in no pickup of foam concentrate

- Air leaks at fittings that cause loss of suction

- Improper cleaning of proportioning equipment that results in clogged foam passages

- Partially closed nozzle control that results in a higher nozzle pressure

- Too long a hose lay on the discharge side of the eductor

- Kinked hose

- Nozzle too far above eductor (results in excessive elevation pressure)

Figure 15.42 Compare the eductor and the nozzle to make sure that they are compatible.

Figure 15.43 Attach the eductor to the hose.

Figure 15.44 Place the pick-up tube into the foam concentrate pail.

• Mixing different types of foam concentrate in the same tank, which can result in a mixture too viscous to pass through the eductor

When using other types of foam proportioning equipment, such as apparatus-mounted systems, the driver/operator must follow the operating instructions provided by the foam system or fire pump manufacturer. Because the operation of these systems vary widely, it is not possible to give specific operating directions in a manual of this type.

Foam Application Techniques

Occasionally, the driver/operator may be required to operate a foam handline or master stream on a fire or spill. It is important to use the correct techniques when manually applying foam. If incorrect techniques are used, such as plunging the foam into a liquid fuel, the effectiveness of the foam is reduced. The techniques for applying foam to a liquid fuel fire or spill include the roll-on method, bank-down method, and rain-down method.

Roll-On Method

The roll-on method directs the foam stream on the ground near the front edge of a burning liquid pool (Figure 15.45). The foam then rolls across the surface of the fuel. The driver/operator continues to apply foam until it spreads across the entire surface of the fuel and the fire is extinguished. It may be necessary to move the stream to different positions along the edge of a liquid spill to cover the entire pool. This method is used only on a pool of liquid fuel (either ignited or unignited) on the open ground.

Bank-Down Method

The bank-down method may be employed when an elevated object is near or within the area of a burning pool of liquid or an unignited liquid spill. The object may be a wall, tank shell, or similar structure. The foam stream is directed off the object, allowing the foam to run down onto the surface of the fuel (Figure 15.46). As with the roll-on method, it may be necessary to direct the stream off various points around the fuel area to achieve total coverage and extinguishment of the fuel. This method is used primarily in dike fires and fires involving spills around damaged or overturned transport vehicles.

Rain-Down Method

The rain-down method is used when the other two methods are not feasible because of either the size of the spill area (either ignited or unignited) or the lack of an object from which to bank the foam. It is also the primary manual application technique used on aboveground storage tank fires. This method directs the stream into the air above the fire or spill and allows the foam to float gently down onto the surface of the fuel (Figure 15.47). On small fires, the driver/operator sweeps the stream back and forth over the entire surface of the fuel until the fuel is completely covered and the fire is extinguished. On large fires, it may be more effective for the driver/operator to direct the stream at one location to allow the foam to take effect there and then work its way out from that point.

Figure 15.46 The bank-down method of foam application.

Figure 15.47 The rain-down method of foam application.

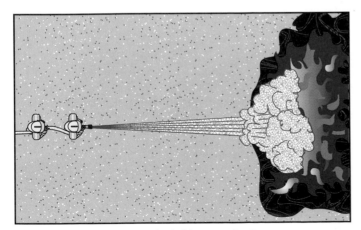

Figure 15.45 The roll-on method of foam application.

Chapter 16

Apparatus Testing

This chapter provides information that will assist the reader in meeting the following job performance requirements from NFPA 1002, *Standard on Fire Apparatus Driver/Operator Professional Qualifications*, 1998 edition. Particular portions of the job performance requirements (JPRs) that are met in this chapter are noted in bold text.

2-2.2 Document the routine tests, inspections, and servicing functions, given maintenance and inspection forms, so that all items are checked for proper operation and deficiencies are reported.

(a) *Requisite Knowledge:* Departmental requirements for documenting maintenance performed, **understanding the importance of accurate record keeping**.

(b) *Requisite Skills:* **The ability to use tools and equipment and complete all related departmental forms.**

3-1.1 Perform the specified routine tests, inspections, and servicing functions specified in the following list in addition to those contained in the list in 2-2.1, given a fire department pumper and its manufacturer's specifications, so that the operational status of the pumper is verified.

- **Water tank and other extinguishing agent levels (if applicable)**
- **Pumping systems**
- **Foam systems**

(a) *Requisite Knowledge:* Manufacturer specifications and requirements, policies, and procedures of the jurisdiction.

(b) *Requisite Skills:* The ability to use hand tools, **recognize system problems, and correct any deficiency noted according to policies and procedures.**

6-1.1 Perform the specified routine tests, inspections, and servicing functions specified in the following list, in addition to those contained in 2-2.1, given a wildland fire apparatus and its manufacturer's specifications, so that the operational status is verified.

- **Water tank and/or other extinguishing agent levels (if applicable)**
- **Pumping systems**
- **Foam systems**

(a) *Requisite Knowledge:* Manufacturer specifications and requirements, policies, and procedures of the jurisdiction.

(b) *Requisite Skills:* The ability to use hand tools, **recognize system problems, and correct any deficiency noted according to policies and procedures.**

8-1.1 Perform routine tests, inspections, and servicing functions specified in the following list, in addition to those specified in the list in 2-2.1, given a fire department mobile water supply apparatus, so that the operational readiness of the mobile water supply apparatus is verified.

- **Water tank and other extinguishing agent levels (if applicable)**
- **Pumping system (if applicable)**
- **Rapid dump system (if applicable)**
- **Foam system (if applicable)**

(a) *Requisite Knowledge:* Manufacturer specifications and requirements, policies, and procedures of the jurisdiction.

(b) *Requisite Skills:* The ability to use hand tools, **recognize system problems, and correct any deficiency noted according to policies and procedures.**

Fire apparatus is tested immediately after its construction — before the purchaser accepts it — to ensure that it performs in the manner for which it was designed. Once it is placed into service, it is tested at least yearly to ensure that it will continue to perform properly under emergency conditions. An organized system of apparatus testing, plus regular maintenance, is the best assurance that apparatus will perform within its design limitations. Furthermore, the insurance industry requires that apparatus be tested in order for the community to receive full credit. This, in turn, affects the insurance rates in the jurisdiction.

Apparatus tests can be grouped into two basic categories: preservice tests and service tests. *Preservice tests* are conducted before the apparatus is placed into service. Usually, the driver/operator is not involved in preservice tests. However, the driver/operator must have a basic understanding of the preservice tests in order to appreciate and understand the service tests. Preservice tests include manufacturer's tests, pump certification tests, and acceptance tests. *Service* tests are conducted on at least a yearly basis while the apparatus is in service. The driver/operator will often be required to perform these tests or at least assist mechanics who are doing them. This chapter concentrates on both types of tests. Important considerations for these tests include correcting net pump discharge pressure for tests, the sequence of tests, equipment needed, safety precautions, and possible causes of trouble during the tests.

Apparatus-mounted foam systems must also be inspected on a regular basis. NFPA 1002 requires that the driver/operator to be able to perform functional tests on foam systems. Most of these tests involve ensuring that the foam system is proportioning the correct amount of foam concentrate into the fire stream. The latter part of this chapter explains several methods for performing these tests.

Preservice Tests

Apparatus equipped with an attack or fire pump undergoes extensive testing before it is put into service. These tests assure the purchaser that the pump and pump components will operate properly under normal use. These tests can be grouped into the categories of manufacturer's tests, certification tests, and acceptance tests. To ensure that they are performed, these tests must be required in the apparatus bid specifications.

NFPA 1901, *Standard for Automotive Fire Apparatus*, and NFPA 1906, *Standard for Wildland Fire Apparatus*, are used as a basis for most apparatus bid specifications. When specifications are written for an apparatus, they should contain a clause that requires them to meet the pertinent chapters of NFPA 1901 or 1906 that apply to that particular apparatus. The clause should state that failure to meet these requirements will be cause for rejecting acceptance of the apparatus.

Generally, fire department personnel are not involved in performing manufacturer's or certification tests. These tests are performed by the manufacturer or Underwriters Laboratories (UL) personnel. Fire department personnel may be involved with acceptance testing. Acceptance testing is commonly performed after the apparatus has been delivered to the purchaser but before final acceptance has been made. The following sections highlight the major points of these preservice tests.

Manufacturer's Tests

If the requirements of NFPA 1901 are included in the apparatus bid specifications, the manufacturer is required to perform two specific tests in addition to the pump certification tests listed in the next section. These two tests are the road test and the hydrostatic test.

Road Test

NFPA requires the following minimum tests to be conducted on the fire apparatus after its construction is complete. The apparatus should be fully loaded in the same manner as it would be once in service. This includes making sure that the water and/or foam tanks are full and the weight of hose and equipment that will be carried on the apparatus are accounted for. The road tests should be conducted in a location and manner that will not violate any applicable traffic laws or motor vehicle codes (Figure 16.1). The test surface should be a flat, dry, paved road surface that is in good condition. At a minimum, the apparatus must meet the following criteria:

- The apparatus must accelerate to 35 mph (56 km/h) from a standing start within 25 seconds. This test must consist of two runs, in opposite directions, over the same surface.

Figure 16.1 The manufacturer's road test should cover a variety of travel conditions.

- The apparatus must achieve a minimum top speed of 50 mph (80 km/h). This requirement may be dropped for specialized wildland apparatus not designed to operate on public roadways.

- The apparatus must come to a full stop from 20 mph (32 km/h) within 35 feet (10.7 m).

- The apparatus parking brake must conform to the specifications listed by the braking system manufacturer.

Beyond these minimums, road tests are centered around the specific needs of each department. These needs are outlined in its apparatus bid specifications. For example, departments that protect hilly jurisdictions may have special requirements for apparatus acceleration, deceleration, and braking abilities on nonlevel surfaces. These special situations are the reason why many jurisdictions prefer to write performance-based specifications as opposed to engineering/equipment specifications. If the specifications are based on desired performance, the purchaser has a stronger position if the apparatus that is delivered does not meet the expectations.

Hydrostatic Test

The *hydrostatic test* determines whether the pump and pump piping can withstand pressures normally encountered during fire fighting operations. Pumps are tested hydrostatically at 250 psi (1 725 kPa) for 3 minutes. The tank fill line, tank-to-pump line, and by-pass line valves should be closed during this test. Discharge valves should be opened and capped. Intake valves should be closed and/or capped (Figure 16.2). The test pressure should be maintained on the system for a minimum of three minutes without the failure of any component of the system.

Pump Certification Tests

Pump certification tests are performed to make sure that the fire pump system operates in the manner for which it was designed after the pump and components are installed on the apparatus chassis. These tests must be conducted by an independent testing organization, such as Underwriters Laboratories (UL). The tests are conducted either at the manufacturer's plant or at the fire department after delivery. These tests assure both the fire department and insurance companies that the apparatus will perform as expected after being placed into service. The results of these tests are required to be stamped into a place that is affixed to the pump panel of the apparatus. Certification tests must be required in the apparatus bid specifications, either by referencing NFPA 1901 or by specific wording, such as:

Figure 16.2 All discharge valves should be open, all intakes should be closed, and the caps should remain on all connections.

"The manufacturer shall submit the completed pumper for pumping performance certification, including tank-to-pump flow rate tests, by Underwriters Laboratories, Inc. A requirement of acceptance is that 'Certification of Inspection for Fire Department Pumper' for the pumper be received by the purchaser from Underwriters Laboratories, Inc."

NFPA 1901 requires the following pump certification tests for apparatus that are equipped with a 750 gpm (3 000 L/min) or larger fire pump:

- Pumping test
- Pumping engine overload test
- Pressure control system test
- Priming device test
- Vacuum test
- Water tank-to-pump flow test

Apparatus equipped with a fire pump whose rated capacity is less than 750 gpm (3 000 L/min) are required to undergo the same certification tests, with the exception of the pumping engine overload test. That test may be omitted for these smaller pumps.

The pump certification tests are performed in a similar manner to the service tests described later in the chapter. The primary differences are duration in which the tests are conducted and minimum criteria for successful completion of the testing. For example, the certification pumping test requires the pump to be operated for 3 hours, while the service pumping test requires only 40 minutes of operation. For more detailed information on pump certification testing, consult NFPA 1901.

Acceptance Testing

Acceptance tests are conducted to assure the purchaser that the apparatus meets bid specifications. These tests are most often performed after the apparatus is delivered to the purchaser. A representative of the manufacturer is present during testing. This ensures that the apparatus will perform in that jurisdiction, as specified. The types of tests and test criteria vary widely with local jurisdiction preference and conditions. The acceptance tests should include another pump test, even if a certification test was performed at the factory. Many instances have been documented where a pump was certified at the factory, only not to perform as desired or rated once it was delivered. The pump test does not have to mirror the certification test but rather may follow the service test procedure listed later in this chapter. This should be proof enough that the certification test was accurate.

If the apparatus fails to perform according to the requirements detailed in the bid specifications, it should be rejected. On occasion, the manufacturer or salesperson may attempt to offer the purchaser alternative "credits," such as chrome wheels or gold-leaf lettering, to compensate for an apparatus that fails to meet all of the performance specifications. It is generally an unwise decision to accept this trade-off. The most important factor in making this purchase is getting an apparatus that performs as needed during an emergency.

An important acceptance testing issue arises when the jurisdiction purchasing the apparatus is located at an altitude in excess of 2,000 feet (610 m) above sea level. In these cases, the pumping engine overload test must be performed during acceptance testing. This ensures that the engine develops the necessary power to operate in the jurisdiction it serves.

Pumper Service Tests

The requirements for fire department service testing are contained in NFPA 1911, *Standard for Service Tests of Fire Pump Systems on Fire Apparatus*. According to this standard, a pumper should be given a service test at least once a year or whenever it has undergone extensive pump or power train repair. These service tests are necessary to ensure that the pumper will perform as it should and to check for defects that otherwise might

go unnoticed until too late. The following sections discuss the site considerations for the testing area as well as information that the driver/operator needs to know to perform the minimum service tests required in NFPA 1911. These tests include the following:

- Engine speed check
- Vacuum test
- Pumping test
- Pressure control test
- Gauge and flow meter test
- Tank-to-pump flow rate test

Site Considerations for Pumper Service Tests

NFPA 1911 says that pumper service tests may be conducted using either a fire hydrant or static body of water as a supply source. Most jurisdictions prefer to draft from a static source during testing because this is a true test of the pump's ability (Figure 16.3). When using a static water supply source, the water must be at least 4 feet (1.2 m) deep. The strainer must be submerged at least 2 feet (0.6 m) below the surface of the water. The surface of the water may be no more than 10 feet (3 m) below the centerline of the pump intake, and 20 feet (6 m) of hard intake hose should be used for drafting during testing (Figure 16.4).

The atmospheric air temperature during testing should be between 0°F and 100°F (-18°C and 38°C). The water being used for testing should be between 35°F and 90°F (2°C and 32°C). The barometric pressure should be at least 29 in.Hg (98.2 kPa), corrected to sea level. This is because a 1-inch (25 mm) drop in barometric pressure reduces a pumper's maximum possible static lift by about 1 foot (0.3 m).

Figure 16.3 It is best to test the pumper from a static water supply source.

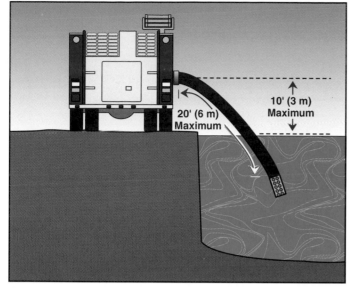

Figure 16.4 Pumps are tested at a 10 foot (3 m) lift through 20 feet (6 m) of intake hose.

Table 16.1 shows the minimum hard intake hose arrangements that are necessary to perform tests on pumps of varying capacities. These figures are good for tests that are performed at altitudes of up to 2,000 feet (610 m) above sea level. Beyond that, it may be necessary to increase intake hose diameter or the number of intake hoses used in order to pump the rated capacity. Altitude affects the pump's performance: lifting ability drops about 1 foot (0.3 m) per 1,000 feet (300 m) increase in altitude; gasoline engines have 3.5 percent less efficiency per 1,000 feet (300 m) increase in altitude.

It is also necessary to lay out a sufficient number of discharge hoses and nozzles to pump the rated capacity of the fire pump; 2½-inch (65 mm) hose is the minimum size hose that may be used for this application. Larger hoses may be used if available. Make sure that whatever hose is used has been tested to ensure that it is capable of withstanding the discharge pressures during the pump test. As you would when service testing the hose, scribe a mark where the hose and couplings meet (Figure 16.5). While the pump testing is proceeding, regularly check the couplings to make sure that the hose is not starting to pull loose of the coupling. If the scribe mark moves more than ⅜-inch (9 mm) away from the coupling, stop the test, and replace the hose (Figure 16.6). Tables 16.2 a and b show the minimum hose and nozzle arrangements needed to discharge sufficient water for various size pumps being tested.

Table 16.1 Required Number and Size of Intake Hoses						
Rated Capacity		Suction Hose Size		Number of Suction Lines	Maximum Lift	
(gpm)	(L/min)	(in.)	(mm)		(ft)	(m)
250	(950)	3	(76)	1	10	(3)
300	(1136)	3	(76)	1	10	(3)
350	(1325)	4	(100)	1	10	(3)
450	(1700)	4	(100)	1	10	(3)
500	(1900)	4	(100)	1	10	(3)
600	(2270)	4	(100)	1	10	(3)
700	(2650)	4	(100)	1	10	(3)
750	(2850)	4½	(113)	1	10	(3)
1000	(3785)	5	(125)	1	10	(3)
1250	(4732)	6	(150)	1	10	(3)
1500	(5678)	6	(150)	2	10	(3)
1750	(6624)	6	(150)	2	8	(2.4)
2000	(7570)	6	(150)	2	6	(1.8)
2250	(8516)	8	(150)	3	6	(1.8)
2500	(9463)	8	(150)	3	6	(1.8)
2750	(10,140)	8	(150)	4	6	(1.8)
3000	(11,356)	8	(150)	4	6	(1.8)

Reprinted with permission from NFPA 1911, *Service Tests of Pumps on Fire Department Apparatus*, Copyright © 1997. National Fire Protection Association, Quincy, MA 02269. This reprinted material is not the complete and official position of the National Fire Protection Association, on the referenced subject which is represented only by the standard in its entirety.

Figure 16.5 Mark the hose before testing.

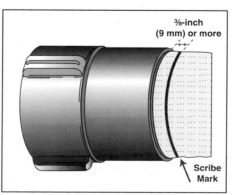

Figure 16.6 If the scribe mark moves more than ⅜-inch (9 mm) away from the coupling, stop the test, and remove the hose from service for repair.

Table 16.2a
Hose and Nozzle Layouts for Pump Tests (U.S.)

Pump Capacity in gpm	Hose and Nozzle Layout (all hose 2½-inch in diameter)
250-350	One 50-foot line with a 1⅛-inch or 1¼-inch nozzle
400-500	One 50-foot line with a 1⅜-inch or 1½-inch nozzle
600-750	Two 100-foot lines with a 1½-inch or 1¾-inch nozzle
1,000	Two or three 100-foot lines with a 2-inch nozzle
1,250	Two 100-foot lines with a 1¾-inch nozzle and one 50-foot line with a 1½-inch nozzle
1,500	Three 100-foot lines with a 2-inch nozzle and one 50-foot line with a 1½-inch nozzle
1,750	Two sets of twin 100-foot lines, each set supplying a 2-inch nozzle
2,000	Two sets of twin 100-foot lines, each set supplying a 2-inch nozzle
2,250	Two sets of three 100-foot lines, each set supplying a 2¼-inch nozzle
2,500	Two sets of three 100-foot lines, each set supplying a 2¼-inch nozzle

Table 16.2b
Hose and Nozzle Layouts for Pump Tests (Metric)

Pump Capacity in L/min	Hose and Nozzle Layout (all hose 65 mm in diameter)
1 000 to 1 400	One 15 m line with a 29 mm or 32 mm nozzle
1 600 to 2 000	One 15 m line with a 35 mm or 38 mm nozzle
2 400 to 3 000	Two 30 m lines with a 38 mm or 45 mm nozzle
4 000	Two or three 30 m lines with a 50 mm nozzle
5 000	Two 30 m lines with a 45 mm nozzle and one 15 m with a 38 mm nozzle
6 000	Three 30 m lines with a 50 mm nozzle and one 15 m with a 38 mm nozzle
7 000	Two sets of twin 30 m lines, each set supplying a 50 mm nozzle
8 000	Two sets of twin 30 m lines, each set supplying a 50 mm nozzle
9 000	Two sets of three 30 m lines, each set supplying a 57 mm nozzle
10 000	Two sets of three 30 m lines, each set supplying a 57 mm nozzle

Correcting Net Pump Discharge Pressure for the Tests

The pump service tests are to be conducted at 150, 165 (overload), 200, and 250 psi (1 000 kPa, 1 150 kPa, 1 350 kPa, and 1 700 kPa) net pump discharge pressure. Net pump discharge pressure is the total work done by the pump to get the water into, through, and out of the pump. When at draft, the net pump discharge pressure is more than the pressure shown on the discharge gauge. Therefore, when the tests are conducted, the allowances for friction loss in the hard intake hose and the height of lift must be taken into account. Table 16.3 gives the friction loss allowances for the various sizes of hard intake hose that may be used for pump testing.

The allowances for friction loss are then used to determine the correct pump discharge pressure during each test. The following formulas are used to make this calculation:

Equation Z (U.S.)

$$\text{Pressure Correction} = \frac{\text{lift (ft.) + intake hose friction loss}}{2.3}$$

Equation AA (metric)

$$\text{Pressure Correction} = \frac{\text{lift (m) + intake hose friction loss}}{0.1}$$

Example 1

Assume that a 1,000 gpm pumper is being tested. The lift is 9 feet through 20 feet of 5-inch intake hose. Find the necessary pressure correction for this test.

From Table 16.3, it can be seen that the friction loss allowance for 20 feet (6 m) of 5-inch intake hose is 8.4 feet; therefore:

$$\text{Pressure Correction} = \frac{9 + 8.4}{2.3} = \frac{17.4}{2.3} = 7.56 \text{ or} \approx 8 \text{ psi}$$

Example 2

Assume that a 4 000 L/min pumper is being tested. The lift is 2.7 m through 6 m of 125 mm intake hose. Find the necessary pressure correction for this test.

From Table 16.3, it can be seen that the friction loss allowance for 6 meters of 125 mm intake hose is 2.6 m; therefore:

$$\text{Pressure Correction} = \frac{2.7 + 2.6}{0.1} = \frac{5.3}{0.1} = 53 \text{ kPa}$$

This correction pressure (8 psi [53 kPa]) is subtracted from the desired net pump discharge pressure to determine what the pump's actual discharge pressure can be. For example, 150 psi − 8 = 142 psi (1 000 kPa − 53 kPa = 947 kPa).

Equipment Needed for Service Tests

Equipment for performing service tests must be in good shape and be tested regularly. Gauges can best be tested for accuracy with a deadweight tester, which should be available at a local waterworks. The following equipment is needed to perform the service test for pumpers:

- A gauge to check the pump intake pressure. This gauge should have a range of 30 in.Hg (100 kPa) to zero for a vacuum gauge, or 30 in. Hg vacuum to 150 psi (1 050 kPa) for compound gauges (Figure 16.7).

- A gauge to check the pump discharge. This gauge should be capable of a range from at least 0 to 400 psi (0 kPa to 2 800 kPa).

- Pitot tube with knife edge and air chamber rated at least from 0 psi (kPa) to 160 psi (1 110 kPa) (Figure 16.8) (**NOTE:** This is not needed if a flowmeter is used).

- Straight stream nozzles of correct sizes to match the volumes pumped for the different tests. (**NOTE:** If a flowmeter is used, fog nozzles may be used provided they are rated for the necessary flows).

- Rope, chain, or a test stand for securing test nozzle(s) (Figures 16.9 a and b).

- Revolution counter or hand tachometer.

- Fire department or insurance agency forms.

Recommended, but not necessary, are the following:

- Two 6-foot (2 m) lengths of ¼-inch (7 mm), 300 psi (2 100 kPa) hose with screw fittings. (**NOTE:** these are used to connect the test gauges to the test gauge fittings at the pump operator's panel.)

- Clamp to hold pitot tube to test nozzle.

- Test stand for gauges.

- Thermometer.

- Stopwatch or watch with sweep second hand.

Figure 16.7 External test gauges are used for the pump test.

Figure 16.8 A pitot tube and gauge are used to measure the velocity pressure of the stream of water. This figure may then be used to calculate the volume of flow.

Figure 16.9a This nozzle test stand is held in place by parking a vehicle on top of the two boards at the base of the stand.

Figure 16.9b Some nozzle test stands are permanently set in place.

Table 16.3
Friction Loss in 20 ft (6 m) of Intake Hose, Including Strainer

Flow Rate gpm	(L/min)	3 in. (77 mm) ft water*	m water*	3½ in. (90 mm) ft water*	m water*	4 in. (100 mm) ft water*	m water*	4½ in. (115 mm) ft water*	m water*	5 in. (125 mm) ft water*	m water*
250	(1 000)	5.2 (1.2)	1.6 (0.4)								
175	(700)	2.6 (0.6)	0.8 (0.2)								
125	(500)	1.4 (0.3)	0.4 (0.1)								
300	(1 200)	7.5 (1.7)	2.3 (0.5)	3.5 (0.8)	1.1 (0.3)						
210	(840)	3.8 (0.8)	1.2 (0.3)	1.8 (0.4)	0.6 (0.2)						
150	(600)	1.9 (0.4)	0.6 (0.2)	0.9 (0.2)	0.3 (0.1)						
350	(1 400)			4.8 (1.1)	1.5 (0.3)	2.5 (0.7)	0.8 (0.2)				
245	(980)			2.4 (0.5)	0.8 (0.2)	1.2 (0.3)	0.4 (0.1)				
175	(700)			1.2 (0.3)	0.4 (0.2)	0.7 (0.1)	0.2 (0.05)				
450	(1 800)					4.1 (1.0)	1.2 (0.3)	2.7 (0.4)	0.8 (0.1)		
315	(1 260)					2.0 (0.6)	0.6 (0.2)	1.2 (0.2)	0.4 (0.1)		
225	(900)					1.0 (0.2)	0.3 (0.1)	0.6 (0.1)	0.2 (0.05)		
500	(2 000)					5.0 (1.3)	1.5 (0.4)	3.6 (0.8)	1.1 (0.3)		
350	(1 400)					2.5 (0.7)	0.8 (0.2)	1.8 (0.4)	0.6 (0.2)		
250	(1 000)					1.3 (0.4)	0.4 (0.1)	0.9 (0.3)	0.3 (0.1)		
600	(2 400)					7.2 (1.8)	2.2 (0.5)	5.3 (1.0)	1.6 (0.3)	3.1 (0.6)	0.9 (0.2)
420	(1 680)					3.5 (1.0)	1.1 (0.3)	2.5 (0.5)	0.8 (0.2)	1.6 (0.3)	0.5 (0.1)
300	(1 200)					1.8 (0.4)	0.6 (0.2)	1.3 (0.2)	0.4 (0.1)	0.6 (0.1)	0.3 (0.05)
700	(2 800)					9.7 (2.7)	3	7.3 (1.3)	2.2 (0.4)	4.3 (0.8)	1.3 (0.3)
490	(1 960)					4.9 (1.1)	1.5 (0.4)	3.5 (0.7)	1.1 (0.3)	2.0 (0.4)	0.7 (0.2)
350	(1 400)					2.5 (0.7)	0.8 (0.2)	1.6 (0.3)	0.6 (0.1)	0.9 (0.2)	0.4 (0.1)

Table 16.3 (continued)
Friction Loss in 20 ft (6 m) of Intake Hose, Including Strainer

Flow Rate gpm	(L/min)	4 in. (100 mm) ft water*	m water*	4½ in. (115 mm) ft water*	m water*	5 in. (125 mm) ft water*	m water*	6 in. (150 mm) ft water*	m water*	Two 4½ in. (115 mm) ft water*	m water*	Two 5 in. (125 mm) ft water*	m water*
750	(3 000)	11.4 (2.9)	3.5 (0.9)	8.0 (1.6)	2.4 (0.5)	4.7 (0.9)	1.4 (0.3)	1.9 (0.4)	0.6 (0.1)				
525	(2 100)	5.5 (1.5)	1.6 (0.5)	8.9 (0.8)	1.2 (0.2)	2.3 (0.5)	0.7 (0.2)	0.9 (0.2)	0.3 (0.05)				
375	(1 500)	2.8 (0.7)	0.8 (0.3)	2.0 (0.4)	0.6 (0.1)	1.2 (0.2)	0.4 (0.1)	0.5 (0.1)	0.2 (0.03)				
1000	(4 000)			14.5 (2.8)	4.4 (0.9)	8.4 (1.6)	2.6 (0.5)	3.4 (0.6)	1 (0.2)				
700	(2 800)			7.0 (1.4)	2.0 (0.4)	4.1 (0.8)	1.2 (0.2)	1.7 (0.3)	0.5 (0.1)				
500	(2 000)			3.6 (0.8)	1.1 (0.2)	2.1 (0.4)	0.6 (0.1)	0.9 (0.2)	0.3 (0.05)				
1250	(5 000)					13.0 (2.4)	4 (0.7)	5.2 (0.9)	1.6 (0.3)	5.5 (1.2)	1.7 (0.4)		
875	(3 500)					6.5 (1.2)	2 (0.4)	2.6 (0.5)	0.8 (0.2)	2.8 (0.7)	0.8 (0.2)		
625	(2 500)					3.3 (0.7)	1 (0.2)	1.3 (0.3)	0.4 (0.1)	1.4 (0.3)	0.4 (0.1)		
1500	(6 000)							7.6 (1.4)	2.3 (0.4)	8.0 (1.6)	2.4 (0.5)	4.7 (0.9)	1.4 (0.3)
1050	(4 200)							3.7 (0.7)	1.2 (0.2)	3.9 (0.8)	1.2 (0.3)	2.3 (0.5)	0.7 (0.2)
750	(3 000)							1.9 (0.4)	0.6 (0.1)	2.0 (0.4)	0.6 (0.1)	1.2 (0.2)	0.4 (0.1)
1750	(7 000)							10.4 (1.8)	3.2 (0.6)	11.0 (2.2)	3.4 (0.7)	6.5 (1.2)	2 (0.4)
1225	(4 900)							5.0 (0.9)	1.6 (0.3)	5.3 (1.1)	1.6 (0.3)	3.1 (0.7)	0.9 (0.2)
875	(3 500)							2.6 (0.5)	0.3 (0.2)	2.8 (0.6)	0.8 (0.2)	1.6 (0.3)	0.5 (0.1)
2000	(8 000)									14.5 (2.8)	4.4 (0.9)	8.4 (1.6)	2.6 (0.5)
1400	(5 600)									7.0 (1.4)	2.0 (0.4)	4.1 (0.8)	1.3 (0.3)
1000	(4 000)									3.6 (0.8)	1.1 (0.2)	2.1 (0.4)	0.7 (0.2)
2250	(9 000)											10.8 (2.2)	3.3 (0.7)
1575	(6 300)											5.3 (1.1)	1.6 (0.4)
1125	(4 500)											2.8 (0.5)	0.9 (0.2)
2500	(10 000)											13.0 (2.4)	4 (0.7)
1750	(7 000)											6.5 (1.2)	2 (0.4)
1250	(5 000)											3.3 (0.7)	1 (0.2)

Flow Rate		Intake Hose Size (inside diameter)					
		Two 6 in. (150 mm)		Three 6 in. (150 mm)		8 in. (200 mm)	
gpm	(L/min)	ft water*	m water*	ft water*	m water*	ft water*	m water*
1500	(6 000)	1.9 (0.4)	0.6 (0.1)				
1050	(4 200)	0.9 (0.3)	0.3 (0.05)				
750	(3 000)	0.5 (0.1)	0.2 (0.03)				
1750	(7 000)	2.6 (0.5)	0.8 (0.2)				
1225	(4 900)	1.2 (0.3)	0.4 (0.1)				
875	(3 500)	0.7 (0.2)	0.2 (0.05)				
2000	(8 000)	3.4 (0.6)	1 (0.2)				
1400	(5 600)	1.7 (0.3)	0.5 (0.1)				
1000	(4 000)	0.9 (0.2)	0.3 (0.05)				
2250	(9 000)	4.3 (0.8)	1.3 (0.2)				
1575	(6 300)	2.2 (0.4)	0.6 (0.1)				
1125	(4 500)	1.1 (0.2)	0.3 (0.05)				
2500	(10 000)	5.2 (0.9)	1.6 (0.3)				
1750	(7 000)	2.6 (0.5)	0.8 (0.2)				
1250	(5 000)	1.3 (0.3)	0.4 (0.1)				
3000	(12 000)	7.6 (1.4)	2.3 (0.4)	3.4 (0.6)	1 (0.2)	8.5 (1.6)	2.6 (0.5)
2100	(8 400)	3.7 (0.7)	1.1 (0.2)	1.7 (0.3)	0.5 (0.1)	4 (0.8)	1.2 (0.4)
1500	(6 000)	1.9 (0.4)	0.6 (0.1)	0.9 (0.2)	0.3 (0.05)	1.9 (0.4)	0.6 (0.2)

*Figures in parentheses after feet (meters) indicate increment to be added or subtracted for each 10 ft (3 m) of hose less than or greater than 20 ft (6 m).

Adopted from NFPA 1911, Table 2-3.1.1 (b).

Flowmeters

A *flowmeter*, which reads the flow directly in gallons per minute (L/min), may be used instead of a pitot gauge arrangement to determine the flow from the nozzles. Flowmeters allow much more flexibility and help to complete the tests more quickly. When a flowmeter is used, all the pump tests can be run without shutting down the pump, without changing nozzles, and without having to convert pitot pressure readings to gpm (L/min). If flowmeters are used, they must be calibrated to the manufacturer's specifications.

Safety Precautions During Service Tests

The following safety precautions should always be followed when performing fire pump service testing:

- All personnel should wear protective headgear and hearing protection (if exposed to noise in excess of 90 dB).
- Prevent water hammer: Open and close all valves and nozzles slowly.
- Do not stand over or straddle hose (Figures 16.10 a and b).
- Manipulate the engine throttle slowly. Prevent sudden pressure changes, which can damage equipment and injure personnel.

- Tie down test nozzles and devices securely.
- Cover all open manholes at the test pit.
- Be aware of the location of all personnel in the test area in relation to hoselines.

Engine Speed Check

The first test that should be performed is the engine speed check. The engine speed should be checked under no-load conditions to ensure that it is still running at the

Figure 16.10a Do not straddle or stand directly over the hose being tested.

Figure 16.10b Observe the hose from a safe distance.

same governed speed that it was rated for when the apparatus was new. If it is not running at the correct speed, no further testing should be started until the situation is corrected by a trained mechanic. The engine speed may be checked by the tachometer on the engine and/or a properly calibrated handheld tachometer or revolution counter.

Vacuum Test

The *vacuum test* is performed to check the priming device, pump, and hard intake hose for air leaks. Most agencies prefer to perform this test first because it will be difficult to proceed if the apparatus is unable to hold an appropriate vacuum.

Step 1: Make sure that the pump is completely drained of all water.

Step 2: Inspect all gaskets of intake hose and caps (Figure 16.11).

Step 3: Look for foreign matter in the intake hose. Clean the hose if necessary.

Step 4: Connect 20 feet (6 m) of the correct intake hose to the pump intake connection (check original test records for correct diameter of hose).

Step 5: Cap the free end of the intake hose (Figure 16.12).

Step 6: Make sure that all intake valves are open and the intake connections are tightly capped. As well, all discharge valves should be closed and their caps should be removed.

Step 7: Connect an accurate vacuum gauge (or mercury manometer) to the threaded test-gauge connection on the intake side of the pump (Figure 16.13). **CAUTION:** If the gauge is not connected to the intake side, it will be irreparably damaged.

Step 8: Check oil level of priming pump reservoir (replenish if necessary).

Step 9: Make pump packing glands accessible for checking (raise floorboards or open compartment doors).

Step 10: Run the priming device until the test gauge shows 22 inches (560 mm) of mercury developed. (**NOTE:** Reduce the amount of mercury developed 1 inch [25 mm] for each 1,000 feet [300 m] of altitude.) Information on the effects of altitude may be found in Appendix E.

Step 11: Compare readings of the apparatus intake gauge and test gauge. Record any difference.

Figure 16.12 Place a cap on the free end of the intake hose.

Figure 16.11 Inspect the condition of the gaskets before connecting the intake hose.

Figure 16.13 The test gauges are attached to the test connections on the pump panel.

Step 12: Shut off the engine. Listen for air leaks. No more than 10 inches (250 mm or 33.9 kPa) of vacuum should be lost in 5 minutes. Excessive leaks will affect the results of subsequent tests and should be located and corrected before performing the rest of the test.

If the apparatus is unable to reach 22 inches (560 mm) of mercury, the apparatus should be removed from service and repaired as soon as possible.

Another test that some jurisdictions choose to perform following the dry vacuum test is a test of the condition of the hard intake hose itself. This is an ideal time to perform this test, as many parts of it mirror the dry vacuum test of the pump. It is important to test the condition of the hard intake hose because defects in it may cause further pump test results to be erroneous. If the outside jacket of the hard intake hose is damaged, pressure will be allowed to leak into the hose during drafting. This causes a bubble or blister to form in the inner lining of the hose. The result will be a restricted flow during drafting.

To test the condition of the hard intake hose, use the following procedure:

Step 1: Make sure that the pump is completely drained of all water.

Step 2: Inspect all gaskets of intake hose, and remove the gasket from the intake cap.

Step 3: Look for foreign matter in the intake hose. Clean the hose if necessary.

Step 4: Place a lighted flashlight into the hose, just inside the female coupling, that will be connected to the intake (Figure 16.14).

Step 5: Connect 20 feet (6 m) of the correct intake hose to the pump intake connection (check original test records for correct diameter of hose). Support the hose on drums or saw horses so that it is relatively straight (Figure 16.15).

Step 6: Make sure that all intake valves are open and the intake connections are tightly capped. As well, close all discharge valves and remove their caps.

Step 7: Connect an accurate vacuum gauge (or mercury manometer) to the threaded test-gauge connection on the intake side of the pump. **CAUTION:** If the gauge is not connected to the intake side, it will be irreparably damaged.

Step 8: Check oil level of priming pump reservoir (replenish if necessary).

Step 9: Make pump packing glands accessible for checking (raise floorboards or open compartment doors).

Figure 16.14 Place the lighted flashlight into the hose before connecting it to the pump panel.

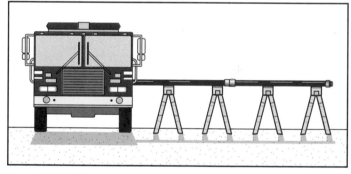

Figure 16.15 The intake hoses should be supported on sawhorses or similar devices.

Step 10: Place the cap gasket against the male end of the hose and then use a sheet of ⅜-inch (9.5 mm) Lexan™ or other hard, clear plastic large enough to cover the entire opening against the gasket. Hold the gasket and plastic in place until the priming device is operated (Figure 16.16).

Step 11: Run the priming device until the test gauge shows 22 inches (560 mm) of mercury developed. (**NOTE:** Reduce the amount of mercury developed 1 inch [25 mm] for each 1,000 feet [300 m] of altitude.)

Step 12: Compare readings of the apparatus intake gauge and test gauge. Record any difference.

Figure 16.16 Place the gasket and plastic sheet against the end of the intake hose. Note that this photo simply shows how that would be done. If an actual test were occurring, the hose would be elevated.

Step 13: Shut off the engine. Listen for air leaks. No more than 10 inches (250 mm or 33.9 kPa) of vacuum should be lost in 5 minutes.

Step 14: Looking through the clear plastic, examine the inside of the hose for any signs of bubbling or liner separation. (These sometimes form slowly).

After the dry vacuum and/or hard intake hose test is completed, prepare the pumper for the remainder of the tests:

Step 1: Open a discharge valve to allow the pressure in the pump to equalize.

Step 2: Replace the cap at the end of the intake hose with the intake strainer.

Step 3: Use standard departmental procedure to tie off intake hose in preparation for drafting, then lower the hose into the water. The strainer should be at least 2 feet (0.6 m) below the surface. The sides and bottom of the strainer should also have at least 2 feet (0.6 m) of water surrounding them (Figure 16.17).

Step 4: Connect the discharge pressure test gauge to the pressure side of the pump at the test fitting on the operator's panel.

Step 5: Connect an adequate number of hoselines to carry the capacity of the pump to the test nozzle. The test nozzle must be the correct size to handle the capacity of the pump. (see Tables 16.4 a and b).

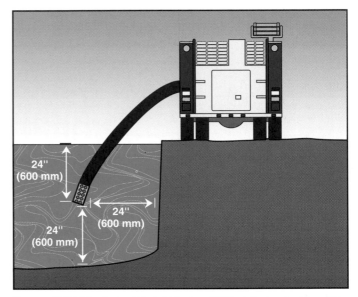

Figure 16.17 It is desired that 24 inches (600 mm) of water should be around all sides of the strainer.

Table 16.4a Flow in gpm from Various Sized Solid Stream Nozzles										
Nozzle Diameter in Inches										
Nozzle Pressure in psi	1	1⅛	1¼	1⅜	1½	1⅝	1¾	1⅞	2	2¼
50	209	265	326	396	472	554	643	740	841	1065
55	219	277	342	415	495	581	674	765	881	1118
60	229	290	357	434	517	607	704	810	920	1168
65	239	301	372	451	537	631	732	843	958	1215
70	246	313	386	469	558	655	761	875	994	1260
75	256	324	399	485	578	678	787	905	1030	1305
80	264	335	413	500	596	700	813	935	1063	1347

Table 16.4b
Flow in L/min from Various Sized Solid Stream Nozzles

Nozzle Pressure in kPa	Nozzle Diameter in mm									
	25	29	32	35	38	42	45	48	50	57
350	791	1 003	1 234	1 499	1 786	2 097	2 434	2 801	3 183	4 031
385	829	1 048	1 294	1 571	1 873	2 200	2 551	2 896	3 334	4 232
420	867	1 098	1 351	1 643	1 957	2 297	2 665	3 066	3 483	4 421
455	905	1 139	1 408	1 707	2 033	2 389	2 771	3 191	3 626	4 600
490	931	1 185	1 461	1 775	2 112	2 480	2 880	3 312	3 763	4 770
525	969	1 226	1 510	1 835	2 188	2 567	2 980	3 425	3 899	4 940
560	1 000	1 000	1 563	1 893	2 256	2 650	3 077	3 539	4 024	5 099

Step 6: Make sure that the nozzle is secured so that it cannot come loose and injure personnel. *NEVER* hold the test nozzle by hand during a test.

Step 7: Connect the pitot gauge and test stand gauges. It is recommended that the pitot gauge be clamped in position at the nozzle.

Pumping Test

The pumping test checks the overall condition of the engine and the pump. To obtain the correct engine and nozzle pressures for the capacity test, some adjustments and readjustments have to be made. All changes must be made slowly to prevent damage to the pump and hose and possible injury to personnel and to allow time for the resulting pressure changes to register on the test gauges. The procedure for the capacity test is as follows:

Step 1: Gradually speed up the pump until the net pump discharge pressure is 150 psi (1 035 kPa), adjusted for intake hose friction loss and altitude. If the pump is a two-stage pump, the transfer valve must be in the VOLUME (PARALLEL) position.

Step 2: Check the flow at the nozzle, using either a pitot gauge or a flowmeter. If the flow is too great, close a valve further. Readjust (lessen) engine speed to correct discharge pressure. If the flow is too low, open a valve further. Readjust (increase) engine speed to correct discharge pressure. (**NOTE**: All these adjustments must be made without the engine speed exceeding 80 percent of its peak.)

Step 3: When both the pump discharge pressure and the volume flowing are satisfactory, the test officially begins. The following readings are made and recorded at the beginning of the test and at 5-minute intervals thereafter until the 20 minutes for the test are over. (**NOTE**: Fluctuations in pressure necessitate more frequent readings.)

- Pump discharge pressure
- Nozzle pressure (or flow)
- Engine tachometer
- Rpm using portable rpm counter
- Engine coolant temperature (optional)
- Oil pressure (optional)
- Automatic transmission fluid temperature (optional)

Step 4: Once the 20-minute capacity test has been completed, the net pump discharge pressure should be increased to 200 psi (1 380 kPa). At this point, the pump should be delivering at least 70 percent of its rated volume capacity. According to NFPA 1911, two-stage pump transfer valves may be in either the VOLUME (PARALLEL) or PRESSURE (SERIES) position for this portion of the test. It is usually best to see the pump certification information (either paperwork or on the data plate attached to the pump panel) to see which position the transfer valve was in during the certification test and use that position at this time. The pump should be allowed to run at this setting for 10 minutes.

Step 5: Once the 200 psi (1 380 kPa) test has been completed, the net pump discharge pressure should be increased to 250 psi (1 725 kPa). At this point, the pump should be delivering at least 50 percent of its rated volume capacity. Two-stage pump transfer valves must be in the PRESSURE (SERIES) position for this portion of the test. The pump should be allowed to run at this setting for 10 minutes.

The following are some additional points to remember while performing the pumping test:

- Hold the pitot gauge with the blade opening in the center of the stream with the tip about one-half the nozzle diameter from the end of the nozzle (Figure 16.18). If the pitot is too close to the nozzle, the reading will be erroneously increased.

- Keep the engine temperature within the proper range.

- Check the oil pressure to be sure that proper engine lubrication is maintained.

- Record any unusual vibration of pump or engine.

- Record any other defect in the performance of pump or engine. Correct minor defects immediately if possible.

Pressure Control Test

The pressure control device(s) should be tested to make sure that they maintain a safe level of pressure on the pump when valves are closed at a variety of discharge pressures. All pressure control devices should be operated according to manufacturer's instructions during this testing. The pressure control test is performed in a three-part sequence as follows:

Part I:

Step 1: Set the fire pump so that it is discharging its rated capacity at a net pump discharge pressure of 150 psi (1 035 kPa).

Step 2: Set the pressure control device to maintain the discharge pressure at 150 psi (1 035 kPa).

Step 3: Once the device is set, close each of the flowing valves, one at a time. Close each valve in no less than 3 seconds and no more than 10 seconds. (**NOTE:** If that valve is closed in less than 3 seconds, damage to the pump, piping, or pressure control device may occur. If it takes longer than 10 seconds to close, it does not provide a realistic test for the pressure control device.)

Step 4: Observe the pump discharge pressure gauge. It should rise no more than 30 psi (207 kPa) when all gauges are closed.

Figure 16.18 The pitot tube should be in the center of the water stream.

Part 2:

Step 1: Set the fire pump so that it is discharging its rated capacity at a net pump discharge pressure of 150 psi (1 035 kPa).

Step 2: Reduce the pumping engine throttle until the net pump discharge pressure drops to 90 psi (620 kPa) with no change to the discharge valve or nozzle setting(s).

Step 3: Set the pressure control device to maintain the discharge pressure at 90 psi (620 kPa).

Step 4: Once the device is set, close each of the flowing valves, one at a time. Close each valve in no less than 3 seconds and no more than 10 seconds. (**NOTE:** If that valve is closed in less than 3 seconds, damage to the pump, piping, or pressure control device may occur. If it takes longer than 10 seconds to close, it does not provide a realistic test for the pressure control device.)

Step 5: Observe the pump discharge pressure gauge. It should rise no more than 30 psi (207 kPa) when all gauges are closed.

Part 3:

Step 1: Set the fire pump so that it is discharging 50 percent of its rated capacity at a net pump discharge pressure of 250 psi (1 725 kPa).

Step 2: Set the pressure control device to maintain the discharge pressure at 250 psi (1 725 kPa).

Step 3: Once the device is set, close each of the flowing valves, one at a time. Close each valve in no less than 3 seconds and no more than 10 seconds. (**NOTE:** If that valve is closed in less than 3 seconds, damage to the pump, piping, or pressure control device may occur. If it takes longer than 10 seconds to close, it does not provide a realistic test for the pressure control device.)

Step 4: Observe the pump discharge pressure gauge. It should rise no more than 30 psi (207 kPa) when all gauges are closed.

Discharge Pressure Gauge and Flowmeter Operational Tests

Check the discharge pressure gauges and flowmeter (if the pump is equipped with one) to make sure that the driver/operator is being given accurate discharge information when the pump is in operation. If these devices are not working properly, it is conceivable that the driver/operator could supply dangerously insufficient or excessive amounts of water to firefighters operating hose streams.

Testing the apparatus pressure discharge gauges is a relatively quick and simple process. Each of the discharges on the apparatus must be capped in order to perform this test properly. This means that preconnected hoselines must be disconnected and caps or closed nozzles screwed onto their discharges (Figure 16.19). Once all the discharges are capped, each discharge valve should be opened slightly. The throttle should then be increased until the test discharge pressure gauge reads 150 psi (1 035 kPa). A quick visual inspection of the master discharge pressure gauge and each individual line discharge gauge should reveal all to be at 150 psi (1 035 kPa) as well. The gauges should then be checked at 200 psi (1 380 kPa) and 250 psi (1 725 kPa) in the same manner. Any gauges that are off by more than 10 psi (70 kPa) should be recalibrated, repaired, or replaced.

Testing discharges equipped with flowmeters is not as simple a process as testing pressure gauges. To test the flowmeter, a hoseline equipped with a solid stream nozzle must be connected to each discharge being tested (they do not all have to be done at once). Refer to Table 16.2 (a or b) to determine appropriate hose and nozzle arrangements for this test. Table 16.5 shows the minimum flow

Table 16.5 Minimum Flow Measuring Points for Flowmeters	
Pipe Size in Inches (mm)	**Test Flow in gpm (L/min)**
1½ (38)	128 (454)
2 (50)	180 (682)
2½ (65)	300 (1 135)
3 (77)	700 (2 650)
4 (100)	1,000 (3 785)

rates that must be achieved for each of the listed pump discharge piping sizes in order for the test to be valid. The actual flow will be calculated using pitot tube readings taken from the discharge of the solid stream nozzle. The flow measured from the nozzle and the reading on the flowmeter should not be off by more than 10 percent. If the margin of error is greater, the flowmeter must be recalibrated, replaced, or repaired.

Tank-To-Pump Flow Test

The tank-to-pump flow test must be conducted on all apparatus that are equipped with a water tank, regardless of size. The purpose of this test is to ensure that the piping between the water tank and pump are sufficient to supply the minimum amount of water specified by NFPA 1901 and the design of the manufacturer. As you will recall from Chapter 10 of this manual, NFPA 1901 states that piping should be sized so that pumpers with a capacity of 500 gpm (1 900 L/min) or less should be capable of flowing 250 gpm (950 L/min) from their booster tank. Pumpers with capacities greater than 500 gpm (1 900 L/min) should be able to flow at least 500 gpm (1 900 L/min). Some departments may specify greater flow rates when the apparatus is ordered from a manufacturer. It is recommended that this test be performed for the higher figure if one greater than the NFPA minimum was specified when the apparatus was ordered. The following test procedure should be used to check the operation of the tank-to-pump line:

Step 1: Make sure that the water tank is filled until it is overflowing.

Step 2: Close the tank fill line, bypass cooling line, and all pump intakes.

Step 3: Attach sufficient hoselines and nozzles to flow the desired discharge rate.

Figure 16.19 The discharge should be capped in some manner.

Step 4: With the pump in gear, open the discharge(s) to which the hose(s) is (are) attached, and begin flowing water.

Step 5: Increase the engine throttle until the maximum consistent pressure is obtained on the discharge gauge.

Step 6: Close the discharge valve, without changing the throttle setting, and refill the tank (usually through the top fill opening or a direct tank fill line). The bypass valve may be temporarily opened during this operation to prevent pump overheating.

Step 7: Reopen the discharge valve, and check the flow through the nozzle using a pitot tube or flowmeter. Adjust the engine throttle if the pressure needs to be brought back to the amount determined in Step 5.

Step 8: Compare the flow rate being measured to the NFPA minimum or the manufacturer's designed rate. If the flow rate is less than this, a problem exists in the tank-to-pump line. Remember that the minimum flow rate should be continuously discharged until at least 80 percent of the capacity of the tank has been emptied.

Reviewing the Test Results

At no time during the tests should the pumping system or pumping engine show signs of overheating, power loss, or any other mechanical problems. All calculations and figures determined during the tests should be recorded so that they may be filed according to departmental record-keeping procedures.

If the fire pump tests to less than 90 percent of its capabilities when it was new and underwent certification testing, two options are available:

- Take the pump out of service and restore it to its designed capabilities (obviously, it will be necessary to test it again after the repairs). This is the preferable option.

- Give the pump a lower rating based on the test results of its actual performance.

Possible Troubles During Service Testing

If the pump fails to meet the requirements of the service test, one or more of the following should be investigated as the probable cause of the problem:

- Transmission in wrong gear

- High gear lockup not functioning (automatic transmission)

- Clutch slipping

- Engine overheating

- Muffler clogged

- Tachometer inaccurate

- Engine governor malfunctioning

- Intake hose too small

- Intake strainer submerged incorrectly (for example, too close to surface, too close to bottom)

- Intake screens clogged

- Wrong strainer is being used for that type of hose

- Lift is higher than 10 feet (3 m)

- Intake hose clogged or inner lining collapsed

- Excessive air leaks at intake side of pump

- Pump impellers are clogged

- Pump or intake hose not fully primed

- Relief valve or pressure governor malfunctioning

- Transfer valve in wrong position

- Inaccurate gauges

- Pitot tube partially clogged

- Nozzle too large

- Seized turbocharger

Every effort should be made to correct any problem that is found. The portion of the test that was unsuccessful should be redone to ensure that the problem has been corrected.

Foam Proportioning Equipment Testing

Like fire pumps, foam proportioning equipment must be tested for proper operation and accuracy before being placed in service and tested on a periodic basis thereafter. Foam proportioning systems and equipment are generally checked for proper operation by one of two methods:

- Testing the foam-to-water solution concentration that the systems and equipment produce

- Testing the rate at which foam concentrate is consumed in proportion to a known flow of water through the system

NFPA 1901 requires foam system accuracy testing to be performed prior to the apparatus being delivered from the manufacturer. It does not *require* foam system testing to be performed on a yearly basis; however, IFSTA recommends it and NFPA 1002 requires the driver/operator to be able to perform this testing. NFPA

1901 does provide four methods for testing a foam proportioning system for calibration accuracy:

- Foam concentrate displacement method
- Foam concentrate pump discharge volume method
- Foam solution refractivity testing
- Foam solution conductivity testing

The following sections detail each of these procedures.

Foam Concentrate Displacement Method

This method measures the accuracy of foam proportioning equipment by checking the volume of foam concentrate that is drawn through the system while it is in operation. The foam system is operated at a predetermined flow using water as a substitute for foam concentrate. The water is drawn from a calibrated tank instead of the normal foam concentrate tank or foam concentrate pails. (**NOTE:** Water has a different viscosity than foam concentrate and may be drawn into the system at a different rate than foam concentrate. The manufacturer of the proportioning system and/or the foam concentrate normally used in the system will be able to provide a correction factor to ensure that the results of testing with water are accurate.) The volume of water drawn from the calibrated tank over a measured period of time is then correlated to the actual percentage of foam concentrate that the system would be proportioning at the test flow rate.

Foam Concentrate Pump Discharge Volume Method

This procedure may be used to test the volume of foam concentrate that is proportioned into the fire stream in some direct injection type foam proportioning systems. As with the previous method, water may be used in place of foam concentrate for testing purposes. With the foam system operating at a predetermined flow, the discharge from the foam concentrate pump is collected in a calibrated container for a specified time period. This volume can then be correlated to the actual percentage of foam concentrate that the system would be proportioning at the test flow rate.

Foam Solution Refractivity Testing

Foam solution refractivity testing is used to test the quality of a foam solution after it has been created by a foam proportioning system. This test method is recommended for protein- and fluoroprotein-based foam solutions. It is not accurate for synthetic-based foams because they typically have a very low refractive index reading. The conductivity test method is used for synthetic-based foams.

The amount of foam concentrate in the solution is measured using a device called a *refractometer*. This device operates on the principle of measuring the velocity of light that travels through a medium. In this case, the medium is the foam solution. The refractometer compares samples of foam solution drawn from the foam system being tested to carefully prepared base reading solutions. Deviations in the foam concentrate content of the foam solution results in different bending of the light beams through the refractometer (Figure 16.20). Because the scale on the refractometer does not reflect the actual foam concentrate proportioning percentage, the results of the tests must be plotted on a graph to be interpreted.

The first step in the testing process is to develop a base calibration curve. This calibration curve is based on the recommended proportioning rate of the foam concentrate being used (usually either 1%, 3%, or 6%). Foam concentrate and water are taken from the system being tested to make the base curve solutions. Three standard solutions are made for each concentration being tested. One solution contains the exact recommended concentration, one solution contains 0.3 percent less concentrate than recommended, and the third contains 0.3 percent more concentrate than recommended. For example, if testing 3% foam, one solution should have exactly 3 percent concentrate in solution, one should have 2.7 percent, and the third should have 3.3 percent.

The following procedure details the preparation of samples for testing 3% foam concentrate. The numbers will be adjusted if 1% or 6% concentrates are tested. The solutions can be mixed as follows:

Step 1: Gently add foam concentrate to each of three labeled, plastic 100 ml or larger graduated bottles. Place 2.7 ml of concentrate into one, 3 ml into the second, and 3.3 ml into the third. A pipette or syringe may be used.

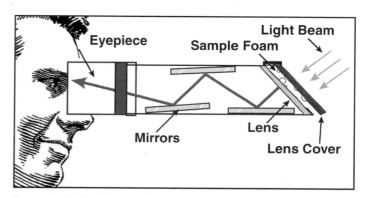

Figure 16.20 Deviations in the foam concentrate content of the foam solution result in different bending of the light beams through the refractometer. *Courtesy of National Foam, Inc.*

Step 2: Fill each bottle with water to the 100 ml mark.

Step 3: Add a plastic stirring bar to each bottle, and cap them tightly.

Step 4: Shake each bottle thoroughly to mix the water and concentrate.

When testing proportioning equipment that can be operated at more than one setting, such as an in-line eductor that has 1%, 3%, and 6% settings, three samples will have to be taken for each and separate charts prepared for each.

After the samples are prepared, a refractive index is taken of each sample. Place a few drops of the sample on the refractometer prism, close the cover plate, and observe the reading. It may take 10 to 20 seconds to get an accurate reading because the refractometer must adjust for temperature fluctuations (Figure 16.21). Each of these readings is then plotted on a piece of graph paper. One axis contains the refractometer reading, and the other contains the proportioning percentage. A line is drawn between the three points to establish a baseline curve.

Once this procedure is complete, samples of the actual foam solution produced by the system being tested are taken. These samples are tested on the refractometer and then plotted on the graph. The results must fall within the parameters previously discussed for the various NFPA standards.

Foam Solution Conductivity Testing

Foam solution conductivity testing is used to check the quality of synthetic-based foams that are produced by foam proportioning equipment and systems. Because synthetic-based foam concentrates are a very light color, the refractivity tests previously discussed are not very accurate for them. Conductivity testing does not rely on the colors of the foam, but rather on their ability to conduct electricity to verify their actual composition.

Conductivity is the ability of a substance to conduct an electrical current. Water and foam concentrate both conduct electricity. When proportioning foam concentrate into water, the conductivity of the resulting foam solution is somewhere in between the figures for plain water and foam concentrate. These figures can be used to measure the amount of foam concentrate in the solution.

There are three methods of performing conductivity testing on foam solution. They are the following:

- Direct reading conductivity testing
- Conductivity comparison testing
- Conductivity calibration curve testing

Each of these are detailed in the following sections.

Direct Reading Conductivity Testing

This method is used when a direct reading conductivity meter is available. The readout on this meter may or may not indicate the actual percentage of foam concentrate in the solution. If the meter does not have an actual percentage reading, a calibration curve will need to be developed. The procedure for developing this curve would be the same as described for refractivity testing.

To perform this test, the meter must first be zeroed in using plain water. This procedure can be done in either one of two ways:

1. Collect a sample of the water that will be used in the test in a container and immerse the sensor head to zero the meter.

2. Mount the sensor directly into the pump discharge line and zero in the meter while flowing water.

Once the meter has been zeroed in, a sample of foam solution from the proportioning system is obtained. As with the water, a reading may be obtained in either a container or the meter may be mounted directly to the pump discharge. The sensor is immersed in the solution, and a reading is taken from the meter. If the meter does not give readings in percentage of foam concentrate in solution, the reading must be plotted on the calibration curve to get the final results.

Conductivity Comparison Testing

This method is used when a conductivity meter that reads out in units of microsiemens per centimeter (ms/cm) is available. The procedure is fairly simple. First, a

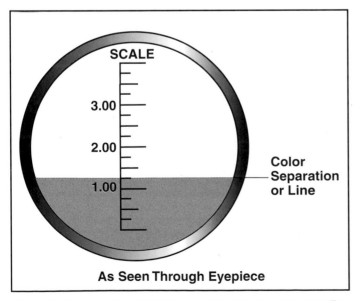

Figure 16.21 It may take 10 to 20 seconds to get an accurate reading because the refractometer must adjust for temperature fluctuations. *Courtesy of National Foam, Inc.*

reading is taken from the plain water that is to be used for the test. Then, a reading is taken from the foam solution produced by the system being tested. The percentage of foam concentrate in the solution can then be determined using the following formula:

$$\% \text{ of concentrate in solution} = \frac{(\text{Conductivity of solution}) - (\text{conductivity of water})}{500}$$

The constant divisor of 500 is used only if the meter that is used is incremented in units of ms/cm. If a different type of meter is used, the divisor will have to be adjusted. The manufacturer of the meter should provide this information.

Conductivity Calibration Curve Testing

Conductivity calibration curve testing is performed by using a handheld, temperature-compensated conduc-tivity meter. The procedure for this method is very similar to that described earlier in this section for refractivity testing. A calibration curve is developed following the same guidelines as those for refractivity testing. The only difference is that in this case the readings will be taken using the conductivity meter. Once the calibration curve has been developed, samples of foam solution from the proportioning equipment being tested may be taken and analyzed. The readings from these tests are then plotted on the calibration curve to determine the percent of concentrate in the solution.

For more detailed information on foam system testing, including testing compressed air foam systems, consult Chapters 18 and 19 of NFPA 1901. IFSTA's **Principles of Foam Fire Fighting** manual also contains detailed information on foam system testing.

ELK GROVE FIRE DEPARTMENT
DAILY VEHICLE PRE-TRIP INSPECTION REPORT

Vehicle No. _____ Mileage _____ Date _____

D01	**VISUAL INSPECTION**		D06	**VEHICLE EXTERIOR**
	General Condition/Body Damage			Lights Operating / Emergency
D02	**ENGINE/ENGINE COMPARTMENT**			Tire / Wheels / Wgs
	Fluid Leakage			Steering
	Belts / Hoses			Fuel Cap / Tank Mount
	Engine Fluids (oil, trans, etc.)		D07	**COMMUNICATIONS EQUIPMENT**
	Batteries			Mobile / Portable
D03	**VEHICLE CAB (INSIDE)**			Mobile Data Terminals
	Forms, Maps, Books, etc.			Cellular Phones
	Fuel Levels			Headsets
	Seat / Seat Belts		D08	**PUMP(S)**
	Glass / Mirrors			Engage / Check Gauges
	Windshield Wipers			Relief / Transfer Valve
	Heater / Defroster			Water Level
	Horn(s) / Sirens			Foam System
D04	**RUN ENGINE**		D09	**EMS EQUIPMENT**
	Check Gauges (oil, temp, amp, etc)			BLS Equipment
D05	**AIR BRAKE SYSTEM**			LP300
	Test Air Leakage Rate			Oxygen Cylinders
	Air Comp. Cut In/Out Pressure		D10	**MISCELLANEOUS EQUIPMENT**
	Test Parking Brake			Water Jugs
	Test Low Pressure Warning			Gas Detection Devices
	Test Service Brakes			Fire Extinguishers
	HYDRAULIC BRAKE SYSTEM			Flare / Cones
	Brake Fluid Level			
	Test Parking Brake			
	Test Brakes			

	OVERALL CONDITION OF VEHICLE			
	Satisfactory			Unsafe To Operate

Remarks: _____

_____

Driver's Signature _____

	Above Defect Need Not Be Corrected For Safe Operation
	Vehicle Taken Out of Service

Captain's Signature _____

	FLEET MANAGEMENT USE ONLY		
	Above Defects Corrected		Above Defects Need Not Be Corrected For Safe Operation

Mechanic's Signature _____

Date _____

White — Maintenance Division   Yellow — Keep In Book

Courtesy of Elk Grove (CA) Fire Department.

Providence Fire Department
Daily Apparatus / Equipment Inspection

Company- _____ Group- _____

Date - _____ Chauffeur- _____

1st Day Check	O.K.	Needs Service
S.C.B.A. - Tank Level, Operation, Harness, P.A.S.S.		
ENGINE - Oil, Coolant, Trans. Power Steering, Brake Fluids / Leaks. Belts / Hoses, Fuel Level Batteries Condition		
TIRES - Tread, Condition Check for Flats (w/ hammer)		
LIGHTS - Headlights, Brake/Tail Signal, Warning Backup/Alarm		
CAB - Clean Interior, Books, Keys Note Paper, Clean Windows Wipers, Seats/Belts		
RADIO/SIREN/HORN - Operation		
E.M.S. EQUIPMENT - Oxygen, Latex Gloves, Infectious Control, BLS Supplies		
OTHER - Handlights, Igloo Cooler		
WASH - Apparatus / Equipment as needed		
PUMPERS - Water Level, Primer, Relief Valve, Transfer Valve, Clean Relief Valve Screen, Hydrant Fittings/Wrenches		
AERIALS / TOWERS - Hydraulic Oil, Level & Operation, PTO, Jacks/Outriggers/Interlocks, Pedestal, Bucket		
RESCUE - Main Oxygen, Stair Chair, Stretcher, Back Boards, Lifepak, ALS Equipment		
OTHER -		

Date - _____ Chauffeur- _____

2nd Day Check	O.K.	Needs Service
S.C.B.A. - Tank Level, Operation, Harness, P.A.S.S.		
HAND TOOLS - Clean And Inspect All Hand Tools		
GENERATORS & INVERTERS - Oil level, Fuel Level, Condition Circuit Breakers Check & Operate Electrical Equipment (Fans, Etc.)		
NOZZLES - Appliances and Fittings Clean and Operate		
EXTINGUISHERS - Pressure, Condition		
PORTABLE PUMPS - Fuel Oil Level		
LIGHTING - Scene Lights, Cords, Reels, Outlet Boxes, Adapters		
HOSE - Preconnects, Booster Hose & Reel Operations, Supply Lines, High Rise Paks, Fittings		
ROPES - Utility, Lifelines, Bags, and Other Equipment		
CHOCKS - Wheel Chocks, Blocks Planks, Shoring, Etc.		
PUMPERS - Water Level, Foam Level & Equipment, Operate Gates & Drains		
AERIALS / TOWERS - Inspect Waterways, Ladder Pipes, Salvage & Overhaul Equip. Lifebelts & Harnesses, Saw - Oil & Fuel Levels, Operate (Per Manuf. Recom.) Spare Fuel, Oil & Blades		
RESCUE - Splints, Hare Traction, KED, MAST, Linen, Extrication & Hand Tools Restraints, Supplies		
OTHER -		

Courtesy of Providence (RI) Fire Department.

Providence Fire Department – Apparatus Defect Report

Date- _____ Co.- _____ Officer- _____ Chauffeur- _____

MAJOR DEFECTS ONLY
Immediate Repair or Change Over Indicated

Apparatus Satisfactory Except as Noted Below Initial-		Defective (Driver)	Repaired (Mechanic)
Brakes–	Soft/Low Pedal		
	Pulls/Grabs		
	Excess Noise/Heat		
	Emergency/Parking Brake		
	Other–		
Electrical–	Total Failure		
	Headlights/Brake Lights		
	All Warning Lights		
	Radio/Siren		
	Charging System		
	Other–		
Engine–	No Power		
	Over Heats		
	Cuts out/Stalls		
	Throttle Controls		
	Rough/Noisy		
	Hard/Doesn't Start		
	Belts/Hoses		
	Other–		
Transm./ Drivetrain–	Does Not Shift		
	Slips		
	Noise/Vibration		
	Other–		
Air System–	No Pressure		
	Slow Pressure Build-up		
	Rapid Loss of Air Pressure		
	Other–		
Suspension–	Spring/Spring Hanger		
	Vehicle Leans (Excessively)		
	Loss/Lack of Control		
	Other–		

Reported To (name) – _____

Time – _____ hrs.

Corrective Action Taken On All Items
(Mechanics Signature) – _____

Corrective Action Taken – Except On Items Listed Below
(Mechanics Signature) – _____

Items Not Repaired – _____

Apparatus Satisfactory Except as Noted Below Initial-			Defective (Driver)	Repaired (Mechanic)
Tires–	Condition			
	Tread			
	Other–			
	Which Tire? (Circle)	Right Left	Front Rear	Inside Outside
Steering–	Hard			
	Pulls			
	Shimmies/Vibrates			
	Loss of/No Control			
	Other–			
Cab Body–	Doors Don't Close/Latch			
	Windshield Wipers			
	Seats/Seat Belts			
	Major Body Damage			
	Other–			
Exhaust–	Filter System			
	Excessive Smoke			
	Fumes			
	Other–			
Aerials– Towers	PTO Won't Engage			
	Jacks/Outriggers			
	Controls			
	Hydraulics			
	Cable/Guides			
	Tiller/Trailer Chassis			
	Other–			
Pump–	Won't Engage			
	Relief Valve			
	No Pressure			
	Other–			

Remarks/Explanations–

Courtesy of Providence (RI) Fire Department.

MINOR DEFECTS
For Preventive Maintenance or When Defect Does NOT Constitute An Immediate Safety Hazard

Apparatus Satisfactory Except as Noted Below Initial-		Defective (Driver)	Repaired (Mechanic)
Elecrical–	Generator		
	Booster/Cable Reel		
	Light Charger		
	Inverter		
	Other–		
Lights–	Warning Lights		
	Tail Lights		
	Marker Lights		
	Compartment Light		
	Panel/Dash Lights		
	Quartz/Spot Lights		
	Other–		
Windows–	Glass Cracked		
	Does not Operate		
	Other–		
Cab/Body–	Body Damage		
	Seats		
	Tool Brackets		
	Hand Rails		
	Other–		
	Location		

Apparatus Satisfactory Except as Noted Below Initial-		Defective (Driver)	Repaired (Mechanic)
Gauges–	Fuel Gauge		
	Compnd./Pressure		
	Line Discharge #		
	Tank Water Level		
	Dash/Panel Gauges		
	Other–		
Leaks–	Pump		
	Tank		
	Gates		
	Booster Reel		
	Fuel (Tan/Clear)		
	Oil (Black)		
	Antifreeze (Green)		
	Transmission (Red)		
	Other–		
Doors–	Sags		
	Binds		
	Loose Handle		
	Latch Sticks		
	Other–		
Mirrors–	Broken		
	Loose		
	Other–		
Pump–	Transfer Valve		
	Priming Pump		
	Other–		

Reported To (name) – _____

Time – _____ hrs.

Corrective Action Taken On All Items
(Mechanics Signature) – _____

Corrective Action Taken – Except On Items Listed Below
(Mechanics Signature) – _____

Items Not Repaired – _____

_____

_____

_____

_____

_____

Remarks/Explanations–

Courtesy of Providence (RI) Fire Department.

Ames Fire Department Apparatus Inspection, Revised 08-11-98 Engine #2

Date _____ Vehicle No._____ Inspected By_____ Shift_____

Pump Hours_____ Engine Miles_____

Marking Code: Okay (\) Repairs Needed (0) Adjustment (x)

Check Fuel Level: Engine_____ Power Blower_____ Chain Saw_____ Port. Light_____

Reg Gas Can_____ Mixed Gas Can 32:1_____

Check Operation of: Engine_____ Power Blower_____ Chain Saw_____ Port. Light_____

ENGINE COMPARTMENT

Check Fluid Levels of: Engine Crankcase:_____ Pump Primer Tank_____ Vent Hole_____

Radiator_____

Check Condition of: Radiator & Heater Hoses_____ Drive Belts_____ Cleanliness_____

CAB

Brake Pedal Travel_____ Park Brake_____ Road Transmission Shift Lever_____

Pump Transmission Shift Lever_____ Battery Master Switch_____ Sirens_____

Dash Instruments and Gauges_____ Radio_____ Portable Radio_____

Air Horns_____ Public Address System_____ Electric Horn_____

Backup Horn_____ Wiper Operations_____ Wiper Blades_____ Windshield Washer_____

Washer Anti-Freeze_____ Cab Cleanliness_____ Glass Cleanliness_____

LIGHTS

Pump Panel_____ Head_____ Tail_____ Stop_____ Turn Signal_____

Backup_____ Hazardous Warning_____ Side Marker_____ Hand Spotlight_____

Clearance_____ Hose Pickup_____ Compartment_____ Hand Lanterns_____

All Emergency Warning Lights_____ Emmitter_____ Halogen Extension Lights_____

BATTERIES: Terminals_____ Fluid Levels_____ Cleanliness_____

BODY

Appearance_____ Body Damage_____ Electrical Outlets_____ Undercarriage_____

Compartment Interiors and Doors_____

CHASSIS

Tires: Tire Pressure L. Front_____ O.L. Dual_____ I.L. Dual_____

R. Front_____ O.R. Dual_____ I.R. Dual_____

Inspect tires for cuts and bruises_____ Check Exhaust System_____

PUMPS

Operate all Drain Valves_____ Primer_____ Booster Tank Condition & Water Level _____

Manually Operate All Controls_____ Throttle_____ Intake_____ Discharge_____

Flow Water Through Pump_____ Pressure Governor_____ Cooling Valve_____

FOAM INDUCTOR SYSTEM

Hale Foam Pro LED Panel Lights_____ Check Operation of A – B Tank Lever_____

Type of Foam in A:_____ Type of Foam in B:_____

Level of Tank A: _____Full _____¾ _____½ _____¼ _____Empty

Level of Tank B: _____Full _____¾ _____½ _____¼ _____Empty

EXTINGUISHERS:

Properly Secured in Mounting Brackets_____ Dry Chemical_____ CO_2_____ H_2O_____

SUPPRESSION EQUIPMENT

Smoke Ejector: Power Blower Operation_____ Cleanliness_____ Oil Level_____

Ropes: Number of Bags_____ Condition_____

Salvage Covers: Number of Covers_____ Number of Hall Runners_____ Condition_____

Hand Tools: Number of Axes_____ Condition of Axes_____ Number of Pike Poles_____

Condition of Pike Poles_____ Number of Hydrant Wrenches_____

All Hand Tools in Proper Location _____

Hydrant Bag: Contents – Spanners, Wrenches, Double Male, Adapters_____

HOSES and ACCESSORIES

Supply Lines: Feet of 5"_____ Feet of 2½"_____ Condition of Load_____

High Rise Bundle (reducer, spanners, nozzle)_____ Feet of 1½"_____

Preconnects: F. Crosslay Condition_____ Feet of 1¾"_____ F. Crosslay Nozzle_____

 R. Crosslay Condition_____ Feet of 1¾"_____ R. Crosslay Nozzle_____

 Front Bumper 1½" Condition_____ Feet of 1½"_____ 1½" Nozzle_____

 Lower Rear 1½" Condition_____ Feet of 1½"_____ 1½" Nozzle_____

 Upper Rear 1¾" Condition_____ Feet of 1¾"_____ 1¾" Nozzle_____

 Rear 2½" Condition_____ Feet of 2½"_____ 2½" Nozzle_____

 Hose Reel_____ Reel Nozzle_____

Nozzles: Number of 2½" Fog_____ 1½" Fog_____ 1¾"_____

Number of 2½" Straight Stream_____ Number of Tips_____ Cellar Nozzle_____

Deck Gun Condition_____

Appliances: Number of Hose Clamps_____ Hose Clamps Operation and Location_____

Number of Wyes_____ Number of Siamese_____ Number of Double Male_____

Number of Double Female_____ Condition of Appliances_____

LADDERS

Ladder Condition_____ Nicks & Burrs_____ Halyards_____ Properly Secured_____

10' Attic Ladder_____ 14' Roof Ladder_____ 24' Extension Ladder_____

Courtesy of Ames (IA) Fire Department.

Daily Check List

	Sun.	Mon.	Tue.	Wed.	Thu.	Fri.
1 Check Compartments						
2 Oil Levels						
3 Fuel Level						
4 Operation of all Lights						
5 Engine Operation						
6 Mobile Radio						
7 Portable Radio						
8 Extra Portable Batteries						
9 Booster Tank Level						
10 Exterior/Interior Appearance						
11 Check Extinguishers						
Signatures						

REMARKS

Officer's Signature _____

Courtesy of Ames (IA) Fire Department.

Weekly Emergency Vehicle Report

Page No. _____

Name of Company _____

Address _____

Emergency Vehicle MFG. _____

Year _____ Serial No. _____ Type _____

Required Tire Pressure: _____

Date Inspection Completed	Inspector	Battery Check	Braking System	Electrical System, Lights & Sirens	Tires & Wheel Lugs	Fuel Level	Oil Level Eng & Hyd	Hydraulic System	Pump Check	Cooling System	Lubrication Pump & Ladder	Engine Check	Booster Tank Level	Doors - Compartment & Cab	Portable Equipment	Special Remarks On Road Test Inspection Use Other Side

Courtesy of Volunteer Fireman's Insurance Services.

REMARKS: (Please itemize procedure taken on unsatisfactory inspection items noted on opposite side)

Inspection Date:	Repair Date:	Comments:	Repairs Completed - By:	Date:

Printed in U.S.A.

Item No. C10:007

Courtesy of Volunteer Fireman's Insurance Services.

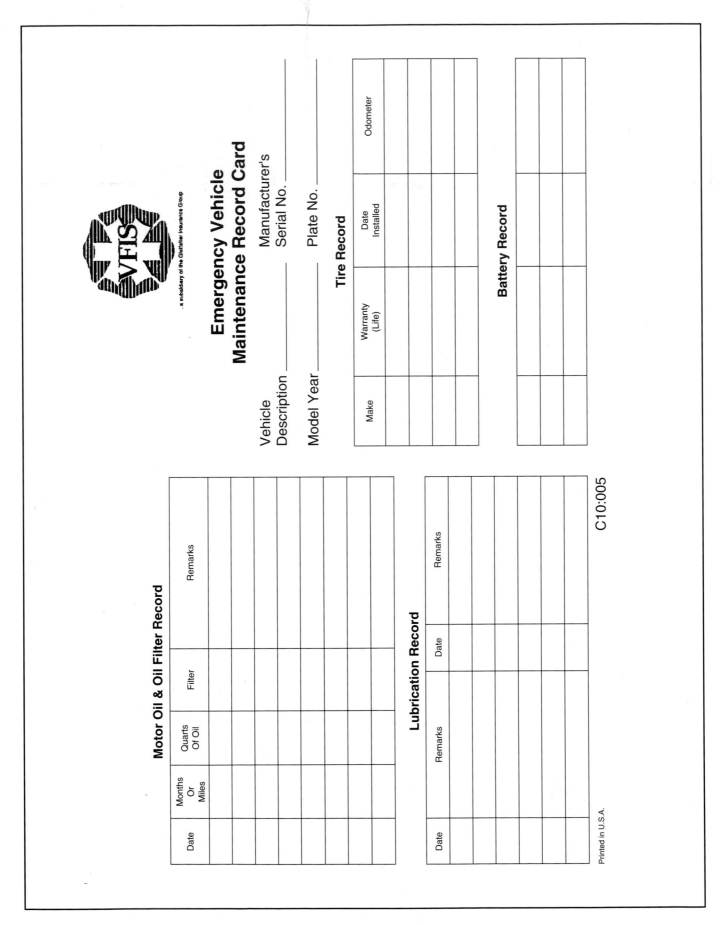

Courtesy of Volunteer Fireman's Insurance Services.

Maintenance And Repair Record

Date	Nature Of Repairs & / Or Maintenance Service	Repaired By	Cost

Emergency Vehicle Driver's Safety Check

VFIS
...a subsidiary of the Glatfelter Insurance Group

Date _____ Odometer Reading _____ Unit No. _____

☐ Pre-Trip Inspection ☐ Post-Trip Inspection

Only Items Checked Require Attention

☐ Gauges - Ammeter, Oil Pressure, Fuel, Water Temperature, Air Pressure or Vacuum
☐ Windshield Wipers
☐ Windshield & Windows
☐ Heater & Defroster
☐ Mirrors
☐ Brakes (Foot & Parking)
☐ Engine Noises
☐ Horn & Sirens
☐ Steering
☐ Vehicle Body
☐ Wheels, Tires, Lugs
☐ Fuel Tank & Cap
☐ Leaks - Water, Fuel, Oil

☐ Head Lights
☐ Tail Lights
☐ Stop Lights
☐ Turn Signals & 4-Way Flasher
☐ Reflectors
☐ Emergency Equipment
    ☐ Other - If Applicable
☐ Clearance Lights
☐ Emergency Warning Lights
☐ Side Marker Lights
☐ Brake Hoses
☐ Compartment Door Locks
☐ Drain Air Tanks of Moisture
☐ Air Systems
☐ Mounted Equipment

Remarks (explain unsatisfactory items noted above) _____

Signature of Driver _____

To Be Completed by Repair Shop

Mechanic's Report (If defects are noted) _____

Signature of Repair Shop
Foreman or Mechanic _____ Date _____

(Use back of form for additional remarks.)

Item No. C10:006

Emergency Vehicle Driver's Safety Check

VFIS
...a subsidiary of the Glatfelter Insurance Group

Date _____ Odometer Reading _____ Unit No. _____

☐ Pre-Trip Inspection ☐ Post-Trip Inspection

Only Items Checked Require Attention

☐ Gauges - Ammeter, Oil Pressure, Fuel, Water Temperature, Air Pressure or Vacuum
☐ Windshield Wipers
☐ Windshield & Windows
☐ Heater & Defroster
☐ Mirrors
☐ Brakes (Foot & Parking)
☐ Engine Noises
☐ Horn & Sirens
☐ Steering
☐ Vehicle Body
☐ Wheels, Tires, Lugs
☐ Fuel Tank & Cap
☐ Leaks - Water, Fuel, Oil

☐ Head Lights
☐ Tail Lights
☐ Stop Lights
☐ Turn Signals & 4-Way Flasher
☐ Reflectors
☐ Emergency Equipment
    ☐ Other - If Applicable
☐ Clearance Lights
☐ Emergency Warning Lights
☐ Side Marker Lights
☐ Brake Hoses
☐ Compartment Door Locks
☐ Drain Air Tanks of Moisture
☐ Air Systems
☐ Mounted Equipment

Remarks (explain unsatisfactory items noted above) _____

Signature of Driver _____

To Be Completed by Repair Shop

Mechanic's Report (If defects are noted) _____

Signature of Repair Shop
Foreman or Mechanic _____ Date _____

(Use back of form for additional remarks.)

Item No. C10:006

Courtesy of Volunteer Fireman's Insurance Services.

DAILY APPARATUS AND EQUIPMENT CHECK

UNIT #	I.D.#	TYPE:	MONTH:

EQUIPMENT	DATE																														
A. VEHICLE	1	2	3	4	5	6	7	8	9	10	11	12	13	14	15	16	17	18	19	20	21	22	23	24	25	26	27	28	29	30	31
1. Engine Crankcase Oil																															
2. Radiator																															
3. Hoses, Radiator Heater																															
4. Fan Belts																															
5. Primer Oil																															
6. Chassis Grease Reservoir																															
7. Seat Belts																															
8. Steering																															
9. Mirrors																															
10. Brakes																															
11. Lights and Warning Signals																															
12. Dashboard Gauges																															
13. Fuel																															
14. Windshield Wipers and Washers																															
15. Windshield Defroster																															
16. Radios																															
17. Sirens																															
18. Horns Air Electric																															
19. Tires and Wheels																															
20. Pump Panel																															
21. Water Tank																															
22. Hose Reel																															
23. Portable Generator																															
24. All Vehicles																															
B. INVENTORY																															
1. Paperwork and Binoculars																															
2. S.C.B.A./S.A.B.A.																															
3. Spare S.C.B.A./S.A.B.A.																															
4. P.A.S.S. Devices																															
5. Oxygen Bottles																															
6. Portable Fire Extinguishers																															
7. High Rise Pack																															
8. Ropes																															
9. Forcible Entry Tools																															
10. Hose Tools																															
11. Hose Appliances and Nozzles																															
12. Foam and Application Equipment																															
13. Traffic Flares and Cones																															
14. Salvage Equipment																															
15. Portable Lights																															
16. Extra Gasoline																															
17. Electric Cords, Adapters, & Fans																															
18. Hose Loads																															
19. Power Saws																															
20. Basic Life Support Equipment																															
21. Ladders																															
22. Life Jackets																															
23. Haz-Maz Bag																															
24. Tool Box																															
25. Hearing Protection																															
C. SPECIAL DEVICES & EQUIP																															
1. Aerial Devices																															
2. Brush Trucks																															
3. Special Operation Vehicles																															
INSPECTED BY:																															

(✔) Accepted   (X) Discrepency   (O) Unable To Check   (M) Missing

Courtesy of Virginia Beach (VA) Fire Department.

DATE	PROBLEM	ACTION TAKEN	DATE CORRECTED

Courtesy of Virginia Beach (VA) Fire Department.

WEEKLY APPARATUS AND EQUIPMENT CHECK

UNIT #	TYPE:																PERIOD ENDING:
EQUIPMENT	JAN APR					FEB MAY					MAR JUN					REMARKS	
	JUL OCT					AUG NOV					SEP DEC						
	1	2	3	4	5	1	2	3	4	5	1	2	3	4	5		
1. Parking Brakes																	
2. Air Brakes																	
3. Air Tank																	
4. Radiator																	
5. Power Steering																	
6. Portable Generator																	
7. Power Saws																	
8. Batteries																	
9. Tires																	
10. Springs																	
11. Drive Train																	
12. Exhaust System																	
14. Seagrave Governor																	
15. Slack Adjusters RF"																	
Slack Adjusters LF"																	
Slack Adjusters RR"																	
Slack Adjusters LR"																	
Slack Adjusters RRR"																	
Slack Adjusters LRR"																	
16. Road Test																	
17. Pump Operation																	
Engage Pump																	
Operate Valves (if necessary)																	
Primer																	
Governor/Relief Valve																	
Auxiliary Cooler																	
Change Over Valve																	
Pump Panel																	
18. Refill Tank & Primer																	
19. Clean Apparatus																	
20. Ladder Truck/Telesquirt																	

(✔) Accepted   (X) Discrepancy   (O) Unable To Check   (M) Missing

Courtesy of Virginia Beach (VA) Fire Department.

DATE	PROBLEM	ACTION TAKEN	DATE CORRECTED

Courtesy of Virginia Beach (VA) Fire Department.

MONTHLY APPARATUS AND EQUIPMENT CHECK

UNIT 3	TYPE							YEAR				
	JAN	FEB	MAR	APR	MAY	JUNE	JULY	AUG	SEPT	OCT	NOV	DEC
1. Doors and Windows												
2. Compartment & Fenders												
3. Emergency Stop												
4. Air Filter												
5. Ladder Maintenance												
6. Clean Equipment												
7. Hard/Soft Suction												
8. Pump Transmission Oil												
9. Hydrostatic Check												
10. Aerial Devices												
11. Seagrave Ladder												
12. Portable Water Pumps												
13. Pump Bearings												
14. Seagrave/Waterous Pump												
15. Change Over Valve												
16. Engine Transmission												
17. Primer Motor (Elec.)												
18. Hose Reel												
19 Steering Box												
20. Rear End												
21. Elec. Pump Shift Oil												
22. Back Flush Pump												
23. Rope Check												
24. Rope Card Filled Out												

(✔) Accepted   (X) Discrepancy   (O) Unable To Check   (M) Missing

Courtesy of Virginia Beach (VA) Fire Department.

DATE	PROBLEM	ACTION TAKEN	DATE CORRECTED

Courtesy of Virginia Beach (VA) Fire Department.

FORM NO. VBFD 77, Rev. 9/89

SUBSEQUENT APPARATUS AND EQUIPMENT CHECK

UNIT #	TYPE						YEAR				
	JAN	FEB	MAR	APR	MAY	JUNE	JULY	AUG	SEPT	OCT	NOV	DEC
QUARTERLY:												
1. Lubricate Grease Fittings												

SEMI-ANNUAL:	MILEAGE/DATE OF COMPLETION	REMARKS
1. Engine Oil		
2. Engine Oil Filters		
3. Auxiliary Generator Oil		
4. Fuel Filters		
5. Ladder Trucks		
6. Air Box Drain Tube		
7. Clean and Polish		
8. Air Filters		
9. Auxiliary Generator		
10. Extra Gasoline		
11. Power Saws		
12. Brush Truck Pump Oil Change		

ANNUAL:	MILEAGE/DATE OF COMPLETION	REMARKS
1. Transmission Oil & Filter		
2. Ladder Trucks		
3. Miscellaneous		
4. Pump Transmission		

(✔) Accepted   (X) Discrepancy   (O) Unable To Check   (M) Missing

Courtesy of Virginia Beach (VA) Fire Department.

DATE	PROBLEM	ACTION TAKEN	DATE CORRECTED

Courtesy of Virginia Beach (VA) Fire Department.

Appliance Test Method

The following procedures may be used to determine the actual friction loss that occurs within portable monitors and turrets.

Portable Monitors

Items required:

- Two 50-foot (15 m) lengths of 2½-inch (65 mm) or larger hose
- One solid stream nozzle capable of flows in range of appliance to be tested
- One pitot gauge

Based on the formulas presented in the text, the basic equation to be used is the following:

PDP = NP + FL + AL + Elev. Press.

Where:

PDP = Pump discharge pressure

FL = Friction loss

AL = Appliance loss

Elev. Press. = Elevation pressure loss

To determine a single variable in the previous equation, it is best to eliminate as many of the other variables as possible. This can be accomplished by making all those variables that might change the same for each test. This allows you to isolate the particular variable that you are trying to determine, which in this case is the appliance loss (AL). Therefore, to determine appliance loss, remove each variable by the following:

- Eliminate friction loss (FL) by using equal flows and equal hose lengths for each test.
- Eliminate elevation pressure loss by conducting the test on level ground.
- Eliminate nozzle pressure (NP) by taking pitot readings at the nozzle and varying the pump discharge pressure to maintain a constant nozzle pressure (and thus a constant flow).

This reduces the basic equation so that now a change in the system (adding the appliance) will have a direct effect on pump discharge pressure.

Two lengths of 2½-inch (65 mm) or larger hose are laid out. A safely secured nozzle is added directly to the end of the hose using appropriate adapters as necessary. The size of the nozzle depends on the amount of flow being tested (Table B.1).

Table B.1 (U.S)
Determining Nozzle Size and Discharge Pressure

Desired Test Flow	Required Nozzle Size	Required Discharge Pressure
340 gpm	1⅛ inch	80 psi
500 gpm	1⅜ inch	80 psi
750 gpm	1¾ inch	70 psi
1,000 gpm	2 inch	72 psi

Table B.1 (Metric)
Determining Nozzle Size and Discharge Pressure

Desired Test Flow	Required Nozzle Size	Required Discharge Pressure
1 350 L/min	32 mm	410 kPa
2 000 L/min	35 mm	620 kPa
3 000 L/min	45mm	520 kPa
4 000 L/min	50 mm	550 kPa

Once the equipment is laid out and connected, the flow is started and pump discharge pressure is increased to the point where nozzle pressure is at the desired pressure (target flow). Note the required pump discharge pressure to achieve this flow. This is pressure PDP_a.

Shut down the flow, and add the portable monitor appliance between the nozzle and the last coupling(s). Once connected, again begin the flow and increase the pump discharge pressure to the point where nozzle pressure (and hence flow) is the same as in the previous step. Note the required pump discharge pressure. This is pressure PDP_b. It should be higher than the first test. To determine the amount of friction loss caused by the monitor, use the following equation:

$PDP_b - PDP_a = AL$

This tells how much friction loss can be expected from that appliance around the flow for which it was tested.

Example B.1
Determine the appliance friction loss in a portable monitor being supplied by 3-inch (77 mm) hose flowing 500 gpm (2 000 L/min).

Solution
For a 500 gpm flow (2 000 L/min), add a 1⅜-inch (35 mm) tip to the hoseline. When the pitot pressure reaches 80 psi (560 kPa), the resulting pump discharge pressure is 102 psi (714 kPa). This is PDP_a.

Now add the portable monitor to be tested. Increase pump discharge pressure until pressure on the nozzle is 80 psi (560 kPa). The new pump discharge pressure is 128 psi (896 kPa). This is PDP_b. Pressure loss for the monitor is as follows:

$$PDP_b - PDP_a = AL$$
$$128 - 102 = 26 \text{ psi}$$
$$896 - 714 = 182 \text{ kPa}$$

Because it is not necessary to determine the exact friction loss in the hose, any layout or number of hoses can be used to deliver the needed flow. The important point is that the number and sizes of hoses used are the same for both tests.

Turrets

To calculate losses through turrets (or portable monitor in the truck-mounted mode), as well as losses between the turret and the pump, an additional item is necessary. A side wall pressure reading piezometer is added between the exit nozzle and the turret. Increase the pressure until desired flow is attained (as determined by a pitot pressure on the nozzle). The pressure loss between the pump and the exit of the turret is shown by the difference between the sidewall piezometer on the exit and the compound pressure gauge in the pump panel. Note that this pressure difference only applies to that particular flow. For other flows, other values must be determined.

IFSTA Friction Loss Calculations (U.S.)

The following tables that include imperial gallons may be used to identify the friction loss for the given flows in imperial gallons per minute. The coefficients used to determine friction loss using U.S. gallons per minute *may not* be used when calculating friction loss using imperial gallons per minute.

Table C.1
¾-Inch Rubber Hose
C-Factor: 1,100

	Flow in gpm				
(U.S. Gallons)	**20**	**30**	**40**	**50**	**60**
(Imperial Gallons)	**17**	**25**	**33**	**42**	**50**
Lay Length Feet	Friction Loss in psi				
50	22	50	88	138	198
100	44	99	176		
150	66	149			
200	88	198			
250	110				
300	132				
350	154				
400	176				

Table C.2
1-Inch Rubber Hose
C-Factor: 150

	Flow in gpm				
(U.S. Gallons)	**20**	**30**	**40**	**50**	**60**
(Imperial Gallons)	**17**	**25**	**33**	**42**	**50**
Lay Length Feet	Friction Loss in psi				
50	3	7	12	19	27
100	6	14	24	38	54
150	9	20	36	56	81
200	12	27	48	75	108
250	15	34	60	94	135
300	18	41	72	113	162
350	21	47	84	131	189
400	24	54	96	150	

Table C.3
1½-Inch Hose
C-Factor: 24

	Flow in gpm					
(U.S. Gallons)	**40**	**60**	**80**	**95**	**125**	**150**
(Imperial Gallons)	**33**	**50**	**67**	**79**	**104**	**125**
Lay Length Feet	Friction Loss in psi					
50	2	4	8	11	19	27
100	4	9	15	22	38	54
150	6	13	23	32	56	81
200	8	17	31	43	75	108
250	10	22	38	54	94	135
300	12	26	46	65	113	162
350	13	30	54	76	131	189
400	15	35	61	87	150	216
450	17	39	69	97	169	
500	19	43	77	108	188	

Table C.4
1¾-Inch Rubber Hose
C-Factor: 15.5

	Flow in gpm				
(U.S. Gallons)	**95**	**125**	**150**	**175**	**200**
(Imperial Gallons)	**79**	**104**	**125**	**146**	**167**
Lay Length Feet	**Friction Loss in psi**				
50	7	12	17	24	31
100	14	24	35	47	62
150	21	36	52	71	93
200	28	48	70	95	124
250	35	61	87	119	155
300	42	73	105	142	186
350	49	85	122	166	
400	56	97	140	190	
450	63	109	157		
500	70	121	174		

Table C.5
2-Inch Rubber Hose
C-Factor: 8

	Flow in gpm				
(U.S. Gallons)	**100**	**125**	**150**	**175**	**200**
(Imperial Gallons)	**83**	**104**	**125**	**146**	**167**
Lay Length Feet	**Friction Loss in psi**				
50	4	6	9	12	16
100	8	13	18	25	32
150	12	19	27	37	48
200	16	25	36	49	64
250	20	31	45	61	80
300	24	38	54	74	96
350	28	44	63	86	112
400	32	50	72	98	128
450	36	56	81	110	144
500	40	63	90	123	160

Table C.6
2½-Inch Hose
C-Factor: 2

Flow in gpm

	175	200	225	250	275	300	350
(U.S. Gallons)	175	200	225	250	275	300	350
(Imperial Gallons)	146	167	187	208	229	250	292

Friction Loss in psi

Lay Length Feet							
50	3	4	5	6	8	9	12
100	6	8	10	13	15	18	25
150	9	12	15	19	23	27	37
200	12	16	20	25	30	36	49
250	15	20	25	31	38	45	61
300	18	24	30	38	45	54	74
350	21	28	35	44	53	63	86
400	25	32	41	50	61	72	98
450	28	36	46	56	68	81	110
500	31	40	51	63	76	90	123
550	34	44	56	69	83	99	135
600	37	48	61	75	91	108	147
650	40	52	66	81	98	117	159
700	43	56	71	88	106	126	172
750	46	60	76	94	113	135	184
800	49	64	81	100	121	144	196
850	52	68	86	106	129	153	208
900	55	72	91	113	136	162	221
950	58	76	96	119	144	171	233
1,000	61	80	101	125	151	180	245
1,050	64	84	106	131	159	189	
1,100	67	88	111	138	166	198	
1,150	70	92	116	144	174	207	
1,200	74	96	122	150	182	216	
1,250	77	100	127	156	189	225	
1,300	80	104	132	163	197	234	
1,350	83	108	137	169	204	243	
1,400	86	112	142	175	212		
1,450	89	116	147	181	219		
1,500	92	120	152	188	227		
1,550	95	124	157	194	234		
1,600	98	128	162	200	242		
1,650	101	132	167	206	250		
1,700	104	136	172	213			
1,750	107	140	177	219			
1,800	110	144	182	225			
1,850	113	148	187	231			
1,900	116	152	192	238			
1,950	119	156	197	244			
2,000	123	160	203	250			

Table C.7
3-Inch Hose
C-Factor: 0.8

	Flow in gpm					
(U.S. Gallons)	250	325	500	750	1,000	1,250
(Imperial Gallons)	208	271	417	625	833	1,041
Lay Length Feet	Friction Loss in psi					
50	3	4	10	23	40	63
100	5	8	20	45	80	125
150	8	13	30	68	120	188
200	10	17	40	90	160	250
250	13	21	50	113	200	
300	15	25	60	135	240	
350	18	30	70	158		
400	20	34	80	180		
450	23	38	90	203		
500	25	42	100	225		
550	28	46	110	248		
600	30	51	120			
650	33	55	130			
700	35	59	140			
750	38	63	150			
800	40	68	160			
850	43	72	170			
900	45	76	180			
950	48	80	190			
1,000	50	85	200			
1,050	53	89	210			
1,100	55	93	220			
1,150	58	97	230			
1,200	60	101	240			
1,250	63	106	250			
1,300	65	110				
1,350	68	114				
1,400	70	118				
1,450	73	123				
1,500	75	127				
1,550	78	131				
1,600	80	135				
1,650	83	139				
1,700	85	144				
1,750	88	148				
1,800	90	152				
1,850	93	156				
1,900	95	161				
1,950	98	165				
2,000	100	169				

Table C.8
3-Inch Hose with 3-Inch Couplings
C-Factor: 0.677

(U.S. Gallons)	Flow in gpm				
	250	350	500	750	1,000
Lay Length Feet	Friction Loss in psi				
50	2	4	8	19	34
100	4	8	17	38	68
150	6	12	25	57	102
200	8	17	34	76	135
250	11	21	42	95	169
300	13	25	51	114	
350	15	29	59	133	
400	17	33	68	152	
450	19	37	76	171	
500	21	41	85		
550	23	46	93		
600	25	50	102		
650	28	54	110		
700	30	58	118		
750	32	62	127		
800	34	66	135		
850	36	70	144		
900	38	75	152		
950	40	79	161		
1,000	42	83	169		
1,050	44	87	178		
1,100	47	91	186		
1,150	49	95			
1,200	51	100			
1,250	53	104			
1,300	55	108			
1,350	57	112			
1,400	59	116			
1,450	61	120			
1,500	63	124			
1,550	66	129			
1,600	68	133			
1,650	70	137			
1,700	72	141			
1,750	74				
1,800	76				
1,850	78				
1,900	80				
1,950	83				
2,000	85				

Table C.9
4-Inch Hose
C-Factor: 0.2

	Flow in gpm					
(U.S. Gallons)	**500**	**750**	**1,000**	**1,250**	**1,500**	**1,750**
(Imperial Gallons)	**417**	**625**	**833**	**1,041**	**1,250**	**1,458**
Lay Length Feet	Friction Loss in psi					
50	3	6	10	16	23	31
100	5	11	20	31	45	61
150	8	17	30	47	68	92
200	10	23	40	63	90	123
250	13	28	50	78	113	153
300	15	34	60	94	135	184
350	18	39	70	109	158	
400	20	45	80	125	180	
450	23	51	90	141		
500	25	56	100	156		
550	28	62	110	172		
600	30	68	120			
650	33	73	130			
700	35	79	140			
750	38	84	150			
800	40	90	160			
850	43	96	170			
900	45	101	180			
950	48	107				
1,000	50	113				
1,050	53	118				
1,100	55	124				
1,150	58	129				
1,200	60	135				
1,250	63	141				
1,300	65	146				
1,350	68	152				
1,400	70	158				
1,450	73	163				
1,500	75	169				
1,550	78	174				
1,600	80	180				
1,650	83					
1,700	85					
1,750	88					
1,800	90					
1,850	93					
1,900	95					
1,950	98					
2,000	100					

Table C.10
4½-Inch Hose
C-Factor: 0.1

Flow in gpm						
(U.S. Gallons)	**500**	**750**	**1,000**	**1,250**	**1,500**	**1,750**
(Imperial Gallons)	**417**	**625**	**833**	**1,041**	**1,250**	**1,458**
Lay Length Feet	**Friction Loss in psi**					
50	1	3	5	8	11	15
100	3	6	10	16	23	31
150	4	8	15	23	34	46
200	5	11	20	31	45	61
250	6	14	25	39	56	77
300	8	17	30	47	68	92
350	9	20	35	55	79	107
400	10	23	40	63	90	123
450	11	25	45	70	101	138
500	13	28	50	78	113	153
550	14	31	55	86	124	168
600	15	34	60	94	135	184
650	16	37	65	102	146	
700	18	39	70	109	158	
750	19	42	75	117	169	
800	20	45	80	125	180	
850	21	48	85	133		
900	23	51	90	141		
950	24	53	95	148		
1,000	25	56	100	156		
1,050	26	59	105	164		
1,100	28	62	110	172		
1,150	29	65	115	180		
1,200	30	68	120			
1,250	31	70	125			
1,300	33	73	130			
1,350	34	76	135			
1,400	35	79	140			
1,450	36	82	145			
1,500	38	84	150			
1,550	39	87	155			
1,600	40	90	160			
1,650	41	93	165			
1,700	43	96	170			
1,750	44	98	175			
1,800	45	101	180			
1,850	46	104	185			
1,900	48	107				
1,950	49	110				
2,000	50	113				

Table C.11
5-Inch Hose
C-Factor: 0.08

	Flow in gpm					
(U.S. Gallons)	750	1,000	1,250	1,500	1,750	2,000
(Imperial Gallons)	625	833	1,041	1,250	1,458	1,666
Lay Length Feet	Friction Loss in psi					
50	2	4	6	9	12	16
100	5	8	13	18	25	32
150	7	12	19	27	37	48
200	9	16	25	36	49	64
250	11	20	31	45	61	80
300	14	24	38	54	74	96
350	16	28	44	63	86	112
400	18	32	50	72	98	128
450	20	36	56	81	110	144
500	23	40	63	90	123	160
550	25	44	69	99	135	176
600	27	48	75	108	147	192
650	29	52	81	117	159	208
700	32	56	88	126	172	224
750	34	60	94	135	184	240
800	36	64	100	144	196	
850	38	68	106	153	208	
900	41	72	113	162	221	
950	43	76	119	171	233	
1,000	45	80	125	180	245	
1,050	47	84	131	189		
1,100	50	88	138	198		
1,150	52	92	144	207		
1,200	54	96	150	216		
1,250	56	100	156	225		
1,300	59	104	163	234		
1,350	61	108	169	243		
1,400	63	112	175	252		
1,450	65	116	181	261		
1,500	68	120	188			
1,550	70	124	194			
1,600	72	128	200			
1,650	74	132	206			
1,700	77	136	213			
1,750	79	140	219			
1,800	81	144	225			
1,850	83	148	231			
1,900	86	152	238			
1,950	88	156	244			
2,000	90	160	250			

Table C.12
Two 2½-Inch Hose
C-Factor: 0.5

Lay Length Feet	Flow in gpm					
(U.S. Gallons)	**500**	**750**	**1,000**	**1,250**	**1,500**	**1,750**
(Imperial Gallons)	**417**	**625**	**833**	**1,041**	**1,250**	**1,458**
	Friction Loss in psi					
50	6	14	25	39	56	77
100	13	28	50	78	113	153
150	19	42	75	117	169	230
200	25	56	100	156	225	
250	31	70	125	195		
300	38	84	150	234		
350	44	98	175			
400	50	113	200			
450	56	127	225			
500	63	141	250			
550	69	155				
600	75	169				
650	81	183				
700	88	197				
750	94	211				
800	100	225				
850	106	239				
900	113					
950	119					
1,000	125					
1,050	131					
1,100	138					
1,150	144					
1,200	150					
1,250	156					
1,300	163					
1,350	169					
1,400	175					
1,450	181					
1,500	188					
1,550	194					
1,600	200					
1,650	206					
1,700	213					
1,750	219					
1,800	225					
1,850	231					
1,900	238					
1,950	244					
2,000	250					

Table C.13
Three 2½-Inch Hose
C-Factor: 0.22

(U.S. Gallons)	Flow in gpm					
	250	**500**	**750**	**1,000**	**1,250**	**1,500**
Lay Length Feet	Friction Loss in psi					
50	1	3	6	11	17	25
100	1	6	12	22	34	50
150	2	8	19	33	52	74
200	3	11	25	44	69	99
250	3	14	31	55	86	124
300	4	17	37	66	103	149
350	5	19	43	77	120	173
400	6	22	50	88	138	198
450	6	25	56	99	155	
500	7	28	62	110	172	
550	8	30	68	121	189	
600	8	33	74	132		
650	9	36	80	143		
700	10	39	87	154		
750	10	41	93	165		
800	11	44	99	176		
850	12	47	105	187		
900	12	50	111	198		
950	13	52	118			
1,000	14	55	124			
1,050	14	58	130			
1,100	15	61	136			
1,150	16	63	142			
1,200	17	66	149			
1,250	17	69	155			
1,300	18	72	161			
1,350	19	74	167			
1,400	19	77	173			
1,450	20	80	179			
1,500	21	83	186			
1,550	21	85	192			
1,600	22	88	198			
1,650	23	91				
1,700	23	94				
1,750	24	96				
1,800	25	99				
1,850	25	102				
1,900	26	105				
1,950	27	107				
2,000	28	110				

Table C.14
One 3-Inch and One 2½-Inch Hose
C-Factor: 0.3

	Flow in gpm					
(U.S. Gallons)	**500**	**750**	**1,000**	**1,250**	**1,500**	**1,750**
(Imperial Gallons)	**417**	**625**	**833**	**1,041**	**1,250**	**1,458**

Lay Length Feet	Friction Loss in psi					
50	4	8	15	23	34	46
100	8	17	30	47	68	92
150	11	25	45	70	101	138
200	15	34	60	94	135	184
250	19	42	75	117	169	
300	23	51	90	141	203	
350	26	59	105	164		
400	30	68	120	188		
450	34	76	135			
500	38	84	150			
550	41	93	165			
600	45	101	180			
650	49	110	195			
700	53	118				
750	56	127				
800	60	135				
850	64	143				
900	68	152				
950	71	160				
1,000	75	169				
1,050	79	177				
1,100	83	186				
1,150	86	194				
1,200	90	203				
1,250	94					
1,300	98					
1,350	101					
1,400	105					
1,450	109					
1,500	113					
1,550	116					
1,600	120					
1,650	124					
1,700	128					
1,750	131					
1,800	135					
1,850	139					
1,900	143					
1,950	146					
2,000	150					

Table C.15
Two 2½-Inch and One 3-Inch Hose
C-Factor: 0.16

(U.S. Gallons)	Flow in gpm					
	500	**750**	**1,000**	**1,250**	**1,500**	**1,750**
Lay Length Feet	Friction Loss in psi					
50	2	5	8	13	18	25
100	4	9	16	25	36	49
150	6	14	24	38	54	74
200	8	18	32	50	72	98
250	10	23	40	63	90	123
300	12	27	48	75	108	147
350	14	32	56	88	126	172
400	16	36	64	100	144	196
450	18	41	72	113	162	
500	20	45	80	125	180	
550	22	50	88	138	198	
600	24	54	96	150		
650	26	59	104	163		
700	28	63	112	175		
750	30	68	120	188		
800	32	72	128	200		
850	34	77	136			
900	36	81	144			
950	38	86	152			
1,000	40	90	160			
1,050	42	95	168			
1,100	44	99	176			
1,150	46	104	184			
1,200	48	108	192			
1,250	50	113	200			
1,300	52	117				
1,350	54	122				
1,400	56	126				
1,450	58	131				
1,500	60	135				
1,550	62	140				
1,600	64	144				
1,650	66	149				
1,700	68	153				
1,750	70	158				
1,800	72	162				
1,850	74	167				
1,900	76	171				
1,950	78	176				
2,000	80	180				

Table C.16
Two 3-Inch Hose
C-Factor: 0.2

	Flow in gpm					
(U.S. Gallons)	**500**	**750**	**1,000**	**1,250**	**1,500**	**1,750**
(Imperial Gallons)	**417**	**625**	**833**	**1,041**	**1,250**	**1,458**
Lay Length Feet	Friction Loss in psi					
50	3	6	10	16	23	31
100	5	11	20	31	45	61
150	8	17	30	47	68	92
200	10	23	40	63	90	123
250	13	28	50	78	113	153
300	15	34	60	94	135	184
350	18	39	70	109	158	
400	20	45	80	125	180	
450	23	51	90	141		
500	25	56	100	156		
550	28	62	110	172		
600	30	68	120	188		
650	33	73	130	203		
700	35	79	140			
750	38	84	150			
800	40	90	160			
850	43	96	170			
900	45	101	180			
950	48	107	190			
1,000	50	113	200			
1,050	53	118				
1,100	55	124				
1,150	58	129				
1,200	60	135				
1,250	63	141				
1,300	65	146				
1,350	68	152				
1,400	70	158				
1,450	73	163				
1,500	75	169				
1,550	78	174				
1,600	80	180				
1,650	83	186				
1,700	85	191				
1,750	88	197				
1,800	90	203				
1,850	93					
1,900	95					
1,950	98					
2,000	100					

Table C.17
Two 3-Inch and One 2½-Inch Hose
C-Factor: 0.12

	Flow in gpm					
(U.S. Gallons)	**500**	**750**	**1,000**	**1,250**	**1,500**	**1,750**
Lay Length Feet	Friction Loss in psi					
50	2	3	6	9	14	18
100	3	7	12	19	27	37
150	5	10	18	28	41	55
200	6	14	24	38	54	74
250	8	17	30	47	68	92
300	9	20	36	56	81	110
350	11	24	42	66	95	129
400	12	27	48	75	108	147
450	14	30	54	84	122	
500	15	34	60	94	135	
550	17	37	66	103	149	
600	18	41	72	113		
650	20	44	78	122		
700	21	47	84	131		
750	23	51	90	141		
800	24	54	96	150		
850	26	57	102			
900	27	61	108			
950	29	64	114			
1,000	30	68	120			
1,050	32	71	126			
1,100	33	74	132			
1,150	35	78	138			
1,200	36	81	144			
1,250	38	84	150			
1,300	39	88				
1,350	41	91				
1,400	42	95				
1,450	44	98				
1,500	45	101				
1,550	47	105				
1,600	48	108				
1,650	50	111				
1,700	51	115				
1,750	53	118				
1,800	54	122				
1,850	56	125				
1,900	57	128				
1,950	59	132				
2,000	60	135				

IFSTA Friction Loss Calculations (Metric)

Table D.1
20 mm Rubber Hose
C-Factor: 1 741

Lay Length Meters	Flow in L/min				
	80	120	160	200	240
	Friction Loss in kPa				
15	167	376	669	1 045	1 504
30	334	752	1 337		
45	501	1 128			
60	669	1 504			
75	836				
90	1 003				
105	1 170				
120	1 337				

Table D.2
25 mm Rubber Hose
C-Factor: 238

Lay Length Meters	Flow in L/min				
	80	120	160	200	240
	Friction Loss in kPa				
15	23	51	91	143	206
30	46	103	183	286	411
45	69	154	274	428	617
60	91	206	366	571	823
75	114	257	457	714	1 028
90	137	308	548	857	1 234
105	160	360	640	1 000	1 439
120	183	411	731	1 142	1 645

Table D.3
38 mm Hose
C-Factor: 38

Lay Length Meters	Flow in L/min				
	240	320	400	500	600
	Friction Loss in kPa				
15	33	58	91	143	205
30	66	117	182	285	410
45	98	175	274	428	616
60	131	233	365	570	821
75	164	292	456	713	1 026
90	197	350	547	855	1 231
105	230	409	638	998	
120	263	467	730	1 140	
135	295	525	821		
150	328	584	912		

Table D.4
45 mm Hose
C-Factor: 24.6

Lay Length Meters	Flow in L/min				
	380	500	600	700	800
	Friction Loss in kPa				
15	53	92	133	181	236
30	107	185	266	362	472
45	160	277	399	542	708
60	213	369	531	723	945
75	266	461	664	904	1 181
90	320	554	797	1 085	1 417
105	373	646	930	1 266	1 653
120	426	738	1 063	1 446	
135	480	830	1 196	1 627	
150	533	923	1 328		

Table D.5
50 mm Hose
C-Factor: 12.7

Lay Length Meters	Flow in L/min				
	400	500	600	700	800
	Friction Loss in kPa				
15	30	48	69	93	122
30	61	95	137	187	244
45	91	143	206	280	366
60	122	191	274	373	488
75	152	238	343	467	610
90	183	286	411	560	732
105	213	333	480	653	853
120	244	381	549	747	975
135	274	429	617	840	1 097
150	305	476	686	933	1 219

Table D.6
65 mm Hose
C-Factor: 3.17

Flow in L/min

Lay Length Meters	700	800	900	1 000	1 100	1 200	1 400
	Friction Loss in kPa						
15	23	30	39	48	58	68	93
30	47	61	77	95	115	137	186
45	70	91	116	143	173	205	280
60	93	122	154	190	230	274	373
75	116	152	193	238	288	342	466
90	140	183	231	285	345	411	559
105	163	213	270	333	403	479	652
120	186	243	308	380	460	548	746
135	210	274	347	428	518	616	839
150	233	304	385	476	575	685	932
165	256	335	424	523	633	753	1 025
180	280	365	462	571	690	822	1 118
195	303	396	501	618	748	890	1 212
210	326	426	539	666	805	959	1 305
225	349	456	578	713	863	1 027	1 398
240	373	487	616	761	921	1 096	1 491
255	396	517	655	808	978	1 164	1 584
270	419	548	693	856	1 036	1 232	1 678
285	443	578	732	903	1 093	1 301	
300	466	609	770	951	1 151	1 369	
315	489	639	809	999	1 208	1 438	
330	513	670	847	1 046	1 266	1 506	
345	536	700	886	1 094	1 323	1 575	
360	559	730	924	1 141	1 381	1 643	
375	582	761	963	1 189	1 438	1 712	
390	606	791	1 001	1 236	1 496		
405	629	822	1 040	1 284	1 553		
420	652	852	1 078	1 331	1 611		
435	676	883	1 117	1 379	1 669		
450	699	913	1 155	1 427			
465	722	943	1 194	1 474			
480	746	974	1 232	1 522			
495	769	1 004	1 271	1 569			
510	792	1 035	1 310	1 617			
525	815	1 065	1 348	1 664			
540	839	1 096	1 387	1 712			
555	862	1 126	1 425				
570	885	1 156	1 464				
585	909	1 187	1 502				
600	932	1 217	1 541				

Table D.7
77 mm Hose with 65 mm Couplings
C-Factor: 1.27

Lay Length Meters	Flow in L/min					
	1 000	1 200	1 400	2 000	3 000	4 000
	Friction Loss in kPa					
15	19	27	37	76	171	305
30	38	55	75	152	343	610
45	57	82	112	229	514	914
60	76	110	149	305	686	1 219
75	95	137	187	381	857	1 524
90	114	165	224	457	1 029	
105	133	192	261	533	1 200	
120	152	219	299	610	1 372	
135	171	247	336	686	1 543	
150	191	274	373	762	1 715	
165	210	302	411	838		
180	229	329	448	914		
195	248	357	485	991		
210	267	384	523	1 067		
225	286	411	560	1 143		
240	305	439	597	1 219		
255	324	466	635	1 295		
270	343	494	672	1 372		
285	362	521	709	1 448		
300	381	549	747	1 524		
315	400	576	784	1 600		
330	419	604	821	1 676		
345	438	631	859			
360	457	658	896			
375	476	686	933			
390	495	713	971			
405	514	741	1 008			
420	533	768	1 045			
435	552	796	1 083			
450	572	823	1 120			
465	591	850	1 157			
480	610	878	1 195			
495	629	905	1 232			
510	648	933	1 269			
525	667	960	1 307			
540	686	988	1 344			
555	705	1 015	1 382			
570	724	1 042	1 419			
585	743	1 070	1 456			
600	762	1 097	1 494			

Table D.8
77 mm Hose with 77 mm Couplings
C-Factor: 1.06

Lay Length Meters	Flow in L/min					
	1 000	1 200	1 400	2 000	3 000	4 000
	Friction Loss in kPa					
15	16	23	31	64	143	254
30	32	46	62	127	286	509
45	48	69	93	191	429	763
60	64	92	125	254	572	1 018
75	80	114	156	318	716	1 272
90	95	137	187	382	859	1 526
105	111	160	218	445	1 002	
120	127	183	249	509	1 145	
135	143	206	280	572	1 288	
150	159	229	312	636	1 431	
165	175	252	343	700	1 574	
180	191	275	374	763	1 717	
195	207	298	405	827		
210	223	321	436	890		
225	239	343	467	954		
240	254	366	499	1 018		
255	270	389	530	1 081		
270	286	412	561	1 145		
285	302	435	592	1 208		
300	318	458	623	1 272		
315	334	481	654	1 336		
330	350	504	686	1 399		
345	366	527	717	1 463		
360	382	550	748	1 526		
375	398	572	779	1 590		
390	413	595	810	1 654		
405	429	618	841	1 717		
420	445	641	873			
435	461	664	904			
450	477	687	935			
465	493	710	966			
480	509	733	997			
495	525	756	1 028			
510	541	778	1 060			
525	557	801	1 091			
540	572	824	1 122			
555	588	847	1 153			
570	604	870	1 184			
585	620	893	1 215			
600	636	916	1 247			

Flow in L/min					
2 000	3 000	4 000	5 000	6 000	7 000

Lay Length Meters	Friction Loss in kPa					
15	18	41	73	114	165	224
30	37	82	146	229	329	448
45	55	124	220	343	494	673
60	73	165	293	458	659	897
75	92	206	366	572	824	1 121
90	110	247	439	686	988	
105	128	288	512	801	1 153	
120	146	329	586	915		
135	165	371	659	1 029		
150	183	412	732	1 144		
165	201	453	805	1 258		
180	220	494	878			
195	238	535	952			
210	256	576	1 025			
225	275	618	1 098			
240	293	659	1 171			
255	311	700	1 244			
270	329	741				
285	348	782				
300	366	824				
315	384	865				
330	403	906				
345	421	947				
360	439	988				
375	458	1 029				
390	476	1 071				
405	494	1 112				
420	512	1 153				
435	531	1 194				
450	549	1 235				
465	567	1 276				
480	586					
495	604					
510	622					
525	641					
540	659					
555	677					
570	695					
585	714					
600	732					

Table D.10
115 mm Hose
C-Factor: 0.167

Lay Length Meters	Flow in L/min					
	2 000	3 000	4 000	5 000	6 000	7 000
	Friction Loss in kPa					
15	10	23	40	63	90	123
30	20	45	80	125	180	245
45	30	68	120	188	271	368
60	40	90	160	251	361	491
75	50	113	200	313	451	614
90	60	135	240	376	541	736
105	70	158	281	438	631	859
120	80	180	321	501	721	982
135	90	203	361	564	812	1 105
150	100	225	401	626	902	1 227
165	110	248	441	689	992	
180	120	271	481	752	1 082	
195	130	293	521	814	1 172	
210	140	316	561	877	1 263	
225	150	338	601	939		
240	160	361	641	1 002		
255	170	383	681	1 065		
270	180	406	721	1 127		
285	190	428	762	1 190		
300	200	451	802	1 253		
315	210	473	842			
330	220	496	882			
345	230	519	922			
360	240	541	962			
375	251	564	1 002			
390	261	586	1 042			
405	271	609	1 082			
420	281	631	1 122			
435	291	654	1 162			
450	301	676	1 202			
465	311	699	1 242			
480	321	721				
495	331	744				
510	341	767				
525	351	789				
540	361	812				
555	371	834				
570	381	857				
585	391	879				
600	401	902				

Table D.11
125 mm Hose
C-Factor: 0.138

Lay Length Meters	Flow in L/min					
	3 000	4 000	5 000	6 000	7 000	8 000
	Friction Loss in kPa					
15	19	33	52	75	101	132
30	37	66	104	149	203	265
45	56	99	155	224	304	397
60	75	132	207	298	406	530
75	93	166	259	373	507	662
90	112	199	311	447	609	795
105	130	232	362	522	710	927
120	149	265	414	596	811	1 060
135	168	298	466	671	913	1 192
150	186	331	518	745	1 014	
165	205	364	569	820	1 116	
180	224	397	621	894	1 217	
195	242	431	673	969		
210	261	464	725	1 043		
225	279	497	776	1 118		
240	298	530	828	1 192		
255	317	563	880	1 267		
270	335	596	932			
285	354	629	983			
300	373	662	1 035			
315	391	696	1 087			
330	410	729	1 139			
345	428	762	1 190			
360	447	795	1 242			
375	466	828				
390	484	861				
405	503	894				
420	522	927				
435	540	960				
450	559	994				
465	578	1 027				
480	596	1 060				
495	615	1 093				
510	633	1 126				
525	652	1 159				
540	671	1 192				
555	689	1 225				
570	708	1 259				
585	727					
600	745					

Lay Length Meters	Flow in L/min					
	1 000	2 000	3 000	4 000	5 000	6 000
	Friction Loss in kPa					
15	12	47	107	189	296	426
30	24	95	213	379	592	852
45	36	142	320	568	888	1 278
60	47	189	426	757	1 184	1 704
75	59	237	533	947	1 479	
90	71	284	639	1 136	1 775	
105	83	331	746	1 326		
120	95	379	852	1 515		
135	107	426	959	1 704		
150	118	473	1 065			
165	130	521	1 172			
180	142	568	1 278			
195	154	615	1 385			
210	166	663	1 491			
225	178	710	1 598			
240	189	757	1 704			
255	201	805				
270	213	852				
285	225	899				
300	237	947				
315	249	994				
330	260	1 041				
345	272	1 089				
360	284	1 136				
375	296	1 184				
390	308	1 231				
405	320	1 278				
420	331	1 326				
435	343	1 373				
450	355	1 420				
465	367	1 468				
480	379	1 515				
495	391	1 562				
510	402	1 610				
525	414	1 657				
540	426	1 704				
555	438	1 752				
570	450					
585	462					
600	473					

Table D.13
Three 65 mm Hose
C-Factor: 0.347

Lay Length Meters	Flow in L/min					
	1 000	2 000	3 000	4 000	5 000	6 000
	Friction Loss in kPa					
15	5	21	47	83	130	187
30	10	42	94	167	260	375
45	16	62	141	250	390	562
60	21	83	187	333	520	750
75	26	104	234	416	651	937
90	31	125	281	500	781	1 124
105	36	146	328	583	911	1 312
120	42	167	375	666	1 041	1 499
135	47	187	422	750	1 171	1 686
150	52	208	468	833	1 301	
165	57	229	515	916	1 431	
180	62	250	562	999	1 561	
195	68	271	609	1 083	1 692	
210	73	291	656	1 166		
225	78	312	703	1 249		
240	83	333	750	1 332		
255	88	354	796	1 416		
270	94	375	843	1 499		
285	99	396	890	1 582		
300	104	416	937	1 666		
315	109	437	984	1 749		
330	115	458	1 031			
345	120	479	1 077			
360	125	500	1 124			
375	130	520	1 171			
390	135	541	1 218			
405	141	562	1 265			
420	146	583	1 312			
435	151	604	1 359			
450	156	625	1 405			
465	161	645	1 452			
480	167	666	1 499			
495	172	687	1 546			
510	177	708	1 593			
525	182	729	1 640			
540	187	750	1 686			
555	193	770				
570	198	791				
585	203	812				
600	208	833				

Table D.14
One 77 mm and One 65 mm Hose
C-Factor: 0.473

Lay Length Meters	Flow in L/min					
	2 000	3 000	4 000	5 000	6 000	7 000
	Friction Loss in kPa					
15	28	54	102	192	364	688
30	57	107	203	384	727	
45	85	161	305	577	1 091	
60	114	215	406	769	1 455	
75	142	268	508	961		
90	170	322	610	1 153		
105	199	376	711	1 345		
120	227	430	813	1 538		
135	255	483	914			
150	284	537	1 016			
165	312	591	1 117			
180	341	644	1 219			
195	369	698	1 321			
210	397	752	1 422			
225	426	805	1 524			
240	454	859	1 625			
255	482	913				
270	511	967				
285	539	1 020				
300	568	1 074				
315	596	1 128				
330	624	1 181				
345	653	1 235				
360	681	1 289				
375	710	1 342				
390	738	1 396				
405	766	1 450				
420	795	1 503				
435	823	1 557				
450	851	1 611				
465	880	1 665				
480	908	1 718				
495	937					
510	965					
525	993					
540	1 022					
555	1 050					
570	1 078					
585	1 107					
600	1 135					

Table D.15
Two 65 mm and One 77 mm Hose
C-Factor: 0.253

Lay Length Meters	Flow in L/min					
	2 000	3 000	4 000	5 000	6 000	7 000
	Friction Loss in kPa					
15	15	34	61	95	137	186
30	30	68	121	190	273	372
45	46	102	182	285	410	558
60	61	137	243	380	546	744
75	76	171	304	474	683	930
90	91	205	364	569	820	1 116
105	106	239	425	664	956	1 302
120	121	273	486	759	1 093	1 488
135	137	307	546	854	1 230	1 674
150	152	342	607	949	1 366	
165	167	376	668	1 044	1 503	
180	182	410	729	1 139	1 639	
195	197	444	789	1 233		
210	213	478	850	1 328		
225	228	512	911	1 423		
240	243	546	972	1 518		
255	258	581	1 032	1 613		
270	273	615	1 093	1 708		
285	288	649	1 154			
300	304	683	1 214			
315	319	717	1 275			
330	334	751	1 336			
345	349	786	1 397			
360	364	820	1 457			
375	380	854	1 518			
390	395	888	1 579			
405	410	922	1 639			
420	425	956	1 700			
435	440	990				
450	455	1 025				
465	471	1 059				
480	486	1 093				
495	501	1 127				
510	516	1 161				
525	531	1 195				
540	546	1 230				
555	562	1 264				
570	577	1 298				
585	592	1 332				
600	607	1 366				

Table D.16
Two 77 mm Hose
C-Factor: 0.316

	Flow in L/min					
	2 000	3 000	4 000	5 000	6 000	7 000
Lay Length Meters	Friction Loss in kPa					
15	19	43	76	119	171	232
30	38	85	152	237	341	465
45	57	128	228	356	512	697
60	76	171	303	474	683	929
75	95	213	379	593	853	1 161
90	114	256	455	711	1 024	
105	133	299	531	830	1 194	
120	152	341	607	948		
135	171	384	683	1 067		
150	190	427	758	1 185		
165	209	469	834			
180	228	512	910			
195	246	555	986			
210	265	597	1 062			
225	284	640	1 138			
240	303	683	1 213			
255	322	725				
270	341	768				
285	360	811				
300	379	853				
315	398	896				
330	417	939				
345	436	981				
360	455	1 024				
375	474	1 067				
390	493	1 109				
405	512	1 152				
420	531	1 194				
435	550	1 237				
450	569					
465	588					
480	607					
495	626					
510	645					
525	664					
540	683					
555	702					
570	720					
585	739					
600	758					

Table D.17
Two 77 mm and One 65 mm Hose
C-Factor: 0.189

	Flow in L/min					
	2 000	3 000	4 000	5 000	6 000	7 000
Lay Length Meters	Friction Loss in kPa					
15	11	26	45	71	102	139
30	23	51	91	142	204	278
45	34	77	136	213	306	417
60	45	102	181	284	408	556
75	57	128	227	354	510	695
90	68	153	272	425	612	833
105	79	179	318	496	714	972
120	91	204	363	567	816	1 111
135	102	230	408	638	919	1 250
150	113	255	454	709	1 021	1 389
165	125	281	499	780	1 123	1 528
180	136	306	544	851	1 225	1 667
195	147	332	590	921	1 327	
210	159	357	635	992	1 429	
225	170	383	680	1 063	1 531	
240	181	408	726	1 134	1 633	
255	193	434	771	1 205	1 735	
270	204	459	816	1 276		
285	215	485	862	1 347		
300	227	510	907	1 418		
315	238	536	953	1 488		
330	249	561	998	1 559		
345	261	587	1 043	1 630		
360	272	612	1 089	1 701		
375	284	638	1 134			
390	295	663	1 179			
405	306	689	1 225			
420	318	714	1 270			
435	329	740	1 315			
450	340	765	1 361			
465	352	791	1 406			
480	363	816	1 452			
495	374	842	1 497			
510	386	868	1 542			
525	397	893	1 588			
540	408	919	1 633			
555	420	944	1 678			
570	431	970	1 724			
585	442	995				
600	454	1 021				

Table E.1
Effects of Altitude on Lift

ALTITUDE FT.	BAROMETER In. Hg.	READING mm Hg.	ATMOS. PRESS. psia	ft. H_2O	GAUGE READING In. Hg.	LIFT FT.	BOILING POINT of H_2O
-1000	31.0	788	15.2	35.2			213.8
-500	30.5	775	15.0	34.6			212.9
0	29.9	760	14.7	33.9	0	0	212.0
+500	29.4	747	14.4	33.3	.95	.6	211.1
+1000	28.9	734	14.2	32.8	1.0	1.1	210.2
1500	28.8	719	13.9	32.1	1.6	1.8	209.3
2000	27.8	706	13.7	31.5	2.1	2.4	208.4
2500	27.3	694	13.4	31.0	2.6	2.9	207.4
3000	26.8	681	13.2	30.4	3.1	3.5	206.5
3500	26.3	668	12.9	29.8	3.6	4.1	205.6
4000	25.8	655	12.7	29.2	4.1	4.7	204.7
4500	25.4	645	12.4	28.8	4.5	5.1	203.8
5000	24.9	633	12.2	28.2	5.0	5.7	202.9
5500	24.4	620	12.0	27.6	5.5	6.3	201.9
6000	24.0	610	11.8	27.2	5.9	6.7	201.0
6500	23.5	597	11.5	26.7	6.4	7.2	200.1
7000	23.1	587	11.3	26.2	6.8	7.7	199.2
7500	22.7	577	11.1	25.7	7.2	8.2	198.3
8000	22.2	564	10.9	25.2	7.7	8.7	197.4
8500	21.8	554	10.7	24.7	8.1	9.2	196.5
9000	21.4	544	10.5	24.3	8.5	9.6	195.5
9500	21.0	533	10.3	23.8	8.9	10.1	194.6
10,000	20.6	523	10.1	23.4	9.3	10.5	193.7
15,000	16.9	429	8.7	19.2	13.0	14.7	184.0

A

Accelerator — Device, usually in the form of a foot pedal, used to control the speed of a vehicle by regulating the fuel supply.

Acceptance Testing (Proof Test) — Preservice tests on fire apparatus or equipment performed at the factory or after delivery to assure the purchaser that the apparatus or equipment meets bid specifications.

Accessibility — Ability of fire apparatus to get close enough to a building to conduct emergency operations.

Accident — Unplanned, uncontrolled event that results from unsafe acts of people and/or unsafe occupational conditions, either of which can result in injury.

Adapter — Fitting for connecting hose couplings with dissimilar threads but with the same inside diameter.

Adjustable Flow Nozzle — Nozzle designed so that the amount of water flowing through the nozzle can be increased or decreased at the nozzle.

Aerial Apparatus — Fire fighting vehicle equipped with a hydraulically operated ladder or elevating platform for the purpose of placing personnel and/or water streams in elevated positions.

Aerial Device — General term used to describe the hydraulically operated ladder or elevating platform attached to a specially designed fire apparatus.

Aerial Ladder — Power-operated (usually hydraulically) ladder mounted on a special truck chassis.

Aerial Ladder Platform — Power-operated (usually hydraulically) ladder with a passenger-carrying device attached to the end of the ladder.

AFFF — Abbreviation for Aqueous Film Forming Foam.

Agent — Generic term used for materials that are used to extinguish fires.

Air-Aspirating Foam Nozzle — Foam nozzle especially designed to provide the aeration required to make the highest quality foam possible; most effective appliance for the generation of low-expansion foam.

Air Cascade System — Three or more large air cylinders, each usually with a capacity of 300 cubic feet (8 490 L), from which SCBA cylinders are recharged.

Air Chamber — Chamber filled with air that eliminates pulsations caused by the operation of piston or rotary-gear pumps.

Air-Supply Unit — Apparatus designed to refill exhausted SCBA air cylinders at the scene of an ongoing emergency.

Ammeter — Gauge that indicates both the amount of electrical current being drawn from and provided to the vehicle's battery.

Angle of Approach — Angle formed by level ground and a line from the point where the front tires of a vehicle touch the ground to the lowest projection at the front of the apparatus.

Angle of Departure — Angle formed by level ground and a line from the point where the rear tires of a vehicle touch the ground to the lowest projection at the rear of the apparatus.

Anti-Electrocution Platform — Slide-out platform mounted beneath the side running board or rear step of an apparatus equipped with an aerial device. This platform is designed to minimize the chance of the driver/operator being electrocuted should the aerial device come in contact with energized electrical wires or equipment.

Apparatus Bay (Apparatus Room) — Area of the fire station where apparatus are parked.

Apparatus Engine — Diesel or gasoline engine that powers the apparatus drive chain and associated fire equipment. Also called Power Plant.

Appliance — Generic term applied to any nozzle, wye, siamese, deluge monitor, or other piece of hardware used in conjunction with fire hose for the purpose of delivering water.

Applicator Pipe — Curved pipe attached to a nozzle for precisely applying water over a burning object.

Aqueous Film Forming Foam (AFFF) — Synthetic foam concentrate that, when combined with water, is a highly effective extinguishing and blanketing agent on hydrocarbon fuels.

ARFF — Acronym for Aircraft Rescue and Fire Fighting.

Articulating Aerial Platform — Aerial device that consists of two or more booms that are attached with hinges and operate in a folding manner. A passenger-carrying platform is attached to the working end of the device.

Articulating Boom — Arm portion of the articulating aerial platform.

Atmospheric Pressure — Pressure exerted by the atmosphere at the surface of the earth due to the weight of air. Atmospheric pressure at sea level is about 14.7 psi (101 kPa). Atmospheric pressure increases as elevation is decreased below sea level and decreases as elevation increases above sea level.

Attack Hose — Hose between the attack pumper and the nozzle(s); also, any hose used in a handline to control and extinguish fire. Minimum size is 1½ inch (38 mm).

Attack Lines — Hoselines or fire streams used to attack, contain, or prevent the spread of a fire.

Attack Pumper — Pumper that is positioned at the fire scene and is directly supplying attack lines.

Automatic Nozzle — Fog stream nozzle that automatically corrects itself to provide a good stream at the proper nozzle pressure.

Average Daily Consumption — Average of the total amount of water used each day during a one-year period.

B

Back Flushing — The cleaning of a fire pump or piping by flowing water through it in the opposite direction of normal flow.

Baffle — Intermediate partial bulkhead that reduces the surge effect in a partially loaded liquid tank.

Balanced Pressure Proportioner — A foam concentrate proportioner that operates in tandem with a fire water pump to ensure a proper foam concentrate-to-water mixture.

Ball Valve — Valve having a ball-shaped internal component with a hole through its center that permits water to flow through when aligned with the waterway.

Bed Ladder Pipe — Nontelescoping section of pipe, usually 3 or 3½ inches (77 mm or 90 mm) in diameter, attached to the underside of the bed section of the aerial ladder for the purpose of deploying an elevated master stream.

Bleeder Valve — Valve on a gate intake that allows air from an incoming supply line to be bled off before allowing the water into the pump.

Booster Apparatus — *See* Brush Apparatus.

Booster Hose — Fabric reinforced, rubber-covered, rubber-lined hose. Booster hose is generally carried on apparatus on a reel and is used for the initial attack and extinguishment of incipient and smoldering fires. Also called Hard Line and Red Line.

Booster Pump — Fire pump used to boost the pressure of the existing water supply within a fixed fire protection system.

Booster Reel — Mounted reel on which booster hose is carried.

Booster Tank — *See* Water Tank.

Bourdon Gauge — Most common device used to measure water system pressures.

Bourdon Tube — Part of a pressure gauge that has a curved, flat tube that changes its curvature as pressure changes. This movement is then transferred mechanically to a pointer on the dial.

Brake Limiting Valve — Valve that allows the vehicle's brakes to be adjusted for the current road conditions.

Braking Distance — Distance the vehicle travels from the time the brakes are applied until it comes to a complete stop.

Breakover Angle — Angle formed by level ground and a line from the point where the rear tires of a vehicle touch the ground to the bottom of the frame at the wheelbase midpoint.

British Thermal Unit (Btu) — Amount of heat energy required to raise the temperature of one pound of water one degree Fahrenheit. One Btu = 1.055 kJ.

Broken Stream — Stream of water that has been broken into coarsely divided drops.

Brush Apparatus — Fire department apparatus designed specifically for fighting wildland fires. Also called Booster Apparatus, Brush Patrol, Brush Pumper, and Field Unit.

Brush Patrol — *See* Brush Apparatus.

Brush Pumper — *See* Brush Apparatus.

Bumper — Structure designed to provide front or rear end protection of a vehicle.

Bumper Line — Preconnected hoseline located on the apparatus bumper.

Butterfly Valve — Type of control valve that uses a flat baffle operated by a quarter-turn handle.

C

Calibrate — To standardize or adjust the increments on a measuring instrument.

Canteen Unit — Emergency vehicle that provides food, drinks, and other rehabilitative services to emergency workers at extended incidents.

Capacity — Maximum ability of a pump or water distribution system to deliver water.

Cascade Air Cylinders — Large air cylinders that are used to refill smaller SCBA cylinders.

Cascade System — Three or more large air cylinders, each usually with a capacity of 300 cubic feet (8 490 L), that are interconnected and from which smaller SCBA cylinders are recharged.

Catch Basin — *See* Portable Tank.

Cellar Pipe — Special nozzle for attacking fires in basements, cellars, and other spaces below the attack level.

Centrifugal Pump — Pump with one or more impellers that utilizes centrifugal force to move the water. Most modern fire pumps are of this type.

Certification Tests — Preservice tests for aerial device, ladder, pump, and other equipment conducted by an independent testing laboratory prior to delivery of an apparatus. These tests assure that the apparatus or equipment will perform as expected after being placed into service.

Certified Shop Test Curves — Results, which are plotted on a graph, of the test performed by the manufacturer on its pump before shipping the pump.

C-Factor — Factor that indicates the roughness of the inner surface of piping or fire hose.

CFM — Abbreviation for cubic feet per minute.

Chamois — Soft pliant leather used for drying furniture and contents or for removing small amounts of water.

Charge — To pressurize a fire hose or fire extinguisher.

Charged Line — Hose loaded with water under pressure and prepared for use.

Chassis — Frame upon which the body of the fire apparatus rests.

Chauffeur — *See* Fire Apparatus Driver/Operator.

Check Valve — Automatic valve that permits liquid flow in only one direction.

Chemical Foam — Foam formed when an alkaline solution and an acid solution unite to form a gas (carbon dioxide) in the presence of a foaming agent that traps the gas in fire-resistive bubbles. Chemical foam is not commonly used today.

Circulating Feed — Fire hydrant that receives water from two or more directions.

Circulating System — *See* Loop System.

Circulation Relief Valve — Small relief valve that opens and provides enough water flow into and out of the pump to prevent the pump from overheating when it is operating at churn against a closed system.

Circulator Valve — Device in a pump that routes water from the pump to the supply to keep the pump cool when hoselines are shut down.

Cistern — Water storage receptacle that is usually underground and may be supplied by a well or rainwater run-off.

Clapper Valve — Hinged valve that permits the flow of water in one direction only.

Class A Foam — Foam specially designed for use on Class A combustibles. Class A foams are essentially wetting agents that reduce the surface tension of water and allow it to soak into combustible materials easier than plain water.

Coefficient of Discharge — Correction factor relating to the shape of the hydrant discharge outlet; used when computing the flow from a hydrant.

Combination Nozzle — Nozzle designed to provide a straight stream and a fog stream.

Commercial Chassis — Truck chassis produced by a commercial truck manufacturer. The chassis is in turn outfitted with a rescue or fire fighting body.

Compound Gauge — (1) Pressure gauge capable of measuring positive or negative pressures. (2) Term used to describe the gauge that measures the intake pressure on a fire pump.

Constant Pressure Relay — Method of establishing a relay water supply utilizing two or more pumpers to supply the attack pumper. This method reduces the need for time-consuming and often confusing fireground calculations of friction loss.

Coupling — Fitting permanently attached to the end of a hose; used to connect two hoselines together or a hoseline to such devices as nozzles, appliances, discharge valves, or hydrants.

D

Dead-End Hydrant — Fire hydrant that receives water from only one direction.

Dead-End Main — Water main that is not looped and in which water can flow in only one direction.

Deck Gun — *See* Turret Pipe.

Deck Pipe — *See* Turret Pipe.

Differential Manometer — Device whose primary application is to reflect the difference in pressures between two points in a system.

Direct Pumping System — Water supply system supplied directly by a system of pumps rather than elevated storage tanks.

Discharge Velocity — Rate at which water travels from an orifice.

Displacement — (1) Volume or weight of a fluid displaced by a floating body of equal weight. (2) Amount of water forced into the pump thus displacing air.

Distribution System — That part of an overall water supply system which receives the water from the pumping station and delivers it throughout the area to be served.

Distributor Nozzle — Nozzle used to create a broken stream that is usually used on basement fires.

Domestic Consumption — Water consumed from the water supply system by residential and commercial occupancies.

Draft — Process of obtaining water from a static source into a pump that is above the source's level. Atmospheric pressure on the water surface forces the water into the pump where a partial vacuum had been created.

Drafting Operation — *See* Draft.

Drafting Pit — Underground reservoir of water from which to draft for pumper testing; usually located at a training center.

Drain Valve — Valve on a pump discharge that facilitates the removal of pressure from a hoseline after the discharge has been closed.

Driver/Operator — *See* Fire Apparatus Driver/Operator.

Driver Reaction Distance — Distance a vehicle travels while a driver is transferring the foot from the accelerator to the brake pedal after perceiving the need for stopping.

Dry-Barrel Hydrant — Fire hydrant that has its opening valve at the water main rather than in the barrel of the hydrant. When operating properly, there is no water in the barrel of the hydrant when it is not in use. These hydrants are used in areas where freezing could occur.

Dry Hoseline — Hoseline without water in it; an uncharged hoseline.

Dry Hydrant — Permanently installed pipe that has pumper suction connections installed at static water sources to speed drafting operations.

Dry Standpipe System — Standpipe system that has closed water supply valves or that lacks a fixed water supply.

Dual Pumping — Operation where a strong hydrant is used to supply two pumpers by connecting the pumpers intake-to-intake. The second pumper receives the excess water not being pumped by the first pumper, which is directly connected to the water supply source. Sometimes incorrectly referred to as tandem pumping.

Dump Line — *See* Waste Line.

E

Eductor — Proportioning device that injects foam concentrate into the water flowing through a hoseline or pipe.

Elevated Master Stream — Fire stream in excess of 350 gpm (1 400 L/min) that is deployed from the tip of an aerial device.

Elevation Loss — *See* Elevation Pressure.

Elevation Pressure — Gain or loss of pressure in a hoseline due to a change in elevation. Also called Elevation Loss.

Elliptical — Describes a large cylindrical, oblong water tank that is used on tankers or tender.

Emergency Truck — Van or similar-type vehicle used to carry portable equipment and personnel.

Engine — Fire department pumper.

Engineer — *See* Fire Apparatus Driver/Operator.

Engine Pressure — *See* Net Pump Discharge Pressure.

External Water Supply — (1) Any water supply to a fire pump from a source other than the vehicle's own water tank. (2) Any water supply to an aerial device from a source other than the vehicle's own fire pump.

F

FDC — Abbreviation for Fire Department Connection.

Feeder Line — *See* Relay-Supply Hose.

Feed Main — Pipe connecting the sprinkler system riser to the cross mains. The cross mains directly service a number of branch lines on which the sprinklers are installed.

Female Coupling — Threaded swivel device on a hose or appliance made to receive a male coupling of the same thread and diameter.

Fender — Exterior body portion of a vehicle adjacent to the front or rear wheels.

FFFP — Abbreviation for film forming fluoroprotein foam.

Field Unit — *See* Wildland Fire Apparatus.

Fill Hose — Short section of hose carried on apparatus equipped with booster tanks to fill the tank from a hydrant or another truck.

Fill Opening — Opening on top of a tank used for filling the tank; usually incorporated in manhole cover.

Fill Site — Location at which tankers/tenders will be loaded during a water shuttle operation.

Film Forming Fluoroprotein Foam (FFFP) — Foam concentrate that is based on fluoroprotein foam technology with aqueous film forming foam (AFFF) capabilities.

Finished Foam — Completed product after the foam solution reaches the nozzle and air is introduced into the solution (aeration). Also simply call *foam*.

Fire Apparatus — Any fire department emergency vehicle used in fire suppression or other emergency situation.

Fire Apparatus Driver/Operator — Firefighter who is charged with the responsibility of operating fire apparatus to, during, and from the scene of a fire operation or any other time the apparatus is in use. The driver/operator is also responsible for routine maintenance of the apparatus and any equipment carried on the apparatus. This is typically the first step in the fire department promotional chain. Also called Chauffeur or Engineer.

Fire Boat — Boat that carries large fire pumps and is capable of supplying boat-mounted master streams or water supply hoselines to land-based fire fighting apparatus. Also called Marine Unit.

Fire Department Connection (FDC) — Point at which the fire department can connect into a sprinkler or standpipe system to boost the water flow in the system. This connection consists of a clappered siamese with two or more 2½-inch (65 mm) intakes or one large-diameter (4-inch [100 mm] or larger) intake. Also called Fire Department Sprinkler Connection.

Fire Department Pumper — Piece of fire apparatus having a permanently mounted fire pump with a rated discharge capacity of 750 gpm (3 000 L/min) or greater. This apparatus may also carry water, hose, and other portable equipment.

Fire Department Sprinkler Connection — *See* Fire Department Connection.

Fire Department Water Supply Officer — Officer in charge of all water supplies at the scene of a fire; duties include placing pumpers at the most advantageous hydrants or other water sources and directing supplementary water supplies, including water shuttles and relay pumping operations. This may also be a permanent, full-time staff position with responsibility for coordinating, with other local agencies, water supply projects of concern to the fire department.

Fire Hydraulics — Science that deals with water in motion as it applies to fire fighting operations.

Fire Pump — Water pump on a piece of fire apparatus.

Fire Stream — Stream of water or other water-based extinguishing agent after it leaves the fire hose and nozzle until it reaches the desired point.

Floating Dock Strainer — Strainer designed to float on top of the water; used for drafting operations. This eliminates the problem of drawing debris into the pump and reduces the required depth of water needed for drafting.

Flowmeter — Mechanical device installed in a discharge line that senses the amount of water flowing and provides a readout in units of gallons per minute (liters per minute).

Flow Pressure — Pressure created by the rate of flow or velocity of water coming from a discharge opening. Also called Plug Pressure.

Flow Test — Tests conducted to establish the capabilities of water supply systems. The objective of a flow test is to establish quantity (gallons or liters per minute) and pressures available at a specific location on a particular water supply system.

Fluoroprotein Foams — Foam concentrates fortified with fluorinated surfactants. These surfactants enable the foam to shed, or separate from, hydrocarbon fuels.

Foam — Extinguishing agent formed by mixing a foam concentrate with water and aerating the solution for expansion; for use on Class A and Class B fires. Foam may be protein, synthetic, aqueous film forming, high expansion, or alcohol type.

Foam Blanket — Covering of foam applied over a burning surface to produce a smothering effect; can be used on nonburning surfaces to prevent ignition.

Foam Concentrate — Raw chemical compound solution that is mixed with water and air to produce foam.

Foam Eductors — Type of foam proportioner used for mixing foam concentrate in proper proportions with a stream of water to produce foam solution.

Foam Proportioner — Device that injects the correct amount of foam concentrate into the water stream to make the foam solution.

Foam Solution — Mixture of foam concentrate and water after it leaves the proportioner but before it is discharged from the nozzle and air is added to it.

Fog Stream — Water stream of finely divided particles used for fire control.

Fold-A-Tank — *See* Portable Tank.

Four-Way Hydrant Valve — Device that permits a pumper to boost the pressure in a supply line connected to a hydrant without interrupting the water flow.

Friction Loss — Loss of pressure created by the turbulence of water moving against the interior walls of the hose or pipe.

Front Bumper Well — Hose or tool compartment built into the front bumper of a fire apparatus.

Front-Mount Pump — Fire pump mounted in front of the radiator of a vehicle and powered off the crankshaft.

G

Gallon — Unit of liquid measure. One U.S. gallon (3.785 L) has the volume of 231 cubic inches (3 785 cubic centimeters). One imperial gallon equals 1.201 U.S. gallons (4.546 L).

Gallons Per Minute (gpm) — Unit of volume measurement used in the U.S. fire service for water movement.

Gaskets — Rubber seals used in fire hose couplings and pump intakes to prevent the leakage of water at connections.

Gated Wye — Hose appliance with one female inlet and two or more male outlets with a gate valve on each outlet.

Gate Valve — Control valve with a solid plate operated by a handle and screw mechanism. Rotating the handle moves the plate into or out of the waterway.

Gauge — Instrument used to show the operating conditions of an appliance or piece of equipment.

Generator — Auxiliary electrical power generating device. Portable generators are powered by small gasoline or diesel engines and generally have 110- and/or 220-volt capacities.

Governor — Built-in pressure-regulating device to control pump discharge pressure by limiting engine rpm.

GPM — Abbreviation for gallons per minute.

Gradability — Ability of a piece of apparatus to traverse various terrain configurations.

Grade — Natural, unaltered ground level.

Gravity System — Water supply system that relies entirely on the force of gravity to propel the water throughout the system. This type of system is generally used in conjunction with an elevated water storage source.

Gravity Tank — Elevated water storage tank for fire protection and community water service. A water level of 100 feet (30 m) provides a static pressure head of 43.4 psi (300 kPa) minus friction losses in piping when water is flowing.

Grid System Water Mains — Interconnecting system of water mains in a crisscross or rectangular pattern.

H

Handline — Small hoselines (2½ inch [65 mm] or less) that can be handled and maneuvered without mechanical assistance.

Hard Intake Hose — Noncollapsible hose that connects a pump to a source of water and is used for drafting.

Hard Sleeve — *See* Hard Intake Hose.

Hard Suction Hose — *See* Hard Intake Hose.

Head — Water pressure due to elevation. For every 1-foot increase in elevation, 0.434 psi is gained (for every 1-meter increase in elevation, 10 kPa is gained). Also called Head Pressure.

Heavy Rescue Vehicle — Large rescue vehicle that may be constructed on a custom or commercial chassis. Additional equipment carried by the heavy rescue unit includes A-frames or gin poles, cascade systems, larger power plants, trench and storing equipment, small pumps and foam equipment, large winches, hydraulic booms, large quantities of rope and rigging equipment, air compressors, and ladders.

Heavy Stream — *See* Master Stream.

High-Pressure Fog — Fog stream operated at high pressures and discharged through small diameter hose.

High-Pressure Hose — Hose leading from the air cylinder to the regulator; may be at cylinder pressure or reduced to some lower pressure.

High-Pressure Nozzle — Fire stream nozzle that is designed to be operated in excess of the 100 psi (700 kPa) to which ordinary fog nozzles are designed.

Hose Bed — Main hose-carrying area of a pumper or other piece of apparatus designed for carrying hose. Also called Hose Body.

Hose Clamp — Mechanical or hydraulic device used to compress fire hose to stop the flow of water.

Hose Control Device — Device used to hold a charged hoseline in a stationary position for an extended period of time.

Hose Couplings — Metal fasteners used to connect fire hose together.

Hydrant Pressure — Amount of pressure being supplied by a hydrant without assistance.

Hydrant Wrench — Specially designed tool used to open or close a hydrant and to remove hydrant caps.

Hydraulics — (1) Operated, moved, or effected by means of water. (2) Of or relating to water or other liquid in motion. (3) Operated by the resistance offered or the pressure transmitted when a quantity of liquid is forced through a comparatively small orifice or through a tube. (4) Branch of fluid mechanics dealing with the mechanical properties of liquids and the application of these properties in engineering.

I

Impeller — Vaned, circulating member of the centrifugal pump that transmits motion to the water.

Impeller Eye — Intake orifice at the center of a centrifugal pump impeller.

Impinging Stream Nozzle — Nozzle that drives several jets of water together at a set angle in order to break water into finely divided particles.

Increaser — Adapter used to attach a larger hoseline to a smaller one. The increaser has female threads on the smaller side and male threads on the larger side.

Indicating Valve — Water main valve that visually shows the open or closed status of the valve.

Initial Attack Apparatus — Fire apparatus whose primary purpose is to initiate a fire attack on structural and wildland fires and support associated fire department actions. Also called minipumpers or midipumpers.

In-Line Eductor — Eductor that is placed along the length of a hoseline.

In-Line Relay Valve — Valve placed along the length of a supply hose that permits a pumper to connect to the valve to boost pressure in the hose.

Intake — Inlet for water into the fire pump.

Intake Pressure — Pressure coming into the fire pump.

Intake Relief Valve — Valve designed to prevent damage to a pump from water hammer or any sudden pressure surge.

Intake Screen — Screen used to prevent foreign objects from entering a pump.

Inverter — Auxiliary electrical power generating device. The inverter is a step-up transformer that converts the vehicle's 12- or 24-volt DC current into 110- or 220-volt AC current.

J

Jet Siphon — Section of pipe or hard suction hose with a 1½-inch (38 mm) discharge line inside that bolsters the flow of water through the tube. The jet siphon is used between portable tanks to maintain a maximum amount of water in the tank from which the pumper is drafting.

Jump Seat — Seats on a fire apparatus that are behind the front seats.

K

Kilopascal (kPa) — Metric unit of measure for pressure; 1 psi = 6.895 kPa, 1 kPa = 0.1450 psi.

Kink — Severe bend in a hoseline that increases friction loss and reduces the flow of water through the hose.

L

Ladder Pipe — Master stream nozzle mounted on the fly of an aerial ladder.

Large Diameter Hose (LDH) — Relay-supply hose of 3½ to 6 inches (90 mm to 150 mm); used to move large volumes of water quickly with a minimum number of pumpers and personnel.

Level I Staging — Used on all multiple-company emergency responses. The first-arriving vehicles of each type proceed directly to the scene, and the others stand by a block or two from the scene and await orders.

Level II Staging — Used on large-scale incidents where greater alarm companies are responding. These companies are sent to a specified location to await assignment.

Lift, Dependable — Height a column of water may be lifted in sufficient quantity to provide a reliable fire flow. Lift may be raised through a hard suction hose to a pump, taking into consideration the atmospheric pressure and friction loss within the hard suction hose. Dependable lift is usually considered to be 14.7 feet (4.5 m).

Lift, Maximum — Maximum height to which any amount of water may be raised through a hard suction hose to a pump.

Lift, Theoretical — Theoretical, scientific height that a column of water may be lifted by atmospheric pressure in a true vacuum. At sea level, this height is 33.8 feet (10 m). The height will decrease as elevation increases.

Light Attack Vehicle — *See* Initial Attack Apparatus.

Light Rescue Vehicle — Small rescue vehicle usually built on a 1-ton or 1½-ton chassis. It is designed to handle only basic extrication and life-support functions and carries only basic hand tools and small equipment.

Loading Site — *See* Fill Site.

Location Marker — Device, such as a reflective marker or flag, used to mark the location of a fire hydrant for quicker identification during a fire response.

Lugging — Condition that occurs when the throttle application is greater than necessary for a given set of conditions. It may result in an excessive amount of carbon particles issuing from the exhaust, oil dilution, and additional fuel consumption. Lugging can be eliminated by using a lower gear and proper shifting techniques.

M

Maintenance — Keeping equipment or apparatus in a state of usefulness or readiness.

Male Coupling — Hose nipple with protruding threads that fits into the thread of a female coupling of the same pitch and appropriate diameter and thread count.

Manifold — (1) Hose appliance that divides one larger hoseline into three or more small hoselines. Also called Portable Hydrant. (2) Hose appliance that combines three or more smaller hoselines into one large hoseline. (3) Top portion of the pump casing.

Manufacturer's Tests — Fire pump or aerial device tests performed by the manufacturer prior to delivery of the apparatus.

Mars Light — Single-beam, oscillating warning light.

Master Stream — Any of a variety of heavy, large-caliber water streams; usually supplied by siamesing two or more hoselines into a manifold device delivering 350 gpm (1 400 L/min) or more. Also called Heavy Stream.

Master Stream Nozzle — Nozzle capable of flowing in excess of 350 gpm (1 400 L/min).

Mattydale Hose Bed — *See* Transverse Hose Bed.

Maximum Daily Consumption — Maximum total amount of water used during any 24-hour interval over a 3-year period.

MDH — *See* Medium Diameter Hose.

Mechanical Foam — Foam that requires the blending of water, foam concentrate, and air to be developed.

Medium Diameter Hose (MDH) — 2½- or 3-inch (65 mm or 77 mm) hose that is used for both fire fighting attack and for relay-supply purposes.

Medium Rescue Vehicle — Rescue vehicle somewhat larger and better equipped than a light rescue vehicle. This vehicle may carry powered hydraulic spreading tools and cutters, air bag lifting systems, power saws, oxyacetylene cutting equipment, ropes and rigging equipment, as well as basic hand equipment.

Midipumper — *See* Initial Attack Apparatus.

Midship Pump — Fire pumps mounted at the center of the fire apparatus.

Minipumper — *See* Initial Attack Apparatus.

Mobile Water Supply Apparatus — Fire apparatus with a water tank of 1,000 gallons (4 000 L) or larger whose primary purpose is transporting water. The truck may also carry a pump, some hose, and other equipment. Also referred to as tenders or tankers.

Monitor — Master stream appliance whose stream direction can be changed while water is being discharged. They can be fixed, portable, or a combination.

Mystery Nozzle — Older style variable gallonage adjustable fog stream nozzle.

N

National Fire Protection Association (NFPA) — Nonprofit educational and technical association located in Quincy, Massachusetts devoted to protecting life and property from fire by developing fire protection standards and educating the public.

National Standard Thread (NST) — Screw thread of specific dimensions for fire service use as specified in NFPA 1963, *Standard for Fire Hose Connections*.

Net Pressure — *See* Net Pump Discharge Pressure.

Net Pump Discharge Pressure (NPDP) — Actual amount of pressure being produced by the pump. When taking water from a hydrant, it is the difference between the intake pressure and the discharge pressure. When drafting, it is the sum of the intake pressure and the discharge pressure. (**NOTE:** Intake pressure is credited for lift and intake hose friction loss and is added to the discharge pressure.) Also called Net Pressure or Engine Pressure.

Nonconforming Apparatus — Apparatus that does not conform to the standards set forth by NFPA standards.

Nozzle — Appliance on the discharge end of a hoseline that forms a fire stream of definite shape, volume, and direction.

Nozzle Pressure — Velocity pressure at which water is discharged from the nozzle.

Nozzle Reaction — Counterforce directed against a person holding a nozzle or a device holding a nozzle by the velocity of water being discharged.

NST — Abbreviation for National Standard Thread.

Nurse Tanker — *See* Nurse Tender.

Nurse Tender — Very large mobile water supply apparatus that is stationed at the fire scene and serves as a portable reservoir rather than as a shuttle tender. Also called Nurse Tanker.

O

Overthrottling — Process of injecting or supplying the diesel engine with more fuel than can be burned.

P

Pattern — Shape of the water stream as it is discharged from a fog nozzle.

PDP — Abbreviation for Pump Discharge Pressure.

Peak Hourly Consumption — Maximum amount of water used during any hour of a day.

Piercing Nozzle — Nozzle with an angled, case-hardened steel tip that can be driven through a wall, roof, or ceiling to extinguish hidden fire. Also called Puncture Nozzle.

Piezometer Tube — Device that uses the heights of liquid columns to illustrate the pressures existing in hydraulic systems.

Piston Pump — Positive-displacement pump using one or more reciprocating pistons to force water from the pump chambers.

Piston Valve — Valve with an internal piston that moves within a cylinder to control the flow of water through the valve.

Pitot Tube — Instrument containing a Bourdon tube that is inserted into a stream of water to measure the velocity pressure of the stream. The gauge reads in units of pounds per square inch (psi) or kilopascals (kPa).

Playpipe — Base part of a three-part nozzle that extends from the hose coupling to the shutoff.

Portable Basin — *See* Portable Tank.

Portable Hydrant — *See* Manifold (1).

Portable Ladder Pipe — Portable, elevated master stream device clamped to the top two rungs of the aerial ladder when needed and supplied by a 3- or 3½-inch (77 mm or 90 mm) fire hose.

Portable Pump — Small fire pump available in several volume and pressure ratings that can be removed from the apparatus and taken to a water supply inaccessible to the main pumper.

Portable Source — Water that is mobile and may be taken directly to the location where it is needed. This may be a fire department tanker or some other vehicle that is capable of hauling a large quantity of water.

Portable Tank — Collapsible storage tank used during a relay or shuttle operation to hold water from water tanks or hydrants. This water can then be used to supply attack apparatus. Also called Catch Basin, Fold-a-Tank, Portable Basin, Portable Water Tank, or Porta-Tank.

Porta-Tank — *See* Portable Tank.

Positive Displacement Pumps — Self-priming pump that moves a given amount of water or hydraulic oil through the pump chamber with each stroke or rotation. These pumps are used for hydraulic pumps on aerial device hydraulic systems and for priming pumps on centrifugal fire pumps.

Pounds Per Square Inch (psi) — U.S. unit for measuring pressure. Its metric equivalent is kilopascals.

Power Plant — *See* Apparatus Engine.

Power Take-Off (PTO) — Rotating shaft that transfers power from the engine to auxiliary equipment.

Preconnect — (1) Attack hose connected to a discharge when the hose is loaded; this shortens the time it takes to deploy the hose for fire fighting. (2) Soft intake hose that is carried connected to the pump intake.

Preservice Tests — Tests performed on fire pumps or aerial devices before they are placed into service. These tests are broken down into manufacturer's tests, certification tests, and acceptance tests.

Pressure — Force per unit area measured in pounds per square inch (psi) or kilopascals (kPa).

Pressure Governor — Pressure control device that controls engine speed and therefore eliminates hazardous conditions that result from excessive pressures.

Pressure Operation — Operation of a two- (or more) stage centrifugal pump in which water passes consecutively through each impeller to provide high pressures at a reduced volume. Also called Series Operation.

Pressure-Relief Device — Automatic device designed to release excess pressure from a fire pump.

Primary Feeder — Large pipes (mains), with relatively widespread spacing, that convey large quantities of water to various points of the system for local distribution to the smaller mains.

Prime — To remove all air from a pump and intake hose in preparation for receiving water under pressure.

Primer — *See* Priming Pump.

Primer Fluid Tank — Tank of fluid used to seal and lubricate the priming pump.

Priming Pump — Small positive-displacement pump used to evacuate air from a centrifugal pump housing and hard suction hose. Evacuating air allows the centrifugal pump to receive water from a static water supply source. Also called Primer.

Private Connection — Connections to water supplies other than the standard municipal water supply system; may include connection within a large industrial facility, a farm, or a private housing development.

Private Hydrant — Hydrant provided on private property or on private water systems to protect private property. Also called Yard Hydrant.

Proportioner — Device used to introduce the correct amount of foam concentrate into a stream of water.

Proportioning Valve — Valve used to balance or divide the air supply between the aeration system and the discharge manifold of a foam system.

Protein Foam — Protein foams are chemically broken down (hydrolyzed) protein solids. The end product of this chemical digestion is protein liquid concentrate.

PSI — Abbreviation for Pounds Per Square Inch.

PSIG — Abbreviation for Pounds Per Square Inch Gauge.

PTO — Abbreviation for Power Take-Off.

Pump and Roll — Ability of an apparatus to pump water while the vehicle is in motion.

Pump Can — Water-filled pump-type extinguisher. Also called Pump Tank.

Pump Capacity Rating — Maximum amount of water a pump will deliver at the indicated pressure.

Pump Charts — Charts carried on a fire apparatus to aid the pump operator in determining the proper pump discharge pressure to use when supplying hoselines.

Pump Discharge Pressure (PDP) — Actual velocity pressure (measured in pounds per square inch) of the water as it leaves the pump and enters the hoseline.

Pump Drain — Drain located at the lowest part of the pump to help remove all water from the pump. This eliminates the danger of damage due to freezing.

Pumper/Tanker — A mobile water supply apparatus equipped with a fire pump. In some jurisdictions, this term is used to differentiate a fire pump equipped mobile water supply apparatus whose main purpose is to attack the fire.

Pumping Apparatus — Fire department apparatus that has the primary responsibility to pump water.

Pump Operator — Firefighter charged with operating the pump and determining the pressures required to operate it efficiently.

Pump Panel — Instrument and control panel located on the pumper.

Puncture Nozzle — *See* Piercing Applicator Nozzle.

Q

Quad — Four-way combination fire apparatus; sometimes referred to as quadruple combination. A quad combines the water tank, pump, and hose of a pumper with the ground ladder complement of a truck company.

Quint — Fire apparatus equipped with a fire pump, water tank, ground ladders, and hose bed in addition to the aerial device.

R

Rapid Intervention Vehicle (RIV) — Small, quick-response fire fighting vehicle carrying at least 600 gallons (2 400 L) of water for producing AFFF, in addition to at least 500 pounds (225 kg) of either dry chemical or Halon 1211.

Reducing Wye — Wye that has two outlets smaller in diameter than the inlet valve. Sometimes called a Leader Line Wye.

Relay — Use of two or more pumpers to move water distances that would require excessive pressures if only one pumper was employed.

Relay Operation — Using two or more pumpers to move water over a long distance by operating them in series. Water discharged from one pumper flows through hoses to the inlet of the next pumper, and so on. Also called Relay Pumping.

Relay Pumping — *See* Relay Operation.

Relay-Supply Hose — Hose between the water source and the attack pumper, laid to provide large volumes of water at low pressure. Also called Feeder Line or Supply Hose.

Relay Valve — Pressure-relief device on the supply side of the pump designed to protect the hose and pump from damaging pressure surges common in relay pumping operations.

Relay Valve, In-Line — Special valve that is inserted in the middle of a long relay hose. This valve allows an additional pumper to connect to the line to boost pressure without having to interrupt the current flow of water.

Relief Valve — Pressure control device designed to eliminate hazardous conditions resulting from excessive pressures by allowing this pressure to bypass to the intake side of the pump.

Rescue Pumper — Specially designed apparatus that combines the functions of both a rescue vehicle and a fire department pumper.

Responder Unit — Emergency medical unit that carries first aid and/or advanced life support equipment but is not equipped for patient transport.

Road Tests — Preservice apparatus maneuverability tests designed to determine the road worthiness of a new vehicle.

Rotary Gear Positive Displacement Pump — Type of positive displacement pump commonly used in hydraulic systems. The pump imparts pressure on the hydraulic fluid by having two intermeshing rotary gears that force the supply of hydraulic oil into the pump casing chamber.

Rotary Vane Pump — Type of positive displacement pump used commonly in hydraulic systems. A rotor with attached vanes is mounted off-center inside the pump housing. Pressure is imparted on the water as the space between the rotor and the pump housing wall decreases.

S

Series Operation — *See* Pressure Operation.

Service Test — Series of tests performed on apparatus and equipment in order to ensure operational readiness of the unit. These tests should be performed at least yearly or whenever a piece of apparatus or equipment has undergone extensive repair.

Sexless Coupling — Coupling with no distinct male or female components. Also called Storz Coupling.

Shutoff Nozzle — Type of nozzle that has a valve or other device for controlling the water supply. Firefighters use it to control water supply at the nozzle rather than at the source of supply.

Siamese — Hose appliance used to combine two or more hoselines into one. The siamese generally has female inlets and a male outlet and is commonly used to supply the hose leading to a ladder pipe.

Single-Stage Centrifugal Pump — Centrifugal pump with only one impeller.

Siphon — Section of hard suction hose or piece of pipe used to maintain an equal level of water in two or more portable tanks.

Small Diameter Hose (SDH) — Hose of ¾ to 2 inches (20 mm to 50 mm) in diameter; used for fire fighting purposes.

Soft Sleeve Hose — Large diameter, collapsible piece of intake hose used to connect a fire pump to a pressurized water supply source; sometimes incorrectly referred to as "soft suction hose."

Solid Stream — Hose stream that stays together as a solid mass as opposed to a fog or spray stream.

Spanner Wrench — Small tool primarily used to tighten or loosen hose couplings; can also be used as a prying tool or a gas key.

Speedometer — Dashboard gauge that measures the speed at which the vehicle is traveling.

Spotter — Firefighter who walks behind a backing apparatus to provide guidance for the driver/operator.

Spotting — Positioning the apparatus in a location that provides the utmost efficiency for operating on the fireground.

Sprinkler Connection — *See* Fire Department Connection.

Staging — Process by which noncommitted units responding to a fire or other emergency incident are stopped at a location away from the fire scene to await their assignment.

Staging Area — (1) Location away from the emergency scene where units assemble and wait until they are assigned a position on the emergency scene. (2) Location on the emergency scene where tools and personnel are assembled before being used or assigned.

Standard Apparatus — Apparatus that conforms to the standards set forth by the National Fire Protection Association standards on fire apparatus design.

Standpipe System—Wet or dry system of pipes in a large single-story or multistory building with fire hose outlets connected to them. The system is used to provide for quick deployment of hoselines during fire fighting operations.

Static Source—Body of water that is not under pressure or in a supply piping system and must be drafted from in order to be used. Static sources include ponds, lakes, rivers, wells, and so on.

Steamer Connection—Large-diameter outlet, usually 4½ inches (115 mm), at a hydrant or the base of an elevated water storage container.

Storz Coupling—Sexless coupling commonly found on large diameter hose.

Strainers—Wire or other metal guards used to prevent debris from clogging the intake hose of fire pumps.

Supply Hose—*See* Relay-Supply Hose.

T

Tachometer—Dashboard or pump panel gauge that measures the engine speed in revolutions per minute (rpm).

Tailboard—Back step of fire apparatus.

Tandem—Two-axle suspension.

Tandem Pumping—Short relay operation in which the pumper taking water from the supply source pumps into the intake of the second pumper. The second pumper boosts the pressure of the water even higher. This method is used when pressures higher than the capability of a single pump are required.

Tanker—(1) *See* Mobile water supply fire apparatus. (2) In ICS terms, tanker refers to a water-transporting fixed-wing aircraft.

Tanker/Pumper—A mobile water supply apparatus equipped with a fire pump. In some jurisdictions, this term is used to differentiate a fire pump equipped mobile water supply apparatus whose main purpose is to shuttle water.

Tanker Shuttle Operation—*See* Water Shuttle Operation.

Telescoping Aerial Platform Apparatus—Type of aerial apparatus equipped with an elevating platform; also equipped with piping systems and nozzles for elevated master stream operations. These apparatus are not meant to be climbed and are equipped with a small ladder that is to be used only for escape from the platform in emergency situations.

Telescoping Boom—Aerial device raised and extended via sections that slide within each other.

Tender—Term used within ICS for a mobile piece of apparatus that has the primary function of supporting another operation. Examples include a water tender that supplies water to pumpers, a fuel tender that supplies fuel to other vehicles, etc. *Also see* Tanker (1) and Mobile Water Supply Apparatus.

Third-Party Testing Agency—Independent agency hired to perform nonbiased testing on a specific piece of apparatus.

Threaded Coupling—Male or female coupling with a spiral thread.

Throttle Control—Device that controls the engine speed.

Total Pressure—Total amount of pressure loss in a hose assembly due to friction loss in the hose and appliances, elevation losses, or any other factors.

Total Stopping Distance—Sum of the driver/operator reaction distance and the vehicle braking distance.

Tower Ladder—Term used to describe a telescoping aerial platform fire apparatus.

Traction—Act of exerting a pulling force.

Tractor-Tiller Aerial Ladder—Aerial ladder apparatus that consists of a tractor power unit and trailer (tiller) section that contains the aerial ladder, ground ladders, and equipment storage areas. The trailer section is steered independently of the tractor by a person called the tiller operator.

Traffic Control Device—Mechanical device that automatically changes traffic signal lights to favor the path of responding emergency apparatus.

Transfer Valve—Valve used for placing multistage centrifugal pumps in volume or pressure operation.

Transverse Hose Bed—Hose bed that lies across the pumper body at a right angle to the main hose bed; designed to deploy preconnected attack hose to the sides of the pumper. Also called Mattydale Hose Bed.

Trash Line—Small diameter, preconnected hoseline intended to be used for trash or other small, exterior fires.

Triple-Combination Pumper—Fire department pumper that carries a fire pump, hose, and a water tank.

Triple Hydrant—Fire hydrant having three outlets, usually two 2½-inch (65 mm) outlets and one 4½-inch (115 mm) outlet.

Turret Pipe—Large master stream appliance mounted on a pumper or trailer and connected directly to a pump. Also called Deck Gun or Deck Pipe.

Two-Stage Centrifugal Pump — Centrifugal pump with two impellers.

UL — Abbreviation for Underwriters Laboratories, Inc.

Ultimate Capacity — Total capacity of a water supply system, including residential and industrial consumption, available fire flow, and all other taxes on the system.

Undercarriage — Portion of a vehicle's frame that is located beneath the vehicle.

Underwriters Laboratories, Inc. (UL) — Independent fire research and testing laboratory.

Unlined Hose — Fire hose without a rubber lining; most frequently used in interior standpipe systems and in wildland fire fighting.

Unloading Site — Place in the water shuttle operation where tankers unload their water into portable tanks. Also called the Dump Site.

Vacuum — Space completely devoid of matter or pressure. In fire service terms, it is more commonly used to describe a pressure that is somewhat less than atmospheric pressure. A vacuum is needed to facilitate drafting of water from a static source.

Valve — Mechanical device with a passageway that controls the flow of a liquid or gas.

Velocity — Speed; the rate of motion in a given direction. It is measured in feet per second (meters per second), miles per hour (kilometers per hour), and so on.

Venturi Principle — When a fluid is forced under pressure through a restricted orifice, there is a decrease in the pressure exerted against the side of the constriction and a corresponding increase in the velocity of the fluid. Because the surrounding air is under greater pressure, it rushes into the area of lower pressure.

Voltmeter — Device used for measuring the voltage existing on an electrical system.

Volume Operation — *See* Parallel Operation.

Volute — Spiral, divergent chamber of a centrifugal pump in which the velocity energy given to water by the impeller blades is converted in to pressure.

Wagon — (1) Water supply piece of apparatus in a two-piece engine company. (2) Special piece of fire apparatus that carries a large quantity of hose.

Wall Hydrant — Hydrant that protrudes through the wall of a building or pump house.

Warning Devices — Any audible or visual devices, such as flashing lights, sirens, horns, or bells, added to an emergency vehicle to gain the attention of drivers of other vehicles.

Warning Lights — Lights on the apparatus designed to attract the attention of other motorists.

Waste Line — Hoseline that is tied off or otherwise secured and is used to handle water in excess of that being used during a relay operation. Also known as a Dump Line.

Water Curtain — Fan-shaped stream of water applied between a fire and an exposed surface to prevent the surface from igniting from radiated heat.

Water Department — Municipal authority responsible for the water supply system in a given community.

Water Distribution System — System designed to supply water for residential, commercial, industrial, and/or fire protection purposes. This water supply is delivered through a network of piping and pressure-developing equipment.

Water Hammer — Force created by the rapid deceleration of water. It generally results from closing a valve or nozzle too quickly.

Water Shuttle Operation — Method of water supply by which tenders/tankers continuously transport water between a fill site and the dump site located near the emergency scene.

Water Supply — Any source of water available for use in fire fighting operations.

Water Supply Pumper — Pumper that takes water from a source and sends it to attack pumpers operating at the fire scene.

Water Tank — Water storage receptacle carried directly on the apparatus. NFPA 1901 specifies that Class A pumpers must carry at least 500 gallons (2 000 L). Also called Booster Tank.

Water Thief — This variation of a wye adapter has three gated outlets, usually two 1½-inch (38 mm) outlets and one 2½-inch (65 mm) outlet. There is a single inlet for 2½-inch (65 mm) or larger hose.

Water Tower — Aerial device primarily intended for deploying an elevated master stream. Not generally intended for climbing operations. Also known as an Elevating Master Stream Device.

Waterway — Path through which water flows within a hose or pipe.

Wet-Barrel Hydrant — Fire hydrant that has water all the way up to the discharge outlets. The hydrant may have separate valves for each discharge or one valve for all the discharges. This type of hydrant is only used in areas where there is no danger of freezing weather conditions.

Wetting Agent — Chemical solution added to water to reduce its surface tension and improve its penetrating ability; detergent is a mild form of wetting agent.

Wet Water — Wetting agent that is introduced to water to reduce the surface tension and improve its penetration qualities.

Wheel Blocks — *See* Chocks.

Wildland Fire Apparatus — A fire apparatus that is designed especially for use in fighting wildland fires. *See* Initial Fire Attack Apparatus and Brush Pumper.

Winch — Pulling tool that consists of a length of steel chain or cable wrapped around a motor-driven drum. These are most commonly attached to the front or rear of a vehicle.

Woven-Jacket Hose — Fire hose constructed with one or two outer jackets woven on looms from cotton or synthetic fibers.

Wye — Hose appliance with one female inlet and two or more male outlets, usually gated.

Y

Yard Hydrant — *See* Private Hydrant.

throttles
  adjusting flow during priming actions 267
  adjusting flow during pumping actions 247, 248
  described 238
  overthrottling 55
  troubleshooting 261, 280
tools. *See* equipment and tools carried on pumpers
torque, peak speed 55
total pressure loss (TPL)
  See also pressure loss
  calculation for standpipe operations 155, 181
  described 141–149, 167–175
  hard intake hoses and 260
  hose layouts and 149–161, 175–187
  total pump discharge compared to 141–142, 167–168
total stopping distance 60–62
trailer mounted pumps 217
training drivers/operators 4, 5, 70, 71–74, 217
transfer valves in pressure or volume positions 210–213, 214, 247, 256–257, 266, 270
troubleshooting
  cativation 277, 279–280
  drafting operations 266, 268–269, 276–280
  foam operations 377–378
  gauges 273, 276
  intake supply lines 280
  pressure 273–274, 276, 280
  priming actions 267, 277–279
  pump capacity 274–275
  during pumper service tests 394
  pumping operations 273–280
  relay operations 280
  relief valves 275–276
  standpipe systems 272
  suction lines 276–277
  tank operations 276
  throttles 280
turret pipes 162, 188
turrets 10, 15, 40, 41, 376
typing apparatus 20

U

United States Bureau of Land Mangement (BLM), high-energy foam proportioning systems 373
United States Department of Transportation (DOT) 6, 45–46
utility and power lines, positioning apparatus away from 80

V

vacuums 260, 262, 296
vacuum test 388–391
valves

See also names of specific valve types
air leaks during priming actions 267
connected to hydrants 251–252, 255
on cooling devices 240
described 225–227
discharge gauges connected to 237–238
in discharge piping 224, 225
inspecting 40, 41
preventing water hammer 117
in water distribution systems 121–122
vaporization, latent heat of 105–106
variable-flow variable-rate injection foam proportioning systems 370–371
velocity 108–113, 114, 116, 117, 136
ventilation systems 26–27, 37
venting during filling operations 335–336, 346
Venturi Principle 366
viscosity 108
volume (parallel) position of transfer valves 210–211, 214, 247, 257, 266, 270
volunteer fire departments 5, 23
volutes 209

W

walls, nozzles driven through 135
warning devices
  backing apparatus 59
  clearing traffic using sirens 60, 66–69
  highway operations and 68, 98–99
  inspection and maintenance 28, 33
  low water tank levels 239, 248
  types of 67
water
  See also drafting operations
  See also hose streams
  characteristics of 104–108
  diagnosing waterflow problems 194
  as extinguishing agent 108
  friction loss and 114–117
  nonpotable 123–124, 264
  pressure and velocity 108–113
water curtain nozzles 135
Water Expansion Systems (WES) and Water Expansion Pumping Systems (WEPS) 373
water hammer 117
water shuttle operations
  apparatus 329, 330–337
  described 329–330
  dump site operations 345–352
  evaluating tender performance 352–354
  fill site operations 330, 340–345
  NFPA 1002 329
  selecting dump and fill sites 336–338
  setting up 336–340
water source supply pumpers, positioning 83–89
water supply lines 103, 104, 125, 126, 249
water supply systems
  See also names of specific system types

described 7–8
distribution systems 120–123
highway operations using 98
moving 118–119
municipal 117–123
positioning apparatus 77–78, 82
pressurized 250–259
private 123–124
processing or treatment facilities 119–120
sources of water 117–118
storing water 118–119
transitioning from tanks to external 248–249, 284
water tanks. *See* tanks
water tenders. *See* tenders (tankers)
water thief appliances 156–159, 182–185
wear rings and packing 213–216
weather
  cisterns freezing 307
  driving in adverse conditions 5, 46–47, 49, 63–64, 66
  idling engine in cold temperatures 56
  snow chains and winterization systems 29, 33, 40, 66, 67
  static water supplies affected by 264, 304, 306, 309
  water shuttle operations and 352
weight transfer 55, 62–63, 64. *See also* brakes
wheels, chocking 243
wildland fire apparatus
  described 12
  equipment and tools carried on pumpers 13–14
  firefighter safety 58
  NFPA 1002 21, 44–45
  positioning 79, 89–93
  pumps used for 206, 217
wildland/urban interface 90
wind direction, positioning apparatus 78
windshields and windows, cleaning and inspecting 24–25, 28, 36–37
wyed hose layouts 151–152, 156–159, 177–178, 182–185
wyes 147, 173, 255

Z

zone system used in hazardous materials incidents 100–101